AF361960

# Modern Approaches to Non-Perturbative QCD and other Confining Gauge Theories

# Modern Approaches to Non-Perturbative QCD and other Confining Gauge Theories

Editor

**Dmitry Antonov**

MDPI • Basel • Beijing • Wuhan • Barcelona • Belgrade • Manchester • Tokyo • Cluj • Tianjin

*Editor*
Dmitry Antonov
Erbprinzenstr. 25,
69126 Heidelberg
Germany

*Editorial Office*
MDPI
St. Alban-Anlage 66
4052 Basel, Switzerland

This is a reprint of articles from the Special Issue published online in the open access journal *Universe* (ISSN 2218-1997) (available at: http://www.mdpi.com).

For citation purposes, cite each article independently as indicated on the article page online and as indicated below:

LastName, A.A.; LastName, B.B.; LastName, C.C. Article Title. *Journal Name* **Year**, *Volume Number*, Page Range.

**ISBN 978-3-0365-3329-2 (Hbk)**
**ISBN 978-3-0365-3330-8 (PDF)**

# Contents

# About the Editor

**Dmitry Antonov** received his Ph.D. in theoretical high-energy physics in 1999, from the Humboldt University of Berlin. From 1999 to 2004, he was working at the Theory Group of the Pisa Section of INFN and, later on, at the Theory Departments of the Humboldt University, Heidelberg University, as well as the Universities of Bielefeld and Lisbon. Those research positions included the fellowships of the European Commission, the Alexander von Humboldt foundation, the German Research Foundation, and the Portuguese Foundation for Science and Technology. He is the author of more than a hundred publications on non-perturbative quantum field theory, including several invited reviews and a monograph.

*Editorial*

# Modern Approaches to Non-Perturbative QCD and Other Confining Gauge Theories

**Dmitry Antonov**

Formerly at Departamento de Física and CFIF, Instituto Superior Técnico, ULisboa, Av. Rovisco Pais, 1049-001 Lisbon, Portugal; dr.dmitry.antonov@gmail.com

**Citation:** Antonov, D. Modern Approaches to Non-Perturbative QCD and Other Confining Gauge Theories. *Universe* **2022**, *8*, 49. https://doi.org/10.3390/universe8010049

Received: 10 January 2022
Accepted: 10 January 2022
Published: 13 January 2022

**Publisher's Note:** MDPI stays neutral with regard to jurisdictional claims in published maps and institutional affiliations.

The primary goal of this Special Issue was to create a collection of reviews on the modern approaches to the problem of quark confinement in QCD. Such approaches include both the microscopic models of the confining Yang–Mills vacuum and the models of the quark–antiquark string. Over the course of this project, the Special Issue also benefited from contributions on other related subjects, such as the topology of baryon-rich matter or a model of the axionic dark matter.

In their broad review [1], Roman Pasechnik and Michal Šumbera provided an outlook on some currently popular scenarios of the confinement phenomenon. The key topics covered by this review include the order parameters for confinement, magnetic order/disorder phase transition, the center-vortex and the monopole models of the Yang–Mills vacuum, as well as realizations of confinement in the gauge-Higgs and Yang–Mills theories, and the phases of QCD matter.

The review [2] by Maria Paola Lombardo is devoted to the subject of topology in dense matter. After a short overview of the status of the corresponding studies at zero density, lattice results for baryon-rich matter were presented. This subject was mostly studied in the two-color QCD and for matter with isospin and chiral imbalances. At high temperatures, some coherent pattern was shown to emerge. Namely, above the critical temperature for superfluidity/superconductivity, the topological susceptibility, as a function of either the isospin or the baryonic chemical potential, turned out to be clearly correlated with the chiral condensate and the confinement-related quantities. This finding holds true also for the chiral chemical potential. In that case, a striking effect, called chiral enhancement, has been found, which is the growth of the chiral condensate with the chemical potential. The same growth turns out to take place also for the topological susceptibility and the string tension.

The review [3] by Michele Caselle starts with a general introduction to the Effective-String-Theory (EST) approach to the description of confinement in the Yang–Mills theory. It further shows that, close to the deconfinement critical temperature, several universal features of confining gauge theories can be accurately described by the EST. Such features include the ratio of the deconfinement critical temperature to the square root of the zero-temperature string tension, the linear increase of the square of the flux-tube width with the interquark distance, and the temperature dependence of the interquark potential. Moreover, close to the deconfinement critical temperature, the behavior of the confining string turns out to be well described by the general principles of conformal invariance and by the Svetitsky–Yaffe dimensional-reduction conjecture. This finding provides further support for the description of confinement by means of the EST.

As mentioned above, this Special Issue contains several reviews and research articles devoted to specific models of the confining Yang–Mills vacuum. In particular, the review [4] by Maria Cristina Diamantini and Carlo Andrea Trugenberger as well as the article [5] by Hideo Suganuma and Hiroki Ohata are devoted to the monopole-based scenario of confinement.

The review by Maria Cristina Diamantini and Carlo Andrea Trugenberger discusses superinsulators (SI), which represent a new topological state of matter that can exist in the vicinity of the superconductor/insulator phase transition. Being dual to superconductors, SI provide a realization of the electric/magnetic duality. The effective field theory that describes SI is governed by the compact Chern–Simons term in (2 + 1)D and the compact BF term in (3 + 1)D. Unlike the superconductor, where the condensate of Cooper pairs leads to the Meissner effect, Cooper pairs in SI form bound states owing to the *dual* Meissner effect, i.e., the monopole-condensate-triggered squeezing of electric fields into the flux tubes. In fact, magnetic monopoles, while elusive as elementary particles, can be realized in certain materials in the form of emergent quasiparticle excitations. The monopole Bose condensate can exist at low temperatures and can manifest itself as a superinsulating state of infinite resistance. The related monopole supercurrents can thus result in the electric counterpart of the Meissner effect, which leads to the linear confinement of Cooper pairs. This way, SI realize one of the mechanisms proposed to explain confinement in the Yang–Mills theory. Furthermore, for SI samples smaller than the width of the confining string, a metallic-like low-temperature behavior of SI has been predicted and experimentally confirmed. It is also predicted that an oblique version of SI can be realized as a pseudogap state of high-temperature superconductors.

In the article by Hideo Suganuma and Hiroki Ohata, the interrelation between the chiral condensate, monopoles, and color-magnetic fields in QCD was studied on the lattice. First, idealized Abelian systems, consisting of a static monopole–antimonopole pair and a magnetic flux without monopoles, were explored. Lattice simulations of the chiral condensate of quasi-massless fermions, coupled to the Abelian gauge field in the mentioned systems, show that this condensate is localized in the vicinity of the magnetic field. Furthermore, by using SU(3) lattice-QCD Monte-Carlo simulations, the Abelian-projected QCD in the Maximal Abelian gauge was studied. The results of these studies show a clear correlation between the chiral condensate, the distribution of monopoles, and the color-magnetic fields of the Abelianized gauge-field configurations. As a statistical indicator, the coefficient measuring the correlation between the chiral condensate and the square of the color-magnetic field in the Abelian-projected QCD, was calculated and found to be approximately equal to 0.8. The same correlation was found to become weaker in the deconfinement phase. Thus, the obtained results show that, similar to what happens in the case of magnetic catalysis, the chiral condensate is locally enhanced by the strong color-magnetic field, which exists in the vicinity of monopoles in the Abelian-projected QCD.

The results of other studies of the interrelation between the dynamics of quarks and the confining dynamics of the Yang–Mills fields were reported in the review [6] by Matteo Giordano and Tamás Kovács, as well as in the articles [7,8] by Manfried Faber, Rudolf Golubich, and their collaborators.

The review by Matteo Giordano and Tamás Kovács is devoted to the Anderson-type localization transition, which affects eigenmodes of the lower part of the Dirac spectrum. Several aspects of this transition were reviewed, mostly by making use of the tools of lattice gauge theory. In particular, the connection of the localization transition with the finite-temperature phase transitions was illustrated. This connection makes the localization transition related to the deconfinement of quarks as well as to the restoration of chiral symmetry, which is spontaneously broken at low temperatures. The review also discusses the universality of the localization transition as well as its connection to the topological excitations of the gauge field, i.e., instantons, and the associated fermionic zero modes. While the review is mostly focused on QCD, it also discusses how the localization transition appears in other gauge models, with different fermionic contents and gauge groups, and in the various space-time dimensions.

The article [7] by Manfried Faber, Rudolf Golubich, and their collaborators, which is devoted to the center-vortex model of the confining Yang–Mills vacuum, discusses back-reaction of quarks on the gauge fields of the model. In particular, it shows that the model reproduces the phenomenological QCD string tension (at interquark distances smaller than the string-breaking distance) also in the presence of dynamical quarks. Their other article [8] suggests a possible resolution for the problems of vortex detection in smooth lattice configurations and discusses recent improvements in the detection of center vortices.

The related review [9] by Luis Esteban Oxman and his collaborators provides an overview of the recent progress achieved by the authors in an analytic derivation of the center-vortex model of the Yang–Mills vacuum. This research program starts with modeling, in the continuum limit, of some of the properties of center vortices that were found in the lattice simulations, and proceeds toward the derivation of the corresponding effective field representations. In particular, when modeling the measure of the center-vortex ensemble, the authors emphasized the importance of the inclusion of the non-oriented center-vortex component and the non-Abelian degrees of freedom. The so-constructed model of percolating center vortices turns out to be capable of reproducing several known important features of confining flux tubes in the SU($N$) Yang–Mills theory.

In their review [10], Ibrahim Burak Ilhan and Alex Kovner discuss the approach aimed at constructing an effective 4D theory that could provide a simple classical picture of the main qualitatively important features of both the Abelian and the non-Abelian gauge theories. This approach starts with ensuring the presence of massless photons, i.e., the Goldstone bosons, in the Abelian theory, and their disappearance in the non-Abelian case, which is happening together with the formation of confining strings between charged states. The suggested formulation avoids the use of vector fields, operating instead with the basic degrees of freedom, which are the scalar fields of a certain non-linear sigma model. The Mark 1 model, discussed in the review, turns out to have a large global symmetry group, with the 2D diffeomorphism invariance in the Abelian limit, which is isomorphic to the group of all canonical transformations in the classical 2D phase space. This symmetry is not present in QED, and it is thus further eliminated by "gauging" this infinite-dimensional global group. By introducing additional modifications to the model (Mark 2), the authors have first proved that the "Abelian" version of such a modified model is equivalent to the theory of a free photon. Achieving the desired properties in the "non-Abelian" regime turns out to be tricky. To this end, the authors introduced a perturbation that led to the formation of confining strings in the Mark 1 model. These strings have somewhat unusual properties, as their profile does not fall off exponentially, away from the center of the string. In addition, the perturbation explicitly breaks the diffeomorphism invariance. The questions of how to preserve this invariance in the gauged model as well as how to obtain realistic confining strings in the Mark 2 model currently remain open.

Last but not least, the article [11] by Janning Meinert and Ralf Hofmann, motivated by the SU(2)$_{\mathrm{CMB}}$-modification of the cosmological model $\Lambda$CDM (where "CMB" stands for Cosmic Microwave Background), considers isolated fuzzy-dark-matter lumps, made of ultralight axion particles with the masses arising due to distinct SU(2) Yang–Mills scales and the Planck mass $M_P$. Unlike the SU(2)$_{\mathrm{CMB}}$-model, the corresponding Yang–Mills theories, which are associated with the three lepton flavors of the Standard Model of particle physics, stay in the confining (zero temperature) phase throughout most of the universe's history. As the universe expands, axionic fuzzy dark matter comprising a three-component fluid undergoes certain depercolation transitions when dark energy (represented by the global axion condensate) is converted into dark matter. The authors extracted the lightest axion mass $m_{a,e} = 0.675 \cdot 10^{-23}$ eV from the well-motivated model fits to observed rotation curves in the low-surface-brightness galaxies (SPARC catalogue). Since the virial mass of an isolated lump solely depends on $M_P$ and the associated Yang–Mills scale, the properties of an $e$-lump predict those of $\mu$- and $\tau$-lumps. As a result, a typical $e$-lump virial mass $\sim 6.3 \cdot 10^{10} M_\odot$ suggests that massive compact objects in galactic centers, such as Sagittarius A$^*$ in the Milky Way, are (merged) $\mu$- and $\tau$-lumps. In addition, $\tau$-lumps may constitute global clusters. If the axial anomaly indeed links leptons with dark matter and the CMB with dark energy, that would demystify the dark universe through a firmly established feature of particle physics.

**Funding:** This Guest Editor's activity received no external funding.



## References


1. Pasechnik, R.; Šumbera, M. Different faces of confinement. *Universe* **2021**, *7*, 330. [CrossRef]
2. Lombardo, M.P. Topological aspects of dense matter: Lattice studies. *Universe* **2021**, *7*, 336. [CrossRef]
3. Caselle, M. Effective string description of the confining flux tube at finite temperature. *Universe* **2021**, *7*, 170. [CrossRef]
4. Diamantini, M.C.; Trugenberger, C.A. Superinsulators: An emergent realisation of confinement. *Universe* **2021**, *7*, 201. [CrossRef]
5. Suganuma, H.; Ohata, H. Local correlation among the chiral condensate, monopoles, and color-magnetic fields in Abelian-projected QCD. *Universe* **2021**, *7*, 318. [CrossRef]
6. Giordano, M.; Kovács, T.G. Localization of Dirac fermions in finite-temperature gauge theory. *Universe* **2021**, *7*, 194. [CrossRef]
7. Dehghan, Z.; Deldar, S.; Faber, M.; Golubich, R.; Höllwieser, R. Influence of fermions on vortices in SU(2)-QCD. *Universe* **2021**, *7*, 130. [CrossRef]
8. Golubich, R.; Faber, M. A possible resolution to troubles of SU(2) center-vortex detection in smooth lattice configurations. *Universe* **2021**, *7*, 122. [CrossRef]
9. Junior, D.R.; Oxman, L.E.; Simões, G.M. From center-vortex ensembles to the confining flux tube. *Universe* **2021**, *7*, 253. [CrossRef]
10. Ilhan, I.B.; Kovner, A. Confinement in 4D: An attempt at classical understanding. *Universe* **2021**, *7*, 291. [CrossRef]
11. Meinert, J.; Hofmann, R. Axial anomaly in galaxies and the dark universe. *Universe* **2021**, *7*, 198. [CrossRef]

Review

# Different Faces of Confinement

Roman Pasechnik [1,2,*] and Michal Šumbera [3]

[1] Department of Astronomy and Theoretical Physics, Lund University, SE-223 62 Lund, Sweden
[2] Departamento de Física, CFM, Universidade Federal de Santa Catarina, C.P. 476, Florianópolis CEP 88040-900, Santa Catarina, Brazil
[3] Nuclear Physics Institute CAS, 25068 Řež, Czech Republic; sumbera@ujf.cas.cz
* Correspondence: Roman.Pasechnik@thep.lu.se

**Abstract:** In this review, we provide a short outlook of some of the current most popular pictures and promising approaches to non-perturbative physics and confinement in gauge theories. A qualitative and by no means exhaustive discussion presented here covers such key topics as the phases of QCD matter, the order parameters for confinement, the central vortex and monopole pictures of the QCD vacuum structure, fundamental properties of the string tension, confinement realisations in gauge-Higgs and Yang–Mills theories, magnetic order/disorder phase transition, among others.

**Keywords:** magnetic disorder; confinement models; lattice QCD; center vortices; magnetic monopoles; quark condensate

**PACS:** 12.38.Aw; 12.38.Gc

## 1. Introduction

Quantum Chromodynamics (QCD) based upon the $SU(3)_c$ gauge theory of colour represents a real-world example of a fundamental Yang–Mills (YM) theory applied to the description of strong interactions and is an organic part of the Standard Model (SM) of particle physics. This theory is extremely successful in predicting various measurable phenomena at particle colliders. The class of phenomena that originate from (or driven by) strong interactions is extremely wide and covers such areas as nuclear physics, hadron physics, physics of quark-gluon plasma, high-temperature and high-density QCD, high-energy particle production and hadronisation. Depending on characteristic length scales, QCD behaves very differently. At short space-time separations, e.g., once we zoom into distances much shorter than the proton radius, QCD appears as a weakly coupled theory that enables a precise Perturbation Theory (PT) analysis. Much of its success has been achieved in this *asymptotic freedom* or ultraviolet (UV) regime where the quark-gluon interaction strength recedes. Thus, success highlights the QCD theory as the correct theory of strong interactions at the fundamental level, precisely matching all the existing observations up to very high momentum transfers reached by the Large Hadron Collider (LHC) so far. However, on the opposite side of length-scales in the infrared (IR) limit, QCD enters entirely different, strongly coupled domain, rendering the PT inapplicable and creating substantial problems for making reliable predictions at intermediate and low momentum transfers, i.e., at large distances. While it is conventionally believed that QCD should remain the correct theory of strong interactions also at large distances, in the so-called *confined regime*, deriving reliable predictions remains a big theoretical challenge. For one of the broadest and comprehensive overviews of many phenomenological and theoretical aspects of QCD and QCD-like gauge theories spanning from IR to UV, from dilute to dense regimes, see Ref. [1].

The *problem of confinement* concerns the strongly coupled sector of QCD composed of interacting coloured partons (quarks and gluons). In virtue of colour confinement, the coloured particles appear to always be trapped (confined) inside colourless composites.

Citation: Pasechnik, R.; Šumbera, M. Different Faces of Confinement. *Universe* **2021**, 7, 330. https://doi.org/10.3390/universe7090330

Academic Editor: Dmitri Antonov

Received: 3 August 2021
Accepted: 1 September 2021
Published: 6 September 2021

**Publisher's Note:** MDPI stays neutral with regard to jurisdictional claims in published maps and institutional affiliations.

The latter emerges as asymptotic states, thus rendering the long-distance regime of hadron physics described by Effective Field Theory (EFT) approaches, such as the chiral PT, as well as a variety of non-perturbative techniques realised in numerical simulations on the lattice. Much of the discussion in the current review is devoted to highlighting main ideas and possible existing ways to address the confinement problem that is known as the main unsolved problem in the SM framework. Despite the major efforts of the research community and tremendous progress made over last few decades, it does not appear to be fully and consistently resolved yet. There are several important subtleties in the formulation of this problem to be discussed in what follows. One of the standard ways of formulating the problem is that there is no complete understanding of why these fundamental degrees of freedom (DoFs) of QCD (or, generically, of any strongly coupled YM theory) do not emerge in the physical spectrum of asymptotic states and how the composite hadrons are dynamically produced starting from the fundamental DoFs in the initial state. In a phenomenological sense, there is a fundamental mismatch between the underlined DoFs of QCD in its short- and long-distance regimes manifest in experimental measurements, and there is not a single consistent theoretical framework that goes beyond the framework of PT and treats both weakly and strongly coupled regimes on the same footing.

For practical purposes, various phenomenological approaches have been proposed that characterise the long-distance effects of QCD absorbing them into universal elements of a given scattering process, such as non-perturbative matrix elements, fragmentation functions or parton distributions. As a commonly adopted picture, a colour-electric flux tube (also known as a colour string) is stretched among the partons produced in a high-energy collision. A string-like picture emerges in the limit of large number of colours already in $D = 1 + 1$ dimensions as has been advocated by t'Hooft back in early 1970s—see, e.g., Ref. [2]. As produced partons move away from each other at large enough distances, those flux tubes fragment into composite particles, such as mesons and baryons, where initial (anti)quarks and gluons get necessarily combined with newly emerged ones from the vacuum into colour-neutral configurations. In a nutshell, the basic problem concerns a first-principle derivation of the long-distance hadron spectrum and dynamics from an underlined strongly coupled gauge theory. More specifically, a successful model of confinement is expected to provide a first-principle dynamical description of the string formation, its basic characteristics and string-breaking effects, also connecting those unambiguously to dynamics of the fundamental DoFs of the underlined gauge theory and deducing the phase structure of the theory at various densities and temperatures. While there are no compelling solutions yet available, there are several distinct approaches to confinement treatment being actively developed in the literature. Not only a large variety of treatments of confinement has hit the literature in past decades but also a proper definition of confinement; what we actually mean by this word posses a notorious difficulty, as was thoroughly discussed in Refs. [3,4]. In this review, we will try to summarise some of the existing attractive treatments of confinement and ideas and why confinement occurs in the way it does in a conceptual and qualitative manner, without pretending to provide an exhaustive overview of all relevant details and corresponding references.

The review is organised as follows. In Section 2, we discuss the basic ingredients of the QCD phase diagram at different temperatures and values of the baryon chemical potential. In Section 3, we provide a brief description of magnetic order/disorder phases and introduce the basic notions of the lattice gauge theory that will be used in follow-up discussions. In Section 4, we overview basic concepts and ideas that lead to different asymptotic behaviours of the Wilson loop VEV as an order parameter for the confining phase. Such distinct properties of QCD scattering amplitudes as the Regge trajectories and the associated picture of a colour string have been outlined in Section 5. In Section 6, we provide a detailed outlook on the complementarity between the Higgs and confining phases and describe such a common feature for both phases as colour confinement. In Section 7, a brief description of the string hadronisation picture realised in the Lund model is given. Section 8 elaborates on why confinement criteria based upon gauge symmetry remnants (un)breaking may be spoiled by gauge-fixing artefacts, highlighting the need for a gauge-invariant description of confinement. Section 9 introduces the basics of the center-symmetry-based confinement criterion and its implications. Section 10 gives a brief outlook on another order parameter of confinement, the Polyakov loop, particularly suitable for confinement description at finite temperatures. In Section 11, yet another important order parameter of confinement probing the vortex structure of the QCD vacuum, the t'Hooft loop, is introduced and the basic features of the center vortices are described. Section 12 elaborates on the most important characteristics of the string tension as the probes for a confining phase. The foundations and implications of the center vortex mechanism of confinement, with its basic tests performed in the literature, have been discussed in Section 13. Section 14 connects the chiral symmetry breaking and the topological charge to the existence of vortex configurations. In Section 15, we briefly describe the Gribov–Zwanziger scenario of confinement, relating it to the non-perturbative behaviour of propagators and describing how a colour string could emerge in this scenario by considering constituent gluons in the gluon chain model. A renown dual superconductivity picture of confinement and the fundamental role of magnetic monopoles have been briefly described in Section 16. A novel generalisation of the confinement criterion applicable in gauge theories with matter in the fundamental representation has been briefly discussed in Section 17. Section 18 highlights an important recent development in understanding the confining property of the gauge-field vacuum and Higgs-confinement transitions via a novel non-local order parameter. A summary and concluding remarks are given in Section 19.

## 2. Phase Structure of QCD Matter

Following the discovery of asymptotic freedom in QCD [5,6], it has been realised that phase transitions in the hot and dense QCD matter between the hadronic (confined) and quark-gluon (deconfined) phases are crucial for understanding the cosmological evolution as well as the state of matter and dynamics of neutron stars [7–14]. Besides, the idea of experimental measurements through heavy-ion collisions has been offered as a tantalising opportunity for explorations of this interesting physics. In those early times, a hypothetical state of QCD matter at characteristic temperatures of around 100 MeV has been envisaged as existing in two possible states of "hadronic plasma" [9] and "quark-gluon plasma" (QGP) [10], with an energy density of order 1 GeV/fm$^3$. Later on, it has been understood that the QCD phase diagram has a much richer structure, particularly, at high baryon number densities, with a lot of important implications for understanding, for instance, neutron star physics as well as heavy-ion collisions at particle colliders.

Strongly interacting QGP was first discovered at RHIC collider in 2005 [15–18] and later has been confirmed at much higher energies at the CERN LHC (for a detailed review, see, e.g., Refs. [19,20] and references therein). In the QGP phase, as the name suggests, the strong interactions between constituents of the plasma, "dressed" light quarks and gluons being its collective excitations, is driven by their $SU(3)_c$ colour charges. For a comprehensive review of early developments and key ideas in the analysis of strongly coupled QCD phenomena and QGP in particular, see, e.g., Ref. [21], while an overview of more recent theoretical and experimental studies can be found in Refs. [19,20,22].

In a weakly interacting QCD gas at very high $T$, the microscopic quark-gluons interactions are relatively weak and should obey the predictions of asymptotic freedom. The leading-order perturbative QCD coupling that determines the strength of QCD interactions at asymptotically short distances,

$$\alpha_s(Q) \simeq \frac{2\pi}{b_0 \ln(Q/\Lambda_{\mathrm{QCD}})}, \qquad b_0 = 11 - \frac{2}{3}N_f, \tag{1}$$

is given in terms of the QCD energy scale $\Lambda_{\mathrm{QCD}} \approx \mathcal{O}(1\,\mathrm{GeV})$, momentum transfer $Q \gg \Lambda_{\mathrm{QCD}}$ and the active quark flavours' number $N_f$. In a perturbative domain of QCD, when going towards shorter distances $l \ll \Lambda_{\mathrm{QCD}}^{-1}$, the colour charge is being diluted compared to the "soft" and non-perturbative domain of QCD at larger distances $l \sim \Lambda_{QCD}^{-1}$, where the charge is being built-up effectively due to the phenomenon called the colour charge "anti-screening" [5,6]. This is quite an opposite effect to what happens in QED. This behaviour of the coupling is demonstrated in Figure 1 (left panel), together with experimentally measured values. As soon as $\alpha_s(Q)$ hits large values entering the strongly coupled (confined) regime at lower $T$, the PT ceases to work such that effective and non-perturbative methods are applied, being, however, often vastly disconnected from the microscopic QCD theory. One could perform a consistent matching of the fundamental QCD to the effective Lagrangian of chiral PT at the "soft" scale $Q \simeq 4\pi f_\pi \simeq 1$ GeV, where both descriptions are expected to be valid and overlap. Such a matching provides a clue about the IR behaviour of $\alpha_S(Q)$ that tends to get "frozen" at the value of $\langle \alpha_S \rangle_{IR} \simeq 0.56$ [23].

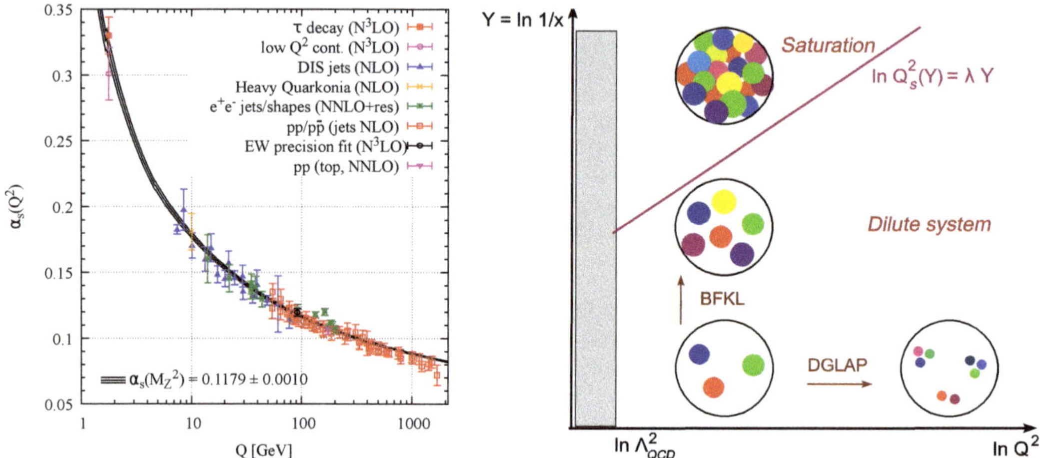


**Figure 1.** On the left panel, the QCD interaction strength $\alpha_s$ as a function of the momentum transfer $Q$ at the next-to-leading order of the PT. The figure is taken from Ref. [24]. On the left panel, the QCD evolution of the characteristic parton (quark and gluon) density and length-scale with respect to rapidity $Y = \ln(1/x)$ and $\ln Q^2$. The figure is taken from Ref. [25].

Besides the weakly coupled short wavelength modes of partonic DoFs with $Q = 2\pi T$ dominating the thermodynamic evolution at very high $T$, the QGP also features long wavelength (non-perturbative) modes, with length scales of $l > T^{-1}$. The latter modes dominate the evolution at not-so-high $T$, forming a liquid and effectively turning QGP into ideal fluid [22,26,27]. The latter fundamental property of QGP has been discovered first at RHIC [15–18] and then confirmed at the LHC. Other effects of such strongly interacting QGP are manifested through a collective flow phenomenon [27] as well as in an effective suppression of high-energy partons transiting through a hot and dense deconfined medium [28,29] (for a review, see Ref. [20] and references therein).

Taking the ratio of the interaction-to-kinetic energy of the QGP constituents and assuming equal contributions from chromo-electric and chromo-magnetic interactions, one introduces the so-called plasma parameter [30]

$$\Gamma \simeq 2\frac{C_{q,g}\alpha_S}{aT}, \qquad C_q = \frac{N_c^2 - 1}{2N_c} = \frac{4}{3}, \qquad C_g = N_c = 3, \tag{2}$$

expressed in terms of the fundamental (quark) and adjoint (gluon) Casimir invariants of $SU(3)_c$, $C_q$ and $C_g$, respectively, and the $T$-dependent average distance between the partons $a$ satisfying $aT \sim d_F^{-1/3}$, where

$$d_F \equiv 2 \times 8 + \frac{3}{4}\Big(3 \times N_f \times 2 \times 2\Big). \tag{3}$$

The latter evolves in $T$ only through $N_f(T)$. Weakly interacting (ideal) plasmas have a very low $\Gamma < 10^{-3}$, while a strongly interacting plasma typically has a much larger $\Gamma \gtrsim 1$. Taking a nearly ideal (weakly coupled) massless QCD gas, for instance, one obtains $\Gamma \sim \alpha_S d_F^{1/3}$ serving as a lower estimate for the plasma parameter as it ignores the partonic interactions in the ideal gas approximation. In a realistic case of QGP created in heavy-ion collisions at RHIC, one finds $T \approx 200$ MeV and $\alpha_S = 0.3$–$0.5$ with only two relevant active flavours, $N_F = 2$, leading to a value of $\Gamma \simeq 1.5$–$6$, indeed being deeply inside the strongly coupled plasma regime.

The QCD evolution of partonic matter in terms of basic kinematic parameters of resolved partons in the medium is illustrated in Figure 1 (right panel). For instance, developing the partonic cascades in typical momentum transfer $Q$, one resolves the partons with a transverse area $1/Q^2$, such that at larger $Q$ and $T \sim Q$, one observes a dilution of the parton density controlled by the DGLAP evolution equations (see for instance Refs. [31,32]). One may also observe how the parton density evolves with energy or, more conveniently, with a fraction of light cone momentum taken by a given radiated parton out of a parent particle, $x = k^+/P^+$. One may visualise the partonic cascade off the initial particle effectively as Brownian-like motion in the transverse plane that can be considered as the Gribov diffusion process in the evolution "time" $Y = \ln(1/x)$. The latter parameter is simply a rapidity difference between the radiated and parent partons, while the diffusion constant is $D \sim \alpha_S$. Such an evolution is controlled by BFKL equations (for more details, see, e.g., Refs. [31,32] and references therein).

The partonic cascade is essentially dominated by soft gluons at high energies or at very small fractions $x \ll 1$, and they are of the same size at a fixed scale $Q$. As soon as the parton scattering cross-section $\sim \alpha_S/Q^2$ multiplied by the probability to find a parton at a given $Q$ with a fraction $x$, $xG_A(x, Q^2)$, becomes of the order of the geometrical cross-section of an area $A$ occupied by the gluons, $\sim \pi R_A^2$, the gluons start to overlap effectively. Due to a repulsive interaction between gluons, however, their occupation number saturates at $f_g \sim 1/\alpha_S$. In particular, this occurs for gluons with transverse momenta below a certain emergent scale $Q_s(x)$, $k_\perp \leq Q_s(x)$, known as a saturation or "close packing" scale [33] (see also Refs. [34,35]),

$$Q_G^2(x) = \frac{\alpha_S(Q_s)}{2(N_c^2 - 1)} \frac{xG_A(x, Q_s^2)}{\pi R_A^2},$$ (4)

thus, representing a fixed point in the parton $x$-evolution. Such a *saturation* phenomenon is rather generic as an analogical scaling of the density $\sim \alpha^{-1}$ characterises various Bose–Einstein condensation phenomena, in particular, those in the Higgs mechanism and in superconductivity [36]. Such a highly coherent gluonic state of matter has properties of a classical field [34] and is known in the literature as the *Colour Glass Condensate* (CGC) [25,35,37] or *glasma* [38].

Indeed, in the path integral formulation of the $SU(N)$ gauge theory, for instance, one sums over all gauge-field configurations weighted with $\exp(-iS_g/\hbar)$, where the action can be written as

$$S_g = -\frac{1}{4g_s^2} \int \mathcal{F}^{\mu\nu,a} \mathcal{F}_{\mu\nu}^a d^4x,$$ (5)

$$A_\mu^a \to \mathcal{A}_\mu^a \equiv g_s A_\mu^a, \quad F_{\mu\nu}^a \to g_s F_{\mu\nu}^a \equiv \mathcal{F}_{\mu\nu}^a = \partial_\mu \mathcal{A}_\nu^a - \partial_\nu \mathcal{A}_\mu^a + f^{abc} \mathcal{A}_\mu^b \mathcal{A}_\nu^c,$$

such that $g_s^2$ multiplies $\hbar$ in the exponent. Here, $f^{abc}$, $(a, b, c) \in \{1, \ldots, N^2 - 1\}$ are the $SU(N)$ structure constants. The path integral would be dominated by the classical configurations for $\hbar \to 0$ (classical limit), which is, therefore, equivalent to taking the weak coupling limit of the theory $g_s^2 \to 0$, where the action is large, $S_g \gg \hbar$, and so is the number of quanta in these configurations, $f_g \sim S_g/\hbar$ [34]. There are certain reasons to believe that such classical-field configurations should describe the state of cold nuclear matter in the initial stages of ultra-relativistic heavy-ion collisions [25,37].

Needless to mention, strongly interacting QCD exhibits a variety of emergent collective effects and phenomena other then those of QGP that are very difficult to understand and to predict starting from the first-principle microscopic theory of QCD. Observable predictions of the hot/dense QCD theory depend on the equation of state (EoS) of compressed nuclear matter, but the latter has not been fully understood yet. This situation is analogical to emergent phenomena in atomic and condensed-matter physics driven by the QED interaction theory at the microscopic level. Notably enough, besides the hadronic and QGP phases, QCD matter features also other distinct phases predicted in various approaches [39,40].

Among important examples of various realisations of confining non-abelian gauge-field dynamics in cosmology are the relaxation phenomena in the real-time cosmological evolution of the QCD vacuum [41] and a possibility of phase transitions in a "dark" strongly coupled $SU(N)$ gauge sectors [42], both potentially testable via the detection of stochastic primordial gravitational-wave spectra in future measurements. The homogeneous gluon condensates in the effective $SU(N)$ theory (such QCD gluodynamics) have also been found to play an important role in the generation of the observable cosmological constant [20,41,43–46]. For a recent review of implications of the quantum YM vacuum for the Dark Energy problem, see Ref. [47].

Systematic explorations of QCD matter at high densities and temperatures, including the search for the critical end point (CEP) in the middle of the phase diagram at $\mu_B \sim$ 0.4 GeV shown in Figure 2, only started about ten years ago. The CEP is located at the end of the first-order phase transition boundary between the hadronic phase and QGP, where a second-order phase transition is predicted to occur. One expects a number of new phenomena in a vicinity of that point [48–51] that have been searched for by the RHIC Beam Energy Scan program.

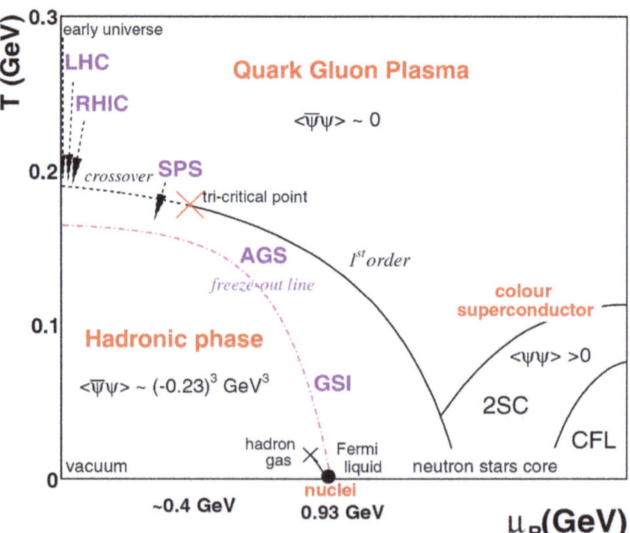


**Figure 2.** An illustration of typical phases of QCD matter that are expected to emerge at various values of the temperature $T$ and the baryonic chemical potential $\mu_B$ associated with $U(1)_B$ breaking (see Ref. [20] and references therein). Accelerators operating at different center-of-mass (c.m.) energies are depicted here.

Currently, a number of different studies of QCD phases in various parts of the $(T, \mu_B)$ diagram are being deployed, both experimentally and theoretically, and a high complexity has started to emerge. Particularly intense are explorations of low $\mu_B \simeq 0$ [52–55] and high $\mu_B \simeq 100$–600 MeV [40,49,51,55] domains, with possible transitions in between, also indicated in Figure 2. Other CEPs may also be expected to emerge such as those for chiral (crossover at low-$T$, not shown in the figure) and nuclear liquid-gas (in the nuclear matter ground-state at nearly-zero $T$ and $\mu_B = 0.93$ GeV) transitions.

More specifically, looking at the QCD phase diagram in Figure 2 along the direction of increasing baryon chemical potential $\mu_B$, we notice that at energies close to the binding energy of bulk nuclear matter, the so-called *cold nuclear matter* phase is found. Interactions between nucleons (quark bound states) may lead to pairing and di-baryon condensation that spontaneously break the $U(1)_B$ baryon number symmetry (see, e.g., Refs. [56,57]). Physically, this also means that the system is in a superfluid (confining) phase. This system has a close analogy, for instance, with liquid helium, where one also finds the Bose–Einstein condensation and Goldstone modes, both associated with the superfluidity property. The same physics emerge in ordinary nuclear matter based upon the nuclear many-body theory, which is applicable at not too large densities.

There is still a substantial lack of knowledge on a transition between the cold nuclear matter and high-density QCD phases, particularly relevant for the physics of neutron stars. Since QCD is asymptotically free, one can go to a very high $\mu_B$ in the quark-matter phase and employ weak-coupling techniques [58]. In this *quark-matter* phase of dense QCD, such calculations predict a nearly Fermi-liquid with residual interactions that lead to pairing among quarks in a gauge-dependent way. This is described by means of a gauge-dependent di-quark condensate $\langle qq \rangle$ playing a role of an order parameter in the dynamical Higgs mechanism such that we deal with a Higgs phase. Indeed, such a di-quark condensate emerges due to long-range attractive forces between the quarks through a Cooper-like pairs' condensation [59,60]. Such a high-density (baryon) superfluid phase where the $SU(3)_c$ gluon field is fully "Higgsed" is known in the literature as a *colour superconductor*[1] (CSC) (for a comprehensive review on key aspects of dense QCD, see Refs. [61,62]). The formation of such Cooper pairs of quarks can be seen in QCD with three massless $u, d, s$ flavours at large baryon number densities featuring the following colour and flavour symmetries' reduction [40,63]

$$SU(3)_c \times SU(3)_R \times SU(3)_L \times 3U(1)_B \to SU(3)_{c+L+R} \times \mathbb{Z}(2) \tag{6}$$

down to a diagonal subgroup $SU(3)_{c+L+R}$. The corresponding symmetry transformations involve a simultaneous "rotation" of colour and flavour group representations known as the colour–flavour locking (CFL). Such a CFL phase is known not to be topologically ordered [64]. Then, in the CFL quark-matter phase, one could also find an order parameter for $U(1)_B$ symmetry breaking (down to $\mathbb{Z}_2$) in analogy to the di-baryon condensate in the nuclear-matter phase—it can be viewed as a cubic power of the di-quark condensate thus being associated with a superfluid flow.

In fact, both quark matter and nuclear matter phases were found to be relevant for the EoS of neutron stars (see, e.g., Refs. [62,65,66]), and the signatures of possible phase transitions might show up in mass-radii relations for neutron stars and gravitational-wave spectra from neutron star collisions. As at high temperatures no baryon number symmetry breaking occurs, one supposedly crosses the line where $U(1)_B$ gets restored when the system heats up. As we noticed above, at low temperatures, both low- and high-density phases have the same order parameter w.r.t. $U(1)_B$ breaking, and one of the fundamental open questions is whether a boundary between the quark-matter (Higgs) and nuclear-matter (confinement) phases actually exists. Following Refs. [67–69], one could consider a simplified picture of pure QCD and include three massless flavours in a maximally symmetric realisation, such that there is no distinction in symmetry realisations between the hadronic phase and asymptotically high-density phase. The latter means there may be no phase transition that is consistent with identical global symmetry realisations in both regimes, t'Hooft anomalies' matching and with smoothly connecting low-lying excitations (see, e.g., Refs. [67,70,71]). Such an assumption has become a working one for many phenomenological studies modelling the EoS for neutron star physics (see, e.g., Ref. [72] and references therein). Below, following the recent results of Ref. [58], one may conclude, however, that the Schäfer–Wilczek conjecture about quark-hadron continuity at large $\mu_B$ may be largely oversimplified. The reality may be even more complex than what emerges in existing theoretical approaches. The basic problem is that there are no well-justified theoretical methods available for the treatment of the strong-coupling regime of QCD, with a non-zero chemical potential, where lattice simulations may not be very reliable.

Finally, yet another QCD phase that is believed to be located somewhere between the chirally restored and confined phases is known as quarkyonic matter [73] that may also have some relevance for neutron star physics [74]. In the limit of large number of colour charges $N_c$, the gluons' contribution scales as $\sim N_c^2$ compared to that of quarks $\sim N_c$ such that this phase is assumed to have energy densities well beyond $\Lambda_{\rm QCD}^4$. Since gluons are bound in glueballs, one ends up with $N_c$ DoFs in this phase.

Let us now turn to a discussion of methods of the lattice gauge theory that became the main tool for explorations of non-perturbative physics in gauge theories and, in particular, QCD in the strongly coupled regime and the associated dynamics of confinement, at least, at not too large chemical potentials.

### 3. Ising Model and Lattice Gauge Theory

To what extent one can expect to derive precision results for low-energy observables from the first-principle QCD theory? A default answer to this question is that we should not expect that, at least, analytically. The collective phenomena that are manifest in the strongly coupled regime of a gauge theory are so complex that none of the existing analytic approaches captures all the relevant dynamics and yields satisfactory results. At the same time, a theory may remain to be correct even if methods of extracting observable information from it are not perfect or suitable. Often though, we start with a simplified model that hopefully captures the same physics as a realistic one, but where we have a better control, and then we abstract the lessons that we learn from such a model back to more complicated theories, such as QCD.

Luckily, a precise and reliable analysis is possible but only numerically. The best available framework so far is the lattice gauge theory providing a first-principle numerical approach for strongly coupled theories, such as QCD. In fact, this framework is often considered as a "numerical experiment" and may be regarded as a black-box whose results need to fit a certain theoretical picture of real underlined physical phenomena and objects providing means to understand those phenomena qualitatively. Whether or not the lattice results fit a particular picture of confinement is an ongoing and long-standing debate in the literature. For relatively recent detailed reviews on non-perturbative physics and the confinement problem, see, e.g., Refs. [3,4,75–77] and references therein. Here and below, we follow the notation adopted in Ref. [3] unless noted otherwise, acknowledging that the latter reference represents one of the most complete, pedagogical and sophisticated reviews available in the literature on what the confinement problem actually is from various perspectives and approaches.

In order to build a consistent picture of confinement, we need to elaborate on such important notions as ordered and disordered systems. One of the simplest examples of the lattice field theory follows the basic principles of statistical mechanics, where the most relevant properties of these systems are readily seen in the Ising model of ferromagnetism. For illustration, consider a simple system—a square ($D = 2$), cubic ($D = 3$) or hypercubic ($D > 3$) array (or *lattice*) of atoms, each with two spin states—in the external magnetic field $h$. This system is described by the Hamiltonian,

$$H = -J \sum_x \sum_{\mu=1}^{D} s(x)s(x+\mu) - h \sum_x s(x), \quad J > 0, \tag{7}$$

where $s(x) = +1$ and $-1$ would correspond to an atom at a point $x$ with spin up and down, respectively, and we denote here the total number of spins as $N$. The probability for a specific configuration of spins, $\{s(x)\}$, at a given temperature $T$, can be written as

$$P_{\{s(x)\}} = \frac{1}{Z} \exp\left[-\frac{H}{kT}\right], \qquad Z = \sum_{\{s(x)\}} \exp[-H/kT]. \tag{8}$$

In the case of zero external field, $h = 0$, the system apparently possesses a *global* $\mathbb{Z}_2$ symmetry w.r.t. transformations

$$s(x) \to s'(x) = \xi s(x), \qquad \xi = \pm 1, \tag{9}$$

such that the mean magnetisation (average spin)

$$\langle s \rangle = \sum_{\{s(x)\}} \frac{P_{\{s(x)\}}}{N} \sum_y s(y) \tag{10}$$

vanishes. This is a system in a so-called *disordered* state.

Assume that the spins in the initial state are aligned. The exact $\mathbb{Z}_2$ symmetry means that at any given temperature, any finite system would end up in a disordered state provided that one waits for long enough for that to occur. This leads to the non-existence of permanent magnets as any alignment of the spins would be destroyed by thermal fluctuations. However, for large $N$, i.e., for macroscopic magnets, the time between sizable fluctuations that could flip a lot of spins would grow exponentially and eventually exceeds the lifetime of the Universe. For non-zero $h$, however, the $\mathbb{Z}_2$ symmetry appears to be explicitly broken, enabling $\langle s \rangle \neq 0$ at any temperature. In this case, the system appears to be in an *ordered* state where a large amount of spins point in the same direction.

Now, consider the magnetisation of a large system in the limit of vanishing $h$. One could show that, in general, this quantity is non-vanishing

$$\lim_{h \to 0} \lim_{N \to \infty} \langle s \rangle \neq 0, \tag{11}$$

yielding the so-called *spontaneous symmetry breaking* (SSB) of the global $\mathbb{Z}_2$ symmetry, which occurs particularly at low temperatures (ordered state). A global symmetry is said to be broken spontaneously when the Hamiltonian and the corresponding equations of motion are symmetric, but the solutions for physical observables (such as the magnetisation introduced above) are not. At high $T$ above a certain critical temperature (Curie temperature), the averaged spin vanishes, and the spin system appears again in a symmetric (disordered) state. Considering the vacuum expectation value (VEV) of a product of two spins, we notice $G(r) \equiv \langle s(0)s(r) \rangle \sim \exp(-r/l)$, i.e., it falls off exponentially with the distance between atoms $r$ in a disordered state, where $l$ is the correlation length. There is a phase transition between the ordered and disordered phases of the system at the Curie temperature for any $D > 1$, while for $D = 1$, the system is in a disordered phase at any $T$. The existence of such phase transitions associated with a global symmetry breaking is a generic property of many different systems and is also manifest in strongly coupled gauge theories, as will be discussed below.

Let us further promote the global $\mathbb{Z}_2$ symmetry to a local one whose transformation parameter depends on the position of the associated DoFs, $\xi(x) = \pm 1$, and can be chosen independently at each site (*gauge transformations*). For this purpose, let us consider the links of the lattice $s_\mu(x)$ along each dimension $\mu = 1 \ldots D$ as dynamical DoFs subjected to the gauge transformation

$$s_\mu(x) \to \xi(x)s_\mu(x)\xi(x + \hat{\mu}), \tag{12}$$

and write down the Hamiltonian of the gauge-invariant Ising model

$$H = -J \sum_x \sum_{\mu=1}^{D-1} \sum_{\nu>\mu}^{D} s_\mu(x)s_\nu(x+\hat{\mu})s_\mu(x+\hat{\nu})s_\nu(x). \tag{13}$$

Thereby, we arrive at the simplest example of the $\mathbb{Z}_2$ lattice gauge theory. In order to describe such systems, one considers observables that are invariant under gauge transformations. A particularly important class of observables can be obtained by taking the VEV of the so-called *Wilson loop*—a product of links on the lattice around a given closed contour $C$ [78],

$$W(C) = \left\langle \Pi_{(x,\mu)\subset C} s_\mu(x) \right\rangle. \tag{14}$$

The Hamiltonian (13) is given by the simplest Wilson loop given by a plaquette, the minimal closed loop on the lattice.

In analogy to the gauged Ising model, in a generic lattice gauge theory described by a certain (discrete or continuous) gauge group $G$, one starts with the Euclidean action where the link variables are the elements of the gauge group. For instance, in the case of a non-abelian group $G \equiv SU(2)$, the group elements in discretized spacetime are

$$U_\mu(x) = e^{iagA_\mu(x)}, \qquad A_\mu(x) = \frac{1}{2}\sigma^a A_\mu^a(x), \tag{15}$$

in terms of the lattice spacing $a$, the gauge coupling $g$, the Pauli spin matrices $\sigma^a$, $a = 1,2,3$, and the $SU(2)$ gauge field $A_\mu^a(x)$. By convention, the link variable $U_\mu(x)$ is associated with a line running from site $x$ on the lattice to a neighbour site $x + \hat{\mu}$ in the positive direction $\mu$. The probability distribution of lattice configurations of the gauge field is found in full analogy to that of the Ising model, namely,

$$P_{\{s(x)\}} = \frac{1}{Z}\exp(-S[U]), \tag{16}$$

where the Euclidean action, also known as the Wilson action,

$$S[U] = -\frac{\beta}{2}\sum_{x,\mu<\nu}\mathrm{Tr}[U_\mu(x)U_\nu(x+\hat{\mu})U_\mu^\dagger(x+\hat{\nu})U_\nu^\dagger(x)] \tag{17}$$

is invariant under local gauge transformations

$$U_\mu(x) \rightarrow G(x)U_\mu(x)G^\dagger(x+\hat{\mu}), \qquad G(x) \subset SU(2). \tag{18}$$

We used the fact that the trace of any $SU(2)$ group element is real. A straightforward extension to the $SU(N)$ gauge theory leads to

$$S[U] = -\frac{\beta}{2N}\sum_{x,\mu<\nu}\left\{\mathrm{Tr}[U_\mu(x)U_\nu(x+\hat{\mu})U_\mu^\dagger(x+\hat{\nu})U_\nu^\dagger(x)] + \text{c.c.}\right\}, \tag{19}$$

with suitably generalised group elements $U_\mu(x)$.

By expanding the latter in powers of $A_\mu^a(x)$, taking $\beta = 2N/g^2$ and turning to the continuum limit of vanishing lattice spacing $a \rightarrow 0$, one arrives at the standard expressions for the action and gauge transformations in Euclidean spacetime

$$S = \frac{1}{2}\int d^4x\,\mathrm{Tr}[F_{\mu\nu}F_{\mu\nu}], \qquad F_{\mu\nu} = \partial_\mu A_\nu - \partial_\nu A_\mu - ig[A_\mu, A_\nu], \tag{20}$$

$$A_\mu(x) \rightarrow G(x)A_\mu(x)G^\dagger(x) - \frac{i}{g}G(x)\partial_\mu G^\dagger(x), \tag{21}$$

in terms of the field strength tensor $F_{\mu\nu}$ and a gauge group element $G(x)$. Here, the repeated indices are summed over as usual.

The formulation of the lattice gauge theory in Euclidean spacetime has quickly become the cornerstone and the main reference for numerical analysis of basic characteristics of the corresponding quantum field theory (QFT) in Minkowski spacetime (such as its low lying spectrum and the static potential). This is due to the single most important fact that the Euclidean formulation of the field theory is conveniently considered as a statistical (not quantum) system whose analysis can be performed using the power of the lattice Monte Carlo methods. For a detailed description of these methods, see, e.g., Ref. [79].

The Euclidean formulation is particularly designed for studies of QFT at finite temperatures in equilibrium and works in Euclidean space with periodic time direction for bosonic fields while fermion fields fulfil antiperiodic boundary conditions in the time direction (for a recent review, see, e.g., Refs. [80,81]). A finite $T$ theory is then constructed from its zero-temperature counterpart by replacing bosonic and fermionic four-momenta $k^\mu$ in Euclidean integrals by $2\pi n T$ and $(2n+1)\pi T$, respectively, and then switching from $k^\mu$ integration to summation over $n$. In a hot medium, an average momentum transfer is given in terms of temperature, $Q = 2\pi T$. The study of thermodynamics and phase transitions is performed in the Hamiltonian formalism starting from the thermal partition function, and the "time" is Euclidean in the path integral formalism from the beginning at any temperature. The order of the deconfinement phase transition in the Euclidean $SU(3)$ lattice gauge theory has been studied in this approach by Monte Carlo methods in Ref. [82].

In the continuum limit, in order to obtain the Minkowski action of the corresponding QFT starting from the thermal theory action in Euclidean spacetime, one conventionally adopts the Wick rotation $t \to -it$ and $A_0 \to iA_0$, relying on the analyticity property of the vector-potential. Then, an assumption that a numerical simulation successfully set up in Euclidean spacetime yields relevant results to the corresponding QFT in Minkowski spacetime would be justified only for smooth transitions between short-distance to long-distance physics enabling analytic (in physical time and in $A_0$) continuations of amplitudes from Minkowski to Euclidean spacetime and backwards. Indeed, such an assumption is violated in the most general case as stated by the so-called Maiani–Testa no-go theorem [83] related to the "failure" of the Wick rotation mentioned above. Indeed, when going out from thermodynamics approaching the study of bound states, the Wick rotation is applicable only to compute static characteristics of the QCD medium, such as vacuum condensates, as well as masses of stable particles that are the minority of the QCD spectrum. Resonances, such as the majority of mesons, charmed and stranged baryons, tetraquarks, pentaquarks, and hadron molecules, are accessible in the Euclidean space only indirectly and only under restrictive assumptions. For more details on the associated problems in the treatment of two-particle systems, see Ref. [84], while a review on the status of three-particle systems can be found, e.g., in Ref. [85].

A manifestation of non-analytic structures (domain walls) in the YM vacuum in physical time has also been discussed recently in the context of the non-stationary background of expanding Universe in Ref. [46]. Such structures were found as attractor cosmological solutions at sufficiently large physical times asymptotically matching the YM dynamics on the Minkowski background. In the essence of the Maiani–Testa theorem, such non-analytic (domain-wall) solutions found in the (nearly) Minkowski background would in general not match the corresponding lattice simulations in Euclidean spacetime, so their implications for confinement are unclear and should be studied separately. As long as such solutions are concerned, one may conjecture that the Euclidean YM field theory predictions match those in Minkowski spacetime only in regions sufficiently far away from the non-analytic phase boundaries. This conjecture, however, requires further in-depth studies of the implications of these novel solutions for confinement dynamics.

Another crucial limitation of Monte Carlo lattice simulations concerns the thermal gauge theory with non-vanishing chemical potential. Indeed, the action becomes complex if the temperature $T$ and the chemical potential $\mu$ are both non-zero, meaning that standard Monte Carlo methods fail in this case (for a thorough review on this issue, see, e.g., Ref. [86]).

In particular, due to the sign problem, the lattice simulations of QCD at $\mu_B > 0$ exhibit difficulties in reproducing the quark-gluon plasma, as observed in heavy-ion collisions, even under an assumption of the thermal equilibrium. The situation becomes even worse when considering the nuclear matter in neutron stars or collapsing black holes at very large densities in the curved spacetime. The way to proceed is to expand the pressure in $\mu_B / T$ and calculate the physical observables as Taylor expansions in this quantity, see, e.g., Ref. [87]. In practice, this requires calculating operators of high order, which are noisy and require very large statistics [88]. Recently, an alternative summation scheme for the equation of state of QCD at finite real chemical potential was proposed in [89], designed to overcome those shortcomings. Using simulations at zero and imaginary chemical potentials, the extracted LO and NLO parameters describing the chemical potential dependence of the baryon density were extrapolated to large real chemical potentials. The proposed expansion scheme converges faster than the Taylor series at a finite density, thus leading to an unprecedented coverage up to $\mu_B / T \leq 3.5$ and to more precise results for the thermodynamic observables.

## 4. Asymptotic Behavior of Large Wilson Loop VEVs

Different phases of a gauge theory are classified based on the behaviour of Wilson loop VEVs at large Euclidean times compared to spacial separations, i.e., $T_E \gg R$. Computing those in Euclidean spacetime provides direct access to the interaction energy between the static field sources in Minkowski QFT when the mass of the sources (and hence the fundamental energy scale of a confining gauge theory) is taken to infinity. Introducing a massive scalar field (a "scalar quark") in an arbitrary representation $r$ to the gauge theory on the $D$-dimensional lattice, the corresponding action

$$S = -\frac{\beta}{N} \sum_p \mathrm{ReTr}[U(p)] - \gamma \sum_{x,\mu} (\phi^\dagger(x) U_\mu^{(r)}(x) \phi^\dagger(x + \hat{\mu}) + \mathrm{c.c.}) + \sum_x (m^2 + 2D)\phi^\dagger(x)\phi(x) \tag{22}$$

is invariant under the gauge transformation of the scalar field: $\phi(x) \to G(x)\phi(x)$, where the link variable is $U_\mu^{(r)}(x)$, and the gauge-field holonomy is $U(p)$ for a given plaquette $p$.

Consider an operator that creates a particle–antiparticle pair in a colour-singlet state at a given time $T_E$ and separation $R$,

$$\mathcal{C}(T_E) = \phi^\dagger(0, T_E) \left[ \Pi_{n=0}^{R-1} U_i^{(r)}(n\hat{i}, T_E) \right] \phi(R\hat{i}, T_E), \tag{23}$$

that also creates a colour-electric flux tube (or string) stretched between the charges. In the limit of heavy static colour-charged sources, $m \gg 1$ in lattice units, the second term in Equation (22) may be considered as a small perturbation, so the string-breaking effect can be neglected to a first approximation. Indeed, as matter fields are very heavy in this limit, it would take an infinite energy to pull them out of the vacuum and to place them on a mass shell in order for them to bind to the sources and hence to screen their charge. This means that one would stretch the flux tube to an infinite length before it can ever break apart, which is, of course, an unrealistic but still useful picture to test the confinement property of the quantum vacuum.

Thus, by integrating out $\phi$ in the functional integral, one finds for the VEV

$$\langle \mathcal{C}(T_E)^\dagger \mathcal{C}(0) \rangle \sim W_r(R, T_E), \tag{24}$$

to the leading order in $1/m^2$ expansion, where

$$W_r(R, T_E) = \langle \mathrm{Tr}[U^{(r)} U^{(r)} \dots U^{(r)}]_C \rangle \equiv \langle \chi_r[U(R, T_E)] \rangle \tag{25}$$

is the VEV of the Wilson loop written in terms of the time-like holonomy $U(R, T_E)$ of the pure gauge theory. Here, the link variables run counter-clockwise on a time-like rectangular contour $C = R \times T_E$, the group character is $\chi_r$, and the sum runs over states with two static charges. In the continuum limit, the corresponding holonomy is given by the path-ordered exponential

$$U(C) = P \exp\left[ ig \oint_C dx^\mu A_\mu(x) \right].$$ 

(26)

Therefore, the Wilson loop (holonomy) operator, in this case, represents a rectangular time-like loop describing the creation, propagation and, finally, destruction of two static quark and antiquark placed at certain fixed spacial points. The time-like links in a given Wilson loop can thus be considered as the worldlines of static heavy charges.

On the other hand, in the operator formalism, one deduces that [3]

$$\langle C(T_E)^\dagger C(0) \rangle \propto \sum_n |c_n|^2 e^{\Delta E_n T_E} \sim e^{-\Delta E_{\min} T_E}, \qquad T_E \to \infty,$$

(27)

where $\Delta E_n$ is the energy of the $n$th excited state above the vacuum, and in the last part of this relation, only the dominant contribution (at large $T_E$) from the minimum-energy eigenstate has been taken into account. In this case, $\Delta E_{\min} = V_r(R)$ corresponds to the energy difference between two static charges, being, in other words, the interaction (static) potential between them $V_r(R)$. Hence, the VEV of the rectangular Wilson loop

$$W_r(R, T_E) \sim e^{-V_r(R) T_E}$$

(28)

is characterised by the potential $V(R)$, which can be inverted as

$$V_r(R) = \lim_{T_E \to \infty} \log\left[ \frac{W_r(R, T_E + 1)}{W_r(R, T_E)} \right].$$

(29)

Now consider, for instance, a planar non-self-intersecting Wilson loop in the $U(1)$ gauge theory, and using the Stokes law, it can be written as

$$U(C) = \exp\left[ ie \oint_C dx^k A_k(x) \right] = \exp\left[ ie \int_C dS_C F_{ij}(x) \right],$$

(30)

where the areal integration represents the magnetic flux and proceeds through the minimal area of the large Wilson loop. Thus, due to the additive nature of the flux, such a planar Wilson loop can be arbitrarily split into a product of smaller loops whose areas add up to the one of the large loop

$$U(C) = \Pi_{i=1}^n U(C_i).$$

(31)

Here, the orientations of the smaller loops are chosen in such a way that neighbouring contours run in opposite directions to each other. In the case of *magnetic disorder*, the magnetic fluxes through smaller loops $C_i$ (e.g., plaquette variables, in the case of smallest loops) are completely uncorrelated, such that the VEV factorises as

$$W_r(C) \equiv \langle U(C) \rangle = \Pi_{i=1}^n \langle U(C_i) \rangle = \exp[-\sigma_r A(C)], \qquad \sigma_r = -\frac{\ln\langle U(C_i) \rangle}{A'},$$

(32)

where $A$ and $A'$ are the larger and smaller Wilson loop areas, respectively.

Assuming the absence of light matter fields that could, in principle, screen the colour charge of the massive sources, and considering a rectangular Wilson loop with $C = R \times T_E$, the magnetically disordered state is characterised by the linear growth of the interaction potential with distance $R$ between the static charges asymptotically,

$$V_r(R) = \sigma_r R + 2V_0,$$

(33)

which represents a potential of a linear string. Here, $V_0$ is interpreted as a self-energy contribution, and $\sigma_r$ has the meaning of the string tension in a given group representation $r$ that does not depend on the subloop area $A'$. For an illustration of the total potential interpolating small-$R$ (Coulomb) and large-$R$ (confining) regimes, see Figure 3. The area-law for the Wilson loop VEV $\sim \exp[-\sigma_r R T_E - 2V_0 T_E]$ is then reproduced for $T_E \gg R$, as expected, or for a generic contour enclosing a large minimal area $A(C)$,

$$W_r(C) \sim \exp[-\sigma_r A(C) - V_0 P(C)], \tag{34}$$

including also a dependence on the perimeter of the contour $P(C)$. Note, the gluon propagator is singular in the UV regime in the continuum limit which generically induces a singular term that is interpreted as a divergent self-energy $V_0$ of the charged particles and antiparticles propagating in the loop. The latter produces a perimeter-law contribution to the large Wilson loop VEV in the above expression. Thus, usually, a kind of smearing of the loop via a superposition of nearby loops is required to regularise the Wilson loop in the continuum limit (see, e.g., Ref. [90]), while on the lattice, such a short-distance regularisation is always implicit.

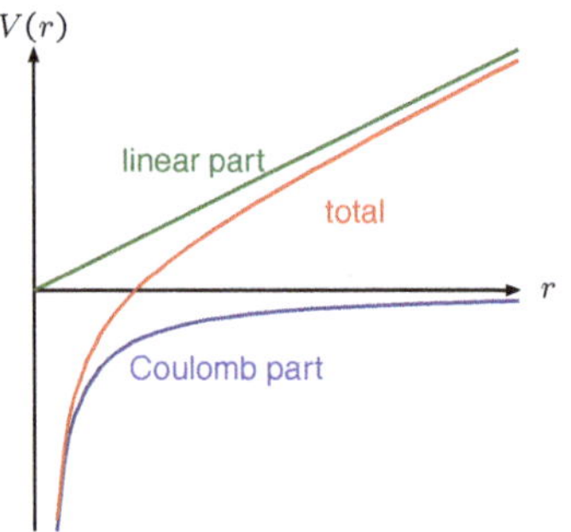


**Figure 3.** An illustration of the total static quark potential as a function of interquark separation.

It is straightforward to show that for any gauge group and $D = 2$, only a magnetically disordered phase is realised, reproducing the area-law falloff due to the absence of a Bianchi constraint on the components of the field strength tensor [91]. It is, however, a much harder problem to prove the area-law falloff of large Wilson loop VEVs in a generic YM theory with a non-trivial center symmetry, which represents the basic confinement problem (for more details, see below). A remarkable property of a Wilson loop is that it characterises vacuum fluctuations of the gauge field, i.e., without the presence of any external sources,

$$W_r(C) = \langle \Psi_0 | \chi_r[U(C)] | \Psi_0 \rangle, \tag{35}$$

with a space-like loop $C$ in terms of the ground-state $\Psi_0$ of the Hamiltonian of the pure gauge theory. As the space-like and time-like loops are related by a Lorentz transformation, one deduces that the potential energy of interaction between static charges is directly connected to the gauge-field vacuum fluctuations in the absence of colour-charged sources.

In $D > 2$ lattice, the Bianchi constraint emerges that correlates the field strength values at neighbour sites so that those no longer fluctuate independently from one point to another [92]. The absence of those correlations among the smallest Wilson loops, the plaquette variables, is the single most important requirement that provides the area-law relation for Wilson loops of arbitrary sizes. For $D > 2$, such correlations disappear, and the area-law is established in the strong-coupling limit only, i.e., in the leading order in $\beta \ll 1$. In the weakly coupled regime $\beta \gg 1$ in $D = 3 + 1$ electrodynamics, this property does not hold, and one recovers the massless phase instead with the potential [93]

$$V(R) = -\frac{g^2(R)}{R} + 2V_0, \tag{36}$$

corresponding to a perimeter-law falloff of the Wilson loop VEV,

$$W(C) \sim \exp[-V_0\, P(C)],\tag{37}$$

where $P(C) = 2T_E$ for a rectangular loop $C$ (with $T_E \gg R$), while the coupling $g(R)$ is a slow function of $R$ that approaches a constant in the Coulomb phase. In non-abelian theories, the magnetically disordered phase has been established for sufficiently large Wilson loops using the non-abelian Stokes law (see, e.g., Refs. [94–101]) and also employing a finite-range behaviour of field strength correlators [102,103]. Let us now briefly discuss one of the most distinctive features of long-range dynamics of QCD associated with Regge trajectories.

## 5. Regge Trajectories and QCD Strings

We have seen that the magnetic disorder phase manifests itself through a linear dependence of the static potential, and this behaviour is inherent to that of a string. What is the nature of such a "colour string" and how is it formed? Which phenomenological implications do such strings may have?

In hadronic scattering processes, the $t$-channel exchanges of QCD resonances are considered to be important at high energies. As suggested by quantum mechanics, a given scattering amplitude can be represented as a series expansion in partial waves,

$$A(k, \cos\theta) = \sum_{l=0}^{\infty} (2l+1) a_l(k) P_l(\cos\theta),\tag{38}$$

in terms of the Legendre polynomials of the first kind and of order $l$, $P_l(\cos\theta)$, the scattering angle $\theta$ and the partial wave amplitudes $a_l$. For a $2 \to 2$ process and particles of equal mass, for instance,

$$\cos\theta = 1 + \frac{2s}{t - 4m^2}.\tag{39}$$

Considering an exchange of a single resonance only, with spin $l_0$ and at large $s \to \infty$, the amplitude behaves as $A(s,t) \propto s^{l_0}$, such that by means of the optical theorem, the corresponding total cross-section, $\sigma_{\text{tot}} \propto s^{l_0-1}$. This result does not work very well against the experimental data for an integer value of $l_0$. The way out is to adopt that there are several resonances being exchanged in the $t$-channel that should all be taken into account. This is consistently done in the formalism of the Regge theory operating with an analytical continuation of partial amplitudes $a_l$ to the complex angular momentum plane (for a thorough discussion of Regge theory principles and applications, see, e.g., Ref. [104]). The poles in this plane are traced out by straight lines known as Regge trajectories, $l = \alpha(t)$, and are associated with particles. The squared mass of an exchanged resonance with spin $l$ corresponds to those $t$ at which $l$ is an integer. As a result of the Regge theory, the asymptotic energy dependence of the scattering amplitude reads

$$A(s,t) \to \beta(t) s^{\alpha(t)}, \qquad s \to \infty.\tag{40}$$

As a striking feature of QCD that has not been observed, e.g., in the electroweak (EW) theory, the Regge trajectories appear to be almost linear functions,

$$\alpha(t) = \alpha(0) + \alpha' t,\tag{41}$$

and one of the big questions is which dynamics could provide such a simple behaviour confirmed experimentally. Namely, hadrons of a given flavour quantum number appear to lie at almost parallel Regge trajectories.

It is clear that such a behaviour must be specific to confining dynamics of QCD. Apparently, the potential that binds the quark and anti-quark together into a meson and rises with the interquark separation linearly should be responsible for such behaviour. One adopts the physical picture of a string stretched between $q$ and $\bar{q}$ as a narrow colour-electric flux tube, which carries the energy $E = \sigma r$, so that one can neglect the quark masses. For simplicity, considering the leading Regge trajectory that maximises $l$ at a given $t$, the flux tube of length $r$ rotates about its center such that its end points move with the speed of light, and

$$\sqrt{t} = \int_{-r/2}^{r/2} dx \frac{\sigma}{\sqrt{1-v_\perp^2}} = \frac{\pi r \sigma}{2}, \tag{42}$$

in terms of the string tension $\sigma$ and the transverse velocity $v_\perp = 2x/r$. Analogously, the angular momentum of such a system

$$l = \int_{-r/2}^{r/2} dx \frac{\sigma v_\perp x}{\sqrt{1-v_\perp^2}} = \frac{\pi r^2 \sigma}{8}, \tag{43}$$

providing us finally with the Regge slope $l/t = 1/2\pi\sigma \equiv \alpha' = $ const. The latter can be extracted by fitting to the experimental data $\alpha' \simeq 0.9\,\text{GeV}^{-2}$, yielding the string tension value of $\sigma \simeq 0.18\,\text{GeV}^2 = 0.91\,\text{GeV/fm}$.

The fundamental question is how non-local string-like objects emerge from the local microscopic parton (quark and gluon) dynamics in QCD. For some peculiar reasons, the gluon field between a static quark and anti-quark gets "squeezed" into a narrow cylindrical domain, whose transverse area is nearly independent on the interquark distance—the main effect of the magnetic disorder phase. In a colour-electric flux tube picture, the energy stored in such a QCD string is proportional to the string tension $\sigma$ that can be found in terms of the colour electric field $E_i^a \equiv F_{0k}^a$ as an integral over the transverse area of the flux tube as [3]

$$\sigma = \frac{1}{2} \int d^2 y_\perp \, (E_i^a(y))^2. \tag{44}$$

Such a string then wildly fluctuates in transverse directions, and the energy of such fluctuations tends to grow with the distance between the static sources. At some critical distance, the strong fluctuations destabilise the flux tube making the longer strings less energetically favourable than the shorter ones. So, instead of indefinitely (and linearly) rising energy stored in a flux tube with its length, one encounters a string breaking effect realised due to the presence of quarks in QCD or, in a general YM theory, matter fields in fundamental representation of the gauge group. Let us elaborate on this point in some more details in what follows.

### 6. Colour Confinement and Higgs-Confinement Complementarity

A traditional and rather generic question one may ask here is what we actually mean by confinement in a gauge theory with and without matter fields that transform in the fundamental representation of the gauge group. As was discussed above, in pure non-abelian gauge theories without dynamical matter fields, the existing attempts to prove confinement consist in demonstrating the area-law dependence of $W(C)$, or equivalently, in showing linear dependence of the static quark potential at large separations[2]. As we will elaborate in more formal details below, confinement in a pure YM theory is associated with an unbroken center symmetry. Thus, the non-perturbative vacuum of QCD or, in general, a non-abelian gauge theory in the range of length-scales where the static potential satisfies a linearly-rising behaviour is considered to be in a *confined phase*.

In the presence of dynamical quarks in the theory, there would not actually be a linear static potential between heavy test quarks at asymptotically large $R$. Indeed, if one attempts to pull them apart, one eventually observes a pair creation (out of the vacuum), thus ending up with the formation of mesons at very large distances. In this picture, such a dynamical quark–antiquark pair creation occurs at the ends of the two shorter strings at the breaking point of the larger one such that the colour charge of the static charges gets effectively screened off. Such a *string breaking* or fragmentation phenomenon in QCD causes the flattening out of the static quark potential at large distances in consistency with the Regge trajectories of QCD and with the vast phenomenology of particle physics processes with hadronic final states. Such a picture has become the cornerstone of hadronisation modelling when long strings loose their stability and decay into shorter strings, yielding the spree of hadrons measurable by experiments at long distances. As we will discuss more later on, no exact center symmetry can be found in such a theory since it generically gets broken by the presence of matter in fundamental gauge-group representation. There are reasons to expect a finite range in intermediate distances where the potential could be seen as approximately linear and hence string-like. Therefore, even as confinement is an unquestionably useful way of thinking about the long-range physics of QCD, it is by far a more complex phenomenon than an assumption about an asymptotically linear static potential associated with unbroken center symmetry.

The phenomenological reality is that coloured quarks and anti-quarks at long distances are always bind together into composite states—mesons and baryons—and do not exist as isolated colour charges. This is realised in an effective string-based hadronisation picture that is proven to work very well phenomenologically in a variety of high-energy scattering processes with hadron final states (see below). The corresponding dynamics have been studied in lattice gauge theory simulations in the strong-coupling regime when matter fields are present in the action [105–107]. The resulting hadrons are automatically colour-neutral and are the true asymptotic states of QCD not the coloured quarks and gluons. Hence, sometimes QCD confinement is naively identified with *colour confinement* (also known as *C*-confinement) due to the colour charge being effectively screened away at large distances by dynamical matter fields such that the coloured partons may only propagate at short distances. However, one must be a little more careful with such an identification. If colour confinement were the only property of the confining phase, than typical Higgs theories (such as the weak interactions' theory in the SM) should also be considered as confining [3], although they do not feature such phenomena as flux tube formation and Regge trajectories [108,109]. This is why "true confinement" appears to be a more complex phenomenon, and, in addition to *C*-confinement, it should also be connected to other distinct properties of the quantum ground state, such as magnetic disorder associated with an unbroken global symmetry [3]. It does appear indeed rather obvious that *C*-confinement always accompanies the magnetic disorder phase, while the opposite may not necessarily be always true [110].

Indeed, consider an even simpler $SU(2)$-invariant gauge-Higgs theory [111], with a Yukawa-type interaction term that can be straightforwardly deduced from Equation (22). Here, the confinement regime is reproduced for small $\beta, \gamma \ll 1$ characterised by the linear rise of the static potential, followed by its flattening at large separations due to string breaking. So, this regime is very similar to the long-range dynamics of real QCD. However, at large values of $\beta, \gamma \gg 1$, one enters the Higgs regime characterised by the presence of massive vector bosons, analogues to those in the EW theory. This is the so-called *massive phase* characterised by a Yukawa-type potential for $T_{\mathrm{E}} \gg R$

$$V(R) = -g^2 \frac{e^{-mR}}{R} + 2V_0, \tag{45}$$

corresponding to a perimeter-law for a generic large planar loop $C$, $W(C) \sim \exp[-V_0 P(C)]$ with $R \gg 1/m$. In fact, in both confinement and Higgs (massive) regimes, the colour field is not detectable far from its source. Indeed, while in the confinement regime, there are only colour-singlets in the physical spectrum of this theory, in the Higgs regime, the gauge forces are the short-range ones, such that one charge screening mechanism transforms into another as the couplings change. This is due to the fact that the gauge-invariant operators in the $SU(2)$ theory that create colour-singlet states in the confinement domain are also responsible for the creation of massive vector bosons in the Higgs domain (for an early discussion on role of the EW theory operators for generation of particle spectra, see, e.g., Ref. [108]), and those states evolve into each other with varying model parameters. Whether this happens continuously or via a first-order phase transition is a subject of ongoing research in the literature, which will be discussed below.

Referring to the EW theory as a particularly important example one should be also very careful about what one actually means by the Higgs phase and the associated Higgs mechanism. Conventionally, the Higgs phase is described in terms of a Mexican-hat shape potential emerging due to the formation of classical scalar fields' (Higgs) condensates in a weakly coupled regime and, as a cause, leading to the spontaneous breaking of a given symmetry. While the gauge symmetry is manifest at the Lagrangian level, due to its spontaneous breakdown by means of the Higgs condensate, it is not a symmetry of solutions of the corresponding equations of motion. Note, however, that it is meaningless to talk about the spontaneous breaking of a gauge symmetry without specifying a certain gauge-fixing condition. Indeed, the Higgs vacuum VEV depends on the gauge choice that we make in practical calculations and can be fixed to any value by an appropriate choice of the gauge, while the actual physical observables and physical states must be gauge-invariant and do not depend on this choice. The gauge symmetry SSB phase cannot be regarded as a true physical system, provided that the gauge symmetries are redundancies of description and cannot actually break spontaneously. The latter is the statement of the so-called Elitzur's theorem [112]. Indeed, according to this theorem, a local gauge symmetry, in variance to less powerful global symmetries, can not break spontaneously such that VEVs of any gauge-noninvariant observables must be zero.

In general, in a gauge theory with fundamental-representation matter fields such as a gauge-Higgs theory, for instance, one typically does not expect to physically identify a *local* order parameter that would distinguish between the Higgs and confinement phases as qualitative descriptions of the corresponding field configurations. If there is no gauge-invariant way to distinguish between these regimes than it would be justified to attribute them to a single phase, as mentioned earlier. A discussion of this issue known as the *Higgs-confinement complementarity* goes back to as early as the late 1970s and early 1980s. In Refs. [109,113,114], by varying parameters in relatively simple lattice gauge-Higgs theories with a global symmetry, analyticity over a set of observables has been rigorously proven when going from a confining regime in the phase diagram to a regime characteristic for the Higgs phase. Although at certain large values of $\beta$, such a phase boundary emerges (see, e.g., Ref. [115]), one can find an analyticity line continuously connecting any two points in the parameter space except $\gamma = 0^3$. In other words, in those models where this is true, there would indeed be no thermodynamical phase transitions (or phase boundaries) along this path that separate the two regimes, suggesting a possible existence of a single, massive phase all along the phase diagram (see Ref. [3] for a more elaborate discussion). Can this statement be applied only for some specific models or is it always true?

This important result, first obtained in specific models, was then conjectured by some of the authors into a kind of "folk theorem" (also known as the Fradkin–Shenker–Banks–Rabinovici theorem), stating that the corresponding conclusion is expected to be always correct. Namely, if there is no local order parameter distinguishing different symmetry realisations, one should probably expect the continuity of phases. There are many examples where such a continuity has indeed been confirmed in simulations such as in transition from low- to high-temperature QCD when turning from physics of dilute gas of hadronic resonances to the physics of quark-gluon plasma (at low $\mu_B$). Indeed, in the Euclidean description of real QCD, there are certain reasons to believe that there is no thermodynamic phase transition that separates these two regimes. However, as will be discussed below, the analyticity conjecture may not actually be always true. As was argued in Ref. [58], considering a discontinuity in a *non-local* order parameter, the Fradkin–Shenker–Banks–Rabinovici theorem does not apply to models where a global symmetry is broken in the same way in both the Higgs and confinement regimes, i.e., where the Higgs fields are charged under global symmetries.

In fact, already in the string-breaking picture of hadronisation, by construction, the gluon vector-potential cannot retain its analyticity and is inherently discontinuous in the effective string-length (or string-time) scale as the string breaks apart, and no gluon field is expected to retain between the daughter strings. Whether or not the observables still remain analytic upon such a string breaking is one of the big questions for confinement models. One interesting example of the analyticity breakdown is associated with the notion of "dense QCD" or QCD at large baryon chemical potentials in the phase with broken $U(1)_B$. We will elaborate on this aspect in the end of this review.

## 7. String Hadronisation and the Lund Model

One of the existing successful realisations of the string hadronisation picture is the so-called Lund string fragmentation model [116] implemented in Monte Carlo event generators widely used in the phenomenology of particle physics, such as Pythia [117,118]. It realises the basic picture of linear confinement described above, where a flux tube is stretched between the colour-charged endpoints of the back-to-back $q\bar{q}$ system that is characterised by the string tension $\sigma \simeq 1$ GeV/fm and the transverse size close to that of the proton, $r_p \simeq 0.7$ fm. In the simplest formulation of the hadronisation model, the quarks at the endpoints are assumed to be massless and to have zero transverse momenta. As the energy transfers between the endpoint quarks and the flux tube, they move along the light cone experiencing the "yo-yo"-type oscillations. As the quarks move apart and pair-creation of dynamical $q\bar{q}$ pairs is enabled, there is non-zeroth probability for the initial "quark-string-antiquark" system to break up into smaller strings. For a simple illustration of this phenomenon, see Figure 4. Ordering the newly produced pairs as $q_i\bar{q}_i$, with $i = 1, \ldots, n-1$, into a chain along the string, depending on the initial energy of $q$ and $\bar{q}$, one eventually ends up with the production of a set of $n$ mesons, $\{q\bar{q}_1, q_1\bar{q}_2, \ldots, q_{n-2}\bar{q}_{n-1}, q_{n-1}\bar{q}\}$ moving along the $x$ axis of the initial string. The $q_i\bar{q}_i$ production vertices with coordinates $(t_i, x_i)$ have a space-like separation, with no unique time-ordering, satisfying the constraint that the produced $i$th meson must be on its mass shell, i.e., $\sigma^2[(x_i - x_{i-1})^2 - (t_i - t_{i-1})^2 = m_i^2]$.

In a more elaborate formulation, quarks have mass $m_q$, while the colour string wildly fluctuates not only in longitudinal but also in transverse directions, and the amplitude of those fluctuations tends to grow with the string length and may eventually destabilise the system causing the string to break up. The transverse momenta $p_\perp$ of the (anti)quarks are then naturally incorporated by giving $q$ and $\bar{q}$ opposite kicks in the transverse plane, with the mean square $\langle p_\perp^2 \rangle = \sigma/\pi \equiv \kappa^2 \simeq (0.25\ \text{GeV})^2$, such that the produced meson receives $\langle p_{\perp\,\text{had}}^2 \rangle = 2\kappa^2$. The virtual (anti)quarks tunnel over a distance $m_\perp/\sigma$, with $m_\perp = \sqrt{m_q^2 + p_\perp^2}$ the transverse quark mass, before they become on-shell, and the tunnelling probability of the produced pair provides an extra Gaussian suppression factor $\exp(-\pi m_\perp^2/\sigma)$.

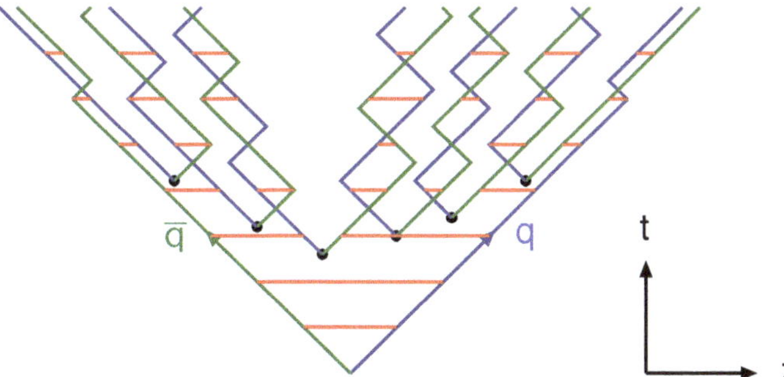


**Figure 4.** An illustration of the string hadronisation picture in the Lund model.

In the framework of the Lund model, a consistent selection of the produced DoFs is performed according to the probability distribution [116],

$$f(z) \sim \frac{(1-z)^a}{z} e^{-bm_\perp^2/z}, \tag{46}$$

implying an equilibrium distribution of the production vertices on the string

$$P(\Gamma) \sim \Gamma^a \, e^{-b\Gamma}, \tag{47}$$

where $\Gamma = \sigma^2(t^2 - x^2)$, $a, b$ are free parameters, and $z$ is the light-cone momentum fraction carried away by a produced meson. The remaining $(1 - z)$ part of the momentum is kept by the string and is then redistributed among other mesons in its subsequent fragmentation. Even though the hadron masses do not enter this approach directly, a good description of the produced particle spectra can be reached with only a few free parameters.

More complicated $q\bar{q}gg\ldots$ topologies can be introduced considering a gluon as a state with separate colour and anticolour indices, well justified in the large-$N_c$ limit [119]. The string then gets stretched between $q$ and $\bar{q}$ as usual, while each of the gluons attach at intermediate points along the string respecting the colour flow that goes in and out of each gluon. Notably, the fragmentation procedure of such a string does not require any extra free parameters [120]. The fact that there is no string that connects $q$ and $\bar{q}$ directly in this case leads to asymmetries in the produced particle spectra in consistency with experimental observations [121]. At last, baryon production can be conceptually tackled by enabling a diquark–antidiquark breaking, e.g., via sequential $q\bar{q}$ production stages (for more details on this mechanism, see, e.g., Refs. [122,123]).

## 8. Gauge Symmetry Remnants and Confinement Criteria

Due to the Elitzur's theorem [112] described above, the phases of a gauge theory cannot be distinguished by means of the breaking of any local gauge symmetry. Thus, there must be an additional, global symmetry whose breaking enables us to identify those phases, at least, when a *local* order parameter is concerned. In the Ising model, the role of such a global symmetry is played by the $\mathbb{Z}_2$ symmetry as we have noticed earlier. Fixing a covariant gauge, in general, does not eliminate the gauge freedom entirely but leaves certain remnant (both dependent and independent on spacetime coordinates) symmetries that can in principle get spontaneously broken since the Elitzur's theorem does not apply to those.

One of the examples of a possible confinement criterion known as the Kugo–Ojima condition [124,125] states that the full residual gauge symmetry in the Landau gauge $\partial^\mu A_\mu^a = 0$ must remain unbroken in order to ensure that the expectation value of the colour charge operator $\langle \psi | Q_a | \psi \rangle$ vanishes in any physical state $\psi$. The spacetime-dependent (but global) part of such a full residual gauge symmetry w.r.t. gauge transformations $A_\mu \to G A_\mu G^\dagger$ in the Landau gauge is known to take the following form [126,127]

$$G(x) = \exp\left(\frac{i}{2}\Xi^a(x)\sigma^a\right), \tag{48}$$

where

$$\Xi^a(x) = \epsilon_\mu^a x^\mu - g\frac{1}{\partial^2}(A_\mu \times \epsilon^\mu)^a + \mathcal{O}(g^2), \tag{49}$$

in terms of a finite number of arbitrary parameters $\epsilon_\mu^a$, and the $SU(2)$ gauge coupling constant $g$. Besides, for confinement to hold yet another spacetime-independent part of the full residual gauge symmetry is required to be unbroken in addition to that in Equation (48). An analogical criterion of confinement has also been formulated in the Coulomb gauge [128,129].

Thus, according to the Kugo–Ojima and Coulomb confinement criteria, the phase boundary between the confining and de-confining regimes of a gauge theory is associated with the boundary between the unbroken and broken full ($x$-dependent and independent) remnants of the gauge symmetry in Landau and Coulomb gauges, respectively. However, a problem highlighted by lattice simulations and demonstrated in Figure 5 is that these criteria predict transitions between confinement and deconfinement phases where actually no such transitions appear in the exact numerical analysis [130].

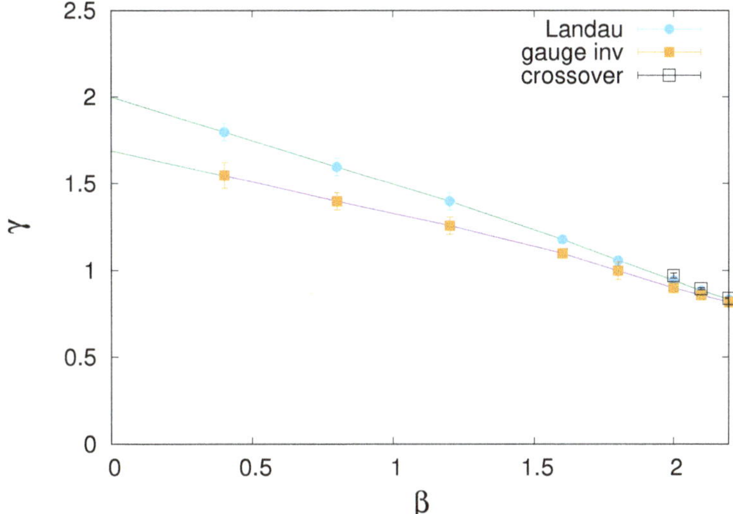

**Figure 5.** Phase boundaries of global gauge symmetry breaking obtained in Landau gauges and in the gauge-invariant approach in the $SU(2)$ gauge-Higgs theory, along with the sharp crossover line at $\beta > 2$. The figure is taken from Ref. [131].

Different remnant symmetries emergent in different gauges break at different values of the couplings, so the resulting phase boundary is in fact gauge-dependent and might indeed emerge even when there is no actual change in the physical state of the system [130]. In order to distinguish a confining state from a non-confining one, one should instead come up with a gauge-invariant criterion whose violation would indicate a true boundary between the magnetic order and disorder states that is the same in any gauge. As was discussed earlier, there is no such criterion in a gauge-Higgs theory. This might indicate that there is no such gauge-invariant separation of phases that can be attributed to a spontaneous breaking of a given symmetry, and the system is in the massive phase characterised by the string-breaking effects and a perimeter-law behaviour of Wilson loop VEVs [3].

There are compelling reasons to believe that the same picture is realised in QCD at very large separations, supported also by lattice simulations. In the gauge-Higgs theory, only in the limit of Higgs decoupling, $\gamma = 0$, the state of magnetic disorder emerges, as indicated by the area-law falloff of large Wilson loops at arbitrary large spacetime separations. The same occurs in the infinite quark mass limit in QCD such that it takes an infinite amount of energy in order to put an infinitely heavy quark–antiquark pair on its mass-shell from the vacuum such that the area-law persists to arbitrarily large string lengths.

### 9. Center Symmetry

So, when one talks about the true (gauge-invariant) separation of phases, one implies a strong first-order (non-analytic) phase transition between the magnetic order (massive) and disorder states that exists at a well-defined (unique!) combination of model parameters in any gauge. Such non-analytic behaviour is associated with a spontaneous breaking of a certain symmetry, and, to comply with the Elitzur's theorem [112], such a symmetry must be global. This type of a symmetry exists and is called the *center symmetry*—a specific subgroup of a given gauge symmetry group, which is defined as a subset of the gauge group elements that commutes with *all* the elements of the gauge group. For instance, the center of the $SU(N)$ gauge symmetry group is its $\mathbb{Z}_N$ subgroup $\{\exp 2\pi in/N\,\hat{1}_N\}$, with $n = 0, \ldots, N-1$.

Each of an infinite number of $SU(N)$ representations can be separated into $N$ possible subsets or *N-alities* depending on the corresponding representation of $\mathbb{Z}_N$ (there are only $N$ of those). Hence, each $SU(N)$ representation is characterised by the N-ality $k$ that is found as the number of boxes in the associated Young tableau mod $N$. In other words, N-ality reflects how a given representation transforms under the center symmetry subgroup of the gauge group. For instance, if for a matrix representation $M[g]$ of an $SU(N)$ group element $g$, $M[zg] = z^k M[g]$ for a center $\mathbb{Z}_N$ element $z$, one says that $g$ belongs to a representation of N-ality $k$ (for a more detailed pedagogical discussion, see, e.g., Ref. [3]). In the lattice formulation, one could show that the action (19) of a pure gauge theory is invariant under the time-like link transformation

$$U_0(\vec{x}, t_0) \rightarrow z U_0(\vec{x}, t_0), \qquad z \subset \mathbb{Z}_N, \tag{50}$$

on a fixed time slice $t = t_0$. This transformation is a particular case of the singular gauge transformation defined on a time-periodic lattice with a period $L_t$ as

$$U_0(\vec{x}, t) \rightarrow G(\vec{x}, t) U_0(\vec{x}, t_0) G^\dagger(\vec{x}, t+1), \tag{51}$$

where $G(\vec{x}, t)$ is a periodic function up to a center symmetry transformation, i.e.

$$G(\vec{x}, L_t + 1) = z^* G(\vec{x}, 1), \tag{52}$$

that also leaves Wilson loops invariant on the lattice. Such a transformation corresponds to an "almost" gauge transformation in the continuum limit,

$$A_\mu(x) \rightarrow G(x) A_\mu(x) G^\dagger(x) - \frac{i}{g} G(x) \partial_\mu G^\dagger(x), \tag{53}$$

where the second term is dropped for $t = L_t$ and for $\mu = 0$ when it turns into a delta-function.

Matter fields in the fundamental representation of the gauge group $SU(N)$, or any other fields with N-ality $k \neq 0$, break the center symmetry $\mathbb{Z}_N$ explicitly if they are not decoupled from the theory—such as the Higgs field for a non-zero coupling $\gamma$ in the example discussed above or the quark sector of real QCD (with $k = 1$) with finite quark masses. Such a breaking, which is also a necessary ingredient of the string hadronisation model (see above), causes the static potential to flatten out instead of growing linearly at asymptotically large distances as the matter fields are, in fact, responsible for the string breaking phenomenon. Gluons or other particles in the adjoint representation having N-ality $k = 0$ do not break the center symmetry so they cannot screen the colour charge of a static source if the latter has a non-zero N-ality. A well-known exception is the $G_2$ gauge symmetry, which has a trivial center subgroup, with a single unit element only, such that the gluons can bind to any source producing a colour-singlet state.

An important criterion of confinement is thus associated with the unbroken center symmetry in a pure YM theory, implying an asymptotically and infinitely rising static quark potential and signalling the area-law falloff of large Wilson line VEVs and hence the presence of the magnetic disorder state. The center symmetry can also be spontaneously broken by thermal effects, i.e., at high temperatures, in pure YM theories, causing the same effect of flattening out the static potential asymptotically as that of the matter fields. Other possible sources of the center symmetry breaking should also be considered in order to reconstruct a full picture of phases in the underlined gauge theory.

## 10. Polyakov Loop

Consider a finite (in space) lattice that is periodic in time. Such a lattice is used, in particular, in quantum statistical mechanics at finite temperatures $T$, where the partition function reads

$$Z = \sum_n \langle n| \exp(-\beta_T H)|n\rangle, \qquad \beta_T = \frac{1}{T}. \tag{54}$$

In the continuum limit of a field theory and in Euclidean time $T_{\mathrm{E}}$, the latter generalises to a path integral

$$Z = \int D\phi(x, 0 \leq T_{\mathrm{E}} < \beta_T)e^{-S}, \tag{55}$$

where the periodic boundary condition in time $\phi(\vec{x},0) = \phi(\vec{x},\beta_T)$ is imposed through an implicit delta-function. Upon lattice regularisation, the temperature is related to the lattice period in time $L_t$ as $T = 1/(L_t a)$, with $a$ being the lattice spacing as usual, and hence, $\beta_T = L_t a$ is the total time extension of the lattice.

While neither the gauge-field action nor Wilson loops are affected by the $\mathbb{Z}_N$ center symmetry transformation (50), the trace of the following holonomy winding in time around the periodic-time lattice, known as the Polyakov loop [132],

$$P(\vec{x}) = \mathrm{Tr}\Pi_{n=1}^{L_t} U_0(\vec{x}, n), \tag{56}$$

is $\mathbb{Z}_N$ non-invariant, i.e., it transforms as $P \to zP$. In the continuum limit, one can represent the Polyakov loop holonomy as follows

$$\mathcal{P}(\vec{x}) = P \exp\left(i\int dt A_0(\vec{x}, T_{\mathrm{E}})\right) = S\, \mathrm{diag}\left[e^{2\pi i\mu_1}, e^{2\pi i\mu_2}, \ldots, e^{2\pi i\mu_N}\right] S^{-1}, \quad \sum_j \mu_j = 0, \tag{57}$$

in terms of an $SU(N)$ matrix $S(x)$ that diagonalises $\mathcal{P}(\vec{x})$.

One can show that in the case of $\mathcal{P}(\vec{x})$ being a center element, all $\mu_j$ are equal, and such a holonomy determines the finite-temperature classical instanton solutions known from Refs. [133,134]. In fact, in center-projected configurations that will be discussed below, the Polyakov loop holonomies $\mathcal{P}(\vec{x})$ are the only center elements. In general, the Polyakov loop is non-trivially charged under $\mathbb{Z}_N$, meaning that its expectation value plays a role of an order parameter for the spontaneous breaking of the center symmetry. Hence, the Polyakov loop is yet another important characteristics of the confined (magnetic disorder) phase of the gauge theory, and vacuum fluctuations of the gauge field are responsible for the formation of this phase and in some ways are associated with the center symmetry.

Indeed, a difference between free energies of two states, one containing a single isolated (heavy) static charge $q$ and the other one defined in a pure gauge theory is as follows

$$e^{-\beta_T F_q} = \frac{Z_q}{Z} \propto \langle P(\vec{x})\rangle, \tag{58}$$

which is obtained by integrating out the massive quark field (in the $m \to \infty$ limit) in the path integral for $Z_q$ over the period of the lattice $0 \leq T_{\mathrm{E}} < \beta_T$. Indeed, $F_q$ for a single quark $q$ would be infinite if $\langle P(\vec{x})\rangle \to 0$, i.e., in the case of unbroken center symmetry. At high temperatures (small $\beta_T$), the center symmetry is in general spontaneously broken such that the isolated charges are described by finite-energy states (deconfining phase). A magnetic disorder-to-order phase transition associated with thermal breaking of the center symmetry is expected to occur at a critical temperature [135].

## 11. t'Hooft Loop and Center Vortices

The singular gauge transformation in the continuum limit (53), unlike the ordinary center symmetry transformation, leaves the action non-invariant. As a result of such a transform, a singular loop of magnetic flux, the so-called *thin center vortex*, is being created.

For instance, as was mentioned earlier, the holonomy for a closed space-like loop $C$ in the $U(1)$ gauge theory

$$U(C) = e^{ie\Phi_B} \tag{59}$$

is given in terms of the magnetic flux $\Phi_B$ through the loop. For a loop winding around a solenoid oriented along the $z$-axis, it is possible that $\Phi_B \neq 0$ even for a zeroth magnetic field along the closed loop, which can be obtained as a result of a singular gauge transformation applied to $A_\mu = 0$ with a discontinuous $G(x)$. If in cylindrical coordinates $\{r, \theta, z, t\}$, the corresponding transformation function $G$ has a discontinuity in $\theta$ for $r > 0$, then

$$U(C) \rightarrow e^{\pm ie\Phi_B} U(C), \tag{60}$$

where $\exp(\pm ie\Phi_B)$ is an element of the $U(1)$ group, the sign $\pm$ depends on the orientation of the loop $C$, such that a singular line of magnetic flux (thin vortex) is produced along the $z$-axis. Instead of the $z$-axis, one could introduce yet another closed contour $C'$ topologically linked to $C$ such that the singular gauge transformation operator $G$ that creates a magnetic flux along $C'$ would satisfy

$$G(\vec{x}(1)) = e^{\pm ie\Phi_B} G(\vec{x}(0)) \tag{61}$$

on the contour $C'$ determined by the parametric equation $\vec{x} = \vec{x}(\zeta)$, with $\zeta = [0, \ldots, 1]$, such that $\vec{x}(1) = \vec{x}(0)$ belong to a surface bounded by $C'$. Upon such a transformation, a Wilson loop $C$ linked to $C'$ appears to transform as in Equation (60). The *winding number* is defined as the number of times a loop goes around a fixed point in $D = 2$, while in $D = 3$, such a topological invariant generalises to the so-called *linking number* that determines the number of times two loops can wind around each other. This can be generalised further on for $D$ dimensions where a loop $C$ links to a $D - 2$ hypersurface $C'$ on which a $(D - 2)$-dimensional thin vortex is created by the corresponding singular gauge transformation, which is discontinuous in the $D - 1$ (Dirac) region bounded by the $D - 2$ hypersurface.

Switching over to the $SU(N)$ YM theory, the $U(1)$ group element that multiplies a transformation operator in Equation (61) should be replaced by a center-group $\mathbb{Z}_N$ element

$$G(\vec{x}(1)) = zG(\vec{x}(0)), \qquad U(C) \rightarrow (z^*)^l U(C), \tag{62}$$

in order for such a transform to create a thin vortex (and hence to affect the action) on the $(D - 2)$-dimensional hypersurface only and not on the Dirac $D - 1$ region that it envelops. Above, the space-like Wilson loop $C$ is topologically linked to the $(D - 2)$-dimensional thin vortex, with the corresponding linking number $l$. Upon quantisation of the non-abelian magnetic flux, its quanta are known in the literature as the *thin center vortices*, while a regularisation of the singular colour-magnetic field by smearing it out in the transverse directions to the $(D - 2)$ hypersurface leads to a vortex with finite thickness or a *thick center vortex*. For a more detailed description of the vortex configurations and properties, see, e.g., Ref. [3] and references therein.

Consider an operator $B(C)$ that creates a thin center vortex at a fixed time $t_0$ along a given loop $C$ in a $D = 3 + 1$ gauge theory [136]. If $C$ and another closed loop $C'$ are topologically linked (with $l = 1$) in a three-dimensional surface, then

$$B(C)U(C') = zU(C')B(C), \qquad z \subset \mathbb{Z}_N, \tag{63}$$

is valid. In this case, the operator $B(C)$ is known as the *t'Hooft loop*. As was demonstrated in Ref. [136], the VEV of a Wilson loop $W(C) \equiv \langle U(C) \rangle$ and a t'Hooft loop $\langle B(C) \rangle$ may satisfy either perimeter-law or area-law falloffs but not simultaneously. Indeed, the confined (magnetic disorder) phase corresponding to an unbroken center symmetry is realised when

$$W(C) \sim e^{-aA(C)} \qquad \Longleftrightarrow \qquad \langle B(C) \rangle \sim e^{-bP(C)}, \qquad a, b > 0, \tag{64}$$

while the opposite case,

$$W(C) \sim e^{-a'P(C)} \qquad \Longleftrightarrow \qquad \langle B(C) \rangle \sim e^{-b'A(C)}, \qquad a', b' > 0, \tag{65}$$

implies a spontaneously broken center symmetry (magnetically-ordered phase). Indeed, the Wilson and t'Hooft loop operators can be considered dual to each other as the first one creates a closed loop of the colour-electric flux, while the second one creates a closed loop of the colour-magnetic flux (thin center vortex) at a fixed time $t$ in both cases.

One could introduce a vortex on a finite lattice in $D = 4$ by replacing $U(p') \to \xi U(p')$ for a given plaquette $p'$ in the $SU(N)$ gauge-field action [137]

$$S = -\frac{\beta}{2N} \Big[ \sum_{p \neq p'} (\mathrm{Tr}[U(p)] + \mathrm{c.c.}) + (\mathrm{Tr}[\xi U(p')] + \mathrm{c.c.}) \Big], \qquad \xi \subset \mathbb{Z}_N, \tag{66}$$

$$U(p') = U_1(x_0, y_0, z, t) U_2(x_0 + 1, y_0, z, t) U_1^{\dagger}(x_0, y_0 + 1, z, t) U_2^{\dagger}(x_0, y_0, z, t), \tag{67}$$

which can be viewed as a change in the periodic boundary conditions, also referred to as twisted boundary conditions. Such a change creates a thick center vortex on the lattice parallel to $(z - t)$-plane, satisfying the ordinary periodic boundary conditions in $z, t$ coordinates.

In the simplest case of $SU(2)$ gauge symmetry, the (magnetic) free energy of a $\mathbb{Z}_2$-center vortex $F_m$ can be found as

$$e^{-F_m} = \frac{Z_-}{Z_+} \tag{68}$$

in terms of the partition functions with ordinary and twisted boundary conditions, $Z_+$ and $Z_-$, respectively, while the free energy of the closed colour-electric flux $F_e$ is

$$e^{-F_e} = 1 - e^{-F_m}. \tag{69}$$

It was shown in Ref. [138] that the VEV of a rectangular Wilson loop $C$ with area $A(C)$ is bounded from above as

$$W(C) \leq [\exp(-F_e)]^{A(C)/(L_x L_y)}. \tag{70}$$

A sufficient condition for the existence of a magnetic-disorder phase, and hence confinement, in terms of the behaviour of the magnetic *vortex free energy* then reads

$$F_m \sim L_z L_t e^{-\kappa L_x L_y}, \tag{71}$$

i.e., it falls off exponentially at a large $L_x L_y$ area, such as $\exp(-F_e) \simeq F_m \ll 1$. Indeed, the latter limit, together with Equation (70), implies an area-law upper bound for a large Wilson loop and, hence, the asymptotic string tension. In Ref. [139], it has been pointed out that quark confinement emerges from a vortex condensate supported by the mass gap.

## 12. Fundamental Properties of the String Tension

One of the fundamental characteristics of confinement is an non-vanishing asymptotic string tension or, equivalently, the asymptotic linearity of the static potential [3,4]. As was proven in Ref. [140], the potential is always convex and is saturated by a straight line from above. At not too large distances, the string tension for a quark in a given representation $r$ of the gauge group interacting with an antiquark can be approximated as

$$\sigma_r = \frac{C_r}{C_F} \sigma_F. \tag{72}$$

This is the property known as the *Casimir scaling*, which is strictly valid in the large-$N$ limit. Here, $\sigma_F$ is the string tension for the defining (fundamental) representation. Such a scaling can be proven in a two-dimensional theory and then to a good precision can be found also in 4D by means of the dimensional reduction [141], supported also by numerical simulations [142]. For a more recent analysis of the Casimir scaling in the $D = 2 + 1$ $SU(N)$ theory in the vortex picture, see Ref. [143]. Asymptotically at very large distances, the Casimir scaling does not hold (apart from $N = 2$ and large-$N$ cases) and can be effective at intermediate distances only.

The dimensional reduction is a specific (approximate) property of the quantum state of the theory $\Psi_0[A]$ emergent at large length-scales. According to this property, a calculation of the VEV of a large Wilson loop $W(R, T)$ in the fundamental representation in a $D = 4$ gauge theory can be sequentially reduced to that in a $D = 3$ theory [144,145] and then down to a $D = 2$ case [91]. In this case,

$$W(R, T) = \langle \mathrm{Tr}[U(C)] \rangle_{D=4} \equiv \langle \Psi_0 | \mathrm{Tr}[U(C)] | \Psi_0 \rangle \simeq \langle \mathrm{Tr}[U(C)] \rangle_{D=3} \simeq \langle \mathrm{Tr}[U(C)] \rangle_{D=2} = e^{-\sigma A(C)}, \tag{73}$$

where the last relation corresponds to the fact that in $D = 2$, the Wilson loop VEVs obey an area-law falloff. For this property to hold in the strong coupling limit, the vacuum functional should take the same form in $D = 2, 3, 4$ at large length-scales:

$$\Psi_0[A] \propto \exp\left[ -\frac{1}{4g_{\mathrm{eff}}^2} \int d^3x \, \mathrm{Tr}[F_{ij}^2] \right]. \tag{74}$$

Note that this form can not be correct at short distances in PT, so it should be regarded as an approximate and generically valid in the non-perturbative regime only. It is also not correct for Wilson loops in the adjoint representation, which follow a perimeter-law, due to the colour screening effect. An elaborate form for the vacuum functional that matches both the dimensional reduction form and the correct free-field limit has been proposed in Ref. [146] predicting the glueball mass spectrum in $D = 2 + 1$ in consistency with the lattice calculations. For other proposals, see, e.g., Refs. [147–151].

Another fundamental property of the string tension, presumably closely related to confinement, is the observation that the string tension depends only on the N-ality of the gauge group representation. For static quarks in the adjoint representation, for instance, gluons screen their charges at large distances, causing the string to break at separations $R$ satisfying $2E < \sigma_A R$, where $E$ is the gluonic energy of the produced "gluelump" state, and $\sigma_A = C_A/C_F \sigma_F$ is the string tension in the adjoint representation valid at intermediate distances. For numerical studies of the adjoint string tensions, see, e.g., Ref. [152]. While the precise form of the N-ality dependence is not known, there are several models widely used in the literature. Among them, for instance, the "Casimir scaling" proposal assumes that the string tension for the lowest dimensional representation ($k$-string tension) behaves asymptotically for the $SU(N)$ gauge theory as

$$\sigma_r = \frac{k(N - k)}{N - 1} \sigma_F. \tag{75}$$

If the true confinement phenomenon implies the formation of an electric flux tube in the form of a quantum Nambu-like string, typical predictions of the string model, such as subleading deviations from linearity of the potential as well as the spectrum of string excitations, should find their evidence in a first principle analysis of the confining gauge theories. In particular, one such prediction is a subleading $1/R$ correction term to the static quark potential emerging due to transverse fluctuations of the string known as the *Lüscher term* [153,154]

$$V(R) = \sigma_r R - \frac{\pi(D - 2)}{24} \frac{1}{R} + \mathrm{const.} \tag{76}$$

such that the VEV of a large rectangular Wilson loop can be generically parameterised as

$$W_r(R, T_E) = \exp[-\sigma_r R T_E + \tau(R + T_E) - \xi(T_E/R + R/T_E) + \eta], \tag{77}$$

where the second term in the exponent is a self-energy contribution that diverges in the continuum limit, as was mentioned above. On the lattice, one may extract the asymptotic string tension $\sigma$ as the following ratio computed at large loop areas

$$-\log\left[\frac{W(R, T_{\mathrm{E}})W(R-1, T_{\mathrm{E}}-1)}{W(R-1, T_{\mathrm{E}})W(R, T_{\mathrm{E}}-1)}\right] \to \sigma \qquad \text{for} \qquad RT_{\mathrm{E}} \to \infty, \tag{78}$$

known as the *Creutz ratio*.

Another property of the Nambu string is that the cross-section area of the string grows logarithmically with the quark separation, the effect known as *roughening* [155,156]. An agreement with Nambu string model predictions was found earlier in the analysis of closed string excitations in the $D = 2 + 1$ $SU(N)$ gauge theory in Ref. [157].

### 13. Center Vortex Mechanism of Confinement

The center vortex mechanism of confinement is strongly supported by the fact that the static potential slope depends only on the N-ality, while N-ality zero (or adjoint) string tensions vanish at asymptotically large distances. Furthermore, when adopting a picture of a pair creation of particles out of the vacuum at a certain distance causing the string to break, one implies a microscopic perturbative language of particle states in a particular configuration. While an extrapolation of perturbative particle states towards large distances may not necessarily work out well in confining theories, an effective particle picture of string breaking is still considered to adequately reflect the reality, at least qualitatively. In proper path integral computations, one sums over all possible field configurations that should provide the same result for the gauge-invariant observables (such as Wilson loop VEVs) as the phenomenologically successful effective particle picture of the string breaking. Ultimately, one would like to find out how the vacuum field fluctuations induce N-ality dependence of the asymptotic string tension and describe colour screening of the static sources [3].

While instantons [158] are saddle points of the classical gauge-field action, vortices are interpreted to be saddle points of the effective one-loop action [159,160] that incorporates the vacuum polarisation effects, and hence have a pronounced fundamental meaning (see also Ref. [4]). Fluctuations of center vortices that can be identified as solitonic objects in typical field configurations are known to give rise to an area law of Wilson loops. A remarkable property is that Wilson loops in different representations but with the same N-ality get the same contributions from center vortices, while loops of N-ality zero are not affected. This follows from the simple fact that the creation of a vortex linked to the loop $C$ affects the loop holonomy of a given N-ality $k$ as $U(C) \to zU(C)$ and its VEV as $W_r(C) \to z^k W_r(C)$ for $z$ from the center group $\mathbb{Z}_N$ of $SU(N)$.

In a more generic case, consider a set of vortices linked to a given loop $C$, with linking numbers $l_{1,2,3,\ldots}$ having the center elements $z_{1,2,3,\ldots}$. Then, the creation of this set modifies the Wilson loop VEV as $W_r(C) \to Z^k(C)W_r(C)$, where $Z(C) = z_1^{l_1} z_2^{l_2} z_3^{l_3} \ldots$. In the vortex picture of confinement [136,159,161,162] (see also Ref. [3] and references therein), the gauge-field vacuum configuration is considered to be a set of vortices superimposed on a non-confining configuration. Then random fluctuations in a number of vortices in the system as well as in their linking numbers to a given Wilson loop $C$ induces the area law dependence of the corresponding Wilson loop VEV. The loop holonomy can be represented in a factorised form $U(C) = Z(C)u(C)$, where $u(C)$ is a contribution from a non-confining background, and $Z(C) \subset \mathbb{Z}_N$ is a center-valued holonomy. Then, the vortex mechanism implies the factorisation of the Wilson loop VEV

$$\langle \chi_r[U(C)] \rangle \simeq \langle Z^k(C) \rangle \langle \chi_r[u(C)] \rangle \simeq \exp[-\sigma_r A(C)] \exp[-\mu_r P(C)]. \tag{79}$$

A detailed proof relies on a weak correlation between $Z(C)$ and $U(C)$, as well as between $Z(C_1)$ and $Z(C_2)$ for any large loops $C, C_{1,2}$, and can be found for instance in Ref. [3]. It manifestly demonstrates that the string tension computed for smaller loops is the same as that for the larger ones provided that the above assumptions hold and $Z(C_i)$ experience independent fluctuations.

Numerical estimates [163] suggest that the thickness of the vortex is close to one fermi, so, in principle, the Wilson loops with an extension below this scale may get affected. As was demonstrated in Refs. [164,165], such a vortex thickness plays an important role for generating the Casimir scaling at intermediate distances. At large distances dominated by large Wilson loops, the N-ality dependence of the linear static potential is reproduced as expected. From this point of view, vortices are non-local objects that represent specific field configurations that lead to an asymptotic string tension as a function of N-ality.

The link configurations $\mathcal{U}_\mu(x) = g(x)z_\mu(x)g^{-1}(x+\hat{\mu})$ that produce $Z(C)$ holonomies can be transformed into the link configurations $z_\mu(x)$ of the $\mathbb{Z}_N$ lattice gauge theory responsible for confinement by means of a specific $SU(N)$ gauge transformation $g(x)$. The thin vortices then have a meaning of excitations of the center-group $\mathbb{Z}_N$ lattice gauge theory. The original link variables $U_\mu(x)$ get separated into a product of center elements $z_\mu(x)$, and the link variables of the non-confining background $V_\mu(x)$ by the $g(x)$ transform

$$U_\mu(x) = g(x)z_\mu(x)V_\mu(x)g^{-1}(x+\mu). \tag{80}$$

The main aim of the vortex mechanism of confinement is to find a specific $g(x)$ for a given non-confining background $V_\mu(x)$, typically assumed to be a small fluctuation about the unity. Locations of center vortices can then be extracted from $z_\mu(x)$ after the above factorisation $U \to zV$ has been performed [166]. One such $g(x)$ transforms the DoFs into a specific gauge known as the direct maximal center gauge where the deviation of the links in the adjoint representation from the identity matrix is minimal, or where the quantity

$$K = \sum_{x,\mu} \mathrm{Tr}[U^A_\mu(x)] = \sum_{x,\mu} \mathrm{Tr}[U_\mu(x)]\mathrm{Tr}[U^\dagger_\mu(x)] - 1, \tag{81}$$

with the adjoint link $U^A_\mu(x)$ is maximal. Locations of center vortices can then be extracted in a dedicated Monte Carlo procedure from the identified center elements $z_\mu(x)$ once the center mapping (projection) $U_\mu(x) \to z_\mu(x)$ has been performed. If the product $Z(p)$ of $z_\mu(x)$ on the projected $\mathbb{Z}_N$ lattice around a plaquette $p$ satisfies $Z(p) \neq 1$, a thin vortex (or $P$-vortex) is then located on that plaquette. The vortex picture of confinement then reduces to a consideration of $P$-vortices as random surfaces percolating through the spacetime volume. Uncorrelated piercings by the $P$-vortices on a given planar surface correspond to uncorrelated large center-projected loops. The numerical procedures, however, may fix the projected lattice to only one out of a large amount of local maxima of the gauge-fixing functional $K$ known as the *Gribov copies* [167], not straight to its global maximum, which is considered to be a problem in several widely used center-gauge fixing approaches.

The problem of Gribov copies is one of the main obstacles for a consistent treatment of the confinement problem. Considering a set of gauge-equivalent configurations of the gauge field known as a gauge orbit and imposing a gauge-fixing condition as a certain hypersurface in a space of gauge field configurations, the Gribov copies can be visualised as many possible intersections of the gauge orbit with the gauge-fixing hypersurface. Summing over all contributions from Gribov copies in a path integral, the latter may actually vanish since those contributions come with opposite signs and may mutually eliminate each other in a given observable. This is the statement of the Neuberger's theorem [168] rendering BRST quantization not well defined in the non-perturbative regime of a gauge theory (see a detailed discussion in Ref. [3]). One possibility is to restrict the functional integral to a subspace of gauge configurations with a positive Faddeev–Popov determinant, the so-called Gribov region, and its boundary also containing the lowest non-trivial eigenmode with zeroth eigenvalue is called the Gribov horizon. An instructive example of such a hypersurface restricted to the Gribov region is the Landau gauge fixing condition that minimises the functional

$$R = -\sum_{x,\mu} \mathrm{Re\,Tr}[U_\mu(x)], \tag{82}$$

such that the corresponding Gribov region consists of all possible minima of $R = R[A]$ for a given gauge orbit. However, various gauge orbits might cross the Gribov region a different number of times leading to different weights assigned to different gauge orbits. A proposal to consider only unique global minima of $R[A]$ functional for each gauge orbit [169] may be very difficult to realise in practical calculations. Furthermore, there is no reason to believe a particular Gribov copy with the global minimum for $R[A]$ is more physical than other local minima. Lattice procedures, in general, assume that a particular choice of a Gribov copy would not make a big difference on the numerical results.

In order to establish a direct connection between the existence of $P$-vortices and a magnetically disordered phase, following the reasoning of Ref. [3], let us first consider whether the center-projected $Z_\mu(x)$ link variables (extracted, for instance, in a maximal center gauge) are responsible for the confinement. For this purpose, it is instructive to consider the VEV of the rectangular $R \times T_{\mathrm{E}}$ Wilson loop, $W(R, T_{\mathrm{E}})$, defined in Equation (77). If such a loop is constructed from $Z_\mu(x)$ links on a center-projected lattice, the corresponding Creutz ratio (78) appears to converge much faster to $\sigma$ than for the unprojected Wilson loop VEV. Already at $R = 2$, the static potential becomes linear—the property of the so-called precocious linearity. The fact that the asymptotic string tensions extracted from center-projected and unprojected Wilson loop VEVs at large $R$ are the same is known as the *center dominance*. There is also an excellent agreement of the Creutz ratios on the center-projected lattice with the well-known predictions of the asymptotic freedom for large $\beta$ (small gauge couplings).

A slow convergence of the Creutz ratio (78) to the string tension at large $R$ in the unprojected case means that we deal with thick vortices linked to large Wilson loops here. The center projection effectively shrinks the thickness of the vortices down to a single lattice spacing, so the linking appears to be relevant already for small center-projected loops. Indeed, as was deduced earlier, $P$-vortex piercings are totally uncorrelated on a planar surface already causing the linearity of the potential at small distances. One naturally wishes to establish that each thin $P$-vortex in the projected configurations matches a thick center vortex in an unprojected lattice in order to prove that the $P$-vortices do not carry artefacts of the gauge fixing procedure and indeed are responsible for the underlined physics of magnetic disorder (and hence confinement).

As thoroughly described in Ref. [3], one way of proving the relevant correlation of $P$-vortices with gauge-invariant observables (unprojected Wilson loops) is to compute a so-called vortex-limited Wilson loop VEV defined as an expectation value of an ordinary unprojected loop holonomy $W_n(C)$ but taken in the ensemble of configurations where the minimal surface area of the loop is pierced by $n$ $P$-vortices. Then, considering for simplicity the $SU(2)$ theory, if the ratios asymptotically behave as $W_n(C)/W_0(C) \to (-1)^n$ provided that $\langle Z(C) \rangle = (-1)^n$ ($-1$ per each vortex piercing), then the procedure of finding thin $P$-vortices on the center-projected lattice effectively locates thick center vortices on the unprojected lattice. This, indeed, has been confirmed by lattice simulations, see, e.g., Ref. [170].

Another test proposed in Ref. [171] suggests to insert a thin vortex found by the center projection operation into a thick vertex on the unprojected lattice and then to check if their disordering effects, due to center dominance, cancel out asymptotically at large distances. Indeed, an explicit calculation shows that this procedure eliminates the string tension and hence the disorder effect. It was also checked in Ref. [172] that the $P$-vortex density is independent on the lattice spacing in the continuum limit, as expected for physical objects. An additional observation of Ref. [173] revealed that the continuum action density appears to be singular at the location of $P$-vortices, which, together with their constant density, signals an intricate cancellation between action and entropy at a surface of infinite action associated with a vortex.

As was discussed above, at finite temperatures $T$ in a time-periodic lattice, the Polyakov loop VEVs determine the quark free energy $F_q$. In the $SU(2)$ gauge theory, at $T > T_c = 220$ MeV, a deconfinement transition occurs when $F_q$ becomes finite and the static quark potential goes flat. However, even at large $T > T_c$, space-like Wilson loops retain their area-law falloff such that vacuum fluctuations inherit some of the key properties of the confined phase.

This observation fits well with the center-vortex mechanism of confinement [3]. At low $T$, due to uncorrelated piercings of the minimal loop areas, one finds $\langle P(\vec{x}) \rangle = 0$ and an exponential falloff of the Polyakov loop correlators for large interquark separation $\langle P(\vec{x}) P(\vec{x} + \vec{R}) \rangle \sim \exp[-\sigma(T) L_t R]$, with $\sigma(T)$—the $T$-dependent string tension of a flux tube stretched between $q$ and $\bar{q}$. Since the vortices running in space-like directions have a finite diameter, as the temperature rises, they get squeezed by the reduced finite lattice extension in time $L_t$ until they effectively stop percolating, eliminating the exponential falloff of the Polyakov loop correlator and hence $\langle P(\vec{x}) \rangle$ is no longer zero [163,174]. The asymptotic behaviour of the space-like Wilson loop, however, is determined by the piercings of center vortices oriented in periodic time (i.e., running in timelike directions), and their cross-section is not limited by a small extension in the time direction at large $T$. Thus, the corresponding $P$-vortices keep percolating on a time slice in the spacial directions such that the exponential falloff of space-like Wilson loops remains unaffected in the deconfined regime [175,176].

As we already discussed above, the center symmetry turns out to be explicitly broken by the dynamical fields in the fundamental representation. The center dominance in the confinement region in the $SU(2)$ gauge-Higgs theory has been tested in Ref. [177]. In a region where the screening effects by the matter fields become important, the center vortices do not disappear but somehow rearrange themselves in order to allow for asymptotically vanishing string tension while still generating a linear slope in the potential at intermediate distances (no signature of linearity has been found in the Higgs region at any scale). In the presence of matter fields, the Dirac volume shrinks and the vortex piercings of the Wilson loop minimal area are expected to become correlated at large distances, but to the best of our knowledge, there is no full consensus on exactly how this occurs.

## 14. Chiral Symmetry Breaking and Topological Charge

The global chiral symmetry of QCD light $u, d$ quark sector $SU(N_f)_\mathrm{R} \times SU(N_f)_\mathrm{L}$ (with the number of flavours, say, $N_f = 2$) is broken spontaneously by the order parameter known as the quark (or chiral) condensate $\langle \bar{q} q \rangle \neq 0$. In addition, it is also broken explicitly by the light current quark mass turning the Goldstone bosons, the pions, into massive pseudo-Goldstone states. Another less known mechanism based upon the linear sigma model of effective quark-meson interactions introduces yet another source of the global chiral symmetry breaking through a linear term in $\sigma$-field proportional to the quark condensate. Such a breaking is also explicit, and as such, it provides an additional finite contribution to the pion mass. A symmetry breaking due to the quark condensation phenomenon is often referred to as *dynamical symmetry breaking* and is considered a baseline for Technicolour models of EW symmetry breaking [178,179] (for a detailed review of existing concepts, see, e.g., Ref. [180]).

As was discussed earlier in the case of Ising model, in order to get a nontrivial value of the order parameter one should perform two limits in a certain order—first, take volume to infinity and then set the quark masses to zero. This procedure leads to the well-known Banks–Casher relation [181] between the chiral condensate as the trace of the quark propagator and the value of the density of the close-to-zero eigenvalues of the Dirac operator characterised by vacuum field configurations. The latter density receives no perturbative contributions, and hence, the dynamical chiral symmetry breaking is an intrinsically non-perturbative phenomenon.

Provided that in real QCD with light quarks, the string tension vanishes asymptotically due to colour-screening and string breaking, the chiral condensate by itself is not tied to the area-law falloff of large Wilson loop VEVs and does not even require the presence of gauge fields, in analogy to the effective Nambu–Jona-Lasinio model [182]. Naively, one might think that these observations indicate no immediate connection between the chiral symmetry breaking mechanism and the confinement phenomenon. As was emphasised in Ref. [183], the low-lying Dirac eigenmodes, which are crucial for chiral symmetry breaking, provide vanishingly small contributions to the string tension and to the Polyakov loop in both confined and deconfined phases. These observations provided no indication of an immediate correspondence between chiral symmetry breaking and confinement.

Interestingly enough though, the critical temperatures of chiral and deconfinement phase transitions appear to be the same or close to each other, as suggested by lattice simulations, motivating a further search for possible hidden connections between the two transitions. In particular, a connection between the Polyakov loop, center symmetry, and the chiral condensate may be due to the fact that, after integrating out fermions, the chiral condensate is basically a complex expectation value of many Wilson loops, including those wrapping around compact dimensions. As was elaborated in detail in Ref. [184], the spectral properties of the Dirac operator are affected by confinement, in particular, causing the correlators of Dirac eigenvector densities to decay exponentially instead of a power law in the deconfined phase. Ultimately, one would need to establish a link between the spectral properties of the Dirac operator in the infrared regime presumably responsible for chiral symmetry breaking with those in the ultraviolet regime tightly connected to confinement.

Remarkably, in Refs. [171,185], it was shown that the chiral condensate vanishes as soon as vortices are removed from the underlined field configurations, while the chiral condensate values are notably larger in center-projected configurations than those on the unmodified lattice. This observation shows that the center vortices are responsible not only for magnetic disorder but also determining the chiral symmetry breaking—thus, both phenomena are tightly connected [186].

It is well known that the axial symmetry $U(1)_A$ of the classical QCD action is broken by the chiral anomaly at the quantum level. The topological charge given by the integral of the divergence of the axial current,

$$Q = \frac{1}{32\pi^2} \int d^4x \epsilon^{\mu\nu\alpha\beta} \, \mathrm{Tr}[F_{\mu\nu} F_{\alpha\beta}], \tag{83}$$

receives contributions from finite action configurations known as instantons [158]. Due to the Atiyah–Singer Index theorem, the integer $Q$ value has a meaning of a difference of numbers of zero modes of the Dirac operator with positive and negative chiralities. The $\eta'$ meson, which would have been a (pseudo-)Goldstone boson of $U(1)_A$ breaking, appears to be way too heavy phenomenologically (above 1 GeV). Its mass is found to be proportional to the topological susceptibility found in the pure gauge theory in the chiral and large-$N_c$ limits, i.e.,

$$m_{\eta'}^2 \simeq \frac{2N_f}{f_\pi^2} \chi, \qquad \chi = \frac{\langle Q \rangle}{V}, \tag{84}$$

—the relation known as the Veneziano–Witten formula [187,188]. Here, $V \to \infty$ is a large volume, and $f_\pi$ is the pion decay constant. For lattice calculations of the topological susceptibility and tests of the Veneziano–Witten formula, see, e.g., Refs. [189,190].

The topological susceptibility $\chi$ is characterised by the vacuum quantum-field fluctuations in a pure gauge theory without any quark fields. Like the chiral condensate, the density of the topological charge may not seem to immediately connect to the IR property of confinement, and naively, one would guess that it may be determined by non-confining configurations, such as instantons in the standard picture. However, as was shown in Ref. [191], a $P$-vortex acquires a fractional topological charge at "writhing" points, and it is possible to get a correct topological susceptability in certain vortex models [192]. Moreover, the results of Ref. [171] actually demonstrate that the topological charge tends to vanish upon vortex removal, while in Ref. [193], it was shown that $\chi$ computed from

$P$-vortices appears to be consistent with the measurements. Therefore, the initial naive guess do appear to be wrong, and confinement plays a crucial role here as well.

Yet another, more recent, test of the vortex mechanism considering the effective quark propagator in the Landau gauge in the following IR form

$$S(k) = \frac{Z(k)}{i\not{k} + M(k)} \tag{85}$$

has been performed in Refs. [186,194] (see also Refs. [3,4] for a pedagogical discussion). With an appropriate smoothing ("cooling") procedure in the $SU(3)$ gauge theory that eliminates short-distance fluctuations, the effective mass $M(k)$ and renormalisation $Z(k)$ functions have been computed for the full, vortex-only and vortex-removed configurations and compared to each other. Removing the vortices causes the mass function to plummet dramatically—see Figure 6—while the full and vortex-only results have appeared to be essentially the same, hence demonstrating a critical role of the vortices in dynamical mass generation and chiral symmetry breaking. The maxima of the action for vortex-only configurations appear to resemble those of instantons, while the number density of those objects is notably similar for the full and vortex-only configurations and by far much larger than that for the vortex-removed case. It seems likely that vortices and instantons are indeed connected in some very non-trivial way.

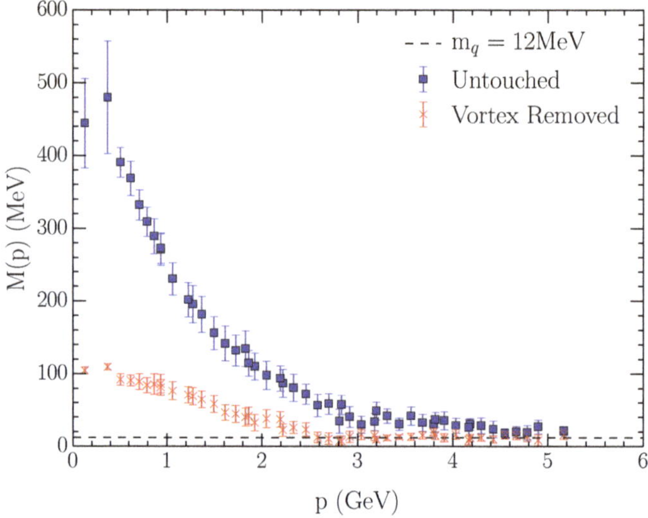


**Figure 6.** The mass function of the effective quark propagator computed accounting for the full and vortex-removed configurations. The figure is taken from Ref. [195].

Remarkably, the center vortices thus appear to describe a number of fundamentally important IR phenomena in non-abelian gauge theories in a gauge-invariant way. Nevertheless, there are also weak points in the vortex mechanism of confinement that require further clarification and, in a perspective, a more complete understanding of, for instance, the Gribov copies problem and a lack of natural explanation of the Lüscher term. Further, a more complete theory of vortices should address these issues, hopefully, on a first-principle basis.

A lack of a perfect consistency of the vortex scenario with full numerical results for the $SU(3)$ gauge theory has also emerged in the literature. For instance, center projection in the $SU(3)$ case yields 2/3 of the asymptotic string action computed on the full lattice [196]. However, consistency has been substantially improved by means of a certain gauge-field smoothing procedure [194], so this may not be regarded as a critical problem.

In order to make the next step in our understanding of the vortex dynamics, it may be enlightening to suggest an EFT of vortices as non-local dynamical objects—fluctuating surfaces—in $D$ dimensions, where all the IR phenomena described above would emerge naturally among its key predictions. Such a theory known as the *random surface model* that resembles a string theory on the lattice has been proposed and elaborated, e.g., in Refs. [191,192,197–199].

In order to build the simplest $D = 4$ action density of vortices in this framework, one considers an extrinsic curvature of the vortex worldsheet multiplied essentially by a single coupling, while the additional Nambu-like string term proportional to the area of the vortex worldsheet appears to be redundant and can be omitted. In the $SU(2)$ version of this model, one assigns $(-1)^n$ to the Wilson loop holonomy for the number of vortex piercings $n$ per minimal loop area, and then one averages it over an ensemble of center vortex configurations. The latter can be generated by Monte Carlo methods for a lattice action density given by the number of cases when a single link is shared by two adjacent orthogonal vortex plaquettes.

In order to compute the topological charge density in this model, for instance, one employs a weighted stochastic procedure of introducing the monopole lines to the surface of each vortex plaquette (see Section 16 below for a brief description of the magnetic monopoles' scenario of confinement). The topological susceptibility appears to be insensitive to the monopole lines' density—a sign of strong predictive power of the model. Besides, the model correctly predicts the emergence of the chiral condensate at $T < T_c$ and the restoration of the chiral symmetry at $T > T_c$, with a critical temperature of the transition $T_c$. A variation in the lattice time extension can provide a temperature dependence, and the second-order deconfinement phase transition has been found. The single dimensionless coupling and the lattice spacing $a$ determine a wide range of long-distance non-perturbative phenomena and were fixed through a matching to the physical $T_c / \sqrt{\sigma}$ and $\sigma / a^2 = (440 \text{ MeV})^2$, in terms of the string tension $\sigma$. Upon such a matching, the temperature-dependent values of $\sigma$, the chiral condensate and $\chi$ are shown to be in agreement with the full theory. Remarkably, in the case of $SU(3)$ gauge theory, the random surface model predicts the electric flux tubes in a form of $Y$-shaped string junctions for baryons (three-quark systems) [200], which is also in agreement with the numerical results of Refs. [201,202].

An alternative EFT approach to dynamics of vortices was suggested in Ref. [162] that is based upon a gauge theory with an adjoint matter field and its gauge-invariant mass term, which provides a mass for the gauge field via the Higgs mechanism. Besides the vortex solutions, it also naturally reveals another type of solutions with magnetic monopoles running along the vortex sheets that are necessary to generate a topological charge.

For a more thorough discussion on the existing vortex-based scenarios, we refer the reader to Ref. [3]. Now, we turn to alternative scenarios of confinement, yet trying to connect them with the existence of vortices whenever possible.

## 15. Gribov-Zwanziger Scenario, Non-Perturbative Propagators and Gluon Chains

Starting from the Coulomb gauge, in Refs. [167,169,203], it has been suggested that very small eigenvalues of the Faddeev–Popov operator that are located close to the Gribov horizon contribute the most to the Coulomb potential $V_C(R)$ and could in principle enhance it to a linear form (see also Ref. [204]). This is the so-called Gribov–Zwanziger scenario of confinement. As it should be for a confined phase, numerical analysis on the lattice demonstrates the linear rise of Coulomb potential $V_C(R)$, which is basically a separation-dependent part of the interaction energy of the physical $q\bar{q}$ state defined as

$$R \to \infty, \qquad V_C(R) \to V(R, T = 0), \qquad V(R, T) = -\frac{d}{dT} \log[G(R, T)],$$

$$G(R, T) = \langle \Psi_{q\bar{q}} | e^{(H - E_0)T} | \Psi_{q\bar{q}} \rangle, \qquad \Psi_{q\bar{q}} = \bar{q}(0) q(R) \Psi_0, \tag{86}$$

with the ground-state of the theory $\Psi_0$, the vacuum energy $E_0$ and with self-energy contribution neglected at large $R$. However, the slope of the extracted Coulomb potential $V_C(R)$ is significantly (for a factor of 2–3, depending on the gauge coupling) larger than that of the static quark potential $V(R) \simeq \lim_{T\to\infty} V(R, T)$ obtained by gauge-invariant methods [129]. Although the latter is in agreement with Zwanziger inequality [205], $V(R) \leq V_C(R)$, the potential is overconfining, prompting discussions in the literature on whether the Coulomb potential in this formulation actually is the full story of confinement or some crucial ingredients are still missing. It is worth mentioning, however, that the asymptotic string tension of the Coulomb potential appears to vanish as soon as vortices are removed from the underlined gauge field configurations, rendering the importance of the vortices for understanding the confinement phenomenon in the Coulomb gauge [206]. Such configurations without vortices in fact behave as perturbations of the free gauge theory, in consistency with expectations.

In the confined phase, the Coulomb self-energy of an isolated static charge $\mathcal{E}$ is expected to be infinite, and the main condition for that reads

$$\mathcal{E} \propto \int d\lambda \left\langle \frac{\rho(\lambda)F(\lambda)}{\lambda} \right\rangle \to \infty, \qquad \lim_{\lambda\to 0} \frac{\rho(\lambda)F(\lambda)}{\lambda} > 0, \qquad F(\lambda) = \langle \phi_\lambda|(-\nabla^2)|\phi_\lambda\rangle, \tag{87}$$

where the first relation relies on the continuum limit of small eigenvalues $\lambda \to 0$ of the Faddeev–Popov operator, with the corresponding eigenstates $\phi_\lambda$ and density of the eigenvalue distribution $\rho(\lambda)$. Using the lattice methods, it was found that [204]

$$\rho(\lambda) \sim \lambda^{0.25}, \qquad F(\lambda) \sim \lambda^{0.38}, \tag{88}$$

yielding a divergent $\mathcal{E} \to \infty$ and hence satisfying the confinement criterion (87). An enhancement of $\rho(\lambda)$ and $F(\lambda)$ close to the Gribov horizon $\lambda \to 0$ seems to be associated with the role of a center vortex ensemble. However, as was advocated in Ref. [129], the Coulomb force appears to be confining also at temperatures above the deconfinement phase transition temperature, which contradicts the fact that a confining potential must be associated with a phase of magnetic disorder.

The linear confining Coulomb potential in the Gribov–Zwanziger scenario can be associated with the instantaneous part of the two-gluon correlator. So, confinement could be effectively considered as an emergent property due to a gluon exchange with a non-perturbative (dressed) gluon propagator. A naive calculation shows that a linear potential may arise if the propagator of the gluon exchange scales with momentum transfer as $\sim 1/k^4$ at $k \to 0$, at least, in one of the possible gauges [207]. One typically attempts to analyse the IR behaviour of the effective gluon and ghost propagators and vertices using the formalism of the Dyson–Schwinger equations following from the disappearance of the functional integral of a total derivative,

$$\left\langle -\frac{\delta S}{\delta \phi_i(x)} + j_i(x) \right\rangle = 0, \tag{89}$$

with subsequent differentiation over the sources $\{j_k\}$. For a review on phenomenological implications of the Dyson–Schwinger approach, see, e.g., Ref. [208] and references therein.

The full gluon and ghost propagators in Euclidean spacetime are conventionally represented in terms of form factors as

$$D_{\mu\nu}^{ab}(k) = \delta^{ab}\left(\delta_{\mu\nu} - \frac{k_\mu k_\nu}{k^2}\right)\frac{Z(k^2)}{k^2}, \qquad G^{ab}(k) = \delta^{ab}\frac{J(k^2)}{k^2}, \tag{90}$$

respectively, such that their IR behaviour, as the virtuality of the exchange vanishes $k^2 \to 0$, is controlled by

$$Z(k^2) \propto (k^2)^{-\kappa_{\rm gl}}, \qquad J(k^2) \propto (k^2)^{-\kappa_{\rm gh}}, \tag{91}$$

where $\kappa_{\mathrm{gh}}$ and $\kappa_{\mathrm{gl}}$ are the so-called IR critical exponents (or anomalous dimensions) to be determined in the calculations.

A necessary condition for the Kugo–Ojima confinement criterion is that the ghost propagator features an enhanced (stronger than $1/k^2$) IR singularity, i.e., $\lim_{k\to 0}[J(k^2)]^{-1} = 0$, known as the horizon condition [209]. The second condition is the vanishing gluon propagator, $\lim_{k\to 0}[Z(k^2)/k^2] = 0$. This is the exactly case for the so-called scaling solution [210–212] that implies a specific relation between $\kappa_{\mathrm{gl}}$ and $\kappa_{\mathrm{gh}}$ in $D$-dimensions [209,210,212,213]

$$\kappa_{\mathrm{gl}} + 2\kappa_{\mathrm{gh}} = -\frac{4 - D}{2}. \tag{92}$$

For the $D = 4$ case, the values are found to be $\kappa_{\mathrm{gh}} \simeq 0.595$ and $\kappa_{\mathrm{gl}} \simeq -1.19$, such that the gluon propagator indeed tends to vanish at $k \to 0$. In order to explain confinement, it was argued in Ref. [214] that the quark-gluon vertex should be sufficiently singular in the long-distance limit, such that its combination with a non-singular gluon propagator gives rise to the confining potential. The scaling solution has been confirmed by a lattice analysis of Ref. [215] in the $SU(2)$ gauge theory in the Landau gauge and only in $D = 2$ dimensions, but it was not observed for $D > 2$ [216,217].

Another well-known solution, the so-called decoupling solution, with

$$\kappa_{\mathrm{gl}} = -1, \qquad \kappa_{\mathrm{gh}} = 0, \tag{93}$$

has been proposed, e.g., in Refs. [218–220]. This solution corresponds to a saturated form of the IR gluon propagator tending to a constant and, hence, effectively decouples from the dynamics in analogy to a massive particle. It is worth noticing here that the non-perturbative gluon propagator does not behave as a propagator for a massive state. Indeed, from numerical simulations, one observes indications of a violation of positivity, in consistency with the fact that no coloured gluons exist in the asymptotic spectrum of a gauge theory that is traditionally connected to gluon confinement [221,222]. Besides, the decoupling solution implies a simple $1/k^2$ pole for the ghost propagator. This solution appears to be favoured by known lattice simulations for $D > 2$, which also indicate a disagreement with the Kugo–Ojima criterion. A more generic criterion for quark confinement applicable in arbitrary gauges relying on the IR behaviour of ghost and gluon propagators has been proposed in Ref. [223].

One would remark here that the primary probe for the magnetic disorder phase is, of course, the area-law falloff of gauge-invariant observables, Wilson loop VEVs, and not the gluon propagator itself, which is not a gauge-invariant object. So one should be extra careful in interpreting the IR behaviour of the propagator in order to avoid spurious results. For recent comprehensive effort to obtain a linear static potential in the framework of Dyson–Schwinger formalism in a Coulomb gauge, see Ref. [224]. A thorough analysis of the Polyakov line VEVs and effective potential based upon the formalism of the Functional Renormalisation Group [225] has been performed in Refs. [226,227], and an agreement with lattice results has been found. However, the search for the area-law dependence of large Wilson loops' VEVs with these methods has not been successful so far.

The picture of strongly collimated colour-electric flux tubes stretched between the colour-charged static sources does not seem to apply to the distribution of colour-electric field in the Coulomb gauge [3]. Indeed, there is a significant long-range dipole contribution to the Coulomb electric field that would cause rather strong van der Waals-type forces between hadrons at large distances. This would immediately contradict to the mass gap existence [139] that requires only short-range forces between composite colour-neutral states. This problem generically emerges in any confinement scenario, such as the Dyson–Schwinger-type approaches where a confining force is associated with a single (dressed) gluon exchange at large distances. While providing a linear potential, such one-gluon exchange scenarios (including the Coulomb confinement one) imply a spread out of the electric field towards large distances, possibly with flux collimation to some extent [228].

A possible development that may eventually address the shortcomings of the Coulomb confinement scenario discussed earlier is to notice that the $q\bar{q}$ state defined in Equation (86) is not necessarily a minimum-energy state of a system containing a single $q\bar{q}$ pair, and lower energy states could in principle be constructed using operators $Q_i^j$—functionals of the lattice links—that effectively create "constituent" coupled gluons as

$$\tilde{\Psi}_{q\bar{q}} = \bar{q}^i(0)Q_i^j q_j(R)\Psi_0,\tag{94}$$

where schematically,

$$Q_i^j = a_0\delta_i^j + a_1 A_i^j + a_2 A_i^k A_k^j + \dots.\tag{95}$$

The resulting state effectively represents a chain of gluons bound by attractive forces, with a $q$ and $\bar{q}$ at the end of the chain, at large $R$, that could, in principle, provide a necessary suppression of the long-range dipole fields. Hence, such a gluon chain may be viewed as a colour-electric flux tube itself [229,230]. Indeed, as $q$ and $\bar{q}$ get separated, more and more constituent gluons get pulled out of the vacuum to minimise the energy of the system [231,232]. This picture rather naturally emerges by expanding the Wilson line stretched between $q$ and $\bar{q}$ in powers of the gluon field and actually implies the absence of dipole fields at large $R$. In the limit of large number of colours $N$ in the the $SU(N)$ theory, such a chain of gluons on a given time slice is dominated by a high-order planar Feynman amplitude that can be, in principle, tackled by analytic methods.

Among remarkable features of the gluon chain model are the Casimir scaling in the leading order of $1/N$ expansion and a subleading $1/N^2$ string breaking effect at some critical length-scale leading to a correct N-ality dependence of the string tension asymptotically. In the case of heavy (static) charges in the adjoint representation of $SU(N)$, for instance, in the limit $N \to \infty$, two gluon chains instead of one are formed between the charges, leading to twice larger adjoint string tension compared to the one in the fundamental representation, i.e., $\sigma_A = 2\sigma_F$. The latter is defined only at intermediate distances but must disappear at asymptotic distances due to colour screening by N-ality zero gluons in the vacuum. Although gluons do not break the center symmetry as such, they take part in the colour screening on the same footing as light quarks in QCD such that both quarks and gluons are absent in the asymptotic spectrum in the virtue of $C$-confinement and the string hadronisation model. This suggests a non-trivial but less explored and speculative possibility that the non-perturbative gauge-field vacuum somehow rearranges itself at large distances in such a way that the center symmetry might get broken somehow even without the presence of matter fields in the fundamental representation[4].

Indeed, pulling the two gluons (or adjoint matter states) apart from each other, eventually the virtual gluons from the QCD vacuum are prompted to bind to the octet-charged sources, yielding colour-singlet states—gluelumps—at asymptotically large distances. Such a gluon colour-screening mechanism is very similar to that driven by dynamical virtual quarks being brought on mass-shell to screen the charge of heavy static quarks as the latter move apart, and the energy accumulated in the string is partially spent for that purpose. So, the colour-screening and hence the string-breaking phenomenon is not particularly sensitive to N-ality but rather to the colour charge itself being the necessary prerequisite for $C$-confinement. While the formation of a flux tube between the two gluons at intermediate distances applies for the confining phase in the strongly coupled regime, $C$-confinement as an asymptotic phenomenon occurs also in the Higgs phase but without the formation of an intermediate flux tube.

Note that an adjoint string breaks via a $1/N^2$ suppressed but very important (at large $R$) interaction between the gluon chains, enabling them to transform into a pair of gluelumps, as described above (see also Ref. [230]). This correctly generalises for sources in an arbitrary gauge group representation giving rise to N-ality dependence of the asymptotic string tension. Such an important string-like property of the gluon chain as the Lüscher term appears due to fluctuations in the gluons' positions on the chain [230]. As was demonstrated in Ref. [233], introducing up two gluons in a chain preserves the linearity of the Coulomb potential, but that is already enough to bring its slope much closer to the true static potential (i.e., obtained by gauge-invariant methods). In this calculation, it was shown that multi-gluon configurations in the chain appear to be increasingly important at large $R$, which also strongly reduces the sensitivity of the results to the lattice volume. This means that the long-range dipole field becomes strongly suppressed, indicating a possible formation of a localised colour flux tube (for more discussion on this aspect, see Ref. [3]).

### 16. Dual Superconductivity and Magnetic Monopoles

As was proposed a long ago in Refs. [132,234–237], the QCD vacuum could be viewed as a "dual" superconductor, an analog of type-II superconductor, where the electric and magnetic fields are interchanged. These studies have pioneered the developments of a beautiful theory of what is sometimes called the *dual superconductor picture of confinement*. In a usual superconductor, one deals with a condensate of electric charges (in fact, bosonic Cooper pairs), and, due to repulsion (or confinement), the magnetic fields get squeezed into magnetic flux tubes with a constant energy density (Abrikosov vortices). In a "dual" superconductor, one instead works with a condensate of magnetic charges known as *magnetic monopoles*, where the electric field of static charges would be squeezed (confined) into electric flux tubes. The latter realisation is what we often regard as ordinary confinement in QCD. Both the static potential of magnetic monopoles in a type-II superconductor and the static potential of colour-electric charges in a "dual" superconductor would rise linearly with the charge separation.

This effect gives rise to a very simple picture of confinement essentially based upon a suitable generalization of the Landau–Ginzburg superconductivity theory. Indeed, starting from relativistic abelian Higgs model

$$S = \int d^D x \left( \frac{1}{4} F_{\mu\nu} F^{\mu\nu} + |D_\mu \phi|^2 + \frac{\lambda}{4} (\phi^\dagger \phi - v^2)^2 \right), \qquad D_\mu \phi = \partial_\mu + ie A_\mu, \tag{96}$$

one recovers the magnetic flux-tube Abrikosov-like solutions dubbed as the Nielsen–Olesen vortices [238]. Attributing a non-trivial winding number $n$ to the Higgs complex phase, a Nielsen–Olesen vortex carries the magnetic flux $2\pi n/e$. In the dual version, such vortex carries an electric flux that confines the electric charges. A particular model, where the dual abelian Higgs model with confinement is realised, is the $N = 2$ supersymmetric YM theory known as the Seiberg–Witten model [239,240] having several distinct types of electric flux tubes. In this model, a continuous set of distinct vacua is spanned by the "moduli" space of certain scalar field operators. Soft supersymmetry breaking then reduces the theory down to an effective $N = 1$ theory, where the confinement of the electric charge is realised due to the condensation of the monopole field and electric flux tube formation. This happens in full analogy to the confinement of the magnetic charge due to magnetic flux tube formation in usual type II superconductors and in the ordinary abelian Higgs model. The duality transformation in the Seiberg–Witten model inverts a certain combination of the effective coupling constant and the $\theta$ angle enabling one to obtain the effective action of light fields at any value of the gauge coupling from the detailed knowledge about the weak-coupling regime of the theory and its infrared singularities (for a detailed review of the underlined concepts and formalism, see, e.g., Refs. [241–243]). Such a duality is due to an exact symmetry of the abelian effective theory manifest at low energies and not of the original $SU(2)$ theory. In fact, this duality is a proper generalisation of the famous electric-magnetic duality of the Dirac formulation of Maxwell electrodynamics (with magnetic monopoles) exchanging the electric charge $q_e$ and its magnetic counterpart $q_m = 2\pi/q_e$. Hence, by means of such a duality transformation, one hopes to learn about strong-coupling (or long-distance) dynamics given from the weak-coupling regime of its dual formulation. In Ref. [244], it was shown that $k$-string tensions in the $SU(N)$ version of the Seiberg–Witten model obey the Sine law

$$\sigma_r = \frac{\sin(\pi k/N)}{\sin(\pi/N)}\sigma_F, \tag{97}$$

being numerically not very different to that of the Casimir scaling (c.f. Equation (75)).

In itself, the superconductivity picture of confinement is an abelian mechanism that has been explored originally by Polyakov [245] in the context of the confinement of electric charges in the $D = 2 + 1$ compact $U(1)$ gauge theory. This theory turns out to be an important starting point to approach QCD confinement. While in the $D = 2 + 1$ case, the compact QED features monopoles (topological excitations), and in $D = 3 + 1$, those monopoles are point-like defects in spacetime, i.e., they are also instantons. Effectively integrating out all the DoFs except monopoles in $D = 2 + 1$ compact QED, it was shown in Refs. [132,237,245] that the action of the monopole gas interacting by means of Coulomb force on the lattice reads

$$S_{\mathrm{m}} = \frac{2\pi^2}{g^2 a}\left[\sum_{i\neq j} m_i m_j G(r_i - r_j) + G(0)\sum_i m_i^2\right], \tag{98}$$

with $i, j = 1 \ldots N$ for $N$ monopoles, and the lattice Coulomb propagator $G$ at large distances behaves as $G \sim 1/4\pi|r_i - r_j|$. A Wilson loop in this approach can be expressed in terms of the monopole density and appears as a current loop that generates its magnetic field being effectively screened away by the (anti)monopoles from the background. Such an effect causes the area-law falloff for the Wilson loop VEVs. Polyakov has explicitly demonstrated that even the arbitrarily low density of these monopoles is sufficient to produce confinement and the mass gap of the theory. This happens in a regime when the entropy related to the size and shape of large Wilson loops wins over the cost in the monopole action for a large loop. The latter effect occurs at any coupling for $D = 3$ QED but only for large enough couplings in the $D = 4$ case.

In the case of YM theories, one needs to extract an abelian subgroup from the gauge group, e.g., by means of an adjoint Higgs field. An important realisation in the case of the $SU(2)$ gauge theory is the Georgi–Glashow model, where in the minimum of the Higgs potential and in unitary gauge, there is a residual $U(1)$ local gauge symmetry. Due to this symmetry, the model exhibits magnetic ('t Hooft–Polyakov) monopoles [246,247] as instanton solutions of the classical equations of motion in $D = 3$ or as static solutions (solitons) in $D = 3 + 1$. The Higgs field that is used to fix the unitary gauge necessarily vanishes at the center of each t'Hooft–Polyakov monopole, making the unitary gauge fixing ambiguous at those sites. The Wilson loop VEVs are then computed in a similar way as was done in compact $D = 3$ QED, resulting in a finite string tension $\sigma \sim \exp(-S_{\mathrm{m}})$ [237]. In $D = 4$, the Georgi–Glashow theory has both confining and non-confining phases; however, stable monopole solutions only exist in the non-confining phase where they do not form a Coulomb plasma.

An important caveat in the $D = 3$ theory is that one cannot simply neglect the effects of $W$ bosons at large distances (and hence in the analysis of confinement) in the long-range effective action. Indeed, the string tensions cannot acquire a correct N-ality dependence without $W$ bosons. The Coulomb monopole gas approximation can be justified in a certain intermediate range below a string-breaking length-scale, where a $W$ bosons carrying two units of electric charge are pair-produced and screen the charges of the static sources, which also possess two units of electric charge. Analogically, the dual abelian Higgs model that ignores the effect of $W$ bosons predicts a wrong N-ality dependence of the Wilson loop VEVs. Thus, it is unable to consistently describe long-range physics of vacuum fluctuations at characteristic distances exceeding the colour screening length-scale. Non-abelian supersymmetric versions of the dual Higgs model have been proposed in a number of existing works yielding specific non-abelian vortex solutions; for a detailed review on these aspects, see, e.g., Refs. [75,248] and references therein.

Dynamical "abelization" of $SU(N)$ gauge fields can be achieved even without an adjoint Higgs field. Instead of using an adjoint Higgs field, another way to extract a Cartan (abelian) subgroup $U(1)^{N-1}$ of $SU(N)$ suggested in Ref. [249] is the so-called *abelian projection*, using a composite operator that transforms like a matter field in the adjoint representation and fixing a gauge in which this operator is diagonal. The same effect emerges also with adjoint fermions fields [250,251] or by adding a trace deformation term to the action [252], and both methods have been successfully explored by lattice simulations (see, e.g., Refs. [253–255]).

The gluons from the coset of abelian projection are charged under $U(1)^{N-1}$, while the monopole condensation would describe their confinement in a way similar to the dual abelian Higgs model. The basic idea then is to look for a specific gauge in which the quantum fluctuations of the $U(1)^{N-1}$-charged gluons are strongly suppressed compared to the fluctuations of "photons" from the Cartan subgroup $U(1)^{N-1}$. In such a gauge called the *maximal abelian gauge* [256], the link variables would be close to a diagonal form. For instance, in the $SU(2)$ gauge theory, this is achieved by means of requiring $\sum \mathrm{Tr}[U_\mu(x)\sigma_3 U_\mu^\dagger(x)\sigma_3]$ to be maximal while leaving the residual $U(1)$ symmetry w.r.t. gauge transformations

$$U_\mu(x) \to e^{i\phi(x)\sigma_3} U_\mu(x) e^{-i\phi(x+\hat{\mu})\sigma_3}. \tag{99}$$

This enables one to decompose $U_\mu(x) = C_\mu(x)u_\mu(x)$, where the $C_\mu(x)$ matrix is expressed in terms of a "matter" field $c_\mu(x)$ with two units of $U(1)$ charge, while the diagonal $u_\mu(x) = \mathrm{diag}(\exp(i\theta_\mu(x)), \exp(-i\theta_\mu(x)))$ is given in terms of the abelian $U(1)$ gauge field $\theta_\mu(x)$, the "photon", coupled to the "matter" field $c_\mu(x)$. One, therefore, obtains the abelian-projected lattice by means of $U_\mu(x) \to \exp(i\theta_\mu(x))$ projection. Note that in the case of the $SU(3)$ theory, the maximal abelian projection is not unambiguously defined, as has been discussed for instance in Ref. [257].

In the *monopole dominance approximation* [258,259], one then replaces the link variables by the monopole links constructed from the Dirac string variables and the Coulomb propagator, and then one computes the VEVs of the Wilson loops over an ensemble of such monopoles. This procedure leads to (almost) the same values for the asymptotic string tensions of the single-charged Wilson loops in the $SU(2)$ lattice gauge theory as in the gauge-invariant approach. Furthermore, the single-charged Polyakov loops computed in the abelian-projected configurations and in the monopole dominance approximation agree with each other and both vanish below the critical temperature of the deconfinement transition, in consistency with expectations. However, these results do not agree for double-charged Polyakov loops. Vanishing VEVs of the latter, and hence the confining disorder, are found in the monopole dominance approximation, which is inconsistent with the charge screening effect that must be in place for double-charged static sources, and for that matter, with the N-ality requirement. This means that in the case of magnetic disorder dominated by abelian gauge field configurations, the abelian flux can not be distributed according to the Coulomb monopole-gas approximation.

The latter problem is not present in the abelian-projected configurations yielding a correct asymptotic behaviour of large Wilson loop VEVs in the fundamental representation. Fixing an abelian projection gauge in the $SU(2)$ gauge theory arranges the monopoles and antimonopoles coupled to each other into a chain with the total monopole flux of $\pm 2\pi$. At a certain fixed time, such a flux can be squeezed into center vortex structures on the abelian-projected lattice [260]—for an illustration of this effect, see Figure 7. Indeed, the numerical analysis that locates both the (anti)monopoles through abelian projection and the center vortices through center projection showed that almost all (anti)monopoles are located on the vortex sheets arranging themselves into alternating order in a chain (for an inspiring discussion, see Refs. [3,4]). The fact that the double-charged (Wilson and Polyakov) loops do not get contributions from linking with such vortices on the abelian-projected lattice is reflected in a vanishing asymptotic string tension in this case, in agreement with the charge screening effect.

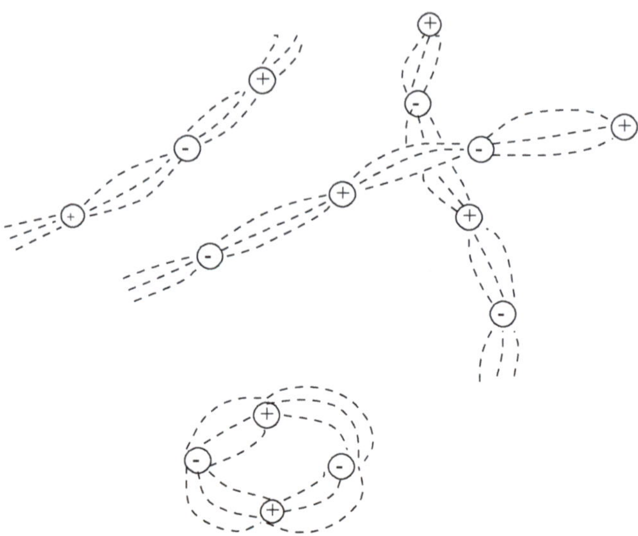

**Figure 7.** The formation of a center vortex through a collimation of the monopole/antimonopole flux. The figure is taken from Ref. [4].

It is instructive to introduce a specific order parameter, the VEV of a monopole creation operator $\langle \mu(\vec{x}) \rangle$, that would signal an emergence of the dual superconducting phase in a non-abelian gauge theory [261,262]. The operator $\mu(\vec{x})$ effectively inserts a monopole configuration at a certain position into the system such that it does not commute with the total magnetic charge operator, and hence its VEV would break the corresponding dual $U(1)$ gauge symmetry, a remnant of the gauge symmetry. According to the monopole condensation mechanism of confinement, the system is in a confining phase if and only if $\langle \mu(\vec{x}) \rangle \neq 0$, while a transition to a non-confining configuration occurs when $\langle \mu(\vec{x}) \rangle \to 0$, which indeed coincides with a more generic numerical analysis in the full theory (also at finite temperatures). There are, however, severe ambiguities such that $\langle \mu(\vec{x}) \rangle$ may also vanish in the absence of any thermodynamic transition to a deconfined phase [263]. Indeed, as was already briefly discussed earlier, the breaking of gauge symmetry remnant cannot be utilised as a correct signature of the magnetic disorder phase.

In a pure non-abelian gauge theory in $D = 4$, classical instanton solutions can not be responsible for magnetic disorder of the vacuum field configurations since their field strength falls off too fast at large distances. However, at finite temperatures, the instanton solutions as saddle points of the Euclidean gauge fields' action called calorons can be relevant for confinement. The latter solutions were found in Refs. [264–266] and are known in the literature as *KvBLL solutions*. They may contain monopole constituents sourcing both electric and magnetic fields, also known as dyons or Bogolmolny–Prasad–Sommerfield (BPS) monopoles [267,268], which can be widely separated. The thermal approach to pure 4D YM theories based upon nonperturbative results on a thermal ground state in the deconfined phase derived from an (anti)caloron ensemble has been thoroughly discussed in Ref. [269] and in references therein. Among important corollaries to this approach is, for instance, the derivation of the 3D critical exponent of the Ising model for the correlation length criticality.

The early work of Ref. [134] made important contributions to understanding the trivial-holonomy calorons in $SU(2)$ Euclidean gauge theory based on Ref. [270], while nontrivial holonomy solutions have been studied, e.g., in Refs. [264,266,271–273]. Considering, for instance, the maximally non-trivial Polyakov loop holonomy $P(\vec{x})$ introduced in $SU(N)$ in Equation (57), where $\mu_j$ are ordered and spaced with a maximal distance from each other

$$\mu_n^{\max} = -\frac{1}{2} - \frac{1}{2N} + \frac{n}{N}, \tag{100}$$

the probability density of calorons in the vacuum would peak at $\mathrm{Tr}P(\vec{x}) = 0$. As was discussed earlier, in the center vortex mechanism of confinement, the vanishing Polyakov loop expectation value computed on an ensemble, where positive and negative fluctuations in vortex configurations cancel out, is a signature of unbroken center symmetry and, hence, that of the confinement property. Notably, in the caloron configurations, the maximally non-trivial Polyakov loop holonomy vanishes by itself before any averaging as the basic property of such configurations.

Due to this property, a system of widely separated dyons, the dyon gas, whose free energy is minimal for $\mathrm{Tr}\, P(\vec{x}) = 0$, has been considered as the basis for the description of the magnetic disorder in YM theories [274]. Indeed, it was shown that the $k$-string tensions extracted from space-like Wilson loops are in agreement with those that determine the asymptotic behaviour of Polyakov loop correlators and follow the Sine law (97). In the high temperature regime, a phase transition to the deconfinement phase occurs with $T_c / \sqrt{\sigma}$ values being in perfect agreement with numerical lattice results. Despite such tremendous success, the path integral measure for the multi-dyon configurations appears to be not positively definite, thus violating the basic property of the exact measure [275]. However, numerical simulations of Ref. [276] with a suitable parameterisation of the integration measure confirmed that the confining static potential indeed emerges in the dyon gas approximation. Thus, one can conclude that the monopole mechanism based upon the caloron classical solutions is one of the most promising scenarios of confinement in non-abelian gauge theories.

There are some critical points to be made regarding the dyon gas picture of confinement neatly summarised in Ref. [3]. The same question as for the monopole Coulomb gas applies also for the dyon gas regarding the asymptotic string tension of double-charged Wilson loops that should disappear due to a screening by gluons. Another question concerns the probability distribution of Polyakov loop holonomies, which is peaked for a vanishing maximally-nontrivial holonomy in the fundamental representation, i.e., $\mathrm{Tr}\, P(\vec{x}) = 0$. If this is indeed true, it implies a negative expectation value of the Polyakov loop in the adjoint representation. However, if the latter is positive, the probability distribution would be peaked at the center-element holonomy, as suggested by the center vortex scenario of confinement. Remarkably, the expectation value of the adjoint Polyakov line in the phase of magnetic disorder has been found to be positive in the $SU(3)$ theory in Ref. [277].

Besides, considering the asymptotic behaviour of double-winding Wilson loops VEVs' (i.e., Wilson loops winding around closed co-planar loops $C_1$ and $C_2$), one reveals a dramatic difference in predictions of the monopole and vortex mechanisms of confinement [278] (see also [3,4]). The vortex scenario provides the "difference-of-areas" law behaviour for such loops, $\sim \exp[-\sigma|A(C_1) - A(C_2)|]$, where the "$-$" sign is due to a vortex linking to the largest loop, correctly reproducing the full lattice results. In this case, the monopole scenario predicts the "sum-of-areas" falloff as $\sim \exp[-\sigma(A(C_1) + A(C_2))]$, which is disfavoured by numerical simulations. The latter observation indicates that the vacuum cannot be in a dual-superconducting state of monopole/dyon plasma. In fact, the "difference-of-areas" law is restored by heavy $W$ bosons that are present in the full YM theory but not in an abelian part of it. As was suggested in Ref. [4], upon integrating out the $W$ bosons' states, one expects the monopole–antimonopole lines to get collimated into $\mathbb{Z}_2$ vortices, as is illustrated in Figure 7, i.e., in a similar fashion to what has been seen on an abelian-projected lattice. This effectively turns the monopole ensemble into a configuration of $\mathbb{Z}_2$ vortices, offering an intricate connection between the two pictures of confinement.

Another observation of Ref. [279] in the case of the $G_2$ gauge theory has suggested that the Polyakov loop expectation value exactly vanishes in the dyon gas picture. The colour screening in $G_2$, however, requires binding a static source in the fundamental representation to a minimum of three gluons, likely leading to a very small, but non-zero, Polyakov loop expectation value that would be difficult to identify numerically [3]. If the colour screening mechanism is a valid approach, the center symmetry breaking at large distances would be manifest in $G_2$, whereas the dyon gas approximation, like the Coulomb gas approximation discussed earlier, might be lacking something relevant in the asymptotic regime.

## 17. Separation-of-Charge Confinement Criterion

A clear symmetry-based distinction between the confining and Higgs phases in a gauge theory with fundamental representation matter fields has been recently proposed in Refs. [110,280,281]. An important generalised criterion of confinement valid in *both* pure YM theories and YM theories with matter in the fundamental representation states that

$$E_V(R) \equiv \langle \Psi_V | H | \Psi_V \rangle - \mathcal{E}_{\text{vac}} \geq E_0(R), \tag{101}$$

with $\Psi_V$ being the $q\bar{q}$ state connected by a Wilson line,

$$\Psi_V \equiv \bar{q}^a(\vec{x}) V^{ab}(\vec{x}, \vec{y}; A) q^b(\vec{y}) \Psi_0, \tag{102}$$

for *any* choice of the gauge bi-covariant non-local operator $V^{ab}(\vec{x}, \vec{y}; A)$. The latter depends only on the gauge field, thus eliminating any possibility for a string breaking by means of dynamical matter fields. This criterion is a necessary and sufficient condition for the *separation-of-charge confinement* (or $S_c$-confinement, for short), which is meaningful only in gauge theories with a non-trivial center symmetry. In Equation (101), $H$ is the Hamiltonian, $\mathcal{E}_{\text{vac}}$ is the vacuum energy, $E_0(R) \sim \sigma R$ at $R \to \infty$ is an asymptotically linear function, which has the meaning of the ground-state energy of the $q\bar{q}$ in a pure $SU(N)$ gauge theory (but not in the one with matter fields), where the above criterion is equivalent to the area-law falloff of Wilson loop VEVs.

In the $SU(2)$ gauge-Higgs theory, the confining phase is found for $\gamma \ll \beta \ll 1$ and $\gamma \ll 1/10$, where the $S_c$-confinement condition (101) is satisfied. However, deeply in the Higgs phase, for other couplings' ranges, this criterion is not fulfilled; hence, we deal with only a weaker $C$-confinement situation there. In Refs. [110,280,281], it has been shown that a transition between the $C$- and $S_c$-confinement phases must take place in the gauge-Higgs theory, and the unbroken custodial symmetry has been found to separate the $S_c$-confining (if not massless) phase from the Higgs phase corresponding to a $C$-confined spin glass state, where the custodial symmetry is actually broken. It would be very interesting to see how such a new concept of $S_c$-confinement can be applied for more realistic theories such as QCD.

## 18. Separating the Higgs and Confinement Phases: Vortex Holonomy Phase

As was discussed earlier, the results of Refs. [109,113,114] state that it is possible to identify a continuous path between the Higgs and confinement regimes where no first-order phase transitions occur like what is believed to happen in physics of high-to-low temperature QCD (smooth crossover) transition and in several other specific models. Indeed, one would naively expect that such a continuity always takes place unless the phases are separated by different realisations of global symmetries.

An important counter-example to this statement has been recently explored in a whole class of models in Refs. [58,64]. Namely, it has been demonstrated that even in the case of a spontaneous global $U(1)$ symmetry breaking in both Higgs and confinement regimes (an analog to the baryon symmetry breaking in dense QCD at low $T$), it is still possible to identify a novel *non-local* order parameter (a vortex holonomy phase) that separates the two phases, leading to a thermodynamical phase transition. This proof provides an important argument against the quark-hadron continuity (Schäfer–Wilczek) conjecture in dense QCD, as mentioned above in Section 2.

For this purpose, the authors of Ref. [58] started with Polyakov's $D = 4$ compact $U(1)$ gauge theory discussed in the previous section, then imposed a single global $U(1)_G$ symmetry (an analog to $U(1)_B$ of QCD) and added three complex scalar fields that effectively mimic dynamical quark fields in QCD, with the following assignments under $U(1) \times U(1)_G$ group

$$\phi^+ = \{+1, -1\}, \qquad \phi^- = \{-1, -1\}, \qquad \phi^0 = \{0, +2\}. \tag{103}$$

Then, the product $\phi^+ \phi^-$ appears to be an analog to the gauge invariant baryon operator, and $\phi^0$ can couple to $\phi^+ \phi^-$ product and can be considered as a baryon interpolating field or a source for baryons. A VEV in $\phi^0$ would bring the theory into a superfluid phase of spontaneously broken $U(1)_G$, such that the global symmetries' realisation is the same in the confinement and Higgs phases. In the monopole-driven confinement picture, the monopoles in this theory would have a finite action that can be UV-completed through an $SU(2)$ symmetry, just as in the original Polyakov's description. Finally, additional discrete $\mathbb{Z}_2$ (charge conjugation and flavour-flip) symmetries have also been introduced and can be considered as analogous to the flavour symmetry in QCD. The effect of monopoles in this model induces an additional term in the Lagrangian [237]

$$V_m(\sigma) \propto e^{-S_I} \cos \sigma, \tag{104}$$

which is a potential for the "dual photon"—a periodic scalar field $\sigma \to \sigma + 2\pi$ related to the field strength by the abelian duality relation

$$F_{\mu\nu} = \frac{ie^2}{2\pi} \epsilon_{\mu\nu\lambda} \partial^\lambda \sigma. \tag{105}$$

As one varies the adjustable mass parameters, three different regimes of the theory emerge. One of them corresponds to the compact $3D$ $U(1)$ theory with heavy scalar quarks and confinement where no symmetries are spontaneously broken ("gapped confined" regime). The second regime features the Higgs mechanism with a non-zero VEV $\langle \phi^+ \phi^- \rangle \neq 0$, such that a cubic $\phi^+ \phi^- \phi^0$ term in the potential drives the condensation of $\phi_0$, $\langle \phi^0 \rangle \neq 0$ and hence spontaneous breaking of $U(1)_G$ (Higgs phase with a single massless Goldstone boson). The third regime has monopole-driven confinement, while $\langle \phi^0 \rangle \neq 0$ spontaneously breaks $U(1)_G$ symmetry, and the heavy charged scalars are very heavy and may be disregarded. A key question here is whether the two phases with spontaneously broken $U(1)_G$ ("Higgs" and "confining") with no distinguishing local order parameter are really two distinct phases or might be continuously connected.

The main claim of Ref. [58] is that these phases are distinct and can be distinguished *only* by a new non-local order parameter that is connected with topological excitations. Physically, both phases with spontaneously broken $U(1)_G$ should be considered as superfluids that have vortices. As a consequence of $U(1)_G$ breaking, the theory possesses a gapless Nambu–Goldstone mode, which is a phase of $\langle \phi^0 \rangle$ condensate, as well as topologically stable vortex excitations when the phase of the $\langle \phi^0 \rangle$ condensate winds around the unit circle when one goes around a given loop. In ordinary superfluids in three spacial dimensions, this provides vortex loops, but in $D = 2 + 1$, vortices act like point particles, the point around which the condensate phase winds. Winding of that phase, in the language of superfluids, is exactly what is called quantized minimal "circulation". The winding number can then be found in terms of a contour integral of the gradient of the phase or simply as

$$w \propto \oint_C \frac{d\langle \phi^0 \rangle}{|\langle \phi^0 \rangle|}. \tag{106}$$

The charge particles in the superfluid phase would then interact with (minimal-energy) vortices through acquiring an Aharonov–Bohm phase as one sends a charged particle into a loop that links with the worldline of the vortex. This phase is measured by the Wilson loop holonomy

$$\Omega(C) = e^{i \oint_C A}, \tag{107}$$

whose expectation value

$$\langle \Omega(C) \rangle \sim e^{-mP(C)}. \tag{108}$$

This means that short-range quantum fluctuations in the phase (with dynamical fundamental representation charges) automatically lead to a perimeter-law decay of the VEV of a large Wilson loop. In fact, this represents the same physics as that of string breaking that turns the area-law behaviour of a pure YM theory into perimeter-law behaviour present in real QCD and in the Higgs phase, as was discussed in detail in previous sections.

Let us consider a Wilson loop expectation value in the presence of a minimal-energy vortex, which can be thought of in terms of a constrained functional integral where one integrates over all the field configurations in the theory but with a constraint forcing the presence of a vortex along some large worldline $C$ in $D = 3$ spacetime. The same short-range physics guarantees it is going to have the same perimeter-law behaviour, but it can have an additional phase factor

$$\langle \Omega(C) \rangle_{w=1} \sim e^{i\Phi} e^{-mP(C)}. \tag{109}$$

In order to extract the phase, one defines the ratio [58]—the *vortex order parameter*,

$$O_\Omega \equiv \lim_{r_v \to \infty} \frac{\langle \Omega(C) \rangle_{w=1}}{\langle \Omega(C) \rangle}, \tag{110}$$

taking the size of the Wilson loop and its separation from the vortex $r_v$ arbitrarily large simultaneously. The symmetries of the theory guarantee that the Wilson loop VEV $\langle \Omega(C) \rangle$ is real, and it can be made positive. While the charge conjugation and reflection symmetries are broken, the flavour-flip symmetry is preserved in the presence of a vortex. The latter also flips the sign of the gauge field and hence conjugates the holonomy guaranteeing that the vortex-constrained expectation value of the Wilson loop is also real but can be either positive or negative, i.e., $O_\Omega = \pm 1$. This means that the vortex order parameter cannot vary smoothly under variations of model parameters when moving between the two phases.

In Ref. [58], it was demonstrated by means of a semi-classical analysis and through the minimization of the vortex energy that in the Higgs phase, the vortex order parameter must be equal to $-1$. Integrating out the heavy charged scalars in the confining phase, one can conclude that the vortex-constrained Wilson loop expectation value hardly knows about the presence of the vortex, thus $O_\Omega = 1$. It was argued in Ref. [58] that if one varies parameters of the theory and at some point the magnetic flux carried by vortices suddenly jumps, which is what the vortex order parameter is really probing, that surely is going to change the core energy density of the vortex. In this case, it does change the probability of having vortex excitations in the ground state wave function, affecting the ground state energy density. In other words, a sudden change in the vortex properties really should be reflected in a genuine thermodynamic phase transition.

Given the close analogies of the considered model with QCD, by construction of the model above, this conclusion may be straightforwardly generalised to a $D = 4$ non-abelian theory with fundamental-representation matter charged under a given global symmetry. This is the case of dense QCD with broken $U(1)_B$, where the $SU(3)$ vortex order parameter can be shown to take two distinct values in two phases [58]

$$O_\Omega \equiv \lim_{r_v \to \infty} \frac{\langle \mathrm{Tr}\Omega(C) \rangle_{w=1}}{\langle \mathrm{Tr}\Omega(C) \rangle} = \begin{cases} e^{2\pi i/3} & \text{CFL}/\text{Higgs phase} \\ +1 & \text{nuclear}/\text{confining phase} \end{cases}, \tag{111}$$

such that the latter $O_\Omega = +1$ is understood as the characteristic signature of confined QCD phase with spontaneously broken baryon symmetry.

This indeed illustrates the main point suggesting that the quark-hadron continuity between the nuclear and quark-matter phases may not hold. Ref. [70] has argued that despite the noted discontinuity in the vortex order parameter, the continuity of phases may still be intact due to a continuous connection between the vortex in the CSC/CFL phase and the corresponding one in the nuclear phase. In response to this claim, Ref. [58] has explicitly proven the existence of thermodynamical phase transition at the interface between the two phases connected to the manifest discontinuity in the non-local vortex order parameter. While the debate about this important issue will likely continue in the literature, it once again reveals the surprising underlined complexity of the non-perturbative QCD vacuum, and the associated approaches to the confinement problem may still not be in their final form. It would be very instructive to find possible connections between the vortex holonomy phase and its discontinuity with the $S_c$-confinement criterion briefly discussed in the previous section, both pursuing the same goal of sharply separating the Higgs and confining phases.

### 19. Summary

To summarise, the mass gap and colour confinement that are already realised in a gauge-Higgs theory may not be connected to an asymptotically rising static potential and Regge trajectories, and, hence, they do not necessarily represent an emergence of the magnetic disorder state. On the other hand, the magnetic disorder and the associated area-law behaviour of Wilson line VEVs imply colour confinement and the mass gap automatically. In this sense, colour confinement and the mass gap only represent a small part of a bigger picture of confinement and should be considered as a consequence of the confined magnetic disorder state and flux tubes formation corresponding to a phase with unbroken non-trivial center symmetry. If there is no non-analytic boundary between the massive and magnetic disorder phases at finite values of coupling constants and at some critical length-scale, i.e., a first-order phase transition, one should talk about a single massive phase at all scales, as, for instance, in the gauge-Higgs theories. The flux tubes formation is only an approximate picture in this case, roughly consistent with reality at some intermediate distances, but it does not necessarily represent an emergence of a new phase.

One of the big questions for real QCD though, i.e., with physical quarks and gluons, which would distinguish it from the EW theory is then whether a magnetic disorder phase really exists within some finite interval of characteristic length-scales that would abruptly transit to a massive phase at asymptotically large length-scales (due to string-breaking), or not. If not, then real QCD would always be considered on the same footing with a gauge-Higgs theory as existing in the massive phase only which is one of the basic options actively discussed in the literature. One thing, however, that distinguishes real QCD from the EW theory is the existence of experimentally observed Regge trajectories in QCD with light quarks that, in fact, may indicate the presence of a non-analytic phase boundary at moderately large distances in QCD in contrast to EW theory, while both would be in the massive phase asymptotically. Numerical values of the coupling constant here should play a decisive role here, and for weak couplings, the magnetic disorder may not emerge at all.

In fact, light "sea" quarks, i.e., with masses way below the confinement energy scale of QCD, emerge due to gluon splitting $G_a \to q\bar{q}$ such that correlated $q\bar{q}$ pairs could be viewed as effective gluons as long as the resolution length-scale is above the wave-length of such a pair. If, at such length-scales, the strong-coupling constant is large enough, one can view physics in such a regime as that of an effective pure YM theory in a magnetically disordered phase with unbroken center symmetry. By pulling $q$ and $\bar{q}$ apart from each other at length-scales larger than the resolution scale, the center symmetry gets effectively broken, and the theory enters the massive phase. In the infinite quark mass limit, however, the string-breaking length-scale grows indefinitely, making the magnetic disorder phase valid for asymptotically large distances.

Depending on the values of the gauge coupling constant, the same can occur in a gauge-Higgs theory in a strongly coupled regime, which is supported by lattice simulations. Thus, we arrive to a radically different phase structure of a gauge theory depending on whether it is in a strongly coupled or in a weakly coupled regime. However, even without matter fields involved, the major problem of confinement remains, namely to understand why pure YM theories with a non-trivial center symmetry in $D \leq 4$ dimensions can only exist in a state of magnetic disorder. Once this key problem is solved, it will become clearer under what conditions realistic theories, such as gauge-Higgs theory or real QCD, may exist and the magnetically disordered phase, and a first-order phase transition towards the massive phase may occur, if at all.

As was elaborated in this review, the phase structure and properties of the quantum QCD vacuum is still under intense explorations, both experimentally and theoretically, numerically and analytically, and is far from its complete and satisfactory description. However, tremendous progress has been made and some basic contours of the fundamental picture of confinement have started to emerge. We do understand a confining phase as an asymptotic magnetic disorder phase with unbroken non-trivial center symmetry that manifests itself through the area-law behaviour of large Wilson loop VEVs and, hence, a linear rise of the corresponding static (string) potential. However, real QCD features such a phase only pre-asymptotically where colour-electric flux tubes exist at not-too-large distances, while they break apart at length-scales beyond an inverse to the lightest meson mass (pion) scale, yielding a massive phase asymptotically. The quanta of the magnetic flux, vortices, have proven to play a crucial role at all stages, from the formation of a flux tube to its breaking. Given the overwhelming qualitative and quantitative evidence collected in vast amounts of studies in the literature, the center vortex mechanism remains among the most favoured scenarios of confinement so far. Other ideas, such as the monopole scenario, highlight the underlined complexity of the confining phase and phase transitions and offer different perspectives but in one way or another connect to the vortex picture. Various order parameters briefly described in this review probe the confining phase and are capable of separating it from a non-confining (Higgs) phase, with the latter remaining under a continuous debate in the literature.

**Author Contributions:** Conceptualization, R.P.; validation, R.P. and M.Š.; formal analysis, R.P.; investigation, R.P.; resources, R.P. and M.Š.; writing—original draft preparation, R.P. and M.Š.; writing—review and editing, R.P. and M.Š.; supervision, R.P.; project administration, R.P.; funding acquisition, R.P. and M.Š. Both authors have read and agreed to the published version of the manuscript.

**Funding:** The research work of R.P. was funded in part by the Swedish Research Council grant, contract number 2016-05996, as well as by the European Research Council (ERC) under the European Union's Horizon 2020 research and innovation programme (grant agreement No 668679). The research work of M.Š. is partially funded by the grants LTT17018 and LTT18002 of the Ministry of Education of the Czech Republic.

**Conflicts of Interest:** The authors declare no conflict of interest. The funders had no role in the design of the study; in the collection, analyses, or interpretation of data; in the writing of the manuscript, or in the decision to publish the results.

## Notes

[1] Real superconductors have observable phenomena such as persistent currents distinguishing the superconducting phase from a normal one. The name "colour superconductor" come out rather misleading in a sense that there are no observable persistent colour-charge currents associated to this phase [58].

[2] One should make a side remark here: considering static charges in the fundamental representation, with a non-zero coupling to the gauge field in the action, automatically implies that the theory is *not* a pure non-abelian gauge theory. Obviously, a pure gauge theory features neither "static" nor "dynamic" quark fields; moreover, as such, the latter fields are not distinguished by the action unless the static ones are made a lot heavier than the dynamic ones. Therefore, any statements about the linear static potential in pure non-abelian gauge theories should be taken with reservations and only makes sense when taking a limit of heavy (static) matter fields that can be effectively integrated out in the corresponding path integral of the theory. However, the latter procedure formally eliminates such heavy charges from asymptotic states of the resulting EFT entirely, making it

impossible to use them as probes for vacuum dynamics and hence confinement in pure gauge theories. So, in practice, one does not eliminate them from the asymptotic states of a gauge theory but rather retains them as heavy sources but with a finite mass.

3.  At $\gamma = 0$, the theory is in the magnetic disorder phase, which cannot be continuously evolved from other regions in parameter space with $\gamma \neq 0$.

4.  By construction, a pure gauge theory does not contain any fundamental-representation charges. So instead of heavy quarks, the use of "constituent gluons" as static colour charges to probe the formation and properties of the flux tubes of finite lengths, colour screening, string breaking mechanism and the phases of the theory would be the most natural approach to study confinement in pure non-abelian theories.

## References


1.  Brambilla, N.; Eidelman, S.; Foka, P.; Gardner, S.; Kronfeld, A.S.; Alford, M.G.; Alkofer, R.; Butenschoen, M.; Cohen, T.D.; Erdmenger, J.; et al. QCD and Strongly Coupled Gauge Theories: Challenges and Perspectives. *Eur. Phys. J. C* **2014**, *74*, 2981. [CrossRef]
2.  't Hooft, G. A Two-Dimensional Model for Mesons. *Nucl. Phys. B* **1974**, *75*, 461–470. [CrossRef]
3.  Greensite, J. *An Introduction to the Confinement Problem*; Springer Nature: Cham, Switzerland, 2020; Volume 972.
4.  Greensite, J. Confinement from Center Vortices: A review of old and new results. *EPJ Web Conf.* **2017**, *137*, 01009. [CrossRef]
5.  Gross, D.J.; Wilczek, F. Ultraviolet Behavior of Nonabelian Gauge Theories. *Phys. Rev. Lett.* **1973**, *30*, 1343–1346. [CrossRef]
6.  Politzer, H.D. Reliable Perturbative Results for Strong Interactions? *Phys. Rev. Lett.* **1973**, *30*, 1346–1349. [CrossRef]
7.  Collins, J.C.; Perry, M.J. Superdense Matter: Neutrons or Asymptotically Free Quarks? *Phys. Rev. Lett.* **1975**, *34*, 1353. [CrossRef]
8.  Cabibbo, N.; Parisi, G. Exponential Hadronic Spectrum and Quark Liberation. *Phys. Lett. B* **1975**, *59*, 67–69. [CrossRef]
9.  Shuryak, E.V. Theory of Hadronic Plasma. *Sov. Phys. JETP* **1978**, *47*, 212–219.
10. Shuryak, E.V. Quark-Gluon Plasma and Hadronic Production of Leptons, Photons and Psions. *Phys. Lett. B* **1978**, *78*, 150. [CrossRef]
11. Freedman, B.A.; McLerran, L.D. Fermions and Gauge Vector Mesons at Finite Temperature and Density. 3. The Ground State Energy of a Relativistic Quark Gas. *Phys. Rev. D* **1977**, *16*, 1169. [CrossRef]
12. Polyakov, A.M. Thermal Properties of Gauge Fields and Quark Liberation. *Phys. Lett. B* **1978**, *72*, 477–480. [CrossRef]
13. Kapusta, J.I. Quantum Chromodynamics at High Temperature. *Nucl. Phys. B* **1979**, *148*, 461–498. [CrossRef]
14. Witten, E. Cosmic Separation of Phases. *Phys. Rev. D* **1984**, *30*, 272–285. [CrossRef]
15. Arsene, I.; Bearden, I.G.; Beavis, D.; Besliu, C.; Budick, B.; Bøggild, H.; Chasman, C.; Christensen, C.H.; Christiansen, P.; Cibor, J.; et al. Quark gluon plasma and color glass condensate at RHIC? The Perspective from the BRAHMS experiment. *Nucl. Phys. A* **2005**, *757*, 1–27. [CrossRef]
16. Back, B.B.; Baker, M.D.; Ballintijn, M.; Barton, D.S.; Becker, B.; Betts, R.R.; Bickley, A.A.; Bindel, R.; Budzanowski, A.; Busza, W.; et al. The PHOBOS perspective on discoveries at RHIC. *Nucl. Phys. A* **2005**, *757*, 28–101. [CrossRef]
17. Adams, J.; Aggarwal, M.M.; Ahammed, Z.; Amonett, J.; Anderson, B.D.; Arkhipkin, D.; Averichev, G.S.; Badyal, S.K.; Bai, Y.; Balewski, J.; et al. Experimental and theoretical challenges in the search for the quark gluon plasma: The STAR Collaboration's critical assessment of the evidence from RHIC collisions. *Nucl. Phys. A* **2005**, *757*, 102–183. [CrossRef]
18. Adcox, K.; Adler, S.S.; Afanasiev, S.; Aidala, C.; Ajitanand, N.N.; Akiba, Y.; Al-Jamel, A.; Alexander, J.; Amirikas, R.; Aoki, K.; et al. Formation of dense partonic matter in relativistic nucleus-nucleus collisions at RHIC: Experimental evaluation by the PHENIX collaboration. *Nucl. Phys. A* **2005**, *757*, 184–283. [CrossRef]
19. Braun-Munzinger, P.; Koch, V.; Schäfer, T.; Stachel, J. Properties of hot and dense matter from relativistic heavy ion collisions. *Phys. Rep.* **2016**, *621*, 76–126. [CrossRef]
20. Pasechnik, R.; Šumbera, M. Phenomenological Review on Quark–Gluon Plasma: Concepts vs. Observations. *Universe* **2017**, *3*, 7. [CrossRef]
21. Kapusta, J.; Muller, B.; Rafelski, J. *Quark-Gluon Plasma: Theoretical Foundations*; Elsevier: Amsterdam, The Netherlands, 2003.
22. Shuryak, E. Strongly coupled quark-gluon plasma in heavy ion collisions. *Rev. Mod. Phys.* **2017**, *89*, 035001. [CrossRef]
23. Fujii, H.; Kharzeev, D. Long range forces of QCD. *Phys. Rev. D* **1999**, *60*, 114039. [CrossRef]
24. Zyla, P.A.; Barnett, R.M.; Beringer, J.; Dahl, O.; Dwyer, D.A.; Groom, D.E.; Lin, C.J.; Lugovsky, K.S.; Pianori, E.; Robinson, D.J.; et al. Review of Particle Physics. *Prog. Theor. Exp. Phys.* **2020**, *2020*, 083C01. [CrossRef]
25. Gelis, F.; Iancu, E.; Jalilian-Marian, J.; Venugopalan, R. The Color Glass Condensate. *Ann. Rev. Nucl. Part. Sci.* **2010**, *60*, 463–489. [CrossRef]
26. Lacey, R.A.; Ajitanand, N.N.; Alexander, J.M.; Chung, P.; Holzmann, W.G.; Issah, M.; Taranenko, A.; Danielewicz, P.; Stoecker, H. Has the QCD Critical Point been Signaled by Observations at RHIC? *Phys. Rev. Lett.* **2007**, *98*, 092301. [CrossRef] [PubMed]
27. Heinz, U.; Snellings, R. Collective flow and viscosity in relativistic heavy-ion collisions. *Ann. Rev. Nucl. Part. Sci.* **2013**, *63*, 123–151. [CrossRef]
28. Adcox, K.; Adler, S.S.; Ajitanand, N.N.; Akiba, Y.; Alexander, J.; Aphecetche, L.; Arai, Y.; Aronson, S.H.; Averbeck, R.; Awes, T.C.; et al. Suppression of hadrons with large transverse momentum in central Au+Au collisions at $\sqrt{s_{NN}}$ = 130-GeV. *Phys. Rev. Lett.* **2002**, *88*, 022301. [CrossRef] [PubMed]

29. Adler, C.; Ahammed, Z.; Allgower, C.; Amonett, J.; Anderson, B.D.; Anderson, M.; Averichev, G.S.; Balewski, J.; Barannikova, O.; Barnby, L.S.; et al. Disappearance of back-to-back high $p_T$ hadron correlations in central Au+Au collisions at $\sqrt{s_{NN}}$ = 200-GeV. *Phys. Rev. Lett.* **2003**, *90*, 082302. [CrossRef]

30. Thoma, M.H. Complex plasmas as a model for the quark-gluon-plasma liquid. *Nucl. Phys. A* **2006**, *774*, 307–314. [CrossRef]

31. Ioffe, B.L.; Fadin, V.S.; Lipatov, L.N. *Quantum Chromodynamics: Perturbative and Nonperturbative Aspects*; Cambridge University Press: Cambridge, UK, 2010.

32. Campbell, J.; Huston, J.; Krauss, F. *The Black Book of Quantum Chromodynamics: A Primer for the LHC Era*; Oxford University Press: Oxford, UK, 2017.

33. Gribov, L.V.; Levin, E.M.; Ryskin, M.G. Semihard Processes in QCD. *Phys. Rep.* **1983**, *100*, 1–150. [CrossRef]

34. Kharzeev, D. Classical chromodynamics of relativistic heavy ion collisions. In Proceedings of the Cargese Summer School on QCD Perspectives on Hot and Dense Matter, Cargese, France, 6–18 August 2001.

35. Berges, J.; Heller, M.P.; Mazeliauskas, A.; Venugopalan, R. Thermalization in QCD: theoretical approaches, phenomenological applications, and interdisciplinary connections. *arXiv* **2020**, arXiv:2005.12299.

36. McLerran, L. A Brief Introduction to the Color Glass Condensate and the Glasma. In Proceedings of the 38th International Symposium on Multiparticle Dynamics, Hamburg, Germany, 15–20 September 2008.

37. McLerran, L.D.; Venugopalan, R. Computing quark and gluon distribution functions for very large nuclei. *Phys. Rev. D* **1994**, *49*, 2233–2241. [CrossRef] [PubMed]

38. Kovner, A.; McLerran, L.D.; Weigert, H. Gluon production from nonAbelian Weizsacker-Williams fields in nucleus-nucleus collisions. *Phys. Rev. D* **1995**, *52*, 6231–6237. [CrossRef] [PubMed]

39. Braun-Munzinger, P.; Wambach, J. The Phase Diagram of Strongly-Interacting Matter. *Rev. Mod. Phys.* **2009**, *81*, 1031–1050. [CrossRef]

40. Fukushima, K.; Hatsuda, T. The phase diagram of dense QCD. *Rep. Prog. Phys.* **2011**, *74*, 014001. [CrossRef]

41. Addazi, A.; Marcianò, A.; Pasechnik, R. Time-crystal ground state and production of gravitational waves from QCD phase transition. *Chin. Phys. C* **2019**, *43*, 065101. [CrossRef]

42. Huang, W.C.; Reichert, M.; Sannino, F.; Wang, Z.W. Testing the Dark Confined Landscape: From Lattice to Gravitational Waves. *arXiv* **2020**, arXiv:2012.11614.

43. Pasechnik, R.; Prokhorov, G.; Teryaev, O. Mirror QCD and Cosmological Constant. *Universe* **2017**, *3*, 43. [CrossRef]

44. Pasechnik, R.; Beylin, V.; Vereshkov, G. Dark Energy from graviton-mediated interactions in the QCD vacuum. *J. Cosmol. Astropart. Phys.* **2013**, *1306*, 011. [CrossRef]

45. Pasechnik, R.; Beylin, V.; Vereshkov, G. Possible compensation of the QCD vacuum contribution to the dark energy. *Phys. Rev.* **2013**, *D88*, 023509. [CrossRef]

46. Addazi, A.; Marcianò, A.; Pasechnik, R.; Prokhorov, G. Mirror Symmetry of quantum Yang-Mills vacua and cosmological implications. *Eur. Phys. J.* **2019**, *C79*, 251. [CrossRef]

47. Pasechnik, R. Quantum Yang-Mills Dark Energy. *Universe* **2016**, *2*, 4. [CrossRef]

48. Stephanov, M.A.; Rajagopal, K.; Shuryak, E.V. Signatures of the tricritical point in QCD. *Phys. Rev. Lett.* **1998**, *81*, 4816–4819. [CrossRef]

49. Gupta, S.; Luo, X.; Mohanty, B.; Ritter, H.G.; Xu, N. Scale for the Phase Diagram of Quantum Chromodynamics. *Science* **2011**, *332*, 1525–1528. [CrossRef]

50. Adamczyk, L.; Adkins, J.K.; Agakishiev, G.; Aggarwal, M.M.; Ahammed, Z.; Ajitan, N.N.; Alekseev, I.; Anderson, D.M.; Aoyama, R.; Aparin, A.; et al. Bulk Properties of the Medium Produced in Relativistic Heavy-Ion Collisions from the Beam Energy Scan Program. *Phys. Rev. C* **2017**, *96*, 044904. [CrossRef]

51. Bzdak, A.; Esumi, S.; Koch, V.; Liao, J.; Stephanov, M.; Xu, N. Mapping the Phases of Quantum Chromodynamics with Beam Energy Scan. *Phys. Rep.* **2020**, *853*, 1–87. [CrossRef]

52. Bellwied, R.; Borsanyi, S.; Fodor, Z.; Günther, J.; Katz, S.D.; Ratti, C.; Szabo, K.K. The QCD phase diagram from analytic continuation. *Phys. Lett. B* **2015**, *751*, 559–564. [CrossRef]

53. Ding, H.T.; Karsch, F.; Mukherjee, S. Thermodynamics of Strong-Interaction Matter from Lattice QCD. In *Quark-Gluon Plasma 5*; Wang, X.N., Ed.; World Scientific: Singapore, 2016.

54. Bazavov, A.; Ding, H.-T.; Hegde, P.; Kaczmarek, O.; Bielefeld, U.; Karsch, F. The QCD Equation of State to $\mathcal{O}(\mu_B^6)$ from Lattice QCD. *Phys. Rev. D* **2017**, *95*, 054504. [CrossRef]

55. Philipsen, O. Constraining the phase diagram of QCD at finite temperature and density. In Proceedings of the 37th International Symposium on Lattice Field Theory, Wuhan, China, 16–22 June 2019.

56. Dean, D.J.; Hjorth-Jensen, M. Pairing in nuclear systems: From neutron stars to finite nuclei. *Rev. Mod. Phys.* **2003**, *75*, 607–656. [CrossRef]

57. Gandolfi, S.; Gezerlis, A.; Carlson, J. Neutron Matter from Low to High Density. *Ann. Rev. Nucl. Part. Sci.* **2015**, *65*, 303–328. [CrossRef]

58. Cherman, A.; Jacobson, T.; Sen, S.; Yaffe, L.G. Higgs-confinement phase transitions with fundamental representation matter. *Phys. Rev. D* **2020**, *102*, 105021. [CrossRef]

59. Barrois, B.C. Superconducting Quark Matter. *Nucl. Phys. B* **1977**, *129*, 390–396. [CrossRef]

60. Bailin, D.; Love, A. Superfluidity and Superconductivity in Relativistic Fermion Systems. *Phys. Rep.* **1984**, *107*, 325. [CrossRef]

61. Alford, M.G.; Schmitt, A.; Rajagopal, K.; Schäfer, T. Color superconductivity in dense quark matter. *Rev. Mod. Phys.* **2008**, *80*, 1455–1515. [CrossRef]
62. Baym, G.; Hatsuda, T.; Kojo, T.; Powell, P.D.; Song, Y.; Takatsuka, T. From hadrons to quarks in neutron stars: A review. *Rep. Prog. Phys.* **2018**, *81*, 056902. [CrossRef]
63. Alford, M.G.; Rajagopal, K.; Wilczek, F. Color flavor locking and chiral symmetry breaking in high density QCD. *Nucl. Phys. B* **1999**, *537*, 443–458. [CrossRef]
64. Cherman, A.; Sen, S.; Yaffe, L.G. Anyonic particle-vortex statistics and the nature of dense quark matter. *Phys. Rev. D* **2019**, *100*, 034015. [CrossRef]
65. Alford, M.; Reddy, S. Compact stars with color superconducting quark matter. *Phys. Rev. D* **2003**, *67*, 074024. [CrossRef]
66. Steiner, A.W.; Reddy, S.; Prakash, M. Color neutral superconducting quark matter. *Phys. Rev. D* **2002**, *66*, 094007. [CrossRef]
67. Schäfer, T.; Wilczek, F. Continuity of quark and hadron matter. *Phys. Rev. Lett.* **1999**, *82*, 3956–3959. [CrossRef]
68. Schäfer, T.; Wilczek, F. Quark description of hadronic phases. *Phys. Rev. D* **1999**, *60*, 074014. [CrossRef]
69. Schäfer, T.; Wilczek, F. Superconductivity from perturbative one gluon exchange in high density quark matter. *Phys. Rev. D* **1999**, *60*, 114033. [CrossRef]
70. Alford, M.G.; Baym, G.; Fukushima, K.; Hatsuda, T.; Tachibana, M. Continuity of vortices from the hadronic to the color-flavor locked phase in dense matter. *Phys. Rev. D* **2019**, *99*, 036004. [CrossRef]
71. Wan, Z.; Wang, J. Higher anomalies, higher symmetries, and cobordisms III: QCD matter phases anew. *Nucl. Phys. B* **2020**, *957*, 115016. [CrossRef]
72. Alford, M.G.; Han, S.; Schwenzer, K. Signatures for quark matter from multi-messenger observations. *J. Phys. G* **2019**, *46*, 114001. [CrossRef]
73. McLerran, L.; Pisarski, R.D. Phases of cold, dense quarks at large N(c). *Nucl. Phys. A* **2007**, *796*, 83–100. [CrossRef]
74. McLerran, L.; Reddy, S. Quarkyonic Matter and Neutron Stars. *Phys. Rev. Lett.* **2019**, *122*, 122701. [CrossRef]
75. Shifman, M. Understanding Confinement in QCD: Elements of a Big Picture. *Int. J. Mod. Phys. A* **2010**, *25*, 4015–4031. [CrossRef]
76. Ogilvie, M.C. Quark Confinement and the Renormalization Group. *Phil. Trans. R. Soc. Lond. A* **2011**, *369*, 2718. [CrossRef] [PubMed]
77. Reinhardt, H. Effective Approaches to QCD. In Proceedings of the 53rd Winter School of Theoretical Physics: Understanding the Origin of Matter from QCD, Karpacz, Poland, 26 February–4 March 2017.
78. Wegner, F.J. Duality in Generalized Ising Models and Phase Transitions Without Local Order Parameters. *J. Math. Phys.* **1971**, *12*, 2259–2272. [CrossRef]
79. DeGrand, T.; Detar, C.E. *Lattice Methods for Quantum Chromodynamics*; World Scientific: Singapore, 2006.
80. Ghiglieri, J.; Kurkela, A.; Strickland, M.; Vuorinen, A. Perturbative Thermal QCD: Formalism and Applications. *Phys. Rep.* **2020**, *880*, 1–73. [CrossRef]
81. Lundberg, T.; Pasechnik, R. Thermal Field Theory in real-time formalism: concepts and applications for particle decays. *Eur. Phys. J. A* **2021**, *57*, 71. [CrossRef]
82. Celik, T.; Engels, J.; Satz, H. The Order of the Deconfinement Transition in SU(3) Yang Mills Theory. *Phys. Lett. B* **1983**, *125*, 411–414. [CrossRef]
83. Maiani, L.; Testa, M. Final state interactions from Euclidean correlation functions. *Phys. Lett. B* **1990**, *245*, 585–590. [CrossRef]
84. Luscher, M. Two particle states on a torus and their relation to the scattering matrix. *Nucl. Phys. B* **1991**, *354*, 531–578. [CrossRef]
85. Hansen, M.T.; Sharpe, S.R. Lattice QCD and Three-particle Decays of Resonances. *Ann. Rev. Nucl. Part. Sci.* **2019**, *69*, 65–107. [CrossRef]
86. Aarts, G. Introductory lectures on lattice QCD at nonzero baryon number. *J. Phys. Conf. Ser.* **2016**, *706*, 022004. [CrossRef]
87. Bollweg, D.; Karsch, F.; Mukherjee, S.; Schmidt, C. Higher order cumulants of net baryon-number distributions at non-zero $\mu_B$. *Nucl. Phys. A* **2021**, *1005*, 121835. [CrossRef]
88. Bazavov, A.; Karsch, F.; Mukherjee, S.; Petreczky, P. Hot-dense Lattice QCD: USQCD whitepaper 2018. *Eur. Phys. J. A* **2019**, *55*, 194. [CrossRef]
89. Borsányi, S.; Fodor, Z.; Guenther, J.N.; Kara, R.; Katz, S.D.; Parotto, P.; Pásztor, A.; Ratti, C.; Szabó, K.K. Lattice QCD equation of state at finite chemical potential from an alternative expansion scheme. *Phys. Rev. Lett.* **2021**, *126*, 232001. [CrossRef]
90. Narayanan, R.; Neuberger, H. Infinite N phase transitions in continuum Wilson loop operators. *J. High Energy Phys.* **2006**, *3*, 64. [CrossRef]
91. Halpern, M.B. Field Strength and Dual Variable Formulations of Gauge Theory. *Phys. Rev. D* **1979**, *19*, 517. [CrossRef]
92. Batrouni, G.G.; Halpern, M.B. String, Corner and Plaquette Formulation of Finite Lattice Gauge Theory. *Phys. Rev. D* **1984**, *30*, 1782. [CrossRef]
93. Intriligator, K.A.; Seiberg, N. Phases of $N = 1$ supersymmetric gauge theories and electric—Magnetic triality. *Nucl. Phys. B Proc. Suppl.* **1996**, *39*, 1. [CrossRef]
94. Arefeva, I. NonAbelian Stokes formula. *Theor. Math. Phys.* **1980**, *43*, 353. [CrossRef]
95. Fishbane, P.M.; Gasiorowicz, S.; Kaus, P. Stokes' Theorems for Nonabelian Fields. *Phys. Rev. D* **1981**, *24*, 2324. [CrossRef]
96. Diakonov, D.; Petrov, V.Y. A Formula for the Wilson Loop. *Phys. Lett. B* **1989**, *224*, 131–135. [CrossRef]
97. Karp, R.L.; Mansouri, F.; Rno, J.S. Product integral formalism and nonAbelian Stokes theorem. *J. Math. Phys.* **1999**, *40*, 6033–6043. [CrossRef]

98. Hirayama, M.; Ueno, M. NonAbelian Stokes theorem for Wilson loops associated with general gauge groups. *Prog. Theor. Phys.* **2000**, *103*, 151–159. [CrossRef]
99. Diakonov, D.; Petrov, V. NonAbelian Stokes theorems in Yang-Mills and gravity theories. *J. Exp. Theor. Phys.* **2001**, *92*, 905–920. [CrossRef]
100. Kondo, K.I.; Taira, Y. NonAbelian Stokes Theorem and Quark confinement in SU(3) Yang-Mills gauge theory. *Mod. Phys. Lett. A* **2000**, *15*, 367–377. [CrossRef]
101. Kondo, K.I.; Taira, Y. NonAbelian Stokes theorem and quark confinement in SU(N) Yang-Mills gauge theory. *Prog. Theor. Phys.* **2000**, *104*, 1189–1265. [CrossRef]
102. Di Giacomo, A.; Dosch, H.G.; Shevchenko, V.I.; Simonov, Y.A. Field correlators in QCD: Theory and applications. *Phys. Rep.* **2002**, *372*, 319–368. [CrossRef]
103. Kuzmenko, D.S.; Shevchenko, V.I.; Simonov, Y.A. The QCD vacuum, confinement and strings in the vacuum correlator method. *Phys. Usp.* **2004**, *47*, 1–15. [CrossRef]
104. Collins, P.D.B. *An Introduction to Regge Theory and High-Energy Physics*; Cambridge Monographs on Mathematical Physics; Cambridge University Press: Cambridge, UK, 2009.
105. Philipsen, O.; Wittig, H. String breaking in nonAbelian gauge theories with fundamental matter fields. *Phys. Rev. Lett.* **1998**, *81*, 4056–4059; Erratum in **1999**, *83*, 2684. [CrossRef]
106. Duncan, A.; Eichten, E.; Thacker, H. String breaking in four-dimensional lattice QCD. *Phys. Rev. D* **2001**, *63*, 111501. [CrossRef]
107. Bernard, C.W.; DeGrand, T.A.; Detar, C.E.; Lacock, P.; Gottlieb, S.A.; Heller, U.M.; Hetrick, J.; Orginos, K.; Toussaint, D.; Sugar, R.L. Zero temperature string breaking in lattice quantum chromodynamics. *Phys. Rev. D* **2001**, *64*, 074509. [CrossRef]
108. Frohlich, J.; Morchio, G.; Strocchi, F. Higgs phenomenon without symmetry breaking order parameter. *Nucl. Phys. B* **1981**, *190*, 553–582. [CrossRef]
109. Fradkin, E.H.; Shenker, S.H. Phase Diagrams of Lattice Gauge Theories with Higgs Fields. *Phys. Rev. D* **1979**, *19*, 3682–3697. [CrossRef]
110. Greensite, J.; Matsuyama, K. Confinement criterion for gauge theories with matter fields. *Phys. Rev. D* **2017**, *96*, 094510. [CrossRef]
111. Lang, C.B.; Rebbi, C.; Virasoro, M. The Phase Structure of a Nonabelian Gauge Higgs Field System. *Phys. Lett. B* **1981**, *104*, 294. [CrossRef]
112. Elitzur, S. Impossibility of Spontaneously Breaking Local Symmetries. *Phys. Rev. D* **1975**, *12*, 3978–3982. [CrossRef]
113. Osterwalder, K.; Seiler, E. Gauge Field Theories on the Lattice. *Ann. Phys.* **1978**, *110*, 440. [CrossRef]
114. Banks, T.; Rabinovici, E. Finite Temperature Behavior of the Lattice Abelian Higgs Model. *Nucl. Phys. B* **1979**, *160*, 349–379. [CrossRef]
115. Bonati, C.; Cossu, G.; D'Elia, M.; Di Giacomo, A. Phase diagram of the lattice SU(2) Higgs model. *Nucl. Phys. B* **2010**, *828*, 390–403. [CrossRef]
116. Andersson, B.; Gustafson, G.; Ingelman, G.; Sjostrand, T. Parton Fragmentation and String Dynamics. *Phys. Rep.* **1983**, *97*, 31–145. [CrossRef]
117. Sjostrand, T.; Mrenna, S.; Skands, P.Z. Pythia 6.4 Physics and Manual. *J. High Energy Phys.* **2006**, *5*, 26. [CrossRef]
118. Sjöstrand, T.; Ask, S.; Christiansen, J.R.; Corke, R.; Desai, N.; Ilten, P.; Mrenna, S.; Prestel, S.; Rasmussen, C.O.; Skands, P.Z. An introduction to Pythia 8.2. *Comput. Phys. Commun.* **2015**, *191*, 159–177. [CrossRef]
119. 't Hooft, G. A Planar Diagram Theory for Strong Interactions. *Nucl. Phys. B* **1974**, *72*, 461. [CrossRef]
120. Sjostrand, T. Jet Fragmentation of Nearby Partons. *Nucl. Phys. B* **1984**, *248*, 469–502.
121. Andersson, B.; Gustafson, G.; Sjostrand, T. How to Find the Gluon Jets in e+ e− Annihilation. *Phys. Lett. B* **1980**, *94*, 211–215. [CrossRef]
122. Andersson, B.; Gustafson, G.; Sjostrand, T. A Model for Baryon Production in Quark and Gluon Jets. *Nucl. Phys. B* **1982**, *197*, 45–54. [CrossRef]
123. Andersson, B.; Gustafson, G.; Sjostrand, T. Baryon Production in Jet Fragmentation and Y Decay. *Phys. Scr.* **1985**, *32*, 574. [CrossRef]
124. Kugo, T.; Ojima, I. Local Covariant Operator Formalism of Nonabelian Gauge Theories and Quark Confinement Problem. *Prog. Theor. Phys. Suppl.* **1979**, *66*, 1–130. [CrossRef]
125. Kugo, T. The Universal renormalization factors Z(1)/Z(3) and color confinement condition in nonAbelian gauge theory. In Proceedings of the International Symposium on BRS Symmetry on the Occasion of Its 20th Anniversary, Kyoto, Japan, 18–22 September 1995.
126. Hata, H. Restoration of the Local Gauge Symmetry and Color Confinement in Nonabelian Gauge Theories. *Prog. Theor. Phys.* **1982**, *67*, 1607. [CrossRef]
127. Hata, H. Restoration of the local gauge symmetry and color confinement in nonabelian gauge theories. II. *Prog. Theor. Phys.* **1983**, *69*, 1524–1536. [CrossRef]
128. Marinari, E.; Paciello, M.L.; Parisi, G.; Taglienti, B. The String tension in gauge theories: A Suggestion for a new measurement method. *Phys. Lett. B* **1993**, *298*, 400–404. [CrossRef]
129. Greensite, J.; Olejnik, S.; Zwanziger, D. Coulomb energy, remnant symmetry, and the phases of nonAbelian gauge theories. *Phys. Rev. D* **2004**, *69*, 074506. [CrossRef]
130. Caudy, W.; Greensite, J. On the ambiguity of spontaneously broken gauge symmetry. *Phys. Rev. D* **2008**, *78*, 025018. [CrossRef]

131. Greensite, J.; Matsuyama, K. On the distinction between color confinement, and confinement. *arXiv* **2018**, arXiv:1811.01512.
132. Polyakov, A.M. Compact Gauge Fields and the Infrared Catastrophe. *Phys. Lett. B* **1975**, *59*, 82–84. [CrossRef]
133. Harrington, B.J.; Shepard, H.K. Thermodynamics of the Yang-Mills Gas. *Phys. Rev. D* **1978**, *18*, 2990. [CrossRef]
134. Harrington, B.J.; Shepard, H.K. Periodic Euclidean Solutions and the Finite Temperature Yang-Mills Gas. *Phys. Rev. D* **1978**, *17*, 2122. [CrossRef]
135. McLerran, L.D.; Svetitsky, B. Quark Liberation at High Temperature: A Monte Carlo Study of SU(2) Gauge Theory. *Phys. Rev. D* **1981**, *24*, 450. [CrossRef]
136. 't Hooft, G. On the Phase Transition Towards Permanent Quark Confinement. *Nucl. Phys. B* **1978**, *138*, 1–25. [CrossRef]
137. 't Hooft, G. A Property of Electric and Magnetic Flux in Nonabelian Gauge Theories. *Nucl. Phys. B* **1979**, *153*, 141–160. [CrossRef]
138. Tomboulis, E.T.; Yaffe, L.G. Finite temperature SU(2) lattice gauge theory. *Commun. Math. Phys.* **1985**, *100*, 313. [CrossRef]
139. Cornwall, J.M. Dynamical Mass Generation in Continuum QCD. *Phys. Rev. D* **1982**, *26*, 1453. [CrossRef]
140. Bachas, C. Convexity of the Quarkonium Potential. *Phys. Rev. D* **1986**, *33*, 2723. [CrossRef] [PubMed]
141. Ambjorn, J.; Olesen, P.; Peterson, C. Stochastic Confinement and Dimensional Reduction. 1. Four-Dimensional SU(2) Lattice Gauge Theory. *Nucl. Phys. B* **1984**, *240*, 189–212. [CrossRef]
142. Bali, G.S. Casimir scaling of SU(3) static potentials. *Phys. Rev. D* **2000**, *62*, 114503. [CrossRef]
143. Junior, D.R.; Oxman, L.E.; Simões, G.M. 3D Yang-Mills confining properties from a non-Abelian ensemble perspective. *J. High Energy Phys.* **2020**, *1*, 180. [CrossRef]
144. Greensite, J.P. Calculation of the Yang-Mills Vacuum Wave Functional. *Nucl. Phys. B* **1979**, *158*, 469–496. [CrossRef]
145. Greensite, J.P. Large Scale Vacuum Structure and New Calculational Techniques in Lattice SU($N$) Gauge Theory. *Nucl. Phys. B* **1980**, *166*, 113–124. [CrossRef]
146. Leigh, R.G.; Minic, D.; Yelnikov, A. On the Glueball Spectrum of Pure Yang-Mills Theory in 2+1 Dimensions. *Phys. Rev. D* **2007**, *76*, 065018. [CrossRef]
147. Karabali, D.; Kim, C.j.; Nair, V.P. On the vacuum wave function and string tension of Yang-Mills theories in (2+1)-dimensions. *Phys. Lett. B* **1998**, *434*, 103–109. [CrossRef]
148. Karabali, D.; Nair, V.P.; Yelnikov, A. The Hamiltonian Approach to Yang-Mills (2+1): An Expansion Scheme and Corrections to String Tension. *Nucl. Phys. B* **2010**, *824*, 387–414. [CrossRef]
149. Reinhardt, H.; Feuchter, C. On the Yang-Mills wave functional in Coulomb gauge. *Phys. Rev. D* **2005**, *71*, 105002. [CrossRef]
150. Feuchter, C.; Reinhardt, H. Variational solution of the Yang-Mills Schrodinger equation in Coulomb gauge. *Phys. Rev. D* **2004**, *70*, 105021. [CrossRef]
151. Greensite, J.; Olejnik, S. Dimensional Reduction and the Yang-Mills Vacuum State in 2+1 Dimensions. *Phys. Rev. D* **2008**, *77*, 065003. [CrossRef]
152. Kratochvila, S.; de Forcrand, P. Observing string breaking with Wilson loops. *Nucl. Phys. B* **2003**, *671*, 103–132. [CrossRef]
153. Luscher, M. Symmetry Breaking Aspects of the Roughening Transition in Gauge Theories. *Nucl. Phys. B* **1981**, *180*, 317–329. [CrossRef]
154. Alvarez, O. The Static Potential in String Models. *Phys. Rev. D* **1981**, *24*, 440. [CrossRef]
155. Luscher, M.; Munster, G.; Weisz, P. How Thick Are Chromoelectric Flux Tubes? *Nucl. Phys. B* **1981**, *180*, 1–12. [CrossRef]
156. Hasenfratz, A.; Hasenfratz, E.; Hasenfratz, P. Generalized Roughening Transition and Its Effect on the String Tension. *Nucl. Phys. B* **1981**, *180*, 353–367. [CrossRef]
157. Athenodorou, A.; Bringoltz, B.; Teper, M. The Closed string spectrum of SU(N) gauge theories in 2+1 dimensions. *Phys. Lett. B* **2007**, *656*, 132–140. [CrossRef]
158. Belavin, A.A.; Polyakov, A.M.; Schwartz, A.S.; Tyupkin, Y.S. Pseudoparticle Solutions of the Yang-Mills Equations. *Phys. Lett. B* **1975**, *59*, 85–87. [CrossRef]
159. Ambjorn, J.; Olesen, P. A Color Magnetic Vortex Condensate in QCD. *Nucl. Phys. B* **1980**, *170*, 265–282. [CrossRef]
160. Diakonov, D.; Maul, M. Center vortex solutions of the Yang-Mills effective action in three and four dimensions. *Phys. Rev. D* **2002**, *66*, 096004. [CrossRef]
161. Nielsen, H.B.; Olesen, P. A Quantum Liquid Model for the QCD Vacuum: Gauge and Rotational Invariance of Domained and Quantized Homogeneous Color Fields. *Nucl. Phys. B* **1979**, *160*, 380–396. [CrossRef]
162. Cornwall, J.M. Quark Confinement and Vortices in Massive Gauge Invariant QCD. *Nucl. Phys. B* **1979**, *157*, 392–412. [CrossRef]
163. Kovacs, T.G.; Tomboulis, E.T. Computation of the vortex free energy in SU(2) gauge theory. *Phys. Rev. Lett.* **2000**, *85*, 704–707. [CrossRef] [PubMed]
164. Faber, M.; Greensite, J.; Olejnik, S. Casimir scaling from center vortices: Towards an understanding of the adjoint string tension. *Phys. Rev. D* **1998**, *57*, 2603–2609. [CrossRef]
165. Greensite, J.; Langfeld, K.; Olejnik, S.; Reinhardt, H.; Tok, T. Color Screening, Casimir Scaling, and Domain Structure in G(2) and SU(N) Gauge Theories. *Phys. Rev. D* **2007**, *75*, 034501. [CrossRef]
166. Del Debbio, L.; Faber, M.; Giedt, J.; Greensite, J.; Olejnik, S. Detection of center vortices in the lattice Yang-Mills vacuum. *Phys. Rev. D* **1998**, *58*, 094501. [CrossRef]
167. Gribov, V.N. Quantization of Nonabelian Gauge Theories. *Nucl. Phys. B* **1978**, *139*, 1. [CrossRef]
168. Neuberger, H. Nonperturbative BRS Invariance and the Gribov Problem. *Phys. Lett. B* **1987**, *183*, 337–340. [CrossRef]

169. Zwanziger, D. Renormalization in the Coulomb gauge and order parameter for confinement in QCD. *Nucl. Phys. B* **1998**, *518*, 237–272. [CrossRef]

170. Faber, M.; Greensite, J.; Olejnik, S. Direct Laplacian center gauge. *J. High Energy Phys.* **2001**, *11*, 053. [CrossRef]

171. de Forcrand, P.; D'Elia, M. On the relevance of center vortices to QCD. *Phys. Rev. Lett.* **1999**, *82*, 4582–4585. [CrossRef]

172. Engelhardt, M.; Langfeld, K.; Reinhardt, H.; Tennert, O. Interaction of confining vortices in SU(2) lattice gauge theory. *Phys. Lett. B* **1998**, *431*, 141–146. [CrossRef]

173. Gubarev, F.V.; Kovalenko, A.V.; Polikarpov, M.I.; Syritsyn, S.N.; Zakharov, V.I. Fine tuned vortices in lattice SU(2) gluodynamics. *Phys. Lett. B* **2003**, *574*, 136–140. [CrossRef]

174. de Forcrand, P.; von Smekal, L. 't Hooft loops, electric flux sectors and confinement in SU(2) Yang-Mills theory. *Phys. Rev. D* **2002**, *66*, 011504. [CrossRef]

175. Engelhardt, M.; Langfeld, K.; Reinhardt, H.; Tennert, O. Deconfinement in SU(2) Yang-Mills theory as a center vortex percolation transition. *Phys. Rev. D* **2000**, *61*, 054504. [CrossRef]

176. Langfeld, K.; Tennert, O.; Engelhardt, M.; Reinhardt, H. Center vortices of Yang-Mills theory at finite temperatures. *Phys. Lett. B* **1999**, *452*, 301. [CrossRef]

177. Greensite, J.; Olejnik, S. Vortices, symmetry breaking and temporary confinement in SU(2) gauge-Higgs theory. *Phys. Rev. D* **2006**, *74*, 014502. [CrossRef]

178. Weinberg, S. Implications of Dynamical Symmetry Breaking. *Phys. Rev. D* **1976**, *13*, 974–996. Erratum in **1979**, *19*, 1277–1280. [CrossRef]

179. Susskind, L. Dynamics of Spontaneous Symmetry Breaking in the Weinberg-Salam Theory. *Phys. Rev. D* **1979**, *20*, 2619–2625. [CrossRef]

180. Hill, C.T.; Simmons, E.H. Strong Dynamics and Electroweak Symmetry Breaking. *Phys. Rep.* **2003**, *381*, 235–402; Erratum in **2004**, *390*, 553–554. [CrossRef]

181. Banks, T.; Casher, A. Chiral Symmetry Breaking in Confining Theories. *Nucl. Phys. B* **1980**, *169*, 103–125. [CrossRef]

182. Nambu, Y.; Jona-Lasinio, G. Dynamical Model of Elementary Particles Based on an Analogy with Superconductivity. 1. *Phys. Rev.* **1961**, *122*, 345–358. [CrossRef]

183. Suganuma, H.; Doi, T.M.; Iritani, T. Analytical formulae of the Polyakov and Wilson loops with Dirac eigenmodes in lattice QCD. *Prog. Theor. Exp. Phys.* **2016**, *2016*, 013B06. [CrossRef]

184. Gattringer, C. Linking confinement to spectral properties of the Dirac operator. *Phys. Rev. Lett.* **2006**, *97*, 032003. [CrossRef] [PubMed]

185. Alexandrou, C.; de Forcrand, P.; D'Elia, M. The Role of center vortices in QCD. *Nucl. Phys. A* **2000**, *663*, 1031–1034. [CrossRef]

186. Trewartha, D.; Kamleh, W.; Leinweber, D. Evidence that center vortices underpin dynamical chiral symmetry breaking in SU(3) gauge theory. *Phys. Lett. B* **2015**, *747*, 373–377. [CrossRef]

187. Witten, E. Current Algebra Theorems for the U(1) Goldstone Boson. *Nucl. Phys. B* **1979**, *156*, 269–283. [CrossRef]

188. Veneziano, G. U(1) Without Instantons. *Nucl. Phys. B* **1979**, *159*, 213–224. [CrossRef]

189. Del Debbio, L.; Giusti, L.; Pica, C. Topological susceptibility in the SU(3) gauge theory. *Phys. Rev. Lett.* **2005**, *94*, 032003. [CrossRef] [PubMed]

190. Cichy, K.; Garcia-Ramos, E.; Jansen, K.; Ottnad, K.; Urbach, C. Non-perturbative Test of the Witten-Veneziano Formula from Lattice QCD. *J. High Energy Phys.* **2015**, *9*, 20. [CrossRef]

191. Engelhardt, M. Center vortex model for the infrared sector of Yang-Mills theory: Topological susceptibility. *Nucl. Phys. B* **2000**, *585*, 614. [CrossRef]

192. Engelhardt, M. Center vortex model for the infrared sector of SU(3) Yang-Mills theory: Topological susceptibility. *Phys. Rev. D* **2011**, *83*, 025015. [CrossRef]

193. Bertle, R.; Engelhardt, M.; Faber, M. Topological susceptibility of Yang-Mills center projection vortices. *Phys. Rev. D* **2001**, *64*, 074504. [CrossRef]

194. Trewartha, D.; Kamleh, W.; Leinweber, D. Connection between center vortices and instantons through gauge-field smoothing. *Phys. Rev. D* **2015**, *92*, 074507. [CrossRef]

195. Kamleh, W.; Leinweber, D.B.; Trewartha, D. Center vortices are the seeds of dynamical chiral symmetry breaking. *arXiv* **2017**, arXiv:1701.03241.

196. Langfeld, K. Vortex structures in pure SU(3) lattice gauge theory. *Phys. Rev. D* **2004**, *69*, 014503. [CrossRef]

197. Engelhardt, M.; Reinhardt, H. Center vortex model for the infrared sector of Yang-Mills theory: Confinement and deconfinement. *Nucl. Phys. B* **2000**, *585*, 591–613. [CrossRef]

198. Engelhardt, M. Center vortex model for the infrared sector of Yang-Mills theory: Quenched Dirac spectrum and chiral condensate. *Nucl. Phys. B* **2002**, *638*, 81–110. [CrossRef]

199. Quandt, M.; Reinhardt, H.; Engelhardt, M. Center vortex model for the infrared sector of SU(3) Yang-Mills theory—Vortex free energy. *Phys. Rev. D* **2005**, *71*, 054026. [CrossRef]

200. Engelhardt, M. Center vortex model for the infrared sector of SU(3) Yang-Mills theory—Baryonic potential. *Phys. Rev. D* **2004**, *70*, 074004. [CrossRef]

201. Alexandrou, C.; de Forcrand, P.; Jahn, O. The Ground state of three quarks. *Nucl. Phys. B Proc. Suppl.* **2003**, *119*, 667–669. [CrossRef]

202. Takahashi, T.T.; Suganuma, H. Detailed analysis of the gluonic excitation in the three-quark system in lattice QCD. *Phys. Rev. D* **2004**, *70*, 074506. [CrossRef]
203. Zwanziger, D. Vanishing of zero momentum lattice gluon propagator and color confinement. *Nucl. Phys. B* **1991**, *364*, 127–161. [CrossRef]
204. Greensite, J.; Olejnik, S.; Zwanziger, D. Center vortices and the Gribov horizon. *J. High Energy Phys.* **2005**, *05*, 070. [CrossRef]
205. Zwanziger, D. No confinement without Coulomb confinement. *Phys. Rev. Lett.* **2003**, *90*, 102001. [CrossRef] [PubMed]
206. Greensite, J.; Olejnik, S. Coulomb energy, vortices, and confinement. *Phys. Rev. D* **2003**, *67*, 094503. [CrossRef]
207. West, G.B. Confinement, the Wilson Loop and the Gluon Propagator. *Phys. Lett. B* **1982**, *115*, 468–472. [CrossRef]
208. Eichmann, G. Hadron phenomenology in the Dyson-Schwinger approach. *J. Phys. Conf. Ser.* **2013**, *426*, 012014. [CrossRef]
209. Zwanziger, D. Nonperturbative Landau gauge and infrared critical exponents in QCD. *Phys. Rev. D* **2002**, *65*, 094039. [CrossRef]
210. Fischer, C.S.; Pawlowski, J.M. Uniqueness of infrared asymptotics in Landau gauge Yang-Mills theory. *Phys. Rev. D* **2007**, *75*, 025012. [CrossRef]
211. Alkofer, R.; Huber, M.Q.; Schwenzer, K. Infrared singularities in Landau gauge Yang-Mills theory. *Phys. Rev. D* **2010**, *81*, 105010. [CrossRef]
212. Fischer, C.S.; Pawlowski, J.M. Uniqueness of infrared asymptotics in Landau gauge Yang-Mills theory II. *Phys. Rev. D* **2009**, *80*, 025023. [CrossRef]
213. Lerche, C.; von Smekal, L. On the infrared exponent for gluon and ghost propagation in Landau gauge QCD. *Phys. Rev. D* **2002**, *65*, 125006. [CrossRef]
214. Alkofer, R.; Fischer, C.S.; Llanes-Estrada, F.J.; Schwenzer, K. The Quark-gluon vertex in Landau gauge QCD: Its role in dynamical chiral symmetry breaking and quark confinement. *Ann. Phys.* **2009**, *324*, 106–172. [CrossRef]
215. Maas, A. Two and three-point Green's functions in two-dimensional Landau-gauge Yang-Mills theory. *Phys. Rev. D* **2007**, *75*, 116004. [CrossRef]
216. Cucchieri, A.; Mendes, T. Constraints on the IR behavior of the gluon propagator in Yang-Mills theories. *Phys. Rev. Lett.* **2008**, *100*, 241601. [CrossRef]
217. Bogolubsky, I.L.; Ilgenfritz, E.M.; Muller-Preussker, M.; Sternbeck, A. Lattice gluodynamics computation of Landau gauge Green's functions in the deep infrared. *Phys. Lett. B* **2009**, *676*, 69–73. [CrossRef]
218. Boucaud, P.; Leroy, J.P.; Yaouanc, A.L.; Micheli, J.; Pene, O.; Rodriguez-Quintero, J. IR finiteness of the ghost dressing function from numerical resolution of the ghost SD equation. *J. High Energy Phys.* **2008**, *6*, 12. [CrossRef]
219. Aguilar, A.C.; Binosi, D.; Papavassiliou, J. Gluon and ghost propagators in the Landau gauge: Deriving lattice results from Schwinger-Dyson equations. *Phys. Rev. D* **2008**, *78*, 025010. [CrossRef]
220. Dudal, D.; Gracey, J.A.; Sorella, S.P.; Vandersickel, N.; Verschelde, H. A Refinement of the Gribov-Zwanziger approach in the Landau gauge: Infrared propagators in harmony with the lattice results. *Phys. Rev. D* **2008**, *78*, 065047. [CrossRef]
221. Fischer, C.S.; Maas, A.; Pawlowski, J.M. On the infrared behavior of Landau gauge Yang-Mills theory. *Ann. Phys.* **2009**, *324*, 2408–2437. [CrossRef]
222. Cucchieri, A.; Mendes, T.; Taurines, A.R. Positivity violation for the lattice Landau gluon propagator. *Phys. Rev. D* **2005**, *71*, 051902. [CrossRef]
223. Braun, J.; Gies, H.; Pawlowski, J.M. Quark Confinement from Color Confinement. *Phys. Lett. B* **2010**, *684*, 262–267. [CrossRef]
224. Cooper, P.; Zwanziger, D. Schwinger-Dyson Equations in Coulomb Gauge Consistent with Numerical Simulation. *Phys. Rev. D* **2018**, *98*, 114006. [CrossRef]
225. Wetterich, C. Exact evolution equation for the effective potential. *Phys. Lett. B* **1993**, *301*, 90–94. [CrossRef]
226. Fister, L.; Pawlowski, J.M. Confinement from Correlation Functions. *Phys. Rev. D* **2013**, *88*, 045010. [CrossRef]
227. Marhauser, F.; Pawlowski, J.M. Confinement in Polyakov Gauge. *arXiv* **2008**, arXiv:0812.1144.
228. Chung, K.; Greensite, J. Coulomb flux tube on the lattice. *Phys. Rev. D* **2017**, *96*, 034512. [CrossRef]
229. Tiktopoulos, G. Gluon Chains. *Phys. Lett. B* **1977**, *66*, 271–275. [CrossRef]
230. Greensite, J.; Thorn, C.B. Gluon chain model of the confining force. *J. High Energy Phys.* **2002**, *02*, 014. [CrossRef]
231. Greensite, J.; Szczepaniak, A.P. Coulomb string tension, asymptotic string tension, and the gluon chain. *Phys. Rev. D* **2015**, *91*, 034503. [CrossRef]
232. Greensite, J.; Szczepaniak, A.P. Constituent gluons and the static quark potential. *Phys. Rev. D* **2016**, *93*, 074506. [CrossRef]
233. Greensite, J.; Olejnik, S. Constituent Gluon Content of the Static Quark-Antiquark State in Coulomb Gauge. *Phys. Rev. D* **2009**, *79*, 114501. [CrossRef]
234. Nambu, Y. Strings, Monopoles and Gauge Fields. *Phys. Rev. D* **1974**, *10*, 4262. [CrossRef]
235. 't Hooft, G. *High Energy Physics*; Editorice Compositori: Bologna, Italy, 1975.
236. Mandelstam, S. Vortices and Quark Confinement in Nonabelian Gauge Theories. *Phys. Rep.* **1976**, *23*, 245–249. [CrossRef]
237. Polyakov, A.M. Quark Confinement and Topology of Gauge Groups. *Nucl. Phys. B* **1977**, *120*, 429–458. [CrossRef]
238. Nielsen, H.B.; Olesen, P. Vortex Line Models for Dual Strings. *Nucl. Phys. B* **1973**, *61*, 45–61. [CrossRef]
239. Seiberg, N.; Witten, E. Electric—magnetic duality, monopole condensation, and confinement in $N = 2$ supersymmetric Yang-Mills theory. *Nucl. Phys. B* **1994**, *426*, 19–52; Erratum in **1994**, *430*, 485–486. [CrossRef]
240. Seiberg, N.; Witten, E. Monopoles, duality and chiral symmetry breaking in $N = 2$ supersymmetric QCD. *Nucl. Phys. B* **1994**, *431*, 484–550. [CrossRef]

241. Gomez, C.; Hernandez, R. Electric—Magnetic Duality and Effective Field Theories; Advanced School on Effective Theories: Granada, Spain, 1995.
242. Bilal, A. *Duality in N = 2 Susy SU(2) Yang-Mills Theory: A Pedagogical Introduction to the Work of Seiberg and Witten*; NATO Advanced Study Institute on Quantum Fields and Quantum Space Time: Cargèse, France, 1997; pp. 21–43.
243. D'Hoker, E.; Phong, D.H. Lectures on supersymmetric Yang-Mills theory and integrable systems. In Proceedings of the 9th CRM Summer School: Theoretical Physics at the End of the 20th Century, Banff, MB, Canada, 27 June–10 July 1999; pp. 1–125.
244. Douglas, M.R.; Shenker, S.H. Dynamics of SU(N) supersymmetric gauge theory. *Nucl. Phys. B* **1995**, *447*, 271–296. [CrossRef]
245. Polyakov, A.M. *Gauge Fields and Strings*; Contemporary Concepts in Physics; Harwood Academic Publishers: Chur, Switzerland, 1987; Volume 3.
246. 't Hooft, G. Magnetic Monopoles in Unified Gauge Theories. *Nucl. Phys. B* **1974**, *79*, 276–284. [CrossRef]
247. Polyakov, A.M. Particle Spectrum in Quantum Field Theory. *JETP Lett.* **1974**, *20*, 194–195.
248. Shifman, M.; Yung, A. Supersymmetric Solitons and How They Help Us Understand Non-Abelian Gauge Theories. *Rev. Mod. Phys.* **2007**, *79*, 1139. [CrossRef]
249. 't Hooft, G. Topology of the Gauge Condition and New Confinement Phases in Nonabelian Gauge Theories. *Nucl. Phys. B* **1981**, *190*, 455–478. [CrossRef]
250. Unsal, M. Abelian duality, confinement, and chiral symmetry breaking in QCD(adj). *Phys. Rev. Lett.* **2008**, *100*, 032005. [CrossRef] [PubMed]
251. Shifman, M.; Unsal, M. QCD-like Theories on R(3) × S(1): A Smooth Journey from Small to Large r(S(1)) with Double-Trace Deformations. *Phys. Rev. D* **2008**, *78*, 065004. [CrossRef]
252. Unsal, M.; Yaffe, L.G. Center-stabilized Yang-Mills theory: Confinement and large N volume independence. *Phys. Rev. D* **2008**, *78*, 065035. [CrossRef]
253. Cossu, G.; Hatanaka, H.; Hosotani, Y.; Noaki, J.I. Polyakov loops and the Hosotani mechanism on the lattice. *Phys. Rev. D* **2014**, *89*, 094509. [CrossRef]
254. Bergner, G.; Piemonte, S.; Ünsal, M. Adiabatic continuity and confinement in supersymmetric Yang-Mills theory on the lattice. *J. High Energy Phys.* **2018**, *11*, 092. [CrossRef]
255. Bonati, C.; Cardinali, M.; D'Elia, M.; Giordano, M.; Mazziotti, F. Reconfinement, localization and thermal monopoles in $SU(3)$ trace-deformed Yang-Mills theory. *Phys. Rev. D* **2021**, *103*, 034506. [CrossRef]
256. Kronfeld, A.S.; Laursen, M.L.; Schierholz, G.; Wiese, U.J. Monopole Condensation and Color Confinement. *Phys. Lett. B* **1987**, *198*, 516–520. [CrossRef]
257. Stack, J.D.; Tucker, W.W.; Wensley, R.J. The Maximal Abelian gauge, monopoles, and vortices in SU(3) lattice gauge theory. *Nucl. Phys. B* **2002**, *639*, 203–222. [CrossRef]
258. Shiba, H.; Suzuki, T. Monopoles and string tension in SU(2) QCD. *Phys. Lett. B* **1994**, *333*, 461–466. [CrossRef]
259. Stack, J.D.; Neiman, S.D.; Wensley, R.J. String tension from monopoles in SU(2) lattice gauge theory. *Phys. Rev. D* **1994**, *50*, 3399–3405. [CrossRef]
260. Ambjorn, J.; Greensite, J. Center disorder in the 3-D Georgi-Glashow model. *J. High Energy Phys.* **1998**, *5*, 4. [CrossRef]
261. Del Debbio, L.; Di Giacomo, A.; Paffuti, G. Detecting dual superconductivity in the ground state of gauge theory. *Phys. Lett. B* **1995**, *349*, 513–518. [CrossRef]
262. Di Giacomo, A.; Lucini, B.; Montesi, L.; Paffuti, G. Color confinement and dual superconductivity of the vacuum. 1. *Phys. Rev. D* **2000**, *61*, 034503. [CrossRef]
263. Greensite, J.; Lucini, B. Is Confinement a Phase of Broken Dual Gauge Symmetry? *Phys. Rev. D* **2008**, *78*, 085004. [CrossRef]
264. Kraan, T.C.; van Baal, P. Periodic instantons with nontrivial holonomy. *Nucl. Phys. B* **1998**, *533*, 627–659. [CrossRef]
265. Kraan, T.C.; van Baal, P. Exact T duality between calorons and Taub—NUT spaces. *Phys. Lett. B* **1998**, *428*, 268–276. [CrossRef]
266. Lee, K.M.; Lu, C.H. SU(2) calorons and magnetic monopoles. *Phys. Rev. D* **1998**, *58*, 025011. [CrossRef]
267. Bogomolny, E.B. Stability of Classical Solutions. *Sov. J. Nucl. Phys.* **1976**, *24*, 449.
268. Prasad, M.K.; Sommerfield, C.M. An Exact Classical Solution for the 't Hooft Monopole and the Julia-Zee Dyon. *Phys. Rev. Lett.* **1975**, *35*, 760–762. [CrossRef]
269. Hofmann, R. *The Thermodynamics of Quantum Yang–Mills Theory*; World Scientific: Singapore, 2011.
270. 't Hooft, G. Computation of the Quantum Effects Due to a Four-Dimensional Pseudoparticle. *Phys. Rev. D* **1976**, *14*, 3432–3450; Erratum in *Phys. Rev. D* **1978**, *18*, 2199. [CrossRef]
271. Nahm, W. Selfdual Monopoles and Calorons. In Proceedings of the 12th International Colloquium on Group Theoretical Methods in Physics, Trieste, Italy, 5–11 September 1983.
272. Garland, H.; Murray, M.K. Kac-Moody Monopoles and Periodic Instantons. *Commun. Math. Phys.* **1988**, *120*, 335–351. [CrossRef]
273. Nahm, W. A Simple Formalism for the BPS Monopole. *Phys. Lett. B* **1980**, *90*, 413–414. [CrossRef]
274. Diakonov, D.; Petrov, V. Confining ensemble of dyons. *Phys. Rev. D* **2007**, *76*, 056001. [CrossRef]
275. Bruckmann, F.; Dinter, S.; Ilgenfritz, E.M.; Muller-Preussker, M.; Wagner, M. Cautionary remarks on the moduli space metric for multi-dyon simulations. *Phys. Rev. D* **2009**, *79*, 116007. [CrossRef]
276. Gerhold, P.; Ilgenfritz, E.M.; Muller-Preussker, M. An SU(2) KvBLL caloron gas model and confinement. *Nucl. Phys. B* **2007**, *760*, 1–37. [CrossRef]

277. Gupta, S.; Huebner, K.; Kaczmarek, O. Renormalized Polyakov loops in many representations. *Phys. Rev. D* **2008**, *77*, 034503. [CrossRef]
278. Greensite, J.; Höllwieser, R. Double-winding Wilson loops and monopole confinement mechanisms. *Phys. Rev. D* **2015**, *91*, 054509. [CrossRef]
279. Diakonov, D. Topology and confinement. *Nucl. Phys. B Proc. Suppl.* **2009**, *195*, 5–45. [CrossRef]
280. Greensite, J.; Matsuyama, K. What symmetry is actually broken in the Higgs phase of a gauge-Higgs theory? *Phys. Rev. D* **2018**, *98*, 074504. [CrossRef]
281. Greensite, J.; Matsuyama, K. Higgs phase as a spin glass and the transition between varieties of confinement. *Phys. Rev. D* **2020**, *101*, 054508. [CrossRef]

 *universe*

*Review*

# Topological Aspects of Dense Matter: Lattice Studies

Maria Paola Lombardo

Istituto Nazionale di Fisica Nucleare, Sezione di Firenze, 50019 Sesto Fiorentino, FI, Italy; lombardo@fi.infn.it

**Abstract:** Topological fluctuations change their nature in the different phases of strong interactions, and the interrelation of topology, chiral symmetry and confinement at high temperature has been investigated in many lattice studies. This review is devoted to the much less explored subject of topology in dense matter. After a short overview of the status at zero density, which will serve as a baseline for the discussion, we will present lattice results for baryon rich matter, which, due to technical difficulties, has been mostly studied in two-color QCD, and for matter with isospin and chiral imbalances. In some cases, a coherent pattern emerges, and in particular the topological susceptibility seems suppressed at high temperature for baryon and isospin rich matter. However, at low temperatures the topological aspects of dense matter remain not completely clear and call for further studies.

**Keywords:** QCD; topology; lattice field theory; dense matter; phase transitions

## 1. Introduction

In broad outline, the general framework of this review is Quantum Chromodynamics (QCD) and its several phases and critical phenomena depending on temperature, baryonic, isospin and chiral densities. At high temperatures the matter is in a plasma phase—the Quark-Gluon Plasma. At lower temperatures and increasing baryonic density, one encounters nuclear matter first, then a transition to a dense deconfined phase of quarks and gluons. This phase of matter is realised in the interior of neutron stars, extremely compact stellar objects produced in the supernova explosions. In this extreme environment several exotic phases can be realised. The recent observation of gravitational wave signals originating from the merging of two neutron stars has triggered further interest in the theoretical investigation in this direction, see e.g., Ref. [1].

Current and planned experiments have the capability of exploring the phase diagram of strong interactions. Ab-initio lattice studies [2,3] have produced results at non-zero baryon density, at rather large temperature in QCD, at non-zero isospin and chiral densities, and in entire phase diagram for two-color QCD, which is protected by the sign problem by the Pauli-Gürsey symmetry. The focus of most studies is on chiral and confining properties, and only a limited subset has addressed topology.

The interplay of chiral symmetry, confinement and topology may well depend on the details of the microscopic dynamics, which in turn is affected by matter density. In vacuum, chiral symmetry breaking occurs via a space-homogeneous condensate. At high temperature this is known to dissolve, while for low temperatures and high-density different pairing phenomena result in a rich, and still not entirely explored phase diagram [4–6]. In particular, Ref. [6] suggests a distinct different behaviour of the topological susceptibility at high temperatures and zero density, and high densities and low temperatures. A simplified view of the phase diagram from Ref. [5] can be seen in Figure 1. Non-homogeneous phases [7,8], predicted and only recently observed [9] in simple models at finite density, may well have different confining and topological properties.

A non-zero density—be it due to baryon, isospin or chiral imbalances—is then an important probe for the interplay of chiral symmetry, confinement and topology, and may shed some light on its general aspects.

**Citation:** Lombardo, M.P. Topological Aspects of Dense Matter: Lattice Studies. *Universe* **2021**, *7*, 336. https://doi.org/10.3390/universe7090336

Academic Editor: Dmitri Antonov

Received: 31 July 2021
Accepted: 8 September 2021
Published: 9 September 2021

**Publisher's Note:** MDPI stays neutral with regard to jurisdictional claims in published maps and institutional affiliations.

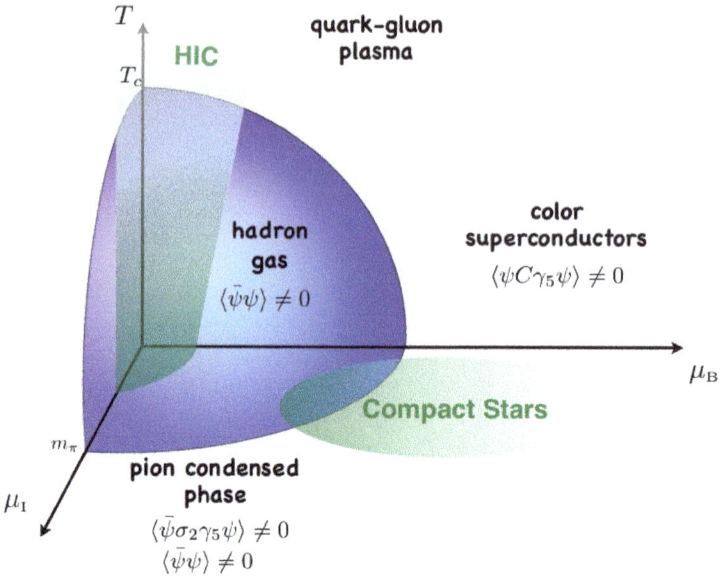



**Figure 1.** A schematic view of the phase diagram of QCD in the baryon, isospin chemical potential, and temperatures, from Ref. [5]. For two-color QCD the diagram would look similar, but the zero-temperature baryon and isospin critical chemical potential are the same, and the two dense phases are characterised by diquark and pion condensation.

## 2. Topology and Strong Interactions

The dynamics of gauge field is a fascinating aspect of strong interactions. Asymptotic freedom, and the self-interacting nature of the gluons, are reflected by the structure of the gauge sector of the theory. A configuration of gauge fields may have a topological content, measured by the topological charge density $q(x)$:

$$q(x) \equiv \frac{g^2}{32\pi^2} F^a_{\mu\nu} \tilde{F}^{\mu\nu}_a \tag{1}$$

Indeed (see e.g., Ref. [10]) $Q \equiv \int q(x)d^4x$ equals the Chern–Pontryagin index or winding number of gauge fields. It can only assume integer values, thus identifying the topological class to which the gauge configuration belongs.

The QCD Lagrangian may be coupled to the topological charge density

$$\mathcal{L} = \mathcal{L}_{QCD} + \theta \frac{g^2}{32\pi^2} F^a_{\mu\nu} \tilde{F}^{\mu\nu}_a, \tag{2}$$

Experiments on the electric dipole moment of the neutron $d_n$ place limits on the value of $\theta$ parameter. The limits follow from the relation between $d_n$ and $\theta$. QCD sum rules give $d_n = 2.4 \times 10^{-16}\theta$ e cm [11] and chiral perturbation theory gives $d_n = 3.3 \times 10^{-16}\theta$ e cm [12]. The most recent experimental measure [13] of the neutron electric dipole moment is $d_n = (0.0 \pm 1.1 \,(\text{stat}) \pm 0.2 \,(\text{sys})) \times 10^{-26}$ e cm, which may be interpreted as an upper limit $|d_n| < 1.8 \times 10^{-26}$ e cm at a 90% C.L. Combining the experimental limit with the relations mentioned above, one arrives at the bound $\theta < 0.5 \times 10^{-10}$. This anomalously small value of $\theta$ leads to the hypothesis of an axion field which would force $\theta$ to vanish dynamically [14,15]. tealIn parallel, and not further discussed in this review, the possibility of a very small, but non-zero nEDM remains open. Such result would indicate physics beyond the standard model [16] and it is a subject of an active theoretical investigation, see e.g., Ref. [17].

### 2.1. Topology, from Low to High Temperatures

The close and challenging interplay of topology, chiral and axial symmetry and gauge field dynamics—which we will briefly review below—has motivated investigations at high temperature, well before considering high density. In short summary, the topological susceptibility drops at the high temperature chiral phase transition, although it is still under debate whether a partial axial restoration coincides or not with chiral restoration, see e.g., Ref. [18] for a recent review, including a full set of references.

The behaviour of topology has been extensively studied at high temperature on the lattice [18–22]: it decreases in the plasma and at very high temperatures follows the predictions of the dilute instanton gas, DIGA (which we will briefly introduce in the next subsection), in which only configurations with zero, or unit, topological charges are possible. The results are briefly summarised in Figure 2, from Ref. [18].

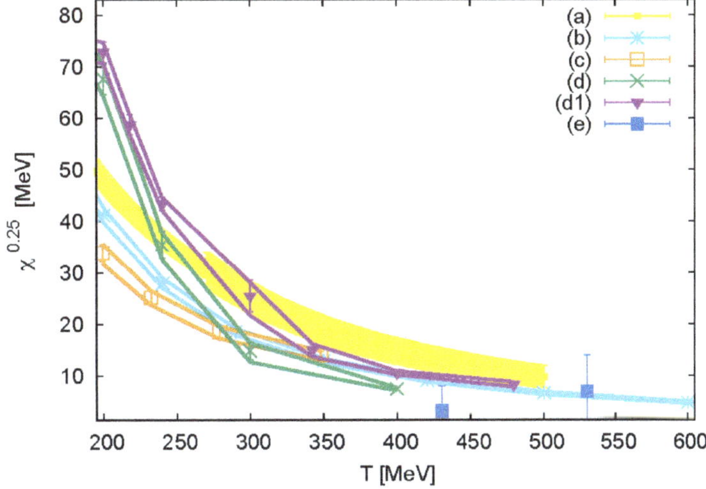



**Figure 2.** The fourth root of topological susceptibility versus the temperature in full QCD. (a) shows the gluonic results from Ref. [21]. (b) shows the tabulated results from Ref. [23]. (c) Ref. [24]; these results are rescaled from a higher pion mass (d) and (d1) show the results from Ref. [22] obtained by rescaling from the two lightest masses $m_\pi = 220, 260$ MeV. (e) the results from Ref. [20], where a careful continuum extrapolation with a conservative error estimate was performed. From Ref. [18].

### 2.2. Symmetries of QCD, and Topology

At a classical level, and in the massless limit, $\mathcal{L}_{QCD}$ has a global $U(N) \times U(N) \equiv SU(N) \times SU(N) \times U(1)_B \times U(1)_A$ symmetry, where $N$ is the number of light flavours. The most natural scenario compatible with the pions and $K$ mesons spectrum is the spontaneous breaking of the $SU(3) \times SU(3)$ symmetry of $\mathcal{L}_{QCD}$ in the three-flavour massless limit, while the other flavours are massive and do not participate in the chiral dynamics. In this scenario, Chiral Perturbation Theory predicts the masses of the mesons and baryons made by the physical up, down and strange quarks. The condensate formed in this breaking would also break the $U(1)_A$ symmetry: hence, the $\eta'$ should follow the same fate as the other mesons, while it is distinctly heavier.

The way out is breaking explicitly the $U(1)_A$ symmetry [25]: since the topological charge appears in the divergence of the $U(1)_A$ current $J_5^\mu$

$$\partial_\mu J_5^\mu = 2N_f Q + 2\sum_{i=1}^{N_f} m_i \bar{\psi}_i \gamma_5 \psi_i \tag{3}$$

the breaking of the axial symmetry may be achieved by a non-trivial topology which leads to the non-conservation of the current. In this way the large mass of the $\eta'$ affords a direct evidence of the non-trivial topology of the vacuum, responsible for the explicit $U(1)_A$ breaking [26–28]. The $\eta'$ carries thus interesting information on the anomalous component and on topology: as anticipated in Ref. [29] the $\eta'$ should be on the same footing as the other mesons in the plasma, once the anomalous component disappears. Indeed, this was verified on the lattice in [30], see Figure 3.

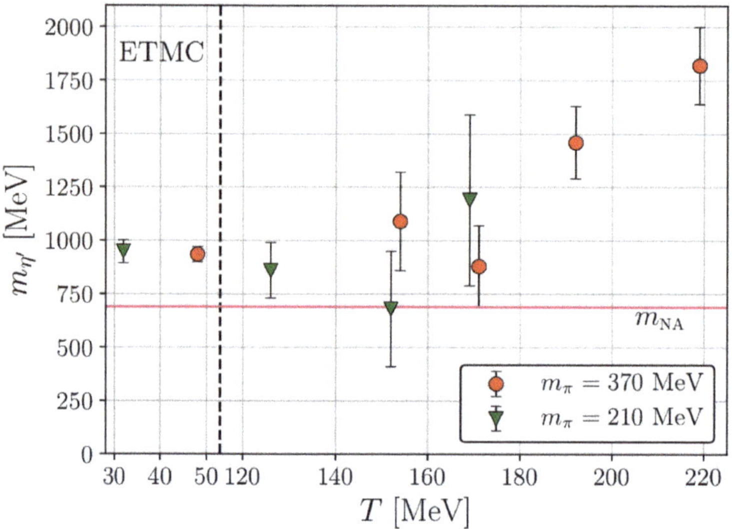


**Figure 3.** The mass of the $\eta'$ as a function of the temperature in QCD: the $\eta'$ mass approaches the mass of a (unphysical) $\bar{s}s$ meson at the transition, signaling the suppression of the anomalous component due to topological fluctuations. From Ref. [30].

Additionally, the spectrum results concurs in indicating that the topological susceptibility is greatly reduced in the plasma.

One interesting case, much studied numerically, is two-color QCD, which enjoys an enlarged chiral symmetry [31]: quarks and antiquarks belong to equivalent representation of the color group. As a consequence of that, the ordinary chiral symmetry of QCD $SU(N) \times SU(N)$ is enlarged to $SU(2N)$. Thanks to this symmetry the theory does not have a sign problem at non-zero baryon density: intuitively, baryon and isospin density are the same for the two-color world.

A further symmetry is the isospin symmetry: a global transformation, an $SU(2)$ rotation in flavour space (QCD interactions are flavour-blind). It acts on up and down quarks and $\mathcal{L}_{QCD}$ is invariant for identical or vanishing masses. In reality this symmetry is explicitly broken by the (small) mass difference between up and down quarks. Isospin breaking has consequences also on topology, since chiral perturbation theory [32] predicts, at leading order (LO),

$$\chi_{LO} = \frac{z}{1+z^2} m_\pi^2 f_\pi^2 \tag{4}$$

with $z = m_u/m_d$. We would not pursue these aspects—we will always consider $m_u = m_d$. We will instead consider the effect on topology of an isospin density [33], artificially induced by an appropriate chemical potential- more precisely, by its third component, $I_3$. The motivation for this is to understand the regime of finite density of a conserved charge (the isospin), aiming at the observation of the transition from hadronic to quark degrees of freedom at zero/low temperatures.

In nuclear matter and in astrophysics isospin imbalance is very important, however in real world the baryonic density is much larger than the isospin one, $\mu_I \ll \mu_B$. In Ref. [33] as well as in lattice studies [34–40], the authors consider an idealization with $\mu_I \neq 0, \mu_B = 0$. Such a system is unstable under weak interactions which do not conserve isospin. Nonetheless, in this purely QCD world this system can still give us relevant information, also considering that it is free from the sign problem [31], and we will discuss lattice results below.

This brief discussion on symmetries leads us to consider three different chemical potentials: two of which associates with conserved charges, $\mu_B$ and $\mu_I$, the third associated with a the non-conserved axial current, $\mu_5 \equiv (\mu_R - - - \mu_L)/2$, which couples to the current $\bar{\psi}_i \gamma_5 \psi_i$ see Equation (3).

### 2.3. Conserved Charges $\mu_B$ and $\mu_I$

Let us consider again the phase diagram in the $\mu_B, \mu_I, T$ space, from Ref. [5], see Figure 1: at high temperature-chiral symmetry is restored. In the limit of zero temperatures isospin and baryochemical potential induce the transition to a dense phase, characterised by different pairing phenomena, see e.g., Refs. [4,5]. A standard scenario predicts the transition when the chemical potential equals the mass of the lowest state carrying the corresponding charge: $\mu_B^c \simeq m_N$, and $\mu_I^c \simeq m_\pi$, with $m_N$ and $m_\pi$ being the nucleon and pion mass, respectively. The two dense phases are characterised by different pairing phenomena in real QCD, and are probably separated by a low temperature phase transition at in the $\mu_B, \mu_I$ plane, studied in chiral perturbation theory [41].

In two-color QCD, due to the already mentioned fact that the baryons of the theory are diquarks, the phase diagram in Figure 1 would look the same in the two directions, with the same thresholds along baryon and isospin chemical potential. The two condensed phases, which are significantly different in QCD, are now the same and have both (colorless) condensates: diquark and pion condensates in the $\mu_B$ and $\mu_I$ directions.

### 2.4. Instantons and Zero Modes

One possible way to discuss the different properties of the phases of strong interactions at zero and non-zero density vis-a-vis topology is to consider the behaviour of instantons [42]. Instantons are classical solutions to the Euclidean equations of motion: localized regions of space-time (typical sizes are 1/3 fm) , with very strong gluonic fields, and characterised by a topological quantum number. It turns out that there are important differences in the instanton behaviour at zero and non-zero densities and temperatures [6,43,44]. At low temperatures one expects a random instanton ensemble, accounting for chiral breaking. At high temperatures one expects instanton-anti-instanton pairs, eventually behaving at high temperatures like a dilute gas, described by the Dilute Instanton Gas Approximation, DIGA, in which only configurations with zero, or unit, topological charges are possible, while at finite density one may expect instanton chains [43]. The reason for this [43] is that either at finite temperature and finite density the quark propagation in time direction is favored over space-like propagation, the latter being suppressed by $e^{-\pi Tr}$ and $e^{i\mu r}$, respectively. In general, the fermion determinant generates strong correlations among instantons. Because of this, the random instanton ensemble, responsible for chiral breaking, dissolves into clusters oriented in the time direction: the already mentioned instanton-anti-instanton pairs at high temperature (eventually turning into the dilute instanton gas) and instanton chains at high density.

The details may be found in Ref. [44], where the authors discuss the role of instantons in the three major phases of strong interactions: the hadronic phase, the color superconductor phase, and the quark-gluon plasma phase. In brief, the reasoning starts from Equation (3) which shows a remarkable connection between gauge fields and fermions. This is made more transparent by the Atiyah–Singer index theorem [45,46]:

$$\frac{1}{32\pi^2}\,\varepsilon_{\mu\nu\rho\sigma}\int \mathrm{Tr}[F^{\mu\nu}(x)F^{\rho\sigma}(x)]\,d^4x = n_+ - n_- . \tag{5}$$

We see that the topological charge 'counts' the number of zero modes of the massless Dirac operator with positive and negative chirality $n_\pm$. The Atiyah–Singer theorem works also at finite density, however, the nature of the zero modes are now different, since there are extra states appearing at the Fermi surface. In QCD this interaction leads to the formation of diquark (colored) Cooper pairs, to BCS instability and to color superconductivity—these are the phases depicted in Figure 1. In two-color QCD, diquark pairs are stable as they are color neutral and a diquark condensate is formed. The different phases of strong interactions can then be characterised by instanton dynamics, and an important point is that pair dynamics should always predominate as temperature increases, at any chemical potential. One would then expect some significant changes in topology at fixed chemical potential when increasing temperature, as well as changes when fixing the temperature and increasing the chemical potentials.

*2.5. Detecting Topology—The Chiral Magnetic Effect and $\mu_5$*

From a phenomenological point of view, the Atiyah–Singer theorem opens the way to the possibility of a direct observation of topology in experiments, see e.g., Refs. [47–49]. In fact, the gluons do not carry conserved charges which could be directly measured. But we can still 'see' topological fluctuations in the quark sector. These observations are at the root of the discovery of the so-called Chiral Magnetic Effect (CME) [49]: electric charge separation in the presence of an external magnetic field that is induced by the chirality imbalance. This striking effect could be observed in heavy ion collisions and in condensed matter experiments and its prediction has spawned a significant experimental activity. Many reviews are available, see e.g., [48] and we will not discuss further the CME here, as our focus is equilibrium studies of topology and their signatures in the phase diagram. However, it is important to underscore that it is indeed the CME that has called the attention of the lattice community on the chiral chemical potential, and has motivated the numerical analysis at equilibrium which we will discuss later. teal It is important to notice that the chiral chemical potential $\mu_5$ couples to the chiral charge density operator $\psi^\dagger\gamma_5\psi$ which is not conserved because of the chiral anomaly. So it is not on the same footing as the baryon chemical potential or the isospin chemical potential. $\mu_5$ cannot be generated in thermodynamic equilibrium, topological fluctuations will wash it out: $\mu_5$ is just an external coupling able to generate a chiral imbalance [50–54], and it may require renormalization in the ultraviolet [50].

## 3. Lattice Results—Topology and Dense Matter

In the previous Section we have argued that one may expect some significant differences in topology in the different phases, and have shown a summary of results at zero density. Here we will review the current lattice results for topology in dense matter, with different temperatures.

A very brief technical note before proceeding: let us remind ourselves that the different densities we have discussed are realised by adding the appropriate zeroth component of the current to the Lagrangian, while on Euclidean lattices the temperature is the reciprocal of the time extent of the lattice, $T = 1/N_t a$, $a$ being the lattice spacing and $N_t$ the number of sites in the temporal direction. The results on topology are so far limited to the topological susceptibility, the fluctuations of the topological charge. Details on the rich and highly technical subject of lattice topology may be found in Ref. [55].

### 3.1. Baryon Density

These studies were mostly carried out for two-color QCD, since this is free from the sign problem. As discussed above, the dynamics may be significantly different from that of QCD, especially concerning the nature of the pairing at high density. However, the hope is that the main features of instanton dynamics which are at the heart of this discussion remain valid—although of course this is subject to verification.

Studies of two-color matter have been performed by several groups [34,56–64]. teal In cold matter, these studies have confirmed that baryonic matter forms at an onset $\mu_o = m_\pi/2$, whereupon diquarks start populating the vacuum. If the dynamics favor pairing, they could condense: diquark condensation at low temperatures, above $\mu_o$ is consistently observed in lattice studies. However, the dependence of the diquark condensate on the chemical potential is still unsettled, leading to different hypotheses on the nature of the phase about $\mu_o$: some studies find consistence with chiral perturbation theory indicating a BEC phase [64], followed from a transition to free behaviour, interpreted as a crossover to BCS. Others find compatibility with a free quark behaviour [65] immediately above $\mu_o$, at largish masses, and non-conclusive results for lower masses [65]. According to the same study, early signals for deconfinement for chemical potentials $\mu \simeq 1.1 m_\pi$ [61] should be interpreted with care. In brief, the issue of the nature of the dense phase above $\mu_o$ is subtle, and under investigation. The very existence of an onset at $\mu_o$ and low temperatures is instead uncontroversial, and we will concern ourselves with the behaviour of topology past this onset, in comparison with the observations at high temperatures.

One first study of topology was carried out in two-color QCD with eight flavours of staggered fermions [58]—this choice may be surprising as the theory in the continuum limit is known to be within the conformal window of QCD, see e.g., Refs. [66–68]. However, the coupling was strong enough to break chiral symmetry at zero temperature. In this condition one may study the phase diagram—similarly, for instance, to what one would do in lattice strong coupling electrodynamics. One important caveat, of course, is that topology is poorly defined at strong coupling: the very nature of the topology require the continuum limit [55]—on a discrete system the barriers among different topological sectors are finite, as the system may be easily deformed. On the lattice, this produces the so-called dislocations, which artificially increase topological fluctuations. To mitigate, at least partially, this effect in Ref. [58] the analysis is restricted to finite temperature, which is realised with finer lattices. In addition to that, the lattice configurations were subjected to smoothing—a local coarse graining—designed to suppress artifacts. The results of [58] indicate that gluon dynamics, chiral symmetry and topology are interrelated in the region of temperatures $0.3 < T/T_c < 0.4$. The behaviour is exemplified in Figure 4: to appreciate the correlation among different observables the diagrams show the derivatives with respect to the chemical potential of the Polyakov loop, the chiral condensate and the topological susceptibility. The coincidence of the peaks, signaling the onset of the superfluid phase, is quite clear.

Obviously, these earlier investigations called for more studies, including lower temperatures. Of particular interest would be the study of topology in the superfluid phase, characterised by a diquark condensate, as discussed at the beginning of this Section.

A study dedicated to topology in the cold phase, i.e., accessing the superfluid region of two-color QCD, found interesting differences between two and four flavours [61], see Figure 5. The topological susceptibility was measured on two different gauge field ensembles. The first used a $12^3 \times 24$ lattice and $N = 2$ flavours of Wilson fermions. The second ensemble used the same system size, and $N = 4$. The two ensembles have similar pion masses in lattice units, $m_\pi a = 0.68$, and may be considered fairly 'cold': the onset for the superfluid phase should then be $\mu_o a \simeq 0.34$. A summary of the results is offered by Figure 5: interestingly, apparently the topology in the two-flavour theory is insensitive to the chemical potential, while the results in the four-flavour model may even suggest an increase of the topological susceptibility. The authors issue a caveat though: also, in this case one may fear important discretization effects. Barring these, the observation is that other thermodynamic studies revealed that the two-flavour model is weakly coupled above the onset [60], while in contrast, the four-flavour model appears to be strongly coupled [62]. So, there is the possibility that the raise of the topological susceptibility in the four-flavour model above the onset does indeed reflect the strongly coupled nature of the theory.

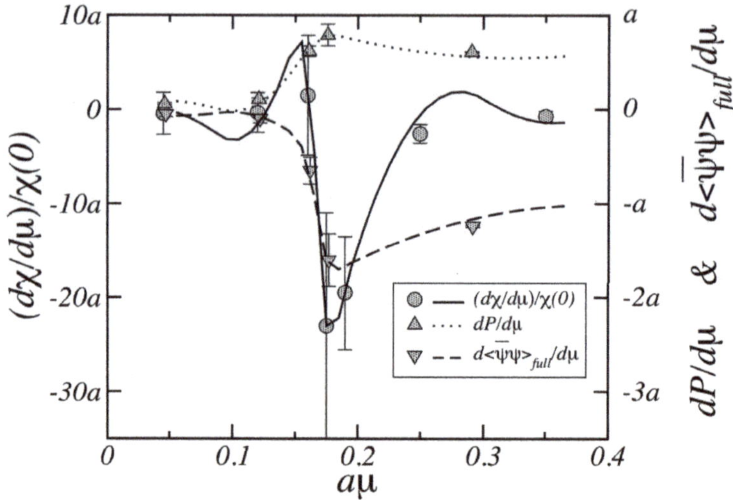

**Figure 4.** The correlation among topology, confinement and chiral symmetry as seen from the $\mu_B$ derivatives of the topological susceptibility, the Polyakov loop and the chiral condensate in two-color QCD, on a hot lattice. From Ref. [58].

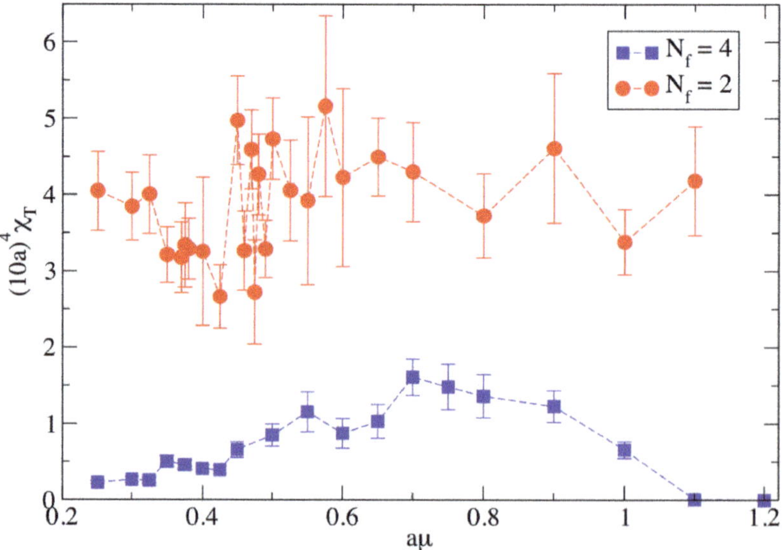

**Figure 5.** Topological susceptibility versus chemical potential in two-color QCD for two and four flavours, from Ref. [61], in a cold lattice; the expected $\mu_o$ in lattice units is $\mu_o a \simeq 0.34$.

A recent paper [64] analysed the temperature dependence of the results at high chemical potential using the same setup at low and high temperature, but for the number of time slices, which controls the temperature itself. At a temperature of about $0.45T_c$, where $T_c$ is the chiral transition temperature at zero chemical potential, the topological susceptibility is found to be almost constant for all the values of chemical potential, from the hadronic to superfluid phase. In contrast, for a temperature of about $0.89T_c$ the topological susceptibility becomes small as the hadronic phase changes into the quark-gluon plasma phase. The results, shown in Figure 6, indicate a significant temperature effect, which changes the behaviour of the topological susceptibility from constant with $\mu_B$ (in the cold phase) to decreasing with $\mu$ in the hot phase, the latter observation in agreement with Ref. [57].

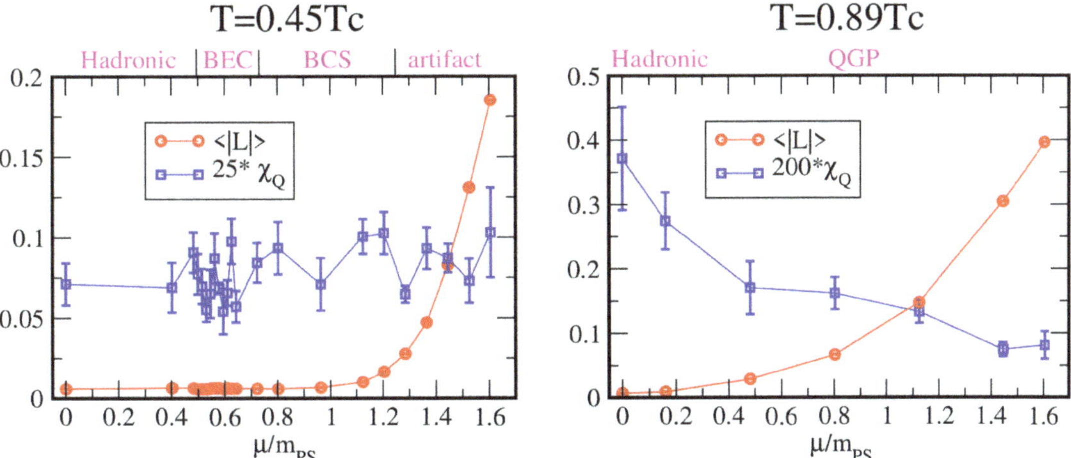


**Figure 6.** Polyakov loop and topological susceptibility in two-color QCD, in a cold (**left**) and in hot lattice, from Ref. [64]. We would like to highlight here the lack of sensitivity of the topological susceptibility on the threshold for the Polyakov loop at $\mu_o$ at $T = 0.45T_c$ (**left**), to be contrasted with the (anti) correlated behaviour of the same observables in the Quark-Gluon Plasma (**right**). Please note that the identification of the diquark dense phase with a BEC phase followed by a BCS one is still under debate [65].

The results of [34] are obtained on a $32^4$ lattice, which is described as a cold one. In simulations for $N = 2$ a clear correlation between chiral condensate and topological susceptibility emerged Figure 7. Accepting that these are simulations on cold lattices, there is an apparent contradiction with the scenario of [62,64], see Figure 6, left, as well Figure 5, left. However, the physical temperature is, according to the estimates of the paper $T = 140$ MeV [34], so it could well be that one is effectively observing a transition to a Quark-Gluon Plasma, albeit in a finite volume. Indeed the diquark condensate remains smaller than the condensate, indicating a different behaviour from the cold transition. This may offer a solution to this apparent puzzle. A clearer conclusion may be reached by performing simulations for chemical potentials in the dense phase, and varying temperatures: one may observe a transition from a phase with large topological susceptibility to a phase with suppressed susceptibility. Lacking those simulations, for the time being the results remain to some extent puzzling.

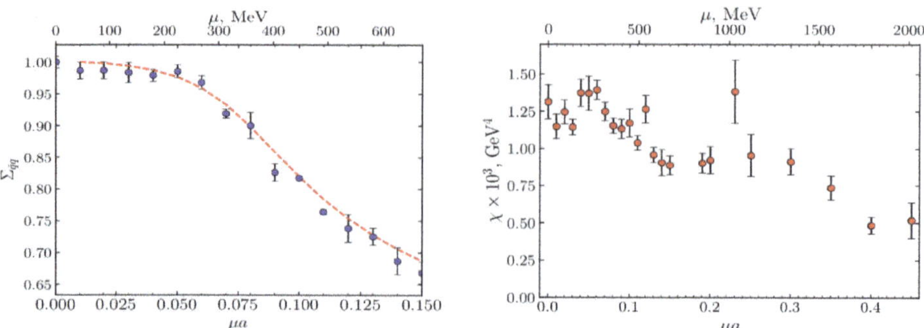

**Figure 7.** Chiral condensate and topological susceptibility as a function of baryochemical potential in two-color QCD, from Ref. [34].

*3.2. Isospin Density*

The phase diagram at finite density of isospin, introduced in Section 2 and shown in Figure 8, has been studied on the lattice by various authors [36,38,69,70]. An interesting feature is that the critical line $T = T(\mu_I)$ has a very small slope—it is almost horizontal. So simulations performed at fixed temperature varying $\mu_I$ are very likely crossing the pion condensation line unless the temperature is really close to $T_c$.

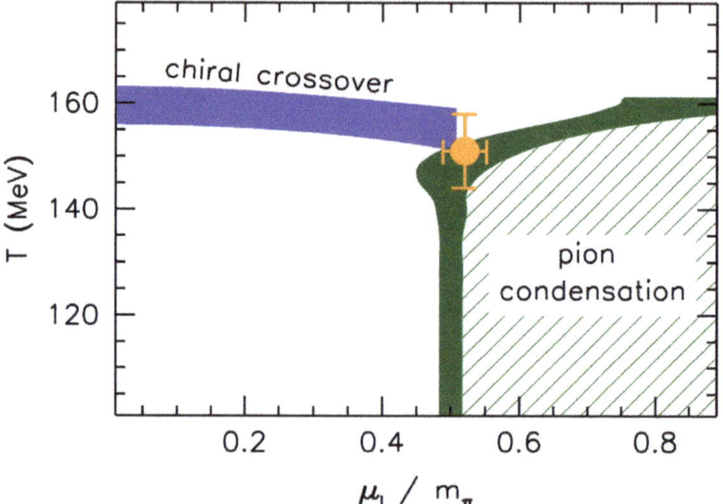

**Figure 8.** Lattice results for the phase diagram of QCD in the temperature-chemical potential for isospin plane, from Ref. [36].

Topology was studied in Ref. [71]: the authors perform simulations in full QCD with staggered fermions on $24^3 \times 6$ lattices, with a similar setup as the one used in Ref. [72] to study a $8^4$ lattice. In Ref. [72] the transition to the condensed phase was clearly observed, with $\mu_I^c \simeq m_\pi/2$ On the $24^3 \times 6$ lattice the pion mass was set at its physical value, corresponding to $m_\pi a = 0.2$, leading to an expected critical isospin chemical potential $\mu_I^c a = 0.1$. Topology was then studied by analysing the zero and non-zero modes of the overlap Dirac operator, which has an exact chiral symmetry, and it is thus particularly suited for this analysis. The eigenvalue distributions were obtained for $\mu_I = 0.5, 1.5\mu_{I_c}$

i.e., below and above the isospin phase transition, and they are remarkably similar, see Figure 9.

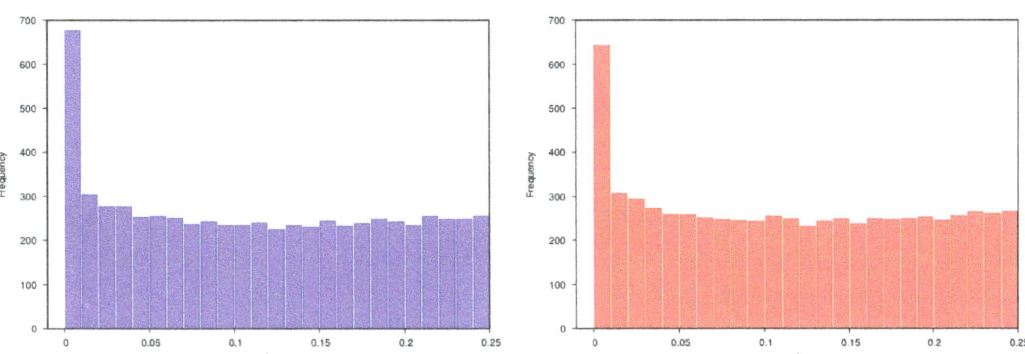

**Figure 9.** Eigenvalue spectrum of the Overlap Dirac operator for QCD with a physical pion mass, on $24^3 \times 6$ lattice for $\mu_I = 0.5, 1.5\mu_{I_c}$, left and right diagram. From Ref. [71].

As with what was observed at finite baryon density, and low temperatures, in two-color QCD, apparently topology in cold systems does not change in dense matter. This would be consistent with the predictions of Ref. [6].

### 3.3. Chiral Density

Early lattice studies of chiral density were performed having in mind a toy model for the chiral magnetic effect in heavy ion collisions [52]. One first systematic study of the phase diagram at equilibrium appeared in Ref. [73]. Even if QCD with a chiral density does not have a sign problem, these first studies were performed for two-color QCD, for the sake of simplicity, economy of computational resources , and possible comparison with results in two-color QCD with a magnetic field. The resulting phase diagram—which confirms the prediction of model studies—is reproduced in Figure 10. The larger extent of the hadronic phase—$T_c$ increases with $\mu_5$—reflects the so-called chiral catalisys [53]—the enhancement of the chiral condensate due to chiral imbalance, which pushes the critical temperatures towards higher values when increasing $\mu_5$. Details and a rich list of references may be found in a recent review [54].

Lattice studies of topology and confinement with a chiral imbalance have been performed in QCD in Ref. [51], using the tree level improved Symanzik gauge action and staggered fermions with two flavours of dynamical quarks. Four different pion masses were explored: $m_\pi = (563, 762, 910)$ MeV. It was found that the model follows the chiral perturbation theory prediction [74] $\rho_5 = \Lambda_{QCD}\mu_5$: the chiral density depends linearly on the chiral chemical potential, there are no thresholds. The study reveals that the topological susceptibility increases with the chiral chemical potential, much in the same way as the chiral condensate did in the two-color study. Moreover, also the string tension increases, in a rather correlated way. This was interpreted [51] as a signal that the chiral chemical potential leads to larger fluctuations of the chiral density and, due to the anomaly, to larger topological fluctuations in QCD: the chiral chemical potential enhances topological fluctuations which in turn are related to the strength of confinement as seen from the string tension, see Figure 11.

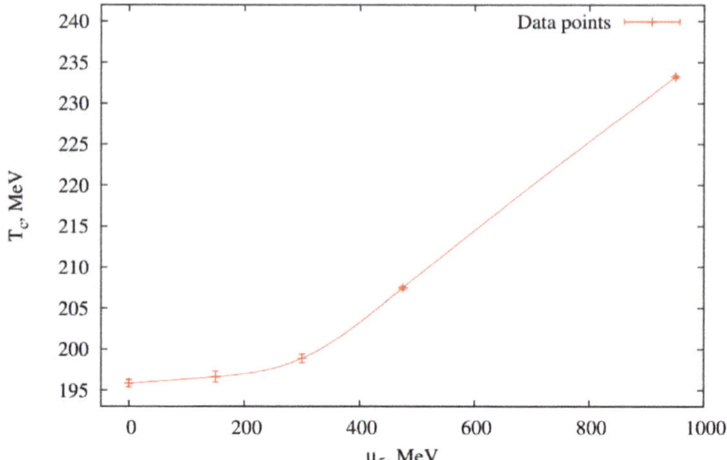

**Figure 10.** The phase diagram of QCD (two-color) in the temperature-chiral chemical potential plane; note the enlargement of the hadronic phase due to the enhancement of chiral breaking. From Ref. [73].

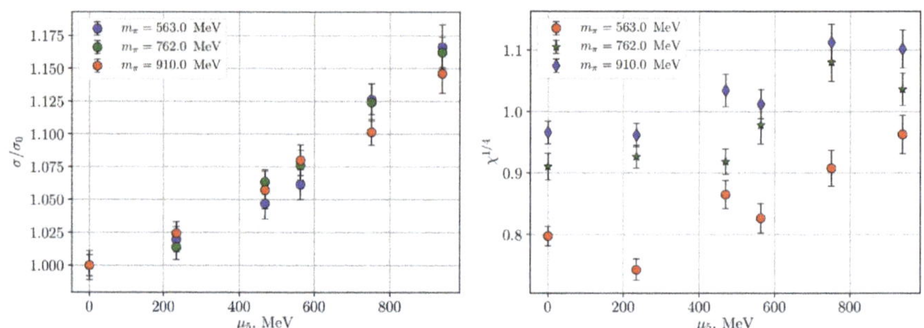

**Figure 11.** String tension (**left**) and topological susceptibility (**right**) in QCD as a function of the chiral chemical potential, and different pion masses. From Ref. [51].

## 4. Summarising

Topology and dense matter are studied independently, and intensively on the lattice. However, studies of the topological aspects in dense matter are still relatively scarce. We have reviewed the available information for a baryon rich matter, mostly coming from two-color QCD, for isospin dense matter, and for a chiral imbalance, hoping to highlight some clear trend, and to contribute to identify the open issues.

Actually, none of the systems considered here is completely realistic: finite baryon density may well be realised in experiments; however, on the lattice one must use (unphysical) theories free from the sign problem to access to cold, dense phase. Dense, baryon-less isospin matter would be unstable under weak interactions. Additionally, chiral imbalance does not exist at equilibrium even in strong interactions. Yet, these studies may add to our understanding of the phases of strong interactions, and , in some cases, pose some new challenge. Indeed, one of the main points to be addressed—the interrelation among topology, confinement, chiral symmetry—remains to large extent unsolved. These ambiguities are partly due to lattice artifacts—they are highlighted in all the works we have reviewed, partly to the different setup of the simulations.The ambiguities may be particularly severe in systems with a chiral chemical potential, which are to some extent artificial, and where a continuum limit is not well defined. Very few studies are available in which dense matter has been explored as a function of temperature within the same model.

Nonetheless, at least at high temperature, some coherent pattern emerges: a tentative conclusion is that above the critical temperature for superfluidity/superconductivity, the topological susceptibility as a function of chemical potential, either isospin and baryonic, is well correlated with the chiral condensate and the signals for confinement. The same remains true for the chiral chemical potential: in that case, there is a striking effect called chiral enhancement: the chiral condensate grows with chemical potential—and the same is true for topological susceptibility and string tension.

At low temperatures, however, the results for the topological susceptibility are not entirely settled: in some cases, a similar behaviour as high temperature has been reported; in other cases a sensitivity to the number of flavours has been observed; other studies conclude for the insensitivity of the topological susceptibility to the matter density. The latter observation would indeed be consistent with the analysis of Ref. [6]. The results in cold systems often call for further investigations and better control of the lattice artifacts, and this may be particularly true for studies with a chiral chemical potential. This said, a feature which seems to be well established is the insensitivity of topological susceptibility to dense matter. However, if instantons' chains were indeed realised in cold and dense matter, there should be some signatures in topological observables. Should one think that the behaviour of topological susceptibility invalidates this picture, or, rather, that topological susceptibility is simply not able to capture it?

**Funding:** This work is partially supported by STRONG-2020, a European Union's Horizon 2020 research and innovation programme under grant agreement No. 824093.

**Conflicts of Interest:** The author declares no conflict of interest.

## References


1. Dexheimer, V.; Constantinou, C.; Most, E.R.; Jens Papenfort, L.; Hanauske, M.; Schramm, S.; Stoecker, H.; Rezzolla, L. Neutron-Star-Merger Equation of State. *Universe* **2019**, *5*, 129. [CrossRef]
2. Ratti, C. Lattice QCD and heavy ion collisions: A review of recent progress. *Rept. Prog. Phys.* **2018**, *81*, 084301. [CrossRef] [PubMed]
3. Guenther, J.N. Overview of the QCD phase diagram: Recent progress from the lattice. *Eur. Phys. J. A* **2021**, *57*, 136. [CrossRef]
4. Rajagopal, K.; Wilczek, F. The Condensed matter physics of QCD. *Front. Part. Phys. Handb. QCD* **2000**, *11*, 2061–2151 .
5. Mannarelli, M. Meson condensation. *Particles* **2019**, *2*, 411–443. [CrossRef]
6. Pisarski, R.D.; Rennecke, F. Multi-instanton contributions to anomalous quark interactions. *Phys. Rev. D* **2020**, *101*, 114019. [CrossRef]
7. McLerran, L.; Redlich, K.; Sasaki, C. Quarkyonic Matter and Chiral Symmetry Breaking. *Nucl. Phys. A* **2009**, *824*, 86–100. [CrossRef]
8. Buballa, M.; Carignano, S. Inhomogeneous chiral condensates. *Prog. Part. Nucl. Phys.* **2015**, *81*, 39–96. [CrossRef]
9. Buballa, M.; Kurth, L.; Wagner, M.; Winstel, M. Regulator dependence of inhomogeneous phases in the (2+1)-dimensional Gross-Neveu model. *Phys. Rev. D* **2021**, *103*, 034503. [CrossRef]
10. Jackiw, R.W. Axial anomaly. *Scholarpedia* **2008**, *3*, 7302. [CrossRef]
11. Pospelov, M.; Ritz, A. Theta vacua, QCD sum rules, and the neutron electric dipole moment. *Nucl. Phys. B* **2000**, *573*, 177–200. [CrossRef]
12. Pich, A.; de Rafael, E. Strong CP violation in an effective chiral Lagrangian approach. *Nucl. Phys. B* **1991**, *367*, 313–333. [CrossRef]
13. Abel, C.; Afach, S.; Ayres, N.J.; Baker, C.A.; Ban, G.; Bison, G.; Bodek, K.; Bondar, V.; Burghoff, M.; Chanel, E.; et al. Measurement of the permanent electric dipole moment of the neutron. *Phys. Rev. Lett.* **2020**, *124*, 081803. [CrossRef] [PubMed]
14. Peccei, R.D.; Quinn, H.R. CP Conservation in the Presence of Instantons. *Phys. Rev. Lett.* **1977**, *38*, 1440–1443. [CrossRef]
15. Peccei, R.D.; Quinn, H.R. Constraints Imposed by CP Conservation in the Presence of Instantons. *Phys. Rev. D* **1977**, *16*, 1791–1797. [CrossRef]
16. Pospelov, M.; Ritz, A. Electric dipole moments as probes of new physics. *Ann. Phys.* **2005**, *318*, 119–169. [CrossRef]
17. Alexandrou, C.; Athenodorou, A.; Hadjiyiannakou, K.; Todaro, A. Neutron electric dipole moment using lattice QCD simulations at the physical point. *Phys. Rev. D* **2021**, *103*, 054501. [CrossRef]
18. Lombardo, M.P.; Trunin, A. Topology and axions in QCD. *Int. J. Mod. Phys. A* **2020**, *35*, 2030010. [CrossRef]
19. Borsanyi, S.; Dierigl, M.; Fodor, Z.; Katz, S.D.; Mages, S.W.; Nogradi, D.; Redondo, J.; Ringwald, A.; Szabo, K.K. Axion cosmology, lattice qcd and the dilute instanton gas. *Phys. Lett.* **2016**, *752*, 175–181. [CrossRef]
20. Bonati, C.; D'Elia, M.; Martinelli, G.; Negro, F.; Sanfilippo, F.; Todaro, A. Topology in full QCD at high temperature: A multicanonical approach. *J. High Energy Phys.* **2018**, *11*, 170. [CrossRef]

21. Petreczky, P.; Schadler, H.; Sharma, S. The topological susceptibility in finite temperature QCD and axion cosmology. *Phys. Lett. B* **2016**, *762*, 498–505. [CrossRef]

22. Burger, F.; Ilgenfritz, E.-M.; Lombardo, M.P.; Trunin, A. Chiral observables and topology in hot QCD with two families of quarks. *Phys. Rev. D* **2018**, *98*, 094501. [CrossRef]

23. Borsányi, S.; Fodor, Z.; Guenther, J.; Kampert, K.H.; Katz, S.D.; Kawanai, T.; Kovacs, T.G.; Mages, S.W.; Pasztor, A.; Pittler, F.; et al. Calculation of the axion mass based on high-temperature lattice quantum chromodynamics. *Nature* **2016**, *539*, 69–71. [CrossRef]

24. Taniguchi, Y.; Kanaya, K.; Suzuki, H.; Umeda, T. Topological susceptibility in finite temperature ( 2+1 )-flavor QCD using gradient flow. *Phys. Rev. D* **2017**, *95*, 054502. [CrossRef]

25. Weinberg, S. The U(1) Problem. *Phys. Rev. D* **1975**, *11*, 3583–3593. [CrossRef]

26. Vecchia, P.D.; Veneziano, G. Chiral Dynamics in the Large n Limit. *Nucl. Phys. B* **1980**, *171*, 253–272. [CrossRef]

27. Vecchia, P.D.; Giannotti, M.; Lattanzi, M.; Lindner, A. Round Table on Axions and Axion-like Particles. *PoS Confin.* **2019**, *2018*, 034.

28. Veneziano, G. U(1) Without Instantons. *Nucl. Phys. B* **1979**, *159*, 213–224. [CrossRef]

29. Kapusta, J.I.; Kharzeev, D.; McLerran, L.D. The Return of the prodigal Goldstone boson. *Phys. Rev. D* **1996**, *53*, 5028–5033. [CrossRef] [PubMed]

30. Kotov, A.Y.; Lombardo, M.P.; Trunin, A.M. Fate of the $\eta'$ in the quark gluon plasma. *Phys. Lett. B* **2019**, *794*, 83–88. [CrossRef]

31. Alford, M.G.; Kapustin, A.; Wilczek, F. Imaginary chemical potential and finite fermion density on the lattice. *Phys. Rev. D* **1999**, *59*, 054502. [CrossRef]

32. Gorghetto, M.; Villadoro, G. Topological Susceptibility and QCD Axion Mass: QED and NNLO corrections. *J. High Energy Phys.* **2019**, *2019*, 033. [CrossRef]

33. Son, D.T.; Stephanov, M.A. QCD at finite isospin density. *Phys. Rev. Lett.* **2001**, *86*, 592–595. [CrossRef]

34. Astrakhantsev, N.; Braguta, V.V.; Ilgenfritz, E.M.; Kotov, A.Y.; Nikolaev, A.A. Lattice study of thermodynamic properties of dense QC$_2$D. *Phys. Rev. D* **2020**, *102*, 074507. [CrossRef]

35. Brandt, B.B.; Cuteri, F.; Endrődi, G.; Schmalzbauer, S. The Dirac spectrum and the BEC-BCS crossover in QCD at nonzero isospin asymmetry. *Particles* **2020**, *3*, 80–86. [CrossRef]

36. Brandt, B.B.; Endrodi, G.; Schmalzbauer, S. QCD phase diagram for nonzero isospin-asymmetry. *Phys. Rev. D* **2018**, *97*, 054514. [CrossRef]

37. Brandt, B.B.; Cuteri, F.; Endrodi, G.; Schmalzbauer, S. Exploring the QCD phase diagram via reweighting from isospin chemical potential. *PoS LATTICE* **2019**, *2019*, 189.

38. Braguta, V.V.; Kotov, A.Y.; Nikolaev, A.A. Lattice Simulation Study of the Properties of Cold Quark Matter with a Nonzero Isospin Density. *JETP Lett.* **2019**, *110*, 1–4. [CrossRef]

39. Detmold, W.; Orginos, K.; Shi, Z. Lattice QCD at non-zero isospin chemical potential. *Phys. Rev. D* **2012**, *86*, 054507. [CrossRef]

40. Cea, P.; Cosmai, L.; D'Elia, M.; Papa, A.; Francesco Sanfilippo The critical line of two-flavor QCD at finite isospin or baryon densities from imaginary chemical potentials. *Phys. Rev. D* **2012**, *85*, 094512. [CrossRef]

41. Toublan, D.; Kogut, J.B. Isospin chemical potential and the QCD phase diagram at nonzero temperature and baryon chemical potential. *Phys. Lett. B* **2003**, *564*, 212–216. [CrossRef]

42. Schäfer, T.; Shuryak, E.V. Instantons in qcd. *Rev. Mod. Phys.* **1998**, *70*, 323–425. [CrossRef]

43. Rapp, R.; Schäfer, T.; Shuryak, E.V.; Velkovsky, M. Diquark Bose condensates in high density matter and instantons. *Phys. Rev. Lett.* **1998**, *81*, 53–56. [CrossRef]

44. Rapp, R.; Schäfer, T.; Shuryak, E.V.; Velkovsky, M. High density QCD and instantons. *Ann. Phys.* **2000**, *280*, 35–99. [CrossRef]

45. Atiyah, M.F.; Singer, I.M. The Index of elliptic operators. 5. *Ann. Math.* **1971**, *93*, 139–149. [CrossRef]

46. Atiyah, M.F.; Singer, I.M. Dirac Operators Coupled to Vector Potentials. *Proc. Natl. Acad. Sci. USA* **1984**, *81*, 2597–2600. [CrossRef]

47. Bzdak, A.; Esumi, S.; Koch, V.; Liao, J.; Stephanov, M.; Xu, N. Mapping the Phases of Quantum Chromodynamics with Beam Energy Scan. *Phys. Rept.* **2020**, *853*, 1–87. [CrossRef]

48. Kharzeev, D.E.; Levin, E.M. Color Confinement and Screening in the $\theta$ Vacuum of QCD. *Phys. Rev. Lett.* **2015**, *114*, 242001. [CrossRef] [PubMed]

49. Kharzeev, D.E. The chiral magnetic effect and anomaly-induced transport. *Prog. Part. Nucl. Phys.* **2014**, *75*, 133–151. [CrossRef]

50. Ruggieri, M.; Chernodub, M.N.; Lu, Z. Topological susceptibility, divergent chiral density, and phase diagram of chirally imbalanced QCD medium at finite temperature. *Phys. Rev.* **2020**, *102*, 014031. [CrossRef]

51. Astrakhantsev, N.Y.; Braguta, V.V.; Kotov, A.Y.; Kuznedelev, D.D.; Nikolaev, A.A. Lattice study of QCD at finite chiral density: Topology and confinement. *Eur. Phys. J. A* **2021**, *57*, 15. [CrossRef]

52. Yamamoto, A. Chiral magnetic effect in lattice qcd with a chiral chemical potential. *Phys. Rev. Lett.* **2011**, *107*, 031601. [CrossRef] [PubMed]

53. Braguta, V.V.; Kotov, A.Y. Catalysis of dynamical chiral symmetry breaking by chiral chemical potential. *Phys. Rev.* **2016**, *93*, 105025. [CrossRef]

54. Yang, L.; Luo, X.; Segovia, J.; Zong, H. A Brief Review of Chiral Chemical Potential and Its Physical Effects. *Symmetry* **2020**, *12*, 2095. [CrossRef]

55. Müller-Preussker, M. Recent results on topology on the lattice (in memory of Pierre van Baal). *PoS LATTICE* **2015**, *2014*, 003.

56. Hands, S.; Kogut, J.B.; Lombardo, M.; Morrison, S.E. Symmetries and spectrum of SU(2) lattice gauge theory at finite chemical potential. *Nucl. Phys. B* **1999**, *558*, 327–346. [CrossRef]

57. Alles, B.; D'Elia, M.; Giacomo, A.D. Topological susceptibility at zero and finite T in SU(3) Yang-Mills theory. *Nucl. Phys. B* **1997**, *494*, 281–292; Erratum in **2004**, *679*, 397–399. [CrossRef]
58. Alles, B.; D'Elia, M.; Lombardo, M.P. Behaviour of the topological susceptibility in two colour QCD across the finite density transition. *Nucl. Phys. B* **2006**, *752*, 124–139. [CrossRef]
59. Lombardo, M.; Paciello, M.L.; Petrarca, S.; Taglienti, B. Glueballs and the superfluid phase of Two-Color QCD. *Eur. Phys. J. C* **2008**, *58*, 69–81. [CrossRef]
60. Hands, S.; Kim, S.; Skullerud, J. A Quarkyonic Phase in Dense Two Color Matter? *Phys. Rev. D* **2010**, *81*, 091502. [CrossRef]
61. Hands, S.; Kenny, P. Topological Fluctuations in Dense Matter with Two Colors. *Phys. Lett. B* **2011**, *701*, 373–377. [CrossRef]
62. Hands, S.; Kenny, P.; Kim, S.; Skullerud, J. Lattice Study of Dense Matter with Two Colors and Four Flavors. *Eur. Phys. J. A* **2011**, *47*, 60. [CrossRef]
63. Astrakhantsev, N.Y.; Bornyakov, V.G.; Braguta, V.V.; Ilgenfritz, E.M.; Kotov, A.Y.; Nikolaev, A.A.; Rothkopf, A. Lattice study of static quark-antiquark interactions in dense quark matter. *J. High Energy Phys.* **2019**, *2019*, 171. [CrossRef]
64. Iida, K.; Itou, E.; Lee, T. Two-colour QCD phases and the topology at low temperature and high density. *J. High Energy Phys.* **2020**, *2020*, 181. [CrossRef]
65. Boz, T.; Giudice, P.; Hands, S.; Skullerud, J. Dense two-color QCD towards continuum and chiral limits. *Phys. Rev. D* **2020**, *101*, 074506. [CrossRef]
66. Appelquist, T.; Ratnaweera, A.; Terning, J.; Wijewardhana, L.C.R. The Phase structure of an SU(N) gauge theory with N(f) flavors. *Phys. Rev. D* **1998**, *58*, 105017. [CrossRef]
67. Appelquist, T.; Sannino, F. The Physical spectrum of conformal SU(N) gauge theories. *Phys. Rev. D* **1999**, *59*, 067702. [CrossRef]
68. Orlando, D.; Reffert, S.; Sannino, F. Charging the Conformal Window. *Phys. Rev. D* **2021**, *103*, 105026. [CrossRef]
69. Brandt, B.B.; Endrodi, G. QCD phase diagram with isospin chemical potential. *PoS LATTICE* **2016**, *2016*, 039.
70. Bornyakov, V.G.; Nikolaev, A.A.; Rogalyov, R.N.; Terentev, A.S. Gluon Propagators in 2 + 1 Lattice QCD with Nonzero Isospin Chemical Potential. *arXiv* **2021**, arXiv:2102.07821.
71. Bali, G.S.; Endrodi, G.; Gavai, R.V.; Mathur, N. Probing the nature of phases across the phase transition at finite isospin chemical potential. *arXiv* **2016**, arXiv:1610.00233.
72. Endrődi, G. Magnetic structure of isospin-asymmetric qcd matter in neutron stars. *Phys. Rev.* **2014**, *90*, 094501. [CrossRef]
73. Braguta, V.V.; Goy, V.A.; Ilgenfritz, E.M.; Kotov, A.Y.; Molochkov, A.V.; Muller-Preussker, M.; Petersson, B. Two-Color QCD with Non-zero Chiral Chemical Potential. *J. High Energy Phys.* **2015**, *2015*, 094. [CrossRef]
74. Espriu, D.; Nicola, A.G.; Vioque-Rodríguez, A. Chiral perturbation theory for nonzero chiral imbalance. *J. High Energy Phys.* **2020**, *2020*, 062. [CrossRef]

MDPI

*Review*

# Effective String Description of the Confining Flux Tube at Finite Temperature

**Michele Caselle**

Department of Physics, University of Turin & INFN, Via Pietro Giuria 1, I-10125 Turin, Italy; michele.caselle@unito.it

**Abstract:** In this review, after a general introduction to the Effective String Theory (EST) description of confinement in pure gauge theories, we discuss the behaviour of EST as the temperature is increased. We show that, as the deconfinement point is approached from below, several universal features of confining gauge theories, like the ratio $T_c/\sqrt{\sigma_0}$, the linear increase of the squared width of the flux tube with the interquark distance, or the temperature dependence of the interquark potential, can be accurately predicted by the effective string. Moreover, in the vicinity of the deconfinement point the EST behaviour turns out to be in good agreement with what was predicted by conformal invariance or by dimensional reduction, thus further supporting the validity of an EST approach to confinement.

**Keywords:** Lattice Gauge Theories; Effective String Theories

**Citation:** Caselle, M. Effective String Description of the Confining Flux Tube at Finite Temperature. *Universe* **2021**, *7*, 170. https://doi.org/10.3390/universe7060170

Academic Editor: Dmitri Antonov

Received: 29 April 2021
Accepted: 26 May 2021
Published: 30 May 2021

**Publisher's Note:** MDPI stays neutral with regard to jurisdictional claims in published maps and institutional affiliations.



## 1. Introduction

One of the most powerful tools we have for studying the non-perturbative behaviour of confining Yang–Mills theories is the so called "Effective String Theory" (EST) in which the confining flux tube joining together a quark-antiquark pair is modeled as a thin vibrating string [1–5]. As explicitly stated in its definition, this model is only an effective large distance description of the flux tube and not an exact non-perturbative solution of the Yang–Mills theory; however, due to the peculiar features of the string action, it turns out to be a highly predictive effective model, whose results can be successfully compared with the most precise existing Monte Carlo simulations in Lattice Gauge Theories (LGTs). Besides its predictive power EST is also interesting from a theoretical point of view, since it is a perfect laboratory to test more refined nonperturbative descriptions of Yang–Mills theories, guess new hypotheses, and drive our understanding of the role of string theory in this game.

EST also plays an important role from a phenomenological point of view since it can be used to model the glueball spectrum of QCD [6] or to model the large distance (non-perturbative) part of the interquark potential in heavy quarkonia [3,4] or the onset of the deconfinement transition [7].

The simplest, Lorentz invariant, EST is the Nambu–Goto model [1,2] which will be the main subject of this review. The Nambu–Goto action can be considered in this framework as a first order approximation of the actual EST describing the non-perturbative behavior of the Yang–Mills theory. The most interesting result of the last ten years of EST studies is that this first order approximation works remarkably well and agrees within the errors, in the large distance limit, with almost all the existing Monte Carlo simulations for all the confining models that have been studied (with only a few exceptions [8,9]). We shall see below that this is not by chance and that it is instead a direct consequence of the peculiar nature of the EST and of the strong constraining power of Lorentz invariance in this context.

This also explains the impressive universality of the infrared regime of confining gauge theories (with only a mild dependence on the number of space–time dimensions, exactly as predicted by the Nambu–Goto model), which show essentially the same behaviour for the interquark potential, the deconfinement temperature and the glueball spectrum. This universality of LGT results, which holds for models as different as the three dimensional gauge Ising model and the four dimensional SU(3) Yang–Mills theory was in the past one of the major puzzles in the Lattice community and is now understood only as a side effect of the impressive effectiveness of the Nambu–Goto approximation. It is thus only an apparent universality and all the details on the gauge group are expected to be encoded in the higher order EST corrections beyond Nambu–Goto. It is thus clear why a lot of efforts have been devoted in these last years to the identification and modellization of these non-universal higher order corrections. The hope is that, since they depend on the particular type of confining gauge theory, they could shed some light on specific non-perturbative properties of the theory, for instance on the non-pertubative degrees of freedom (say, instatons or monopoles) driving confinement in the model.

Due to the asymptotic large-distance nature of the EST expansion, the optimal regime in which one can observe higher order terms is for short interquark separations, much smaller than the length of the Polyakov loops. In the string language this is known as the "open string channel".

While the above choice is the one which is more often used, it is easy to see, looking at the explicit expression of the EST partition function that the opposite choice (the "closed string channel" in string language) in which the short direction is the one with periodic boundary conditions, is better suited to observe higher order corrections. In the language of Lattice Gauge Theories this is the high temperature regime of the theory in which the temperature is just below the deconfinement transition. This is exactly the limit in which we are interested in this review.

While there are already several good general reviews on EST (see for instance [10–12]), the goal of this paper is to focus specifically on the EST behaviour in the high-$T$ regime and to discuss how our understanding of Lattice Gauge Theory in this limit can help us to constrain and check EST. In particular an important reason of interest of this limit is that, in LGTs with a second order deconfinement transition, several non trivial results on EST can be obtained using renormalization group arguments of the type discussed in [13]. This approach, which allows to map a $(d + 1)$ dimensional LGT into a suitably chosen $d$ dimensional spin model, will be one of the main focus of the present review.

In this review we shall mainly focus on two observables: The interquark potential and the flux tube width which in the past years played a major role in the progress of our understanding of EST properties. In particular, the review is organized as follows. Section 2 will be devoted to a brief introduction to Lattice Gauge Theories. In Section 3 we shall discuss the main properties of EST (and in particular of the Nambu–Goto action) with a particular focus on the high-$T$ behaviour. Then in Section 4 we shall compare EST predictions for the interquark potential in the high temperature regime with Monte Carlo simulations. In Section 5 we shall address the important issue of the width of the flux tube and discuss its behaviour in the high-$T$ regime. Section 6 will be devoted to a few concluding remarks.

## 2. A Brief Summary of LGTs

The natural context in which we can see the EST at work is in the confining phase of Lattice Gauge Theories (LGT) where EST is expected to describe the large distance behaviour of the confining flux tube joining a quark antiquark pair. Thus in order to fix notations and to better understand the physics behind EST it is useful to briefly discuss a few basic notions of LGTs. We refer the interested reader to the book [14] for a more detailed introduction to LGTs.

The partition function of a gauge theory in $D$ spacetime dimensions with gauge group $G$ regularized on a lattice is

$$Z = \int \prod dU_\mu(\vec{x}, t) \exp\{-\beta \sum_p \operatorname{Re} \operatorname{Tr}(1 - U_p)\}, \tag{1}$$

where $U_\mu(\vec{x}, t) \in G$ is the link variable at the site $(\vec{x}, t) = (x_1, .., x_{D-1}, t)$ in the direction $\mu$ and $U_p$ is the product of the links around the plaquette $p$.

We shall denote in the following with $N_t$ ($N_s$) the lattice size in the time (space) direction and assume for simplicity $N_s$ to be the same for all the space-like directions. We shall use $d$ to denote the number of space-like directions in the lattice; thus $D = d + 1$. To simplify notations we shall fix the lattice spacing $a$ to 1 and neglect it in the following.

As it is well known the link variable $U_\mu(\vec{x}, t)$ is not gauge invariant and only the traces of ordered products of link variables along closed paths are gauge invariant.

The simplest choice is the Wilson loops

$$W(\gamma) \;=\; \mathrm{Tr} \prod_{(\vec{x},t)\in\gamma} U_\mu(\vec{x}, t) \tag{2}$$

where the product is assumed to be ordered along the path $\gamma$. If we choose the path $\gamma$ to be a rectangle of size $R \times L$ (with $L$ along the Euclidean "time" direction and $R$ along one of the space directions) then it is possible to relate the expectation value of $W(R \times L)$ to the interquark potential as:

$$V(R) \;=\; -\lim_{L\to\infty} \frac{1}{L} \log\langle W(R, L)\rangle. \tag{3}$$

The idea behind this definition is that we may think of $\langle W(R, L)\rangle$ as the free energy due to the creation at the time $t_0$ of a quark and an antiquark pair which are instantaneously moved at a distance $R$ from each other, keep their position for a time $L$ and finally annihilate at the instant $t_0 + L$.

A confining LGT will be characterized by a linearly rising potential and thus, according to Equation (3) we expect for the Wilson loop an "area law" of this type

$$\langle W(R, T)\rangle \;\sim\; e^{-\sigma_0 RT + p(R+T) + k}. \tag{4}$$

The area term is responsible for confinement while the perimeter and constant terms are non universal contributions related to the discretization procedure. The physically important quantity is the coefficient of the area term which represents the lattice estimate of the string tension.

*2.1. Finite Temperature LGTs*

It is important at this point to stress that Equation (3) above, only defines the so called zero temperature interquark potential and accordingly $\sigma_0$ is the zero temperature string tension. If one is interested in the finite temperature behaviour of the interquark potential and in the possible presence of a deconfinement transition at some finite temperature $T_c$ the lattice regularization prescription must be modified.

The lattice regularization of a generic Quantum Field Theory (QFT) at a non-zero, finite temperature $T$ can be obtained by imposing periodic boundary conditions in the time direction for the bosonic field (and antiperiodic for fermionic ones). With this choice a lattice of size $(N_s a)^d (N_t a)$ represents the regularized version of a system of finite volume $V = (N_s a)^d$ at a finite temperature $T = 1/N_t a$. Even if in the rest of the paper, having set $a = 1$, we shall systematically use $N_t$ and $N_s$ as a shorthand notation for $N_t a$ and $N_s a$, in the above formulas we restored the lattice spacing to emphasize the correct dimensions of the physical quantities we are defining. The compactified "time" direction at this point does not have any longer the meaning of time (recall that we are describing a system at equilibrium in the canonical ensemble) but its size $N_t a$ is instead a measure of the inverse temperature of the system $N_t a = 1/T$.

As a consequence, in a finite temperature setting, the Wilson loop cannot be related any more to the interquark potential as we did above. Fortunately we have a different way to construct a quantity with the same physical interpretation. In a finite temperature setting one can define a new class of gauge invariant observables which are usually called Polyakov loops.

A Polyakov loop $P(\vec{x})$ is the trace of the ordered product of all time-like links with the same space-like coordinates; this loop is closed owing to the periodic boundary conditions in the time direction:

$$P(\vec{x}) = \mathrm{Tr} \prod_{z=1}^{N_t} U_t(\vec{x}, z). \tag{5}$$

In a pure LGT the Polyakov loop has a deep physical meaning, since its expectation value is related to the free energy of a single isolated quark. Hence the fact that the Polyakov loop acquires a non-zero expectation value can be considered as a signature of deconfinement and the Polyakov loop is thus the order parameter of such a deconfinement transition.

The value $\beta_c(N_t)$ of this deconfinement transition in a lattice of size $N_t = 1/T$ in the compactified time direction can be used to define a new physical observable $T_c$ which is obtained by inverting $\beta_c(N_t)$. We obtain in this way, for each value of $\beta$, the lattice size in the time direction (which we shall call in the following $N_{t,c}(\beta)$) at which the model undergoes the deconfinement transition and from this the critical temperature $T_c(\beta) \equiv 1/N_{t,c}(\beta)$ as a function of $\beta$.

### 2.2. The Finite Temperature Interquark Potential

In a finite temperature setting the interquark potential can be extracted by looking at the correlations of Polyakov loops in the confined phase. The correlation of two loops $P(x)$ at a distance $R$ and at a temperature $T = 1/N_t$ (which we denote with the subscript $N_t$ in the expectation value) is given by

$$\langle P(x)P^\dagger(x+R) \rangle_{N_t} \equiv e^{-\frac{1}{T}V(R,T)} = e^{-N_t V(R,T)}, \tag{6}$$

where we consider the free energy $V(R, N_t)$ as a proxy for the interquark potential at a finite temperature $T$

$$V(R,T) = -\frac{1}{N_t} \log \langle P(x)P^\dagger(x+R) \rangle_{N_t}. \tag{7}$$

If we assume also for this correlator an area law similar to the one discussed above for the Wilson loop:

$$\langle P(x)P^\dagger(x+R) \rangle_{N_t} \sim e^{-\sigma(T)N_t R}, \tag{8}$$

then we find again a confining behaviour for the interquark potential. In the above equation $\sigma(T)$ denotes this time the finite temperature string tension. As we shall see below $\sigma(T)$ is a decreasing function of $T$ and vanishes exactly at the deconfinement point [15,16]. Following the definitions of the previous section, we may identify $V(R)$ with the $T \to 0$ limit of $V(R,T)$ and $\sigma_0$ with the $T \to 0$ limit of $\sigma(T)$.

It is interesting to notice that the observable Equation (6) is similar to the expectation value of an ordinary Wilson loop except for the boundary conditions, which are in this case fixed in the space directions and periodic in the time direction. The resulting geometry is that of a cylinder, which is topologically different from the rectangular geometry of the Wilson loop.

### 2.3. Center Symmetry and the Polyakov Loop

The major consequence of the periodic boundary conditions in the time direction is the appearance of a new global symmetry of the action, with symmetry group the center $C$ of the gauge group (i.e., $\mathbf{Z}_N$ if the gauge group is $SU(N)$).

This symmetry can be realized, for instance, by acting on all the timelike links of a given space-like slice with the same element $W_0$ belonging to the center of the gauge group.

$$U_t(\vec{x}, t) \to W_0\, U_t(\vec{x}, t) \qquad \forall\, \vec{x}, \quad t \text{ fixed}. \tag{9}$$

it is easy to see that the Wilson action is invariant under such transformation. while the Polyakov loop transforms as:

$$P(\vec{x}) \rightarrow W_0\, P(\vec{x}); \tag{10}$$

thus it is a natural order parameter for this symmetry. It will acquire a non zero expectation value if the center symmetry is spontaneously broken.

Thus we see that the Polyakov loop is at the same time the order parameter of the deconfinement transition and of the center symmetry: The deconfinement transition in a pure lattice gauge theory coincides with the center symmetry breaking phase transition. In the deconfined phase the center symmetry is spontaneously broken while in the confining phase it is conserved.

### 2.4. The Svetitsky–Yaffe Conjecture

The peculiar role played by the Polyakov loops in the above discussion, suggests to use some kind of effective action for the Polyakov loops, integrating out the spacelike links of the model, to study the deconfinement transition and, more generally, the physics of finite temperature LGT. Such a construction corresponds in all respects to a "dimensional reduction": Starting from a $(d+1)$ dimensional LGT we end up with an effective action for the Polyakov loops which will be a $d$ dimensional spin model with global symmetry the center of the original gauge group.

While the explicit construction of such an effective action may be cumbersome and can be performed only as a strong coupling expansion, some general insight on the behaviour of the model can be deduced by simple renormalization group arguments [13] .

Indeed, even if as a result of the integration over the original gauge degrees of freedom we may expect long range interactions between the Polyakov loops, it can be shown [13] that these interactions decrease exponentially with the distance. Thus, if the phase transition is continuous, in the vicinity of the critical point the fine details of the interactions can be neglected, and the model will belong to the same universality class of the simplest spin model, with only nearest neighbour interactions, sharing the same symmetry breaking pattern. For instance, the deconfinement transition of the $SU(2)$ LGT in $(2+1)$ and $(3+1)$ dimensions, which in both cases is continuous, belong to the same universality class as, respectively, the two dimensional and the three dimensional Ising models.

This mapping has several important consequences:

(a)   The ordered (low temperature) phase of the spin model corresponds to the deconfined (high temperature) phase of the original gauge theory. This is the phase in which both the Polyakov loop, in the original LGT, and the spin, in the effective spin model, acquire a non-zero expectation value.

(b)   As for the operator content of the two models, the Polyakov loop is mapped into the spin operator, while the plaquette is mapped into the energy operator of the effective spin model. Accordingly, the Polyakov loop correlator in the confining phase, from which we extract the interquark potential, is mapped into the spin–spin correlator of the disordered, high temperature phase of the spin model

(c)   Thermal perturbations from the critical point in the original gauge theory, which are driven by the plaquette operator, are mapped into thermal perturbation of the effective spin model which are driven by the energy operator. Notice, however, the change in sign: An increase in temperature of the original gauge theory corresponds to a decrease of the temperature of the effective spin model.

A major consequence of this correspondence is that, in the vicinity of the deconfinement point, the behaviour of the interquark potential is strongly constrained and thus it represents, as we shall see, a powerful tool to test the predictions of the effective string model.

*2.5. EST Versus LGT: The Roughening Transition*

As we mentioned above, a confining interquark potential implies an area law for the Wilson loop (at zero temperature) or for the correlator of two Polyakov loops (at finite temperature). A nice feature of the lattice regularization is that such an area law naturally arises from a strong coupling expansion of these observables. Order by order in the strong coupling parameter $\beta$, the expectation value of a Wilson loop (or of a Polyakov loops correlator) is described by the sum over all the possible surfaces bordered by the Wilson loop with a weight proportional to their area. As it is well known this expansion diverges at the so called "roughening point" [3,4,17,18], well before the values of $\beta$ for which a continuum limit of the lattice regularization can be approached. This roughening transition is due to the vanishing of the stiffness of the strong coupling surfaces and has a very insightful explanation from an EST point of view. The vanishing of the surface stiffness ensures that the surfaces bordered by the Wilson loop can freely fluctuate as actual continuum-like surfaces and that they are not any more anchored to the crystallographic planes of the lattice and can thus be described by a set of $(D-2)$ real degrees of freedom representing their transverse displacement from the Wilson loop plane [17,18]. Upon quantization these transverse coordinates will become the $(D-2)$ bosonic degrees of freedom of the EST description which we shall discuss in the next section [3,4]. These massless quantum fluctuations delocalize the flux tube which acquires a nonzero width, which diverges logarithmically as the interquark distance increases [19]. We shall discuss in detail this issue in Section 5.

We may summarize all these observations by saying that the LGT regularization strongly supports an Effective String Theory description of confinement. We shall devote the next section to a precise formulation of this EST.

## 3. Effective String Description of the Interquark Potential

Even if a rigorous proof of quark confinement in Yang–Mills theories is still missing, there is little doubt that confinement is associated to the formation of a thin string-like flux tube [1–5], which generates, for large quark separations, a linearly rising confining potential.

This picture is strongly supported by the lattice regularization of Yang–Mills theories where, as we have seen in the previous section, the vacuum expectation value of Polyakov loops correlators is given by a sum over certain lattice surfaces which can be considered as the world-sheet of the underlying confining string.

This picture led Lüscher and collaborators [3,4], more than forty years ago, to propose that the dynamics of the flux tube for large interquark distances could be described by a free massless bosonic field theory in two dimensions.

$$S[X] = S_{cl} + S_0[X] + \dots, \tag{11}$$

where the classical action $S_{cl}$ describes the usual perimeter-area term, $X$ denotes the two-dimensional bosonic fields $X_i(\xi_1, \xi_2)$, with $i = 1, 2, \dots, D-2$, describing the transverse displacements of the string with respect the configuration of minimal energy, $\xi_1, \xi_2$ are the coordinates on the world-sheet and $S_0[X]$ is the Gaussian action

$$S_0[X] = \frac{\sigma_0}{2} \int d^2\xi (\partial_\alpha X \cdot \partial^\alpha X) \tag{12}$$

We are assuming an Euclidean signature for both the worldsheet and the target space.

This is the first example of an effective string action and, as we shall see below, it is actually nothing else than the large distance limit of the Nambu–Goto string written in the so called "physical gauge".

This free Gaussian action can be easily integrated, leading in the $T \to 0$ ($N_t \to \infty$) limit to a correction to the linear quark-anti-quark potential, known as Lüscher term [3,4]

$$V(R) = \sigma_0 R + c - \frac{\pi(D-2)}{24R} + O(1/R^2). \tag{13}$$

We shall neglect from now on the constant $c$ which is related to the perimeter term discussed in the previous section.

It is instructive to look at this correction for finite values of $N_t$. Thanks to the Gaussian nature of the action the integration can be easily performed also for finite values of $N_t$, for instance using the $\zeta$ function regularization, leading to the following result [20–22]:

$$V(R, T) = \sigma_0 R + \frac{D-2}{N_t} \log(\eta(q)), \tag{14}$$

where $\eta$ denotes the Dedekind eta function (see the Appendix A):

$$\eta(\tau) = q^{\frac{1}{24}} \prod_{n=1}^{\infty} (1 - q^n) \quad ; \quad q = e^{2\pi i \tau} \quad ; \quad \tau = i\frac{N_t}{2R}. \tag{15}$$

To understand the meaning of this result it is useful to expand it in the two limits $R \ll N_t$ and $R \gg N_t$.

$R \ll N_t$, low temperature

$$V(R, T) = \sigma_0 R + \left[ -\frac{\pi}{24R} + \frac{1}{N_t} \sum_{n=1}^{\infty} \log(1 - e^{-\pi n N_t/R}) \right](D-2), \tag{16}$$

$R \gg N_t$, high temperature

$$V(R, T) = \sigma_0 R + \frac{D-2}{N_t} \left[ -\frac{\pi R}{6N_t} + \frac{1}{2} \log \frac{2R}{N_t} + \sum_{n=1}^{\infty} \log(1 - e^{-4\pi n R/N_t}) \right]. \tag{17}$$

From a string point of view these limits correspond to the open and closed string channels, respectively. They are related by a modular transformation $\tau \to -1/\tau$

$$\eta\left(e^{-2\pi i/\tau}\right) = \sqrt{-i\tau}\,\eta\left(e^{2\pi i\tau}\right). \tag{18}$$

which is known as open-closed string duality.

In the LGT language the two limits correspond, respectively, to the low temperature and the high temperature limits where, obviously, with high temperature we mean a value of $T$ large, but still below the deconfinement temperature, so that a confining flux tube still exists between the quark and the antiquark and an EST picture is still a valid description of the infrared behaviour of the theory.

It is interesting to see that the EST corrections have a completely different behaviour in the two regimes.

At low temperature we find a rather mild correction, which is dominated by the Lüscher term mentioned above (the first term in Equation (16)) while the remaining terms vanish in the $N_t \to \infty$ limit.

On the contrary at high temperature we find that the dominant term is linear in $R$ and gives a large correction, which increases as the temperature increases, and counteracts the string tension.

$$V(R, T) \sim \left(\sigma_0 - \frac{\pi(D-2)}{6N_t^2}\right) R. \tag{19}$$

We shall see below that, if one studies the whole Nambu–Goto action this correction represents only the first term of an infinite set of corrections which can be resummed as follows

$$V(R, T) \sim \sigma_0 \sqrt{1 - \frac{\pi(D-2)}{3\sigma_0 N_t^2}}\, R \equiv \sigma(T)R \tag{20}$$

where we have introduced a temperature dependent string tension $\sigma(T)$ defined as:

$$\sigma(T) = \sigma_0 \sqrt{1 - \frac{\pi(D-2)}{3\sigma_0 N_t^2}} \tag{21}$$

Intuitively, what is happening in this regime is that the fluctuations induced by the temperature tend to reduce the confining force of the flux tube. As the temperature increases, fluctuations get stronger and stronger and finally, at the deconfinement point, the flux tube is destroyed by the fluctuations and there is no more a confining potential between the quark and the antiquark.

It is exactly this finite temperature regime the main focus of the present review, and it is clear now the reason of this choice: In this regime string effects are magnified and can be more easily compared with numerical simulations.

### 3.1. The Nambu–Goto Action

It is easy to see that the free Gaussian action discussed above cannot be a consistent effective string description of the flux tube since it does not fulfill the constraints imposed by the Lorentz invariance of the original gauge theory (we shall discuss this issue in more detail below). The simplest possible EST fulfilling these constraints is the well known Nambu–Goto action [1,2]. As we shall see below the free Gaussian action of Equation (12) is actually the first term of the large distance expansion of the NG action. This explains why, notwithstanding its lack of consistency, its predictions, and in particular the Lüscher term, were initially found in good agreement with LGT simulations of several different gauge models [23–30], and why, with the improvement of LGT simulations, this agreement was later shown to hold only for for Polyakov loops correlators with large separations and higher order corrections (in particular the next to leading Nambu–Goto term that we shall discuss below) started to be detected [31–46]. Notice that some of the first studies on EST were actually performed in the three dimensional Ising model. In these simulations instead of the interquark potential one studies the interface free energy which is also described by the EST, but with different boundary conditions. In particular this is the case of the following papers: [23,31,36,45].

In the Nambu–Goto model [1,2], the string action $S_{\mathrm{NG}}$ is :

$$S_{\mathrm{NG}} = \sigma_0 \int_\Sigma d^2\xi \sqrt{g} \, , \tag{22}$$

where $g \equiv \det g_{\alpha\beta}$ and

$$g_{\alpha\beta} = \partial_\alpha X_\mu \, \partial_\beta X^\mu \tag{23}$$

is the induced metric on the reference world-sheet surface $\Sigma$ and, as above, we denote the worldsheet coordinates as $\xi \equiv (\xi^0, \xi^1)$. This term has a simple geometric interpretation: It measures the area of the surface spanned by the string in the target space and is thus the natural EST realization of the sum over surfaces weighted by their area in the rough phase of the LGT model which we discussed above. This action has only one free parameter: The string tension $\sigma_0$ which has dimension $(\text{length})^{-2}$. Once this is fixed, say, by a fit to the large distance behaviour of the lattice data at zero temperature, there are no more free degrees of freedom in the model which is thus, as we shall see, highly predictive.

In order to perform calculations with the Nambu–Goto action one has first to fix its reparametrization invariance. The standard choice is the so called "physical gauge". In this gauge the two worldsheet coordinates are identified with the longitudinal degrees of freedom of the string: $\xi^0 = X^0$, $\xi^1 = X^1$, so that the string action can be expressed as a function only of the $(D-2)$ degrees of freedom corresponding to the transverse displacements, $X^i$, with $i = 2, \ldots, (D-1)$ which are assumed to be single-valued functions of the worldsheet coordinates. We shall comment below on the problems of this gauge fixing choice, but let us assume it for the moment and let us see what are the consequences.

With this gauge choice the determinant of the metric can be written as

$$
\begin{aligned}
g \;=\;& 1 + \partial_0 X_i \partial_0 X^i + \partial_1 X_i \partial_1 X^i \\
& + \partial_0 X_i \partial_0 X^i \partial_1 X_j \partial_1 X^j - (\partial_0 X_i \partial_1 X^i)^2
\end{aligned}
\tag{24}
$$

and the Nambu–Goto action can then be written as a low-energy expansion in the number of derivatives of the transverse degrees of freedom of the string which, by a suitable redefinition of the fields, can be rephrased as a large distance expansion. The first few terms in this expansion are

$$
S = S_{\mathrm{cl}} + \frac{\sigma_0}{2} \int d^2 \xi \left[ \partial_\alpha X_i \cdot \partial^\alpha X^i + \frac{1}{8}(\partial_\alpha X_i \cdot \partial^\alpha X^i)^2 - \frac{1}{4}(\partial_\alpha X_i \cdot \partial_\beta X^i)^2 + \dots \right],
\tag{25}
$$

and we see, as anticipated, that the first term of the expansion is exactly the Gaussian action of Equation (12). From a Quantum Field Theory point of view the free Gaussian action is the two dimensional Conformal Field Theory (CFT) of the $D-2$ free bosons which represent the transverse degrees of freedom.

Remarkably enough, it can be shown that all the additional terms in the expansion beyond the Gaussian one combine themselves so as to give an exactly integrable, irrelevant perturbation of the Gaussian term [47], driven by the $T\bar{T}$ operator of the $D-2$ free bosons [48].

Thanks to this exact integrability, the partition function of the model can be written explicitely [49,50]. The explicit expression for the partition function was actually found even before this $T\bar{T}$ study, first by using the constraints imposed by the open-closed string duality [51] and then using a d-brane formalism [52]. For the Polyakov loop correlator in which we are interested here (similar expressions can be obtained also for the other relevant geometries: The Wilson loop [42] and the interface [53]), the expression in $D$ space–time dimensions is, using the notations of [51,52]:

$$
\langle P(x)^*P(y)\rangle = \sum_{n=0}^{\infty} w_n \frac{2R\sigma_0 N_t}{E_n} \left(\frac{\pi}{\sigma_0}\right)^{\frac{1}{2}(D-2)} \left(\frac{E_n}{2\pi R}\right)^{\frac{1}{2}(D-1)} K_{\frac{1}{2}(D-3)}(E_n R)
\tag{26}
$$

where $R$ denotes, as above, the interquark distance $R = |x - y|$, $w_n$ the multiplicity of the closed string states which propagate from one Polyakov loop to the other, and $E_n$ their energies which are given by

$$
E_n = \sigma_0 N_t \sqrt{1 + \frac{8\pi}{\sigma_0 N_t^2}\left[-\frac{1}{24}(D-2) + n\right]}.
\tag{27}
$$

At large distance the correlator is dominated by the lowest state

$$
E_0 = \sigma_0 N_t \sqrt{1 - \frac{\pi(D-2)}{3\sigma_0 N_t^2}} = \sigma(T)N_t.
\tag{28}
$$

where $\sigma(T)$ is the finite temperature string tension defined in Equation (21).

The weights $w_n$ can be easily obtained from the expansion in series of $q$ of the infinite products contained in the Dedekind functions which describes the large-$R$ limit of Equation (26) (see reference [52] for a detailed derivation):

$$
\left(\prod_{r=1}^{\infty} \frac{1}{1 - q^r}\right)^{D-2} = \sum_{k=0}^{\infty} w_k q^k.
\tag{29}
$$

For $D = 3$ we have simply $w_k = p_k$, the number of partitions of the integer $k$, while for $D > 3$ these weights can be straightforwardly obtained from combinations of the $p_k$.

These weights diverge exponentially as $n$ increases; in particular we have:

$$w_n \sim \exp\left(\pi\sqrt{\frac{2(D-2)n}{3}}\right).$$
(30)

Again, it is easy to see that the large distance expansion of Equation (26) exactly matches the free Gaussian result of Equation (14).

### 3.2. The Nambu–Goto Action at Finite Temperature

Looking at Equation (26) we see that the NG partition function coincides with a a collection of free particles of mass $E_n$ and multiplicity $w_n$ in $D-1$ dimensions. In the large distance limit only the lowest of these masses survives and the Polyakov loop correlator is described by an expression of this type

$$\langle P(x)^*P(y)\rangle \sim \left(\frac{1}{R}\right)^{\frac{1}{2}(D-3)} K_{\frac{1}{2}(D-3)}(E_0 R)$$
(31)

Remarkably enough this is exactly what we would expect from the Renormalization Group analysis of Section 2.4. In fact, if we interpret the Polyakov loop as a spin of a $D-1$ dimensional spin model with global symmetry the center of the gauge group, then, if the symmetry group is discrete (like for instance for the $SU(2)$ or $SU(3)$ LGTs), in the symmetric phase of the model the spin–spin correlator is described by an isolated pole in the Fourier space, which, when transformed back to the coordinate space becomes exactly the expression of Equation (31). In this interpretation, the mass $E_0$ becomes the inverse of the correlation length $\zeta$ of the system. We thus find (see Equation (28)).

$$\frac{1}{\zeta} = \sigma_0 N_t \sqrt{1 - \frac{\pi(D-2)}{3\sigma_0 N_t^2}} = \frac{\sigma_0}{T}\sqrt{1 - \frac{\pi(D-2)T^2}{3\sigma_0}}$$
(32)

It is interesting to look at the large distance expansion of the interquark potential in this regime. As anticipated the dominant term is linear in $R$ and is proportional to the finite temperature string tension $\sigma(T)$. On top of this we have a set of subleading corrections, (encoded in the asymptotic expansion of the modified Bessel function $K_{\frac{D-3}{2}}$) which represent a specific signature of the Nambu–Goto action.

Using the large distance expansion of the modified Bessel function $K_n(z)$:

$$K_n(z) = \sqrt{\frac{\pi}{2z}}e^{-z}\left[1 + \frac{4n^2 - 1}{8z} + \frac{16n^4 - 40n^2 + 9}{128z^2} + \mathcal{O}(z^{-3})\right]$$
(33)

and the definition of the interquark potential in Equation (7) we find, using Equation (31)

$$V(R, N_t) \sim RTE_0 + \frac{T(D-2)}{2}\ln R + \frac{T(1-(D-3)^2)}{8RE_0} \cdots$$
(34)

where we dropped an irrelevant additive constant, and neglected terms which are suppressed by higher powers of $(RT)^{-1}$. In the following we shall mainly study models in $D = 3$ dimensions. In this case the above expression becomes:

$$V(R, N_t) \sim R\sigma(T) + \frac{T}{2}\ln R + \frac{T^2}{8R\sigma(T)} \cdots$$
(35)

In the framework of the Nambu–Goto approximation one can also derive an estimate of the critical temperature $T_{c,NG}$ measured in units of the square root of the string tension $\sqrt{\sigma_0}$ [7,54,55]

$$\frac{T_{c,NG}}{\sqrt{\sigma_0}} = \sqrt{\frac{3}{\pi(D-2)}}$$
(36)

given by the value of the ratio $\frac{T_{c,NG}}{\sqrt{\sigma_0}}$ for which the lowest mass $E_0$ vanishes. We can thus rewrite the energy levels as a function of $T/T_{c,NG}$ as

$$E_n = \frac{(D-2)\pi T_{c,NG}^2}{3T}\sqrt{1 - \frac{T^2}{T_{c,NG}^2}\left[1 - \frac{24n}{D-2}\right]}. \tag{37}$$

In this framework the correlation length can be written as:

$$\xi(T) = \frac{3T}{(D-2)\pi T_{c,NG}^2}\frac{1}{\sqrt{1 - \frac{T^2}{T_{c,NG}^2}}}. \tag{38}$$

which diverges as expected at the critical point. This result is particularly interesting from a conceptual point of view since it makes explicit in which sense the Nambu–Goto action is an approximation of the "correct" effective string action. The critical index that we find: $\nu = 1/2$ is the typical signature of the mean field approximation. We know from the Svetitsky-Yaffe analysis that this cannot be the correct answer and that the critical index should instead be that of the symmetry breaking phase transition of the $(D-1)$ dimensional spin model with symmetry group the center of the original gauge group. For instance, for the $(3+1)$ dimensional $SU(2)$ model we expect to find $\nu = 0.6299709(40)$ [56,57] which is the value for the three dimensional Ising model or for the $(2+1)$ $SU(2)$ LGT we expect $\nu = 1$, which is the cirtical index for the 2D Ising model. Besides this anomalous dimension, the major effect of the mean field approximation is, as usual, a shift in the critical temperature. Indeed, while the Nambu–Goto prediction for the deconfinement temperature is in four dimensions $T_{c,NG} = \sqrt{\frac{3\sigma_0}{2\pi}} \sim= 0.691\sqrt{\sigma_0}$ the actual value for the $SU(2)$ deconfinement transition is slightly larger: $T_c/\sqrt{\sigma_0} = 0.7091(36)$ [58]. The fact that this shift is so small is another evidence of the goodness of the Nambu–Goto approximation. It is interesting to notice that this agreement holds for all the LGTs which have been studied [58–62], both in $(2+1)$ and in $(3+1)$ dimensions, with the only exception of the 3D $U(1)$ model [63], for which, in fact, a different EST is expected [8,64], with a dominant contribution from the extrinsic curvature term.

### 3.3. Beyond Nambu–Goto

We have seen from the above analysis that the Nambu–Goto action alone cannot be the end of the story. Finding hints of the correct EST action beyond the Nambu–Goto term is one of the major open challenges in this context. We shall devote this subsection and the following to a brief discussion of this issue.

As a starting point let us notice that, from an effective action point of view, there is no reason to constrain the coefficients of the higher order terms in Equation (25) to the values displayed there. In principle, one should instead assume the most general form for such an effective action

$$S = S_{cl} + \frac{\sigma_0}{2}\int d^2\xi\left[\partial_\alpha X_i \cdot \partial^\alpha X^i + c_2(\partial_\alpha X_i \cdot \partial^\alpha X^i)^2 + c_3(\partial_\alpha X_i \cdot \partial_\beta X^i)^2 + \dots\right], \tag{39}$$

and then fix the coefficients order by order using Monte Carlo simulations or experimental results. However, one of the most interesting results of the last few years is that the $c_i$ coefficients are not arbitrary, but must satisfy a set of constraints to enforce the Poincarè invariance of the lattice gauge theory in the $D$ dimensional target space. These constraints were first obtained by comparing the string partition function in different channels, using the open-closed string duality [51,65]. It was later realized [66–70] that they could be directly obtained as a consequence of the Poincaré symmetry of the underlying Yang–Mills theory. A similar result, for the first few coefficients of the EST, was obtained also in the Polchinski-Strominger [5] formalism in [71,72] (See also [73–77] for a debate on these results and a discussion on the extension of this analysis to higher orders and its interplay with conformal invariance).

In fact, even though the $SO(D)$ invariance of the original theory is spontaneously broken by the formation of the classical string configuration around which one is expanding, the effective action should still respect this symmetry through a non-linear realization in terms of the transverse fields $X_i$ [66–70]. These non-linear constraints induce a set of recursive relations among the coefficients of the expansion, which strongly constrain the coefficients $c_i$. In particular, it can be shown that the terms with only first derivatives coincide with the Nambu–Goto action to all orders in the derivative expansion [78] and that the first correction with respect to the Nambu–Goto action appears at order $1/R^7$ in the large $R$ expansion. This explains why the Nambu–Goto model has been so succesfull over these last forty years to describe the infrared behaviour of confining gauge theories despite its simplicity and why the deconfinement temperature predicted by Nambu–Goto is so close to the one obtained in Monte Carlo simulations.

This argument can be better understood looking at the original string action, before fixing the reparametrization invariance from a geometric point of view. In this framework the effective action is obtained by the mapping

$$X^\mu : \mathcal{M} \to \mathbb{R}^D, \qquad \mu = 0, \cdots, D-1 \tag{40}$$

of the two-dimensional surface describing the worldsheet of the string $\mathcal{M}$ into the (flat) $D$-dimensional target space $\mathbb{R}^D$ of the gauge theory and then imposing the constraints due to Poincaré and parity invariance of the original theory. This approach was discussed in detail in reference [11]. The first few terms of the action compatible with these constraints must be combinations of the geometric invariants which can be constructed from the induced metric $g_{\alpha\beta} = \partial_\alpha X^\mu \partial_\beta X_\mu$. These terms can be classified according to their "weight", defined as the difference between the number of derivatives minus the number of fields $X^\mu$ (i.e., as their energy dimension). Due to invariance under parity, only terms with an even number of fields should be considered. The first term of this expansion, which is also the only term of weight zero, corresponds, as we mentioned above, to the Nambu–Goto action

$$S_{\text{NG}} = \sigma_0 \int d^2\zeta \sqrt{g}, \tag{41}$$

At weight two, two new contributions appear:

$$S_{2,\mathcal{R}} = \gamma \int d^2\zeta \sqrt{g}\mathcal{R}, \tag{42}$$

$$S_{2,K} = \alpha \int d^2\zeta \sqrt{g}K^2, \tag{43}$$

where $\alpha$, and $\gamma$ are two new free parameters, $\mathcal{R}$ denotes the Ricci scalar constructed from the induced metric, and $K$ is the extrinsic curvature, defined as $K = \Delta(g)X$, with

$$\Delta(g) = \frac{1}{\sqrt{(g)}}\partial_a[\sqrt{(g)}g^{ab}\partial_b] \tag{44}$$

the Laplacian in the space with metric $g_{\alpha\beta}$. In principle the new free parameters $\alpha$, and $\gamma$ should be fixed, as we did for $\sigma_0$ by comparing with Monte Carlo simulations. However this process is simplified by the observation that the term proportional to $\mathcal{R}$ is a topological invariant in two dimensions and, since in the long-string limit in which we are interested one does not expect topology-changing fluctuations, its contribution can be neglected [11]. On the other hand, the term in Equation (43) which contains $K^2$ leads to quantum corrections which decrease exponentially with the interquark distance [8] and are thus negligible unless the ratio between the coefficient of the $K^2$ term and the string tension grows to infinity in the continuum limit and this seems to occur only in very few models like, for instance, the $d = 3$ $U(1)$ model [8]. In these cases an Effective String Theory model, which combines Nambu–Goto and extrinsic curvature was proposed long ago in [79,80]. The resulting EST is usually known as "rigid string". We shall comment on this issue in the last section of the review.

At weight four, two new combinations can be constructed and correspondingly two new parameters appear, leading in the open string channel (i.e., in the low $T$ regime) to the $1/R^7$ correction mentioned above. Notice, however, that also these new parameters are not completely free and can be constrained using a bootstrap type of analysis [81] in the framework of the S-matrix approach pioneered by [47]. As above they should in principle be fixed by comparing with Monte Carlo estimates of the potential; however, their contributions appear at such a high level that they are very difficult to detect even with the most precise numerical simulations.

*3.4. Beyond Nambu–Goto: The Boundary Term*

Another term which must be considered beyond the Nambu–Goto one is the so called "boundary term". This term has an origin different from those discussed above. It is due to the presence of the Polyakov loops at the boundary of the correlator. The classical contribution associated to this correction is the constant term $c$ which appears in the potential and that we have systematically neglected in the previous analysis. Beyond this classical term we may find quantum corrections due to the interaction with the flux tube. The main result in this context is that also these terms are strongly constrained by Lorentz invariance

The first boundary correction compatible with Lorentz invariance is [82]:

$$b_2 \int d\xi_0 \left[ \frac{\partial_0 \partial_1 X \cdot \partial_0 \partial_1 X}{1 + \partial_1 X \cdot \partial_1 X} - \frac{(\partial_0 \partial_1 X \cdot \partial_1 X)^2}{(1 + \partial_1 X \cdot \partial_1 X)^2} \right]. \tag{45}$$

with an arbitrary, non-universal coefficient $b_2$. The lowest order term of the expansion of Equation (45) is:

$$S_{b,2}^{(1)} = b_2 \int d\xi_0 (\partial_0 \partial_1 X)^2 \tag{46}$$

The contribution of this term to the interquark potential was evaluated in [78] using the zeta function regularization:

$$\langle S_{b,2}^{(1)} \rangle = -b_2 \frac{\pi^3 N_t}{60 R^4} E_4 (e^{-\frac{\pi N_t}{R}}) \tag{47}$$

where $E_4$ denotes the fourth order Eisenstein series (see the Appendix for definitions and properties of these functions). In the standard low temperature ($N_t \gg R$) setting, this amounts to a correction proportional to $1/R^4$ to the interquark potential, which turns out to be the dominant correction term beyond Nambu–Goto in this regime and represents a further obstacle to detect signatures of the "bulk" correction terms discussed in the previous subsection.

Recent high precision Monte Carlo simulations [82–86] allowed to estimate $b_2$ for a few LGTs with remarkable precision. For the $SU(2)$ model in (2 + 1) dimensions a boundary correction was estimated even for the first string excitations [86]. Preliminary results have been also obtained for the (3 + 1) dimensional SU(3) LGT [87,88] where, besides $b_2$, a tentative estimate of the next to leading term $b_4$ is also reported. As expected these values are not any more universal and represent the first hint of the fact that different LGTs are described by different ESTs and that the information on the gauge group of the model and the gluon content of the flux tube is somehow encoded in the effective string model.

At the same time the above discussion shows that this boundary term is the dominant non universal correction beyond Nambu–Goto in this low $T$ regime and it is clear that its presence makes it almost impossible to detect the much weaker signatures of the "bulk" correction terms discussed in the previous subsection.

However, by performing a modular transformation (see the Appendix A) it is easy to see that in the high temperature limit (i.e., $R \gg N_t$) this correction becomes

$$\langle S_{b,2}^{(1)} \rangle = -b_2 \frac{4\pi^3}{15 N_t^3} E_4 (e^{-\frac{4\pi R}{N_t}})$$

(48)

and does not contain a term proportional to $R$ and thus it does not give a correction to the temperature dependent string tension $\sigma(T)$.

This is a second important reason of interest of the high temperature regime which is the focus of this review. In this limit the boundary term does not interfere with the "bulk" EST corrections beyond Nambu–Goto which can thus be directly observed with Monte Carlo simulations.

## 4. Comparison with Monte Carlo Simulations

In the past years the predictions of EST for the interquark potential were tested with Monte Carlo simulations of increasing precision in several different LGTs [23–46]. Most of these tests were performed in the low temperature regime. However, as we have seen in the previous sections, in order to have a complete understanding of EST and to test its consistency under the open–closed string transformation, it is interesting to test EST predictions also in the high temperature regime. This is the goal of this section in which we shall report the results of a few papers in which EST was compared with high-$T$ Monte Carlo simulations.

We shall first discuss in detail, as an example, the $SU(2)$ gauge theory in $(2+1)$ dimensions, which is the simplest non-abelian LGT and allows to reach high precision results with a relatively small amount of computing power. Then, in the last subsection, we shall briefly review the results obtained in other LGTs both in (2 + 1) and in (3 + 1) dimensions.

### 4.1. LGT Observables

Let us first discuss a few combinations of Polyakov loop correlators which are particularly useful to address the comparison between EST predictions and LGT results in the high temperature regime.

To simplfy notations let us define the Polyakov loop correlator as:

$$G(R,T) = \langle P(x) P^\dagger(x+R) \rangle_{N_t}$$

(49)

Following [30,33,89] it is particularly convenient to introduce the following quantities:

$$Q(R,T) = T \ln \frac{G(R,T)}{G(R+1,T)},$$

(50)

$$A(R,T) = R^2 \ln \frac{G(R+1,T)G(R-1,T)}{G^2(R,T)}.$$

(51)

Note that, in the continuum limit $a \to 0$, $Q(R,T)$ tends to the first derivative of $V(R,T)$ with respect to $R$:

$$\lim_{a \to 0} Q = \frac{\partial V}{\partial R},$$

(52)

so that it can be interpreted as a lattice version of (minus) the interquark force. On the other hand, $A(R,T)$ is a dimensionless quantity proportional to the discretized derivative of the force:

$$\lim_{a \to 0} A = -\frac{R^2}{T} \frac{\partial^2 V}{\partial R^2}.$$

(53)

These quantities are the finite temperature version of the observables introduced in [30]. In particular $Q(R, T)$ coincides in the low-$T$ limit with the "force" $F(R)$ of [30] while $A(R, T)$ is related to the "central charge" $c(R)$ of [30] as follows

$$A(R, T) = \frac{2}{RT} c(R). \tag{54}$$

Using Equation (35) we may estimate the large-$R$ limit of these two observables for $D = 3$ LGTs in the framework of the Nambu–Goto effective string model:

$$Q(R, T) \simeq \sigma(T) + \frac{T}{2R} - \frac{T^2}{8\sigma(T)R^2} + \cdots \tag{55}$$

$$A(R, T) \simeq \frac{1}{2} - \frac{T}{4\sigma(T)R} + \cdots \tag{56}$$

The constraints on the EST discussed in Section 3.3 tell us that these expressions for $Q$ and $A$ should be universal and should hold for any LGT (except, as usual, the 3d $U(1)$ LGT). Corrections to the EST beyond Nambu–Goto should only affect higher order terms (the dots in the above Equations (55) and (56), and are expected to affect the finite temperature string tension $\sigma(T)$ only with corrections of the order of $(T/T_c)^7$. In the next section we shall compare this prediction with Monte Carlo simulations.

*4.2. The Su(2) LGT in (2 + 1) Dimensions*

The $(2 + 1)$ dimensional $SU(2)$ model has been the subject of several numerical efforts in the last years; most of them, however, focused on the low $T$ regime of the model. We shall report here the results of the simulations discussed in [89] which were instead performed at a relatively high $(T = \frac{3}{4}T_c)$ temperature.

The only imput we need to fix our predictions is the zero temperature string tension $\sigma_0$. This can be fixed using for instance the results of [33] which we report here

$$\sqrt{\sigma_0} \simeq \frac{1.324(12)}{\beta} + \frac{1.20(11)}{\beta^2} \tag{57}$$

Simulations were perfomed at $\beta = 9$ for which we have $\sigma_0 = 0.0262(1)$ [33] on a lattice of size $120^2 \times 8$. For this value of $\beta$ the critical tempearture is, almost exactly located at $N_t = 6$ thus this choice of lattice sizes corresponds to a temperature $T = \frac{3}{4}T_c$. Polyakov loop correlators were measured up to the distance of $R = 19$ lattice spacings. We report for completeness the results of the simulations in Table 1 and refer the interested reader to [33] for more details on the simulation settings and on the fitting protocol.

**Table 1.** Results for $Q(r, T)$, as a function of the interquark distance $R$, for the $(2 + 1)$ dimensional SU(2) model at $T = 3T_c/4$, taken from [89].

| $R$ | $Q$ | $R$ | $Q$ | $R$ | $Q$ |
|---|---|---|---|---|---|
| 2 | 0.037433(46) | 8 | 0.02232(11) | 14 | 0.01971(16) |
| 3 | 0.030958(56) | 9 | 0.02170(12) | 15 | 0.01949(18) |
| 4 | 0.027600(64) | 10 | 0.02117(12) | 16 | 0.01926(19) |
| 5 | 0.025553(72) | 11 | 0.02072(13) | 17 | 0.01906(20) |
| 6 | 0.024154(84) | 12 | 0.02034(15) | 18 | 0.01892(22) |
| 7 | 0.023118(94) | 13 | 0.02000(15) | 19 | 0.01876(24) |

Following Equation (55) the values of $Q(R, T)$ are fitted with:

$$Q(R, T)|_{T=3T_c/4} = s + \frac{b}{R} + \frac{c}{R^2}, \tag{58}$$

and the following best fit values for the parameters are found

$$s = 0.01530(37) \quad b = 0.0668(58) \quad c = -0.087(27)$$

with a reduced $\chi_r^2 = 0.75$.

The universal correction in which we are interested are encoded in the parameter $b$ which according to the analysis discussed in the previous sections should be given by

$$b = \frac{T}{2} = \frac{1}{16} = 0.0625,$$

which turns out to be in remarkable agreement with the result of the fit.

This is further confirmed by the analysis of the $A(R, T)$ values (which can be easily obtained from the data reported in Table 1). These values are fitted with

$$A(R, T)|_{T=3T_c/4} = k - \frac{m}{R}, \tag{59}$$

finding

$$k = 0.528(28), \quad m = -1.09(28), \quad \text{with } \chi_{\text{red}}^2 = 1.6,$$

which is again in perfect agreement with the expected value $k = 1/2$.

From the first fit we can extract the value $\sigma(T) = 0.01530(37)$ for the finite temperature string tension at $T = 1/8$. Using the value $\sigma_0$ reported in Equation (57), we may obtain a "Nambu–Goto" prediction for $\sigma(T)$ using Equation (21), which turns out to be $\sigma_{\text{NG}}(T = 1/8) = 0.01605(6)$, at two standard deviations from the observed value. This indicates, as already observed in [90], that for the (2 + 1) SU(2) LGT the Nambu–Goto string represents a rather good approximation but, as the precision of the simulations improves, small deviations start to be detected. These deviations are the signatures of the $(T/T_c)^7$ term mentioned above.

Finally, using the measured value of $\sigma(T)$ it is possible to obtain predictions for the subleading corrections in the two fits. One find for the $c$ term in the first fit $c_{\text{NG}} \sim -0.1216(5)$ and for $m$ in the second fit $m_{\text{NG}} = -1.946(8)$. Both values are similar to those extracted from the fits, but not compatible within the errors. This small discrepancy agrees in sign and magnitude with the analogous deviations from the Nambu–Goto ansatz observed in [90] and summarized in the coefficient $C_3$ evaluated there. We shall comment on these deviations in the next section.

### 4.3. EST Predictions Versus Monte Carlo Results for Different LGTs

The same analysis was performed in [33] for the (2 + 1) $SU(3)$ and $SU(4)$ lattice gauge theories and, using data obtained in [91], also in the case of the three dimensional Ising gauge model for two different temperatures. We summarize the results for the fits to $Q(R.T)$ in Table 2.

**Table 2.** Results of the fits to $Q(R, T)$ for various LGTs (listed in the first column), together with the expected values for the best fit parameters according to the Nambu–Goto EST.

| Gauge Group | $N_t$ | $T/T_c$ | $s$ | $\sigma_{\text{NG}}(T)$ | $b$ | $b_{\text{NG}}$ | $c$ | $c_{\text{NG}}$ |
|---|---|---|---|---|---|---|---|---|
| SU(2) | 8 | 3/4 | 0.01530(37) | 0.01605(6) | 0.0668(58) | 0.0625 | $-0.087(27)$ | $-0.1216(5)$ |
| SU(3) | 8 | 3/4 | 0.01884(44) | 0.01946(6) | 0.0612(74) | 0.0625 | $-0.063(34)$ | $-0.1003(5)$ |
| SU(4) | 8 | 3/4 | 0.01721(43) | 0.01830(40) | 0.0634(70) | 0.0625 | $-0.063(32)$ | $-0.1070(20)$ |
| $Z_2$ | 12 | 1/2 | 0.01485(2) | 0.01487(6) | 0.0414(8) | 0.04167 | $-0.049(6)$ | $-0.058(1)$ |
| $Z_2$ | 9 | 2/3 | 0.01137(11) | 0.01067(6) | 0.0522(40) | 0.0556 | $-0.076(34)$ | $-0.145(1)$ |

Looking at the table we can make a few interesting observations

- All the models, except the one at the lowest temperature, show deviations in the fitted value of $\sigma(T)$ with respect to the Nambu–Goto prediction. These deviations are the signatures of the terms beyond Nambu–Goto which must be included in the EST action which we discussed in the previous section. They are exactly those needed to match the critical index of the deconfinement transition which in this case is $\nu = 1$ instead of the Nambu–Goto value $\nu = 1/2$.
- The universal constant $b$ is always compatible with the theoretical expectation. This represents a remarkable consistency check of the whole EST construction.
- The constant $c$ shows the same trend for all the models: It is similar to the expected Nambu–Goto value, but always slightly smaller in magnitude. Most likely this deviation is due to the fact that in the fit we are neglecting higher terms, and indeed the first of them, the one proportional to $1/R^3$ , due to the expansion of the modified Bessel function has the opposite sign with respect to the $1/R^2$ one and may explain the decrease in magnitude of $c$.

Similar results are found fitting the $A(R, T)$ function (see [33] for further details).

A similar analysis was also performed for the SU(2) model in (3 + 1) dimensions [92] and (with a different set of observables) for SU(3) in [93]. In both cases two different temperatures were tested and a good agreement with the Nambu–Goto predictions was found for the lower one, while deviations were detected for the higher one, pointing to the possible presence of terms in the EST beyond the Nambu–Goto one. Besides its physical relevance, this extension to (3 + 1) dimensions is also interesting because the interquark potential, as can be seen in Equation (34), shows a non-trivial dependece on the number of space time dimensions which is precisely confirmed by the numerical simulations at the lowest temperatures.

As a matter of fact this type of corrections in the (3 + 1) dimensional SU(3) models were already observed more than twenty years ago when the first high precision determinations of $\sigma(T)$ were obtained [15,16]. The behaviour of $\sigma(T)$ was very similar to the one predicted by the Nambu–Goto action, but with small deviations in the vicinity of the deconfinement point. Thes deviations led to a non-zero, even if small, value of the string tension $\sigma(T_c)$ at the critical point which had the effect of transforming the second order phase transition predicted by the Nambu–Goto model into the first order deconfinement phase transition of the SU(3) (3 + 1) dimensional model.

## 5. Width of the Confining Flux Tube at High Temperature

One of the most intriguing features of the EST picture of confinement is the logarithmic increase of the square width $w^2(R)$ of the flux tube as a function of the interquark distance $R$ [19].

$$\sigma_0 w^2(R) = \frac{1}{2\pi} \log \frac{R}{R_c} \tag{60}$$

where $R_c$ is known as "intrinsic width" and sets the scale of the logarithmic growth.

This logarithmic growth, which is commonly referred to as the "delocalization" of the flux tube was discussed for the first time many years ago by Lüscher, Münster and Weisz in [19] but it required several years of efforts before it could be observed in lattice simulations. The first numerical results were obtained in abelian models [94–100] where, thanks to duality, simulations can be performed more easily and later the flux tube width was studied also in non abelian LGTs [87,88,101–111].

An important issue in this context is to understand the fate of the flux tube width as the deconfinement transition is approached from below. It is important to stress that delocalization and deconfinement are two deeply different conditions of the flux tube. As we have seen in Section 2, deconfinement is characterized by the vanishing of the string tension $\sigma(T)$ and, accordingly, of the flux tube. The delocalization of the flux tube instead coincides with the onset of the rough phase. Delocalization is a typical quantum effect. It is a consequence of the Mermin-Wagner theorem which imposes the restoring in the continuum limit of the translational symmetry for the fluctuations of the flux tube in the transverse directions. Intutively it amounts to say that we cannot fix deterministically the trajectory of the flux tube but may only describe it as a probabilty distribution. It is important to stress that, even if delocalized, the flux tube fully keeps its confining function. The quantum fluctuations which drive the delocalization also influence the confining potential (as the presence of Lüscher term indicates) but do not destroy it.

While the behaviour of the string tension $\sigma(T)$ as the deconfinement temperature $T_c$ is approached from below is rather well understood, much less is known on the behaviour of the flux tube thickness in this regime. This is an important issue from a physical point of view since the interplay between delocalization and deconfinement could strongly influence the transition from hadrons to free quarks as $T_c$ is approached.

Similarly to what we did for the interquark potential, also this problem can be addressed by performing a modular transformation of the low temperature result. This was done in [112] in the case of the free Gaussian action (i.e., the first order in the perturbative expansion of the Nambu–Goto effective string) leading in the large $R$ limit to the following result:

$$\sigma_0 w_{lo}^2 = \frac{R}{4N_t} + \frac{1}{2\pi} \log \frac{N_t}{L_c} - \frac{1}{\pi} e^{-2\pi \frac{R}{N_t}} + \cdots \tag{61}$$

where $L_c$ is a length scale which plays the role in this limit of the intrinsic width of Equation (60) and the suffix $lo$ is added to emphasize that this is only the leading order (Gaussian) approximation of the true flux tube width.

We see that the large $R$ behaviour of the square width changes completely and becomes linear (with a coefficient $\frac{1}{4\sigma_0 N_t}$) instead of logarithmic. This behaviour holds in principle for any temperature $T$, but as $T$ decreases it requires larger and larger values of $R$ to be observed. Similarly it is possible to show that for any fixed value of $R$ the square width smoothly converges toward the expected logarithmic behaviour as $T$ decreases. The threshold between the two behaviours is $R \sim 1/T$.

In [112] this prediction was tested with a set of high precision Monte Carlo simulations of the 3D gauge Ising models and only a partial agreement with Equation (61) was found. For all the temperatures studied in [112] $w^2(R)$ was indeed a linearly increasing function of $R$. However the coefficient of this linear behaviour was in general larger than the one predicted by the effective string (except for the smallest temperature values) and, what is more important, it seemed to diverge as the deconfinement point was approached (while the coefficient $\frac{1}{4\sigma_0 N_t}$ converges instead to a finite value at the deconfinement point).

This discrepancy tells us that as the deconfinement transition is approached the leading order approximation gets worse and worse and that, similarly to what happens for the interquark potential, higher order terms must be included. The problem is that, while for the interquark potential we have the exact solution to all orders, for the flux tube width only the next to leading order is known [101,113,114]. This correction goes in the right direction but is not enough to fill the gap between numerical data and theoretical expectations.

We shall see in the next section that the Svetitsky-Yaffe conjecture offers a powerful tool to address this issue when one approaches the deconfinement transition and allows to guess the resummation to all orders of the flux tube width for the Nambu–Goto effective string. The complete answer for the leading term linear in $R$ turns out to be [115–117]

$$w^2(R) = \frac{1}{4\sigma(T)} RT \tag{62}$$

where $\sigma(T)$ is the temperature dependent string tension of Equation (21).

By expanding this expression in powers of $T/T_c$ it is easy to see that both the leading order $w_{lo}^2$ and the next to leading order of [101,113,114] fully agree with Equation (62).

The results for the Ising model of [112] agree with Equation (62) and a few years later, the same behaviour was observed with a set of high precision simulations in the $(3+1)$ dimensional $SU(3)$ model [104].

To understand the origin of Equation (62) we should first define the LGT observables which allows to evaluate the flux tube width, then address their dimensionally reduced version, according to the Svetitsky-Yaffe projection and finally evaluate these expectation values using the S-matrix approach. Let us address these issues step by step.

### 5.1. Definition of the Flux Tube Thickness

In a finite temperature setting the lattice operator which is used to evaluate the flux through a plaquette $p$ of the lattice is:

$$\langle \phi(p; P, P') \rangle_{N_t} = \frac{\langle PP'^\dagger U_p \rangle_{N_t}}{\langle PP'^\dagger \rangle_{N_t}} - \langle U_p \rangle_{N_t} \tag{63}$$

where $P$, $P'$ are two Polyakov loops separated by $R$ lattice spacings and $U_p$ is the operator associated with the plaquette $p$. Different possible orientations of the plaquette $p$ measure different components of the flux. In the following we shall neglect this dependence which plays no role in our analysis. The only information that we need is the position of the plaquette. Let us define

$$\langle \phi(p; P, P') \rangle_{N_t} = \left\langle \phi(\vec{h}; R, N_t) \right\rangle$$

where $\vec{h}$ denotes the displacement of $p$ from the $P\,P'$ plane. In each transverse direction, the flux density shows a Gaussian like shape (see for instance Figure 2 of [94]). The width of this Gaussian $w$ is the quantity which is usually denoted as "flux tube thickness":

$$w^2(R, N_t) = \frac{\sum_{\vec{h}} \vec{h}^2 \left\langle \phi(\vec{h}; R, N_t) \right\rangle}{\sum_{\vec{h}} \left\langle \phi(\vec{h}; R, N_t) \right\rangle} \tag{64}$$

This quantity depends on the number of transverse dimensions and on the bare gauge coupling $\beta$. Once $\beta$ is fixed the only remaining dependences are on the interquark distance $R$ and on the inverse temperature $N_t$. By tuning $N_t$ we can thus study the flux tube thickness near the deconfinement transition.

### 5.2. Dimensional Reduction and the Svetitsky–Yaffe Approach

As we have seen in Section 2.4. In the vicinity of the deconfinement transition the physics of a $(d+1)$ LGT can be described using an effective model in which the spacelike links are integrated out and the only remaining degrees of freedom are the Polyakov loops. The simplest examples of this effective mapping are the $(2+1)$ $SU(2)$ LGT and the $(2+1)$ Ising gauge model which have the same center $Z_2$ and are thus both mapped into the 2D spin Ising model. We shall use this case as an example in the following to simplify the discussion. Using the correspondences discussed in Section 2.4 it is possible to construct the dimensionally reduced projection of the operator which measures the flux tube thickness which turns out to be a suitable ratio of three and two point correlators of the spin and energy operators (see [115] for a detailed discussion of this mapping). In the particular case of the 2D Ising model that we are using as an example this combination is:

$$\frac{\langle \sigma(x_1)\epsilon(x_2)\sigma(x_3) \rangle}{\langle \sigma(x_1)\sigma(x_3) \rangle} \tag{65}$$

to be evaluated in the high temperature phase and in zero magnetic field. Since we are interested in the large distance behaviour of these correlators we can use the so-called Form Factors approach (see [118] for an introduction to Form Factors and their application in the context of the 2D Ising model without magnetic field).

A straightforward calculation leads to the following expression for the flux distribution [115]

$$P(R, y) = \frac{2\pi R}{4y^2 + R^2} \frac{e^{-m\sqrt{4y^2 + R^2}}}{K_0(mR)}. \tag{66}$$

where $y$ denotes the transverse direction, $K_0$ is the modified Bessel function of order 0, $m$ is the mass of the 2D Ising model and a large $mR$ limit is assumed.

From this flux distribution it is easy to extract the square of the flux tube width as the ratio

$$w^2(R) = \frac{\int_{-\infty}^{\infty} dy\, y^2\, P(R,y)}{\int_{-\infty}^{\infty} dy\, P(R,y)} \tag{67}$$

which, setting $x = 2y/R$ amounts to evaluate

$$w^2(R) = \frac{R^2}{4} \frac{\int_{-\infty}^{\infty} dx\, \frac{x^2}{1+x^2} e^{-2mr\sqrt{1+x^2}}}{\int_{-\infty}^{\infty} dx\, \frac{e^{-2mr\sqrt{1+x^2}}}{1+x^2}} \tag{68}$$

These integrals can be evaluated asymptotically in the large $mR$ limit [115,116] leading to the following result:

$$w^2(R) \simeq \frac{1}{4} \frac{R}{m} + \dots . \tag{69}$$

where the dots stay for terms constant or proportional to negative powers of $R$.

The last step in order to compare this result with Equation (62) is to give a meaning to the Ising mass $m$ in terms of LGT quantities.

This can be easily accomplished if we recall that the mass can be obtained from the large $R$ limit of the spin spin correlator, which according to the Svetitsky-Yaffe mapping is the 2D limit of the expectation value of two Polyakov loops at distance $R$. Following Equation (28) we can thus identify

$$m = \sigma(T)N_t \tag{70}$$

from which we immediately obtain the result of Equation (62).

Similar arguments allow to obtain also estimates of the intrinsic width of the model [117].

The above analysis was performed in the case of the Ising model, but the argument is completely general and the derivation of the large distance behaviour holds for any spin model with a gap in the spectrum.

### 6. Open Issues and Concluding Remarks

In this review we focused in particular on the behaviour of the interquark potential and of the flux tube width. There are, however, a few other observables which show a non trivial behaviour at high-$T$ and allow for non-trivial tests of EST. We could not discuss them in detail in this review for lack of space and specific expertise but we briefly mention them here and list a few relevant references which may help the interested reader to deepen the subject.

- **The deconfinement transition as a Hagedorn transition.**
  One of the more interesting consequences of the EST description of confinement is that the deconfinement transition can be interpreted as a Hagedorn transition [119]. This can be understood (using a dual transformation) as a direct consequence of the tachyonic singularity in the interquark potential [7]. This Hagedorn behaviour has relevant consequences on the equation of state of pure gauge theories which can be precisely tested using Monte Carlo simulations. In fact, in pure gauge theories the only massive excitations in the confining phase are glueballs and the equation of state can be accurately modeled in terms of a gas of these massive, non-interacting glueballs. If one assumes a description of glueballs as closed color flux tubes (as for instance in the Isgur-Paton model [6]) then one should expect a Hagedorn-like [119] stringy behaviour of the glueball spectrum and as a consequence a highly non trivial temperature dependence of pressure and entropy across the deconfinement transition. This effect was observed for the first time in reference [120] for the SU(3) Yang–Mills theory in

(3 + 1) dimensions, and later also in SU($N$) theories in (2 + 1) dimensions [121] and in the (2 + 1) dimensional $SU(2)$ [122,123], finding always a very good agreement with the expected Hagedorn behaviour.

- **The spacelike string tension at high Temperature.**
  An interesting open issue in Lattice Gauge Theory is to understand and model the behaviour of the so called "space–like string tension" [124–138] across the deconfinement transition.

  The space–like string tension is extracted from the correlator of space–like Polyakov loops, i.e., Polyakov loops which lay in a space–like plane, orthogonal to the compact time direction $N_t$. Due to their space–like nature these Polyakov loops do not play the role of order parameter of deconfinement and the space–like string tension extracted from them is different from the actual string tension of the model $\sigma(T)$.

  At low temperature the two string tensions coincide but as the temperature increases they behave differently [124–127,139]. As we have seen $\sigma(T)$ decreases as the deconfinement temperature is approached and vanishes at the deconfinement point, while the space–like string tension remains constant and then increases in the deconfined phase [124–126]. The physical reason for this behavior is that the correlator of two space–like Polyakov loops describes quarks moving in a finite temperature environment. It can be shown that what we called space–like string tension is related to the screening masses in hot QCD [128–134] and thus it does not vanish in the deconfined phase.

  An EST description of this behaviour has been recently obtained [140] using the mapping between the Nambu–Goto action and the $T\bar{T}$ deformation of the free bosonic action. An important open issue in this context is to address the interplay of the space–like string tension with the intrinsic width of the flux tube.

- **EST and interfaces.**
  In this review we studied EST in two particular choices of boundary conditions for the world sheet: Wilson loops (rectangular geometry) and Polyakov loop correlators (cylindrical geometry). There is a third important case, the toroidal geometry, which cannot be easily realized in non-abelian LGTs, but is pretty natural in three dimensional abelian gauge theories. These models, thanks to the Kramers–Wannier duality can be mapped into standard three dimensional spin models (the most relevant example being the 3D gauge Ising model which is mapped into the three dimensional Ising spin model). By suitably choosing the boundary conditions of the spin model (for instance: Antiperiodic in the Ising case) in the low temperature phase one can induce the formation of interfaces which can be described by EST with a toroidal world sheet [31,36,141–146]. Interfaces in the spin model are in some sense the dual of the Wilson loops in the gauge model. The partition function of the Nambu–Goto string with this toroidal boundary conditions can be evaluated with the same tools used for the Polyakov loop correlators [53]. The major reason of interest of this setting is the absence of boundary terms. It is thus much easier to study higher order terms of EST and in fact some of the most precise Monte Carlo studies of these terms were obtained using interfaces in the 3D Ising model [45,146]. The analogy of the high temperature regime in this context is obtained by "squeezing" the interface in one direction. From the spin model point of view this is the regime in which one is approaching dimensional reduction from three to two dimensions [147]. A systematic comparison of EST predictions and Monte Carlo simulations in this regime is still lacking and could lead to an interesting and original insight into EST behaviour.

- **Interplay between the EST and the dual superconductor model of confinement.**
  In this review we introduced the EST, following the seminal papers of Lüscher and collaborators, as a tool to describe the behaviour of Wilson loops in LGT beyond the roughening transition. There is, however, a different, interesting, route which may lead to an effective string description of confinement which was proposed long time ago by Nielsen and Olesen [148], 't Hooft [149], Mandelstam [150] and Polyakov [151].

The proposal relies on the description of the QCD vacuum as a coherent state of color magnetic monopoles or, equivalently, as a magnetic (dual) superconductor (for a review see for instance [152–154]). According to this picture the (dual) Meissner effect naturally leads to vortex like structures: The Abrikosov vortices [155] which are very similar to the confining color flux tubes which are described by the EST. A very interesting laboratory to address this picture is the 3D $U(1)$ LGT for which it can be shown, using a duality transformation, that confinement is indeed due to the condensation of monopoles [156]. The remarkable success of this approach led to conjecture that a similar mechanism could drive confinement also in non-Abelian Yang–Mills theories [105–111].

The implicit assumption behind this scenario is that there should exist a duality transformation mapping gauge fields into strings. In the non-Abelian case, such gauge/string duality transformation is in general unknown (a notable exception, however, is given by the holographic correspondence, relating gauge theories and string theories defined in a higher-dimensional spacetime [157–159]), but in the 3D $U(1)$ case Polyakov [160] (see also [153,154,161] for an alternative derivation) was able to give a heuristic proof of this mapping and proposed to describe the free energy of a large Wilson loop with a string action combining both the Nambu–Goto and the extrinsic curvature terms, the so called "rigid string" [79,80].

It is by now clear that this approach leads to an EST *different* from the one discussed in this review [8]. The "rigid string", dominated by the extrinsic curvature term, agrees with the expectation of the dual superconductor model while the one which we discussed in this review has a negligible extrisic curvature term and is dominated by the Nambu–Goto behaviour. The major difference between the two ESTs is in the shape and width of the flux tube [9]. Interestingly this difference is magnified exactly in the high temperature regime [9,64] which is the subject of this review. It would be interesting to pursue this study to better understand the role of the extrinsic curvature term in driving this difference and, more importantly, which one better describes the behaviour of the flux tube in non-abelian LGTs.

As a final remark on this issue, let us stress that the rigid string shows a pretty different behaviour depending on the sign of the extrinsic curvature term. An EST with negative extrinsic curvature was proposed more than twenty years ago in [162–164] and was subsequently thoroughly studied in [153,154,165–168]. Despite the apparent instability due to the negative sign of the curvature term, it can be shown that the string is stabilized by higher order terms in the derivative expansion [165] (for a review, see for instance [153]). In particular, as far as the topic of this review is concerned, the high temperature behaviour of the model was studied in detail in [166,167] and, also in this case, it would be very interesting to test these prediction with high precision Monte Carlo data for non-abelian LGTs.

In the last few years we have witnessed remarkable progress in our understanding of EST; however several important issues are still open, from the identification of EST terms beyond the Nambu–Goto one, to a better understanding of the role and properties of the rigidity term. The main goal of this review was to show that the high-$T$ regime of LGTs is a perfect laboratory to test new ideas in this context and compare them with Monte Carlo simulations. We hope that this review will stimulate further research in this direction.

**Funding:** This research received no external funding.

**Acknowledgments:** We thank D. Antonov and F. Caristo for a careful reading of the draft and for many useful suggestions. We warmly thank M. Billo', F. Gliozzi, M. Hasenbusch, A. Nada, M. Panero and D. Vadacchino for a longlasting fruitful collaboration on the topics discussed in this review.

**Conflicts of Interest:** The authors declare no conflict of interest.

## Appendix A. Useful Formulae

Here are some properties of the modular functions which appear in the text. To simplify notations we shall denote $\tilde{\tau} \equiv -\frac{1}{\tau}$ in the following

The relation with the variables used in the text is:

$$\tau \equiv i\frac{N_t}{2R}\,, \qquad\qquad q \equiv e^{2\pi i\tau} = e^{-\frac{\pi N_t}{R}}\,,$$

$$\tilde{\tau} \equiv -\frac{1}{\tau} = i\frac{2R}{N_t}\,, \qquad\qquad \tilde{q} \equiv e^{2\pi i\tilde{\tau}} = e^{-\frac{4\pi R}{N_t}}\,. \tag{A1}$$

The Dedekind-$\eta$-function is

$$\eta(q) \equiv q^{\frac{1}{24}} \prod_{n=1}^{\infty}(1 - q^n)\,. \tag{A2}$$

The Eisenstein functions are defined as:

$$E_{2k}(q) \equiv 1 + \frac{2}{\zeta(1-2k)} \sum_{n=1}^{\infty} \frac{n^{2k-1}q^n}{1-q^n}\,, \tag{A3}$$

where $\zeta(s)$ denotes the Riemann $\zeta$ function defined as follows:

$$\zeta(s) \equiv \sum_{n=1}^{\infty} n^{-s}\,, \tag{A4}$$

The Eisenstein functions can be expanded as follows:

$$E_2(q) = 1 - 24q - 3\cdot 24q^2 - 4\cdot 24q^3 - 7\cdot 24q^4 - \cdots$$
$$E_4(q) = 1 + 10\cdot 24q + 90\cdot 24q^2 + \cdots \tag{A5}$$

These functions transform as follows under the modular transformation $\tau \to -\frac{1}{\tau}$ (notice the inhomogeneous term in the $E_2$ function):

$$\eta(q) = (-i\tilde{\tau})^{1/2}\eta(\tilde{q}) = \left(\frac{2R}{N_t}\right)^{\frac{1}{2}}\eta(\tilde{q})\,,$$

$$E_2(q) = -\frac{6i}{\pi}\tilde{\tau} + \tilde{\tau}^2 E_2(\tilde{q}) = \frac{12R}{\pi N_t} - \left(\frac{2r}{l}\right)^2 E_2(\tilde{q}) = \frac{12R}{\pi N_t}\left(1 - \frac{\pi R}{3N_t}E_2(\tilde{q})\right)\,,$$

$$E_4(q) = \tilde{\tau}^4 E_4(\tilde{q}) = \left(\frac{2R}{N_t}\right)^4 E_4(\tilde{q})\,. \tag{A6}$$

## References

1. Nambu, Y. Strings, Monopoles and Gauge Fields. *Phys. Rev.* **1974**, *D10*, 4262. [CrossRef]
2. Goto, T. Relativistic quantum mechanics of one-dimensional mechanical continuum and subsidiary condition of dual resonance model. *Prog. Theor. Phys.* **1971**, *46*, 1560. [CrossRef]
3. Luscher, M. Symmetry Breaking Aspects of the Roughening Transition in Gauge Theories. *Nucl. Phys.* **1981**, *B180*, 317. [CrossRef]
4. Luscher, M.; Symanzik, K.; Weisz, P. Anomalies of the Free Loop Wave Equation in the WKB Approximation. *Nucl. Phys.* **1980**, *B173*, 365. [CrossRef]
5. Polchinski, J.; Strominger, A. Effective string theory. *Phys. Rev. Lett.* **1991**, *67*, 1681. [CrossRef]
6. Isgur, N.; Paton, J.E. A Flux Tube Model for Hadrons in QCD. *Phys. Rev. D* **1985**, *31*, 2910. [CrossRef]
7. Olesen, P. Strings, Tachyons and Deconfinement. *Phys. Lett.* **1985**, *B160*, 408. [CrossRef]
8. Caselle, M.; Panero, M.; Pellegrini, R.; Vadacchino, D. A different kind of string. *J. High Energy Phys.* **2015**, *1501*, 105. [CrossRef]
9. Caselle, M.; Panero, M.; Vadacchino, D. Width of the flux tube in compact U(1) gauge theory in three dimensions. *J. High Energy Phys.* **2016**, *1602*, 180. [CrossRef]
10. Bali, G.S. QCD forces and heavy quark bound states. *Phys. Rept.* **2001**, *343*, 1. [CrossRef]
11. Aharony, O.; Komargodski, Z. The Effective Theory of Long Strings. *J. High Energy Phys.* **2013**, *1305*, 118. [CrossRef]

12. Brandt, B.B.; Meineri, M. Effective string description of confining flux tubes. *Int. J. Mod. Phys.* **2016**, *A31*, 1643001. [CrossRef]

13. Svetitsky, B.; Yaffe, L.G. Critical Behavior at Finite Temperature Confinement Transitions. *Nucl. Phys.* **1982**, *B210*, 423. [CrossRef]

14. Montvay, I.; Munster, G. *Quantum Fields on a Lattice, Cambridge Monographs on Mathematical Physics*; Cambridge University Press: Cambridge, UK, 1997; Volume 3.

15. Kaczmarek, O.; Karsch, F.; Laermann, E.; Lutgemeier, M. Heavy quark potentials in quenched QCD at high temperature. *Phys. Rev. D* **2000**, *62*, 034021. [CrossRef]

16. Cardoso, N.; Bicudo, P. Lattice QCD computation of the SU(3) String Tension critical curve. *Phys. Rev. D* **2012**, *85*, 077501. [CrossRef]

17. Hasenfratz, A.; Hasenfratz, E.; Hasenfratz, P. Generalized Roughening Transition and Its Effect on the String Tension. *Nucl. Phys. B* **1981**, *180*, 353. [CrossRef]

18. Itzykson, C.; Peskin, M.E.; Zuber, J.-B. Roughening of Wilson's Surface. *Phys. Lett. B* **1980**, *95*, 259. [CrossRef]

19. Luscher, M.; Munster, G.; Weisz, P. How Thick Are Chromoelectric Flux Tubes? *Nucl. Phys.* **1981**, *B180*, 1. [CrossRef]

20. Dietz, K.; Filk, T. On the renormalization of string functionals. *Phys. Rev.* **1983**, *D27*, 2944. [CrossRef]

21. Alvarez, O. The Static Potential in String Models. *Phys. Rev. D* **1981**, *24*, 440. [CrossRef]

22. Arvis, J. The Exact $q\bar{q}$ Potential in Nambu String Theory. *Phys. Lett.* **1983**, *B127*, 106. [CrossRef]

23. Hasenbusch, M.; Pinn, K. Surface tension, surface stiffness, and surface width of the three-dimensional Ising model on a cubic lattice. *Physica A* **1993**, *192*, 342. [CrossRef]

24. Caselle, M.; Fiore, R.; Gliozzi, F.; Hasenbusch, M.; Provero, P. String effects in the Wilson loop: A High precision numerical test. *Nucl. Phys.* **1997**, *B486*, 245. [CrossRef]

25. Ambjorn, J.; Olesen, P.; Peterson, C. Observation of a String in Three-dimensional SU(2) Lattice Gauge Theory. *Phys. Lett. B* **1984**, *142*, 410. [CrossRef]

26. Ambjorn, J.; Olesen, P.; Peterson, C. Three-dimensional Lattice Gauge Theory and Strings. *Nucl. Phys. B* **1984**, *244*, 262. [CrossRef]

27. Necco, S.; Sommer, R. The N(f) = 0 heavy quark potential from short to intermediate distances. *Nucl. Phys.* **2002**, *B622*, 328. [CrossRef]

28. Juge, K.J.; Kuti, J.; Morningstar, C. Fine structure of the QCD string spectrum. *Phys. Rev. Lett.* **2003**, *90*, 161601. [CrossRef] [PubMed]

29. Lucini, B.; Teper, M. Confining strings in SU(N) gauge theories. *Phys. Rev.* **2001**, *D64*, 105019.

30. Luscher, M.; Weisz, P. Quark confinement and the bosonic string. *J. High Energy Phys.* **2002**, *0207*, 049. [CrossRef]

31. Caselle, M.; Fiore, R.; Gliozzi, F.; Hasenbusch, M.; Pinn, K. Rough interfaces beyond the Gaussian approximation. *Nucl. Phys.* **1994**, *B432*, 590. [CrossRef]

32. Lucini, B.; Teper, M. SU(N) gauge theories in (2 + 1)-dimensions: Further results. *Phys. Rev. D* **2002**, *66*, 097502. [CrossRef]

33. Caselle, M.; Pepe, M.; Rago, A. Static quark potential and effective string corrections in the (2 + 1)-d SU(2) Yang–Mills theory. *J. High Energy Phys.* **2004**, *10*, 005. [CrossRef]

34. Caselle, M.; Hasenbusch, M.; Panero, M. Comparing the Nambu–Goto string with LGT results. *J. High Energy Phys.* **2005**, *3*, 026. [CrossRef]

35. Bringoltz, B.; Teper, M. A Precise calculation of the fundamental string tension in SU(N) gauge theories in 2 + 1 dimensions. *Phys. Lett.* **2007**, *B645*, 383. [CrossRef]

36. Caselle, M.; Hasenbusch, M.; Panero, M. High precision Monte Carlo simulations of interfaces in the three-dimensional ising model: A Comparison with the Nambu–Goto effective string model. *J. High Energy Phys.* **2006**, *0603*, 084. [CrossRef]

37. Dass, N.D.H.; Majumdar, P. String-like behaviour of 4-D SU(3) Yang–Mills flux tubes. *J. High Energy Phys.* **2006**, *10*, 020. [CrossRef]

38. Bringoltz, B.; Teper, M. Closed k-strings in SU(N) gauge theories : 2 + 1 dimensions. *Phys. Lett. B* **2008**, *663*, 429. [CrossRef]

39. Athenodorou, A.; Bringoltz, B.; Teper, M. Closed flux tubes and their string description in D = 2 + 1 SU(N) gauge theories. *J. High Energy Phys.* **2011**, *1105*, 042. [CrossRef]

40. Athenodorou, A.; Bringoltz, B.; Teper, M. Closed flux tubes and their string description in D = 3 + 1 SU(N) gauge theories. *J. High Energy Phys.* **2011**, *1102*, 030. [CrossRef]

41. Caselle, M.; Zago, M. A new approach to the study of effective string corrections in LGTs. *Eur. Phys. J.* **2011**, *C71*, 1658. [CrossRef]

42. Billo, M.; Caselle, M.; Pellegrini, R. New numerical results and novel effective string predictions for Wilson loops. *J. High Energy Phys.* **2012**, *1*, 104. [CrossRef]

43. Mykkanen, A. The static quark potential from a multilevel algorithm for the improved gauge action. *J. High Energy Phys.* **2012**, *12*, 069. [CrossRef]

44. Athenodorou, A.; Teper, M. Closed flux tubes in higher representations and their string description in D = 2 + 1 SU(N) gauge theories. *J. High Energy Phys.* **2013**, *1306*, 053. [CrossRef]

45. Caselle, M.; Costagliola, G.; Nada, A.; Panero, M.; Toniato, A. Jarzynski's theorem for lattice gauge theory. *Phys. Rev.* **2016**, *D94*, 034503. [CrossRef]

46. Athenodorou, A.; Teper, M. Closed flux tubes in D = 2 + 1 SU(N ) gauge theories: Dynamics and effective string description. *J. High Energy Phys.* **2016**, *10*, 093. [CrossRef]

47. Dubovsky, S.; Flauger, R.; Gorbenko, V. Effective String Theory Revisited. *J. High Energy Phys.* **2012**, *1209*, 044. [CrossRef]

48. Caselle, M.; Fioravanti, D.; Gliozzi, F.; Tateo, R. Quantisation of the effective string with TBA. *J. High Energy Phys.* **2013**, *7*, 071. [CrossRef]

49. Aharony, O.; Datta, S.; Giveon, A.; Jiang, Y.; Kutasov, D. Modular invariance and uniqueness of $T\bar{T}$ deformed CFT. *J. High Energy Phys.* **2019**, *1*, 086. [CrossRef]
50. Datta, S.; Jiang, Y. $T\bar{T}$ deformed partition functions. *J. High Energy Phys.* **2018**, *8*, 106. [CrossRef]
51. Luscher, M.; Weisz, P. String excitation energies in SU(N) gauge theories beyond the free-string approximation. *J. High Energy Phys.* **2004**, *7*, 014. [CrossRef]
52. Billo, M.; Caselle, M. Polyakov loop correlators from D0-brane interactions in bosonic string theory. *J. High Energy Phys.* **2005**, *507*, 038. [CrossRef]
53. Billo, M.; Caselle, M.; Ferro, L. The Partition function of interfaces from the Nambu–Goto effective string theory. *J. High Energy Phys.* **2006**, *0602*, 070. [CrossRef]
54. Olesen, P. On the Exponentially Increasing Level Density in String Models and the Tachyon Singularity. *Nucl. Phys. B* **1986**, *267*, 539–556. [CrossRef]
55. Pisarski, R.D.; Alvarez, O. Strings at Finite Temperature and Deconfinement. *Phys. Rev. D* **1982**, *26*, 3735. [CrossRef]
56. Komargodski, Z.; Simmons-Duffin, D.The Random-Bond Ising Model in 2.01 and 3 Dimensions. *J. Phys.* **2017**, *A50*, 154001. [CrossRef]
57. Kos, F.; Poland, D.; Simmons-Duffin, D.; Vichi, A. Precision Islands in the Ising and $O(N)$ Models. *J. High Energy Phys.* **2016**, *8*, 036. [CrossRef]
58. Lucini, B.; Teper, M.; Wenger, U. The High temperature phase transition in SU(N) gauge theories. *J. High Energy Phys.* **2004**, *401*, 061. [CrossRef]
59. Lucini, B.; Teper, M.; Wenger, U. The Deconfinement transition in SU(N) gauge theories. *Phys. Lett. B* **2002**, *545*, 197. [CrossRef]
60. Lucini, B.; Teper, M.; Wenger, U. Properties of the deconfining phase transition in SU(N) gauge theories. *J. High Energy Phys.* **2005**, *2*, 033. [CrossRef]
61. Liddle, J.; Tepe, M. The Deconfining phase transition in D = 2 + 1 SU(N) gauge theories. *arXiv* **2008**, arXiv:0803.2128.
62. Lau, R.; Teper, M. The deconfining phase transition of SO(N) gauge theories in 2 + 1 dimensions. *J. High Energy Phys.* **2016**, *3*, 072. [CrossRef]
63. Borisenko, O.; Chelnokov, V.; Gravina, M.; Papa, A. Deconfinement and universality in the 3D U(1) lattice gauge theory at finite temperature: Study in the dual formulation. *J. High Energy Phys.* **2015**, *9*, 062. [CrossRef]
64. Caselle, M.; Nada, A.; Panero, M.; Vadacchino, D. Conformal field theory and the hot phase of three-dimensional U(1) gauge theory. *J. High Energy Phys.* **2019**, *5*, 068. [CrossRef]
65. Aharony, O.; Karzbrun, E. On the effective action of confining strings. *J. High Energy Phys.* **2009**, *6*, 012. [CrossRef]
66. Meyer, H.B. Poincare invariance in effective string theories. *J. High Energy Phys.* **2006**, *5*, 066. [CrossRef]
67. Aharony, O.; Dodelson, M. Effective String Theory and Nonlinear Lorentz Invariance. *J. High Energy Phys.* **2012**, *1202*, 008. [CrossRef]
68. Gliozzi, F. Dirac-Born-Infeld action from spontaneous breakdown of Lorentz symmetry in brane-world scenarios. *Phys. Rev. D* **2011**, *84*, 027702. [CrossRef]
69. Gliozzi, F.; Meineri, M. Lorentz completion of effective string (and p-brane) action. *J. High Energy Phys.* **2012**, *1208*, 056. [CrossRef]
70. Meineri, M. Lorentz completion of effective string action. *arXiv* **2013**, arXiv:1301.3437.
71. Drummond, J.M. Universal subleading spectrum of effective string theory. *arXiv* **2004**, arXiv:0411017.
72. Dass, N.D.H.; Matlock, P. Universality of correction to Luescher term in Polchinski-Strominger effective string theories. *arXiv* **2006**, arXiv:0606265.
73. Drummond, J.M. Reply to hep-th/0606265. *arXiv* **2006**, arXiv:0608109.
74. Dass, N.D.H.; Matlock, P. Our response to the response hep-th/0608109 by Drummond. *arXiv* **2006**, arXiv:0611215.
75. Dass, N.D.H.; Matlock, P. Covariant Calculus for Effective String Theories. *Indian J. Phys.* **2014**, *88*, 965. [CrossRef]
76. Dass, N.D.H.; Matlock, P.; Bharadwa, Y. Spectrum to all orders of Polchinski-Strominger Effective String Theory of Polyakov-Liouville Type. *arXiv* **2009**, arXiv:0910.5615.
77. Dass, N.D.H. All Conformal Effective String Theories are Isospectral to Nambu–Goto Theory. *arXiv* **2009**, arXiv:0911.3236.
78. Aharony, O.; Field, M. On the effective theory of long open strings. *J. High Energy Phys.* **2011**, *1101*, 065. [CrossRef]
79. Polyakov, A.M. Fine Structure of Strings. *Nucl. Phys. B* **1986**, *268*, 406. [CrossRef]
80. Kleinert, H. The Membrane Properties of Condensing Strings. *Phys. Lett. B* **1986**, *174*, 335. [CrossRef]
81. Miró, J.E.; Guerrieri, A.L.; Hebbar, A.; Penedones, J.a.; Vieira, P. Flux Tube S-matrix Bootstrap. *Phys. Rev. Lett.* **2019**, *123*, 221602. [CrossRef]
82. Billo, M.; Caselle, M.; Gliozzi, F.; Meineri, M.; Pellegrini, R. The Lorentz-invariant boundary action of the confining string and its universal contribution to the inter-quark potential. *J. High Energy Phys.* **2012**, *5*, 130. [CrossRef]
83. Brand, B.B. Probing boundary-corrections to Nambu–Goto open string energy levels in 3D SU(2) gauge theory. *J. High Energy Phys.* **2011**, *2*, 040. [CrossRef]
84. Brandt, B.B. Spectrum of the open QCD flux tube and its effective string description I: 3D static potential in SU(N = 2, 3). *J. High Energy Phys.* **2017**, *7*, 008. [CrossRef]
85. Brandt, B.B. Spectrum of the open QCD flux tube and its effective string description. *arXiv* **2018**, arXiv:1811.11779.
86. Brandt, B.B. Revisiting the flux tube spectrum of 3D SU(2) lattice gauge theory. *arXiv* **2021**, arXiv:2102.06413.

87. Bakry, A.S.; Deliyergiyev, M.A.; Galal, A.A.; Khalaf, A.M.; William, M.K. Quantum delocalization of strings with boundary action in Yang–Mills theory. *arXiv* **2020**, arXiv:2001.02392.

88. Bakry, A.S.; Deliyergiyev, M.A.; Galal, A.A.; Williams, M.K. Boundary action and profile of effective bosonic strings. *arXiv* **2019**, arXiv:1912.13381.

89. Caselle, M.; Feo, A.; Panero, M.; Pellegrini, R. Universal signatures of the effective string in finite temperature lattice gauge theories. *J. High Energy Phys.* **2011**, *1104*, 020. [CrossRef]

90. Athenodorou, A.; Bringoltz, B.; Teper, M. The closed string spectrum of SU(N) gauge theories in 2+1 dimensions. *Phys. Lett.* **2007**, *B656*, 132. [CrossRef]

91. Caselle, M.; Hasenbusch, M.; Panero, M. String effects in the 3-d gauge Ising model. *J. High Energy Phys.* **2003**, *301*, 057. [CrossRef]

92. Bonati, C. Finite temperature effective string corrections in (3+1)D SU(2) lattice gauge theory. *Phys. Lett. B* **2011**, *703*, 376. [CrossRef]

93. Bakry, A.S.; Deliyergiyev, M.A.; Galal, A.A.; Khalil, M.N. On QCD strings beyond non-interacting model. *arXiv* **2020**, arXiv:2001.04203.

94. Caselle, M.; Gliozzi, F.; Magnea, U.; Vinti, S. Width of long color flux tubes in lattice gauge systems. *Nucl. Phys.* **1996**, *B460*, 397. [CrossRef]

95. Zach, M.; Faber, M.; Skala, P. Investigating confinement in dually transformed U(1) lattice gauge theory. *Phys. Rev.* **1998**, *D57*, 123. [CrossRef]

96. Koma, Y.; Koma, M.; Majumdar, P. Static potential, force, and flux tube profile in 4-D compact U(1) lattice gauge theory with the multilevel algorithm. *Nucl. Phys.* **2004**, *B692*, 209. [CrossRef]

97. Panero, M. A Numerical study of confinement in compact QED. *J. High Energy Phys.* **2005**, *5*, 066. [CrossRef]

98. Giudice, P.; Gliozzi, F.; Lottini, S. Quantum broadening of k-strings in gauge theories. *J. High Energy Phys.* **2007**, *1*, 084. [CrossRef]

99. Amado, A.; Cardoso, N.; Bicudo, P. Flux tube widening in compact U (1) lattice gauge theory computed at $T < T_c$ with the multilevel method and GPUs. *arXiv* **2013**, arXiv:1309.3859.

100. Amado, A.; Cardoso, N.; Cardoso, M.; Bicudo, P. Study of compact U(1) flux tubes in 3 + 1 dimensions in lattice gauge theory using GPU's. *Acta Phys. Polon. Supp.* **2012**, *5*, 1129. [CrossRef]

101. Gliozzi, F.; Pepe, M.; Wiese, U.J. The Width of the Confining String in Yang–Mills Theory. *Phys. Rev. Lett.* **2010**, *104*, 232001. [CrossRef]

102. Bakry, A.S.; Leinweber, D.B.; Moran, P.J.; Sternbeck, A.; Williams, A.G. String effects and the distribution of the glue in mesons at finite temperature. *Phys. Rev. D* **2010**, *82*, 094503. [CrossRef]

103. Cardoso, N.; Cardoso, M.; Bicudo, P. Inside the SU(3) quark-antiquark QCD flux tube: Screening versus quantum widening. *Phys. Rev.* **2013**, *D88*, 054504. [CrossRef]

104. Bicudo, P.; Cardoso, N.; Cardoso, M. Pure gauge QCD flux tubes and their widths at finite temperature. *Nucl. Phys. B* **2019**, *940*, 88. [CrossRef]

105. Cardaci, M.S.; Cea, P.; Cosmai, L.; Falcone, R.; Papa, A. Chromoelectric flux tubes in QCD. *Phys. Rev.* **2011**, *D83*, 014502. [CrossRef]

106. Cea, P.; Cosmai, L.; Papa, A. Chromoelectric flux tubes and coherence length in QCD. *Phys. Rev.* **2012**, *D86*, 054501. [CrossRef]

107. Cea, P.; Cosmai, L.; Cuteri, F.; Papa, A. Flux tubes in the SU(3) vacuum: London penetration depth and coherence length. *Phys. Rev.* **2014**, *D89*, 094505. [CrossRef]

108. Cea, P.; Cosmai, L.; Cuteri, F.; Papa, A. Flux tubes at finite temperature. *J. High Energy Phys.* **2016**, *6*, 033. [CrossRef]

109. Cea, P.; Cosmai, L.; Cuteri, F.; Papa, A. Flux tubes in the QCD vacuum. *Phys. Rev. D* **2017**, *95*, 114511. [CrossRef]

110. Baker, M.; Cea, P.; Chelnokov, V.; Cosmai, L.; Cuteri, F.; Papa, A. Isolating the confining color field in the SU(3) flux tube. *Eur. Phys. J. C* **2019**, *79*, 478. [CrossRef]

111. Baker, M.; Cea, P.; Chelnokov, V.; Cosmai, L.; Cuteri, F.; Papa, A. The confining color field in SU(3) gauge theory. *Eur. Phys. J. C* **2020**, *80*, 514. [CrossRef]

112. Allais, A.; Caselle, M. On the linear increase of the flux tube thickness near the deconfinement transition. *J. High Energy Phys.* **2009**, *901*, 073. [CrossRef]

113. Gliozzi, F.; Pepe, M.; Wiese, U.J. The Width of the Color Flux Tube at 2-Loop Order. *J. High Energy Phys.* **2010**, *11*, 053. [CrossRef]

114. Gliozzi, F.; Pepe, M.; Wiese, U.J. Linear Broadening of the Confining String in Yang–Mills Theory at Low Temperature. *J. High Energy Phys.* **2011**, *1*, 057. [CrossRef]

115. Caselle, M.; Grinza, P.; Magnoli, N. Study of the flux tube thickness in 3-D LGT's by means of 2-D spin models. *J. Stat. Mech.* **2006**, *0611*, P11003. [CrossRef]

116. Caselle, M. Flux tube delocalization at the deconfinement point. *J. High Energy Phys.* **2010**, *8*, 063. [CrossRef]

117. Caselle, M.; Grinza, P. On the intrinsic width of the chromoelectric flux tube in finite temperature LGTs. *J. High Energy Phys.* **2012**, *11*, 174. [CrossRef]

118. Yurov, V.P.; Zamolodchikov, A.B. Correlation functions of integrable 2-D models of relativistic field theory. Ising model. *Int. J. Mod. Phys. A* **1991**, *6*, 3419. [CrossRef]

119. Hagedorn, R. Statistical thermodynamics of strong interactions at high-energies. *Nuovo Cim. Suppl.* **1965**, *3*, 147.

120. Meyer, H.B. High-Precision Thermodynamics and Hagedorn Density of States. *Phys. Rev. D* **2009**, *80*, 051502. [CrossRef]

121. Caselle, M.; Castagnini, L.; Feo, A.; Gliozzi, F.; Panero, M. Thermodynamics of SU(N) Yang–Mills theories in 2 + 1 dimensions I—The confining phase. *J. High Energy Phys.* **2011**, *6*, 142. [CrossRef]

122. Caselle, M.; Nada, A.; Panero, M. Hagedorn spectrum and thermodynamics of SU(2) and SU(3) Yang–Mills theories. *J. High Energy Phys.* **2015**, *7*, 143. [CrossRef]
123. Alba, P.; Alberico, W.M.; Nada, A.; Panero, M.; Stöcker, H. Excluded-volume effects for a hadron gas in Yang–Mills theory. *Phys. Rev. D* **2017**, *95*, 094511. [CrossRef]
124. Karkkainen, L.; Lacock, P.; Miller, D.E.; Petersson, B.; Reisz, T. Space-like Wilson loops at finite temperature. *Phys. Lett.* **1993**, *B312*, 173. [CrossRef]
125. Bali, G.S.; Fingberg, J.; Heller, U.M.; Karsch, F.; Schilling, K. The Spatial string tension in the deconfined phase of the (3 + 1)-dimensional SU(2) gauge theory. *Phys. Rev. Lett.* **1993**, *71*, 3059. [CrossRef]
126. Karsch, F.; Laermann, E.; Lutgemeier, M. Three-dimensional SU(3) gauge theory and the spatial string tension of the (3 + 1)-dimensional finite temperature SU(3) gauge theory. *Phys. Lett.* **1995**, *B346*, 94. [CrossRef]
127. Caselle, M.; Fiore, R.; Gliozzi, F.; Guaita, P.; Vinti, S. On the behavior of spatial Wilson loops in the high temperature phase of LGT. *Nucl. Phys.* **1994**, *B422*, 397. [CrossRef]
128. Koch, V. On the temperature dependence of correlation functions in the space-like direction in hot QCD. *Phys. Rev.* **1994**, *D49*, 6063.
129. Ejiri, S. Monopoles and spatial string tension in the high temperature phase of SU(2) QCD. *Phys. Lett.* **1996**, *B376*, 163. [CrossRef]
130. Sekiguchi, T.; Ishigur, K. Abelian spatial string tension in finite temperature SU(2) gauge theory. *Int. J. Mod. Phys.* **2016**, *A31*, 1650149. [CrossRef]
131. Schroder, Y.; Laine, M. Spatial string tension revisited. *PoS* **2006**, *LAT2005*, 180.
132. RBC-Bielefeld Collaboration. The Spatial string tension and dimensional reduction in QCD. *PoS* **2007**, *LATTICE2007*, 204.
133. Cheng, S.M.; Datta, J.; van der Heide, K.; Huebner, F.; Karsch, O.; Kaczmarek, E.; Laermann, J.; Liddle, R.; Mawhinney, D.; Miao, C.; et al. The Spatial String Tension and Dimensional Reduction in QCD. *Phys. Rev.* **2008**, *D78*, 034506. [CrossRef]
134. WHOT-QCD Collaboration. Heavy-quark free energy, debye mass, and spatial string tension at finite temperature in two flavor lattice QCD with Wilson quark action. *Phys. Rev.* **2007**, *D75*, 074501.
135. Alanen, J.; Kajantie, K.; Suur-Uski, V. Spatial string tension of finite temperature QCD matter in gauge/gravity duality. *Phys. Rev.* **2009**, *D80*, 075017. [CrossRef]
136. Andreev, O.; Zakharov, V.I. The Spatial String Tension, Thermal Phase Transition, and AdS/QCD. *Phys. Lett.* **2007**, *B645*, 437. [CrossRef]
137. Andreev, O. The Spatial String Tension in the Deconfined Phase of SU(N) Gauge Theory and Gauge/String Duality. *Phys. Lett.* **2008**, *B659*, 416. [CrossRef]
138. Meyer, H.B. Vortices on the worldsheet of the QCD string. *Nucl. Phys.* **2005**, *B724*, 432. [CrossRef]
139. Caselle, M.; Gliozzi, F.; Vinti, S. On the relation between the width of the flux tube and T(c)**1 in lattice gauge theories. *Nucl. Phys. Proc. Suppl.* **1994**, *34*, 263. [CrossRef]
140. Beratto, E.; Billò, M.; Caselle, M. $T\bar{T}$ deformation of the compactified boson and its interpretation in lattice gauge theory. *Phys. Rev. D* **2020**, *102*, 014504. [CrossRef]
141. Munster, G. Interface Tension in Three-dimensional Systems From Field Theory. *Nucl. Phys.* **1990**, *B340*, 559. [CrossRef]
142. Caselle, M.; Gliozzi, F.; Vinti, S. Finite size effects in the interface of 3-D Ising model. *Phys. Lett.* **1993**, *B302*, 74. [CrossRef]
143. Klessinger, S.; Munster, G. Numerical investigation of the interface tension in the three-dimensional Ising model. *Nucl. Phys.* **1992**, *B386*, 701. [CrossRef]
144. Hoppe, P.; Munster, G. The Interface tension of the three-dimensional Ising model in two loop order. *Phys. Lett.* **1998**, *A238*, 265. [CrossRef]
145. Muller, M.; Munster, G. Profile and width of rough interfaces. *J. Statist. Phys.* **2005**, *118*, 669. [CrossRef]
146. Caselle, M.; Hasenbusch, M.; Panero, M. The Interface free energy: Comparison of accurate Monte Carlo results for the 3D Ising model with effective interface models. *J. High Energy Phys.* **2007**, *709*, 117. [CrossRef]
147. Billo, M.; Caselle, M.; Ferro, L. Universal behaviour of interfaces in 2d and dimensional reduction of Nambu–Goto strings. *Nucl. Phys.* **2008**, *B795*, 623. [CrossRef]
148. Nielsen, H.B.; Olesen, P. Vortex Line Models for Dual Strings. *Nucl. Phys.* **1973**, *B61*, 45. [CrossRef]
149. Hooft, G. Magnetic Monopoles in Unified Gauge Theories. *Nucl. Phys. B* **1974**, *79*, 276. [CrossRef]
150. Mandelstam, S. Vortices and Quark Confinement in Nonabelian Gauge Theories. *Phys.Rept.* **1976**, *23*, 245. [CrossRef]
151. Polyakov, A.M. Particle Spectrum in the Quantum Field Theory. *J. Exp. Theor. Phys. Lett.* **1974**, *20*, 194.
152. Ripka, G. Dual superconductor models of color confinement. *arXiv* **2004**, arXiv:0310102.
153. Antonov, D.; Diamantini, M.C. 3D Georgi-Glashow model and confining strings at zero and finite temperatures. In From Fields to Strings: Circumnavigating Theoretical Physics: A Conference in Tribute to Ian Kogan. *arXiv* **2004**, arXiv:0406272.
154. Antonov, D. Monopole-Based Scenarios of Confinement and Deconfinement in 3D and 4D. *Universe* **2017**, *3*, 50. [CrossRef]
155. Abrikosov, A.A. On the Magnetic properties of superconductors of the second group. *Sov. Phys. J. Exp. Theor. Phys.* **1957**, *5*, 1174.
156. Polyakov, A.M. Quark Confinement and Topology of Gauge Groups. *Nucl. Phys.* **1977**, *B120*, 429. [CrossRef]
157. Maldacena, J.M. The large N limit of superconformal field theories and supergravity. *Adv. Theor. Math. Phys.* **1998**, *2*, 231. [CrossRef]
158. Gubser, S.; Klebanov, I.R.; Polyakov, A.M. Gauge theory correlators from noncritical string theory. *Phys. Lett.* **1998**, *B428*, 105. [CrossRef]

159. Witten, E. Anti-de Sitter space and holography. *Adv. Theor. Math. Phys.* **1998**, *2*, 253. [CrossRef]
160. Polyakov, A.M. Confining strings. *Nucl. Phys. B* **1997**, *486*, 23. [CrossRef]
161. Antonov, D. Various properties of compact QED and confining strings. *Phys. Lett.* **1998**, *B428*, 346. [CrossRef]
162. Orland, P. Extrinsic curvature dependence of Nielsen-Olesen strings. *Nucl. Phys. B* **1994**, *428*, 221. [CrossRef]
163. Sato, M.; Yahikozawa, S. 'Topological' formulation of effective vortex strings. *Nucl. Phys. B* **1995**, *436*, 100. [CrossRef]
164. Kleinert, H.; Chervyakov, A.M. Evidence for negative stiffness of QCD flux tubes in the large-N limit of SU(N). *Phys. Lett. B* **1996**, *381*, 286. [CrossRef]
165. Diamantini, M.C.; Kleinert, H.; Trugenberger, C.A. Strings with negative stiffness and hyperfine structure. *Phys. Rev. Lett.* **1999**, *82*, 267. [CrossRef]
166. Diamantini, M.C.; Trugenberger, C.A. QCD like behaviour of high temperature confining strings. *Phys. Rev. Lett.* **2002**, *88*, 251601. [CrossRef] [PubMed]
167. Diamantini, M.C.; Trugenberger, C.A. Confining strings at high temperature. *J. High Energy Phys.* **2002**, *4*, 032. [CrossRef]
168. Hidaka, Y.; Pisarski, R.D. Zero Point Energy of Renormalized Wilson Loops. *Phys. Rev. D* **2009**, *80*, 074504. [CrossRef]

*Review*

# Superinsulators: An Emergent Realisation of Confinement

**Maria Cristina Diamantini** [1,*,†] **and Carlo A. Trugenberger** [2,†]

1    NiPS Laboratory, INFN and Dipartimento di Fisica e Geologia, University of Perugia, Via A. Pascoli, I-06100 Perugia, Italy
2    SwissScientific Technologies SA, Rue du Rhone 59, CH-1204 Geneva, Switzerland; ca.trugenberger@bluewin.ch
*    Correspondence: cristina.diamantini@pg.infn.it
†    These authors contributed equally to this work.

**Abstract:** Superinsulators (SI) are a new topological state of matter, predicted by our collaboration and experimentally observed in the critical vicinity of the superconductor-insulator transition (SIT). SI are dual to superconductors and realise electric-magnetic (S)-duality. The effective field theory that describes this topological phase of matter is governed by a compact Chern-Simons in (2+1) dimensions and a compact BF term in (3+1) dimensions. While in a superconductor the condensate of Cooper pairs generates the Meissner effect, which constricts the magnetic field lines penetrating a type II superconductor into Abrikosov vortices, in superinsulators Cooper pairs are linearly bound by electric fields squeezed into strings (dual Meissner effect) by a monopole condensate. Magnetic monopoles, while elusive as elementary particles, exist in certain materials in the form of emergent quasiparticle excitations. We demonstrate that at low temperatures magnetic monopoles can form a quantum Bose condensate (plasma in (2+1) dimensions) dual to the charge condensate in superconductors. The monopole Bose condensate manifests as a superinsulating state with infinite resistance, dual to superconductivity. The monopole supercurrents result in the electric analogue of the Meissner effect and lead to linear confinement of the Cooper pairs by Polyakov electric strings in analogy to quarks in hadrons. Superinsulators realise thus one of the mechanism proposed to explain confinement in QCD. Moreover, the string mechanism of confinement implies asymptotic freedom at the IR fixed point. We predict thus for superinsulators a metallic-like low temperature behaviour when samples are smaller than the string scale. This has been experimentally confirmed. We predict that an oblique version of SI is realised as the pseudogap state of high-$T_C$ superconductors.

**Keywords:** monopoles; confinement; topological interactions

**Citation:** Diamantini, M.C.; Trugenberger, C.A. Superinsulators: An Emergent Realisation of Confinement. *Universe* **2021**, *7*, 201. https://doi.org/10.3390/universe7060201

Academic Editor: Maarten Golterman

Received: 22 May 2021
Accepted: 11 June 2021
Published: 17 June 2021

**Publisher's Note:** MDPI stays neutral with regard to jurisdictional claims in published maps and institutional affiliations.

## 1. Introduction

Although extremely successful in describing many aspects of particle physics, the standard model does not explain the mechanism of confinement that binds quarks into hadrons. In 1978, in a Gedanken experiment for quark confinement [1] 't Hooft introduced the idea of a dual superconductor in which, in analogy to the Meissner effect, chromo-electric fields would be squeezed into thin flux tubes with quarks at their ends in a condensate of magnetic monopoles. When quarks are pulled apart, it is energetically favourable to pull out of the vacuum additional quark-antiquark pairs and to form several short strings instead of a long string. As a consequence, colour charge can never be observed at distances above a fundamental length scale, $1/\Lambda_{QCD}$ and quarks are confined. Only colour-neutral hadron jets can be observed in collider events. In this phase, that he called the "extreme opposite" of a superconductor, there is zero quark mobility and, thus, an infinite chromo-electric resistance. He, hence, called this phase a "superinsulator".

In condensed matter, superinsulation emerges in materials that have Cooper pairs and vortices as relevant degrees of freedom. It was originally predicted for Josephson junction arrays (JJA) [2] and then experimentally found in InO superconducting films [3], in Tin films [4,5] and NbTiN films [6]. Superinsulations emerge in all these systems at the insulating side of the superconductor-insulator transition (SIT). The SIT and the nature of the phases that it harbours is determined by the competition between two quantum orders embodied in the topological interactions between charges and vortices (Aharonov-Bohm/Aharonov-Casher (ABC)) and is, thus, a realisation of the field-theoretical Mandelstam't Hooft S-duality [1,7] in a material. A local formulation of such topological interactions requires the introduction of two emergent gauge fields $a_\mu$ and $b_\mu$ (a tensor field $b_{\mu\nu}$ in (3+1) dimensions) coupled to the conserved charge and vortex currents, respectively. The effective field theory that describes this topological phase of matter is a mixed Chern-Simons (CS) field theory in (2+1) dimensions and a BF theory in (3+1) dimensions.

Superinsulators are characterised by an infinite resistance that persists at finite temperatures: charges cannot move even if a voltage (below a critical threshold) is applied. This infinite resistance is due to linear confinement of charges [8] in a condensate of magnetic monopoles (instanton plasma in (2+1) dimensions). This confining mechanism is exactly the mechanism that is realised in compact QED, [9,10], the simplest example of a strongly coupled gauge theory with a massive photon and linear confinement of charges: the vortex Bose condensate constricts electric fields into electric flux tubes that bind Cooper pairs and anti-Copper pairs (Figure 1). The superinsulating state is nothing else than a plasma of magnetic monopoles (instantons) since, in a condensate, vortex number is not conserved. In (3+1), dimensions vortices can be viewed as magnetic filaments connecting magnetic monopoles at their ends [10]. In this one-color version of quantum chromodynamics (QCD) Cooper pairs play the role of quarks.

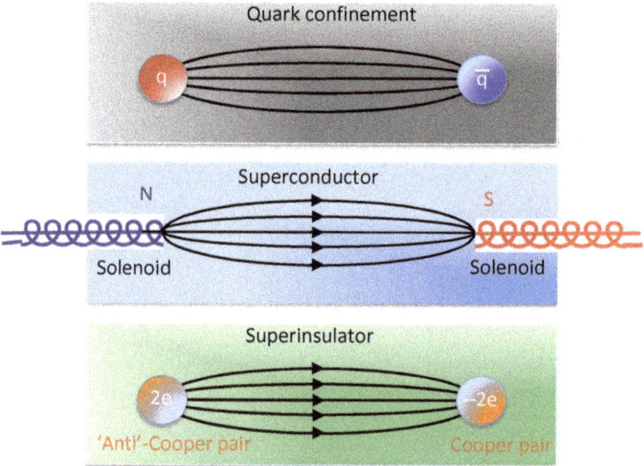


**Figure 1.** Dual Mandelstam't Hooft–Polyakov confinement. From top to bottom: quark confinement by chromo-electric strings; magnetic tube (Abrikosov vortex) that forms in a superconductor between two magnetic monopoles; electric string that forms in a superinsulator between the Cooper pair and anti-Cooper pair. The lines are the force lines for magnetic and electric fields respectively. In all cases the energy of the string (the binding energy) is proportional to the distance between either the monopoles or the charges.

Although the search of magnetic monopoles has been the object of years of efforts [11], they are elusive as elementary particles. In the materials that exhibit superinsulations, instead, magnetic monopoles are present in the form of emergent quasiparticle excitations realising the electric–magnetic symmetry. They behave as quantum particles and at low temperatures, they form a quantum Bose condensate dual to the charge condensate in superconductors. Their supercurrents cause the electric analogue of the Meissner effect and lead to linear confinement of the Cooper pairs [8,12–14]. Magnetic monopoles play, thus, a crucial role in the formation and properties of the superinsulating state. As we will show below, this phase is a phase in which vortex strings with magnetic monopoles at their the endpoints become loose and confine charges. On the contrary, appreciable vortex tension implies that vortices are short and confines monopoles in small dipoles, as shown in Figure 2. Charges are thus liberated.

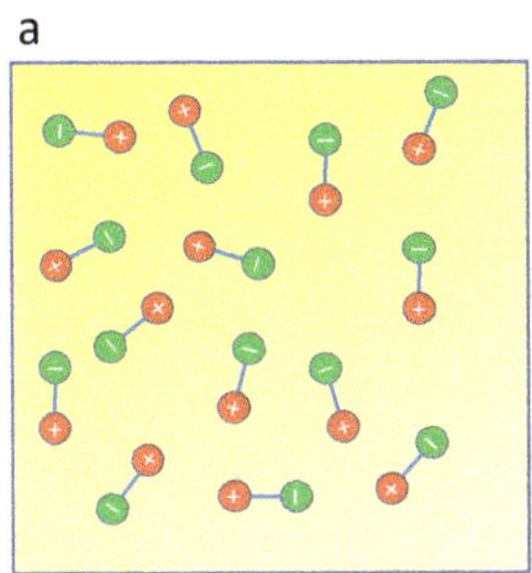
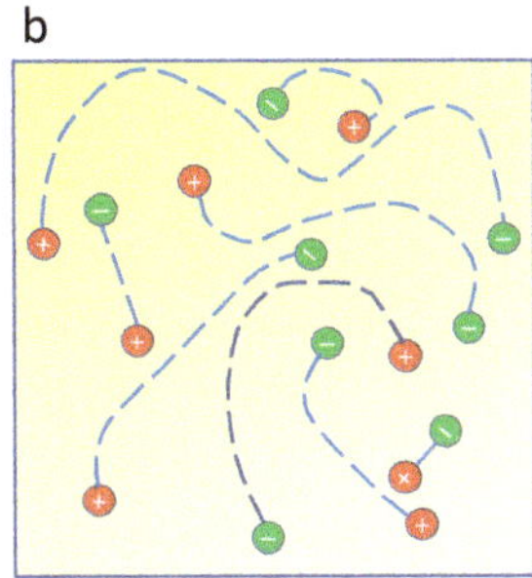

**Figure 2.** Magnetic monopole states at low temperatures. (**a**) Monopoles are confined into small dipoles by the tension of vortices connecting them. (**b**) As the tension vanishes, vortices become loose and magnetic monopoles at their endpoints condense.

A salient feature of QCD is asymptotic freedom, the weakening of the interaction coupling strength at short distances (ultraviolet (UV) limit). At large distances (infrared (IR) limit), the quarks are thought to be confined within hadrons, which are physical observable excitations, by the QCD strings. Quarks themselves cannot be extracted from hadrons and be seen in isolation. The mechanism for the transition from weak quark interactions in the UV regime to confinement and strings in the IR regime remains an open issue. Confinement by strong interactions prevents a direct view on quarks despite that they move nearly free at the small scales. As we will show below, superinsulators, instead, allow for a direct observation of the interior of electric mesons made of Cooper pairs by standard transport measurements. We reveal the transition from the confined to the asymptotic free Cooper pair motion upon decreasing the distance between electrodes, modelling the observation scale.

Pure gauge compact QED in 2D, with only closed string excitations [15] is not renormalisable. However, coupling the action to dynamical matter results in a non-trivial fixed point [16]. The same occurs in our case: deep non-relativistic compact QED is induced by an underlying matter dynamics from which it inherits the corresponding Berezinskii-Kosterlitz-Thouless (BKT) [17–19] fixed point separating an integer topological phase, [20–22] from a confined phase. The CS mass sets the gap for the topological phase, [20–22], that corresponds to a functional first Landau level and consists of an intertwined incompressible fluid of charges and vortices. The confined phase, instead, is a highly entangled vortex condensate in which charges are linearly bound. As we will show below, the effective coupling of the theory will thus flows to small values in the UV limit, and the induced compact QED2 becomes asymptotically free (the theory is actually asymptotically safe, since the critical point is at a finite value of the coupling different from zero but we will use the more familiar term for simplicity's sake here), the BKT transition representing the infrared (IR) confining fixed point.

The review is organised like this. In Section 1, we present the effective gauge theories description of the SIT in (2+1) dimensions and show how the BKT transition arises. We then derive the phase diagram. In Section 2, we will discuss the characteristics of the superinsulating phase, computing the string tensions for the electric strings that bind the Cooper pairs and show how the asymptotic free regime arises. We then generalised the model of the SIT to the (3+1)-dimensional case in Section 3 and compute the phase diagram that essentially coincides with the one in one dimension less. In Section 4, we discuss the characteristic of the superinsulating phase and show that also in (3+1) dimensions this is a confinement phase in which Cooper pairs are bounded by electric flux tubes in a condensate of magnetic monopole. Section 5 is devoted to conclusions.

## 2. 2+1 Dimensions

We will use natural units $c = 1$, $\hbar = 1$, $\epsilon_0 = 1$ but restore physical units when necessary. The infinite-range ABC interaction, embodying the quantum phase acquired either by a charge encircling a vortex or by a vortex encircling a charge, dominates the structure of the critical vicinity of the SIT. The world-lines of elementary charges and vortices are described by:

$$Q_\mu = \Sigma_i \int_{x_q^{(i)}} d\tau \frac{dx_{q\mu}^{(i)}(\tau)}{d\tau} \delta^3\left(x - x_q^{(i)}(\tau)\right),$$

$$M_\mu = \Sigma_i \int_{x_m^{(i)}} d\tau \frac{dx_{m\mu}^{(i)}(\tau)}{d\tau} \delta^3\left(x - x_m^{(i)}(\tau)\right),$$

(1)

where the index i labels the elementary charges and vortices, parametrized by the coordinates $x_q^{(i)}$ and $x_m^{(i)}$, respectively, $n$ is the dimensionless charge ( in our case $n = 2$ to describes Cooper pairs), and Greek subscripts run over the Euclidean three dimensional space encompassing the 2D space coordinates and the Wick rotated time coordinate. ABC phases are encoded in the Gauss linking number between the two curves (1):

$$S_{\text{linking}} = i \int d^3x\, Q_\mu \epsilon_{\mu\alpha\nu} \frac{\partial_\alpha}{-\nabla^2} M_\nu,$$

(2)

where $\epsilon_{\mu\alpha\nu}$ is the completely antisymmetric tensor. To ensure a local formulation of the action (2), one introduces two emergent gauge fields, $a_\mu$ and $b_\mu$ mediating ABC interactions and the topological part of the action takes the form

$$S^{\text{CS}} = \int d^3x \left[ i\frac{n}{2\pi} a_\mu \epsilon_{\mu\alpha\nu} \partial_\alpha b_\nu + i\sqrt{n} a_\mu Q_\mu + i\sqrt{n} b_\mu M_\mu \right].$$

(3)

Equation (3) defines the mixed Chern-Simons (CS) action [23–25] and represents the local formulation of the topological interactions between charges and vortices. Since it contains only one field derivative, it is the dominant contribution to the action at long distances and it is invariant under the gauge transformations $a_\mu \to a_\mu + \partial_\mu \lambda$ and $b_\mu \to b_\mu + \partial_\mu \chi$, reflecting the conservation of the charge and vortex numbers. In this representation $j_\mu = (\sqrt{n}/2\pi)\epsilon_{\mu\alpha\nu}\partial_\alpha b_\mu$ and $\phi_\mu = (\sqrt{n}/2\pi)\epsilon_{\mu\alpha\nu}\partial_\alpha a_\mu$ are the continuous charge and vortex number current fluctuations, while $Q_\mu$ and $M_\mu$ stand for integer point charges and vortices.

The CS kernel has a zero mode [23–25] and needs a regularisation. To this end we will use the next-order terms in the effective action of the SIT that contain two field derivatives and that are gauge invariant. Introducing the dual field strengths $f_\mu = \epsilon_{\mu\alpha\nu}\partial_\alpha b_\mu$ and $g_\mu = \epsilon_{\mu\alpha\nu}\partial_\alpha a_\mu$ and setting $n = 2$ for Cooper pairs, we obtain the action

$$S_{2D} = \int d^3x\, i\frac{1}{\pi} a_\mu \epsilon_{\mu\alpha\nu}\partial_\alpha b_\nu + \frac{1}{2e_v^2\mu}f_0^2 + \frac{\varepsilon}{2e_v^2}f_i^2 + \frac{1}{2e_q^2\mu}g_0^2 + \frac{\varepsilon}{2e_q^2}g_i^2 + i\sqrt{2}a_\mu Q_\mu + i\sqrt{2}b_\mu M_\mu,$$ (4)

where $f_0$ and $g_0$ are the magnetic fields, $f_i$ and $g_i$ the electric fields and $\mu$ is the magnetic permeability and $\varepsilon$ is the electric permittivity which define the speed of light $v = 1/\sqrt{\mu\varepsilon}$ in the material. The two coupling constants $e_q^2$ and $e_v^2$ have canonical dimension [1/length] so, naively the two kinetic terms are infrared-irrelevant. However they are necessary to correctly define the pure CS limit in which the topological mass $m = e_q e_v/2\pi v = O(1/v\lambda_L) \to \infty$ [26,27], where $\lambda_L$ is the London penetration depth in the bulk material. With the two energy scales $e_q^2$ and $e_v^2$ we can define a dimensionless coupling constant $g = e_v/e_q = O(d/(\alpha\lambda_L))$, where d is the thickness of the film and $\alpha = e^2/4\pi$ is the fine structure constant. $g$ plays the role of the conductance in materials. The electric-magnetic duality (charge–vortex symmetry) is given by the action symmetry with respect to the transformation $g \equiv e_v/e_q \leftrightarrow 1/g$. Thus, $g$ is a tuning parameter driving the system across the SIT, and the SIT itself corresponds to $g = g_c = 1$.

To describe the linking number the two compact emergent gauge fields must be compact. To formulate U(1) symmetries we will use a lattice regularisation introducing a lattice of spacing $\ell$. This is not entirely trivial, however, since particular care has to be exercised in the definition of the lattice CS term so that discrete gauge invariance is maintained [2]. To this end we introduce the forward and backward derivatives and shift operators

$$d_\mu f(x) = \tfrac{f(x+\ell\hat{\mu})-f(x)}{\ell} , \qquad S_\mu f(x) = f(x + \ell\hat{\mu}) ,$$
$$\hat{d}_\mu f(x) = \tfrac{f(x)-f(x+\ell\hat{\mu})}{\ell} , \qquad \hat{S}_\mu f(x) = f(x - \ell\hat{\mu}) . \tag{5}$$

We also introduce forward and backward finite differences:

$$\Delta_\mu f(x) = f(x + \ell\hat{\mu}) - f(x) \quad ; \quad \hat{\Delta}_\mu f(x) = f(x) - f(x + \ell\hat{\mu}) . \tag{6}$$

Summation by parts on the lattice interchanges both the two derivatives (with a minus sign) and the two shift operators. Gauge transformations are defined by using the forward lattice derivative. In terms of these operators one can then define two lattice Chern-Simons terms

$$k_{\mu\nu} = S_\mu \epsilon_{\mu\alpha\nu} d_\alpha , \qquad \hat{k}_{\mu\nu} = \epsilon_{\mu\alpha\nu} \hat{d}_\alpha \hat{S}_\nu , \tag{7}$$

where no summation is implied over equal indices. Summation by parts on the lattice interchanges also these two operators (without any minus sign). Gauge invariance is then guaranteed by the relations

$$k_{\mu\alpha} d_\nu = \hat{d}_\mu k_{\alpha\nu} = 0 , \qquad \hat{k}_{\mu\nu} d_\nu = \hat{d}_\mu \hat{k}_{\mu\nu} = 0 . \tag{8}$$

Note that the product of the two Chern-Simons terms gives the lattice Maxwell operator

$$k_{\mu\alpha} \hat{k}_{\alpha\nu} = \hat{k}_{\mu\alpha} k_{\alpha\nu} = -\delta_{\mu\nu} \nabla^2 + d_\mu \hat{d}_\nu , \tag{9}$$

where $\nabla^2 = \hat{d}_\mu d_\mu$ is the 3D Laplace operator.

Integrating out the fictitious gauge fields we obtain an action for the topological excitations alone:

$$S_{\text{top}} = \sum_x \quad v^2 \tfrac{e_q^2}{\ell} Q_\mu \frac{1}{v^4 m^2 - d_0 \hat{d}_0 - v^2 \nabla_2^2} Q_\mu + v^2 \tfrac{e_v^2}{\ell} M_\mu \frac{1}{v^4 m^2 - d_0 \hat{d}_0 - v^2 \nabla_2^2} M_\mu$$
$$+ i \tfrac{2\pi v^6 m^2}{\ell} Q_\mu \frac{k_{\mu\nu}}{(d_0 \hat{d}_0 + v^2 \nabla_2^2)(v^4 m^2 - d_0 \hat{d}_0 - v^2 \nabla_2^2)} M_\mu , \tag{10}$$

where $\nabla_2$ is the 2D spatial Laplacian. The third term in this action describes the lattice version of the topological linking of electric and magnetic strings of width $1/(vm)^2$ and, due to the Dirac quantization condition, at large distances, it reduces to an integer. We will thus drop this term.

The phase diagram is determined by the condensation (or lack thereof) of topological defects. The conditions for the condensation are derived using the standard free energy arguments [28]: the action of the Euclidean field theory model plays the same role as the energy and quantum corrections to the classical action play the same role as the entropy in an equivalent statistical mechanics model in one additional spatial dimension. The ground state of the quantum model corresponds to the minimum of its free energy. Following the standard lattice gauge field theory arguments of [29] we retain only the self-interaction terms in (10)

$$S_{\text{top}} = 2\pi m\ell v G(m\ell v)\left[\frac{e_q}{e_v}\,Q^2 + \frac{e_v}{e_q}\,M^2\right]N\,, \tag{11}$$

where $G(m\ell v)$ is proportional to the diagonal element of the lattice kernel $G(m\ell v, x - y)$ representing the inverse of the operator $(\ell^2/v^2)(m_T^2 v^4 - d_0\hat{d}_0 - v^2\nabla_2^2)$ and we consider strings made of $N$ bonds with integer electric and magnetic quantum numbers $Q$ and $M$. We assign to strings an entropy proportional to their length, being given by $\mu N$ with $\mu \approx \ln(5)$ since, at each step, the non-backtracking strings can choose among 5 possible directions on how to continue. The main contribution to the free energy is thus:

$$F = 2\pi m\ell v G(m\ell v)\left[\frac{e_q}{e_v}\,Q^2 + \frac{e_v}{e_q}\,M^2 - \frac{1}{\eta}\right]N\,, \tag{12}$$

with the dimensionless parameter $\eta$ given by:

$$\eta = \frac{2\pi m\ell v G(m\ell v)}{\mu}\,, \tag{13}$$

which, together with the ratio $g = e_v/e_q$ fully determines the quantum phase structure.

When the energy term in (12) dominates, the free energy is positive and minimized by short closed loop configurations while, when the entropy dominates, the free energy is negative and minimised by large strings and long closed loops giving the following condensation conditions for long strings with integer quantum numbers $Q$ and $M$:

$$\eta\frac{e_q}{e_v}Q^2 + \eta\frac{e_v}{e_q}M^2 < 1\,. \tag{14}$$

If two or more condensations are allowed, one has to choose the one with the lowest free energy. This condition describes the interior of an ellipse with semi-axes

$$\begin{aligned}
r_Q &= \sqrt{\frac{e_v}{e_q}}\sqrt{\frac{1}{\eta}}\,,\\[6pt]
r_M &= \sqrt{\frac{e_q}{e_v}}\sqrt{\frac{1}{\eta}}\,,
\end{aligned} \tag{15}$$

on a square lattice of integer electric and magnetic charges. The ratio $g = e_v/e_q$ determines the ratio of the semi-axes while the parameter $\eta$ sets the overall scale of the ellipse. The quantum phase diagram is found by noting which integer charges lie within the ellipse when the semi-axes and the overall scale are varied:

$$\eta \quad < 1 \to \begin{cases} g > 1\,,\text{electric condensation } = \text{ superconductor}\,,\\ g < 1\,,\text{magnetic condensation } = \text{ superinsulator}\,, \end{cases}$$

$$\eta \quad > 1 \to \begin{cases} g > \eta\,,\text{electric condensation } = \text{ superconductor}\,,\\ \eta > g > \frac{1}{\eta}\,,\text{no condensation } = \text{ Bose metal } = \text{ topologicalinsulator}\,,\\ g < \frac{1}{\eta}\,,\text{magnetic condensation } = \text{ superinsulator}\,. \end{cases}$$

The phase structure is shown in Figure 3.

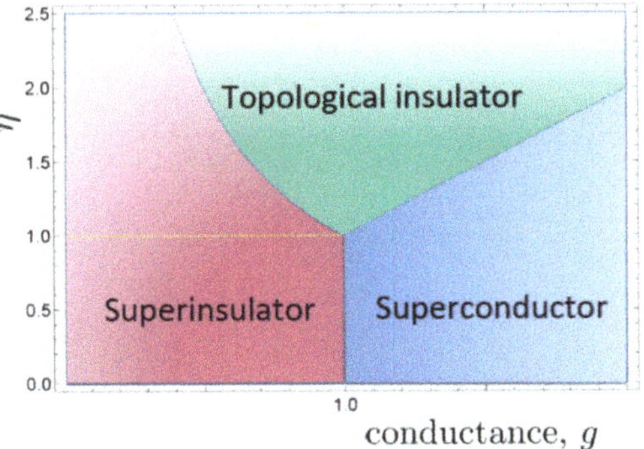


**Figure 3.** The quantum phase diagram of the SIT as a function of the coupling constant $g$. The point $g = 1$, $\eta = 1$ is a tricritical point dominating the phase structure.

A detailed description of all these possible phases can be found in [8,30]. In what follows we will concentrate on the superinsulating phase.

### 3. Superinsulating Phase

To understand the nature of the superinsulating state, we couple the charge current $j_\mu$ to the physical electromagnetic gauge field $A_\mu$ by adding to the action the minimal coupling term $2eA_\mu j_\mu$ and set $Q_\mu = 0$, since charges are dilute, in (4). The effective action $S_{\text{eff}}(A_\mu)$, which gives the electromagnetic response of an ensemble of charges in a superinsulator, is obtained by integrating out the gauge fields $a_\mu$ and $b_\mu$, and summing over the condensed vortices $M_\mu$. The action we obtain is the deep non-relativistic version of Polyakov's compact QED action [10] in which only the electric fields survive

$$S_{\text{top}}\left(M_\mu, A_\mu\right) = \sum_{x,i} \frac{1}{2e_{\text{eff}}^2}\left(\mathcal{F}_i + 2\pi M_i\right)^2, \tag{16}$$

where $e_{\text{eff}}^2$ is the effective coupling constant

$$e_{\text{eff}}^2 = \frac{2\pi^2}{\mu}\frac{1}{\eta g} = e^2 \frac{\pi}{2\mu\eta}\frac{\lambda_L}{d} = e^2 O\left(\frac{\lambda_L}{d}\right). \tag{17}$$

The partition function that we obtain is:

$$Z = Z_0 \cdot Z_{\text{inst.}} = \int_{-\infty}^{+\infty} \mathcal{D}A_\mu\, e^{-\frac{1}{2e_{\text{eff}}^2}\sum_{x,i}\mathcal{F}_i^2} \cdot \sum_{\{m\}} e^{-\frac{2\pi^2}{e_{\text{eff}}^2}\sum_x m\frac{1}{-\nabla_2^2}m}, \tag{18}$$

where $\nabla_2^2$ is the spatial Laplacian instead of the full Laplacian in 3D Euclidean space-time present in the relativistic version of the model. As we will see, this difference has important consequences on the model since the interaction of the monopoles near the SIT, in this case, is logarithmic, $(e_{\text{eff}}^2/2\pi)\ln|\mathbf{x}|$, instead of an inverse linear power of the relativistic model.

The deep non-relativistic limit does not affect, however, the main consequence of Polyakov's original idea [10]: the physics of a superinsulator is governed by the spontaneous proliferation of instantons $M = d_0 M_0 + d_i M_i$, corresponding to magnetic monopoles, so that the vortex number is not conserved in the condensate. These instantons represent quantum tunnelling events by which vortex fluctuations appear and disappear in the condensate. Then, in a mirror analogue to the monopole confinement, i.e., formation of Abrikosov vortices as a result of the Meissner effect in a Cooper pair condensate, the magnetic monopole condensation leads to a dual phenomenon, the emergence of the electric strings [10] mediating confinement of Cooper pairs in superinsulators.

The deep non-relativistic limit plays a crucial role in the shape of the monopoles. The gauge invariance of the $b_\mu$ fields force the constraint $d_\mu M_\mu = 0$ that is satisfied by choosing $d_i M_i = m$ and consequently $d_t M_0 = -m$. $m$ represents instanton quantum tunnelling events in which vortices on the film appear and disappear and their magnetic flux flows in and out isotropically in the four available spatial directions, as it is shown in Figure 4. Another important effects of instantons is that they disorder the system and generate a mass for the photon given by [10]

$$m_\gamma = \frac{8\pi^2}{e_{\text{eff}}^2} z \,, \qquad (19)$$

rendering thus the Coulomb potential screened with a a screening length $\lambda_{\text{el}} = 1/m_\gamma$.

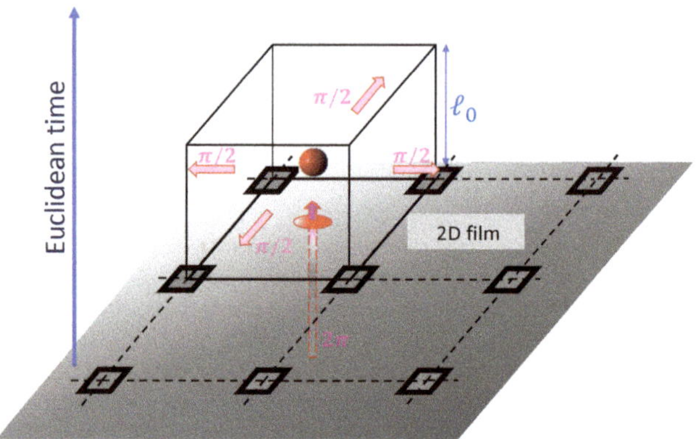


**Figure 4.** A non-relativistic magnetic monopole instanton representing a quantum tunnelling event in which a fundamental vortex of flux $2\pi$ at time $t$ is divided up into four fluxes $\pi/2$ that flow out isotropically in the spatial directions. At the next instant $t + \ell_0$ there is no vortex on the film anymore. For simplicity the condensate islands are represented schematically as in a regular array.

To probe Cooper pair confinement we compute the expectation value of the Wilson loop operator $W(C)$, where $C$ is a closed loop in 3D Euclidean space-time (a factor $\ell$ is absorbed into the gauge field $A_\mu$ to make it dimensionless),

$$\langle W(C) \rangle = \frac{1}{Z_{A_\mu, M_i}} \sum_{\{M_i\}} \int_{-\pi}^{+\pi} \mathcal{D} A_\mu \, e^{-\frac{1}{2e_{eff}^2} \Sigma_x (F_i - 2\pi M_i)^2} e^{i q_{\text{ext}} \Sigma_C A_\mu} \,. \qquad (20)$$

When the loop $C$ is restricted to the plane formed by the Euclidean time and one of the space coordinates, $\langle W(C) \rangle$ measures the potential between two external probe charges $\pm q_{\text{ext}}$. A perimeter law indicates a short-range potential, while an area-law is tantamount to a linear interaction between the probe charges [10] with a new emergent scale represented by the string tension $\sigma$ that gives the strength of the linear potential. We now multiply the Wilson loop operator by 1 in the form $\exp(-i2\pi q_{\text{ext}} M_i)$ on the plaquettes forming the surface $S$ encircled by the loop $C$ and we introduce a unit vector $S_i$ perpendicular to the plaquettes forming the surface $S$.

We then decompose $M_i$ into transverse and longitudinal components, $M_i = M^{\text{T}}_i + M^{\text{L}}_i$ with $M^{\text{T}}_i = \epsilon_{ij} \Delta_j n + \epsilon_{ij} \Delta_j \xi$, $M^{\text{L}}_i = \Delta_i \lambda$, where $\{n\}$ are integers and $\Delta \lambda = \hat{\Delta}_i \Delta_i \lambda = m$. The two sets of integers $\{M_i\}$ are thus traded for one set of integers $\{n\}$ and one set of integers $\{m\}$ representing the magnetic monopoles. The integers $\{n\}$ are used to shift the integration domain for the gauge field $A_\mu$ to $[-\infty, +\infty]$. The real variables $\{\xi\}$ are then also absorbed into the gauge field. The integral over this non-compact gauge field $A_\mu$ gives then the Gaussian fluctuations around the instantons $m$, representing the saddle points of the action. Gaussian fluctuations do not contribute to confinement and, thus, can be neglected. Only the summation over instantons, $\{m\}$, remains:

$$\langle W(C)\rangle = \frac{1}{Z_{\rm m}} \sum_{\{m\}} e^{-\frac{2\pi^2}{e_{\rm eff}^2} \sum_x m_x \frac{1}{-\nabla_2^2} m_x} \, e^{i2\pi q_{\rm ext} \sum_S \hat{\Delta}_i S_i \frac{1}{-\Delta} m_x} \, . \tag{21}$$

For $q_{\rm ext} = 1$, i.e., Cooper pair probes, the sum over instantons gives rise to an area law for the expectation value of the Wilson loop (21) with a string tension given by [31]

$$\sigma = \frac{\sqrt{8}}{\ell_0 \ell} \frac{e_{\rm eff}}{\pi} e^{-\frac{\pi^2}{e_{\rm eff}^2} G_2(0)} \, , \tag{22}$$

where $G_2(0)$ is now the infrared-regularized 2D lattice Coulomb potential at coinciding points. This linear potential is due to a flux tube (string) of electric field connecting Cooper pairs and Cooper holes. This string, with a Cooper pair and a Cooper hole at its endpoints, has a typical width $\lambda_{\rm el}$ [15] and typical length $d_s = 1/\sqrt{\sigma}$ and is the electric equivalent of a strong interaction pion. When one pulls this string by, say, an external voltage, Cooper pairs and Cooper holes start moving apart but there comes a moment where it becomes energetically favourable for the system to pop out a Cooper pair-Cooper hole pair in some intermediate island and to form two short strings. Only neutral states exist asymptotically in this phase of the system and the resistance becomes infinite since charges cannot move anymore. There is, however, a crucial difference with the relativistic case, in which monopoles are always in a plasma phase due to their weak inverse linear interaction. In the deep non-relativistic limit, the interactions between monopoles is logarithmic, as we already pointed out, so, near the SIT they can undergo a confining quantum BKT transition [17–19] for sufficiently strong coupling constants $g$. In fact $e_{\rm eff}^2$ plays the role of the temperature and, from (17), we see that $g$ plays the role of an inverse temperature, so we have the usual XY model: for low values of $g$ instantons are free and charges are confined, while instants undergo a confining transition and become logarithmically confined at $g = g_{\rm cr}$. This quantum BKT transition represents the SIT itself with a transition between the superinsulating phase and the intermediate Bose metal, the bosoic topological insulator phase.

The BKT transition is an infinite-order transition and follows from the observation that the dual formulation of the 2D Coulomb gas is the well known sine-Gordon model. To obtain the Coulomb gas formulation we start from from $Z_{\rm inst}$ (18) and rewrite the Gaussian term in the action for the topological excitations in terms on an auxiliary field as:

$$Z_{\rm inst} = \int_{-\pi}^{+\pi} \mathcal{D}\chi e^{-\sum_{x,i} \frac{e_{eff}^2}{8\pi^2}(\Delta_i \chi_x)^2} \sum_N \frac{z^N}{N!} \sum_{x_1,\dots,x_N} \sum_{m_1,\dots,m_n=\pm 1} e^{i\sum_x m_x(\chi_x + \eta_x)} \, , \tag{23}$$

where

$$z = e^{\frac{-2\pi^2 G(0)}{e_{eff}^2}} \, , \tag{24}$$

is the instanton fugacity and we have adopted the dilute gas approximation in which we consider only $m_x = \pm 1$. $G(0)$ is the infrared-regularised value of the lattice Coulomb kernel at coinciding points. The sums can be now computed, with the result

$$Z_{\rm inst} = \int_{-\infty}^{+\infty} \mathcal{D}\chi \, e^{-\sum_{x,i} \frac{e_{eff}^2}{8\pi^2}(\Delta_i \chi_x)^2 + 2z(1-\cos(\chi_x))} \, , \tag{25}$$

which is nothing else than the partition function of the sine-Gordon model (for a review see [32]) which describes the physics of the planar XY model. Following the results for the XY model [32] we find thus a critical coupling $g_{\rm crit} = (4\pi/\mu)(1/\eta)$ which plays the role of the critical temperature in this quantum BKT transitions. Monopoles and linear confinement of charges can exist only for $g < g_{cr}$, in excellent agreement with the estimate obtained from the crude free energy argument for strings which would correspond to the string entropy value $\mu = 4\pi$.

For the XY model described by (25) re-normalisation group flow (varying the temperature) is expressed best in terms of the two variables

$$u = 1 - \frac{T_{cr}}{T},$$
$$v = 16\pi z \frac{T_{cr}}{T}.$$

(26)

The half line $z = 0$, $T < T_{cr}$ is an half line of infrared fixed points all corresponding to states with bound vortices and differing by a constant representing the initial conditions of the flow equations. In our case, g flows to large values in the IR limit, and the line $z = 0$, $g < g_{cr}$ is a line of confining IR fixed points for the charges. The magnetic monopole instantons cause linear confinement of charges in the superinsulating phase and the granularity scale $\ell$ (lattice spacing) determines the string tension of this linear potential and sets thus also the scale of linearly bound pairs of charges, see (22). The SIT corresponds to an IR Berezinskii-Kosterlitz-Thouless [17–19] fixed point $(g_{cr}, z = 0)$. The BKT re-normalisation flow toward short UV scales implies a decreasing $g$ and an increasing $z$. The confining interaction decreases when flowing towards short scales and, we reach the scale $O(\ell)$ as we will show, charges essentially do not feel any potential anymore, showing what is called *asymptotic freedom*. This phenomenon is typically associated with non-Abelian gauge theories, where it characterises their UV fixed point [33], here it is associated with the sine-Gordon model (and not the compact QED) and describes an IR fixed point so it should be called *asymptotic safety* but here will use the more familiar term asymptotic freedom.

The only evidence for quarks inside hadrons is indirect, through high-energy collision that smash them and create hadron jets in colliders such as LHC. In such experiments, it is impossible to "look inside hadrons" to study the UV to IR confining transition. Here we show that this is, instead, possible in condensed matter superinsulators since the electric interaction is much weaker than the strong interaction and, therefore, the size an electric pion larger than the size of real pion. In superinsulators if the string length scale $d_s$ is large enough so that the regime $\lambda_{el} < d_s$ is realized, one can probe the interior of "superinsulating mesons" by measuring the $IV$ dependencies on samples with linear dimensions $L < \lambda_{string}$. In this case the "interior" interaction at intermediate scales $\lambda_{el} < r < d_s$ is a screened Coulomb potential. This should result in a strong size-dependence of the $I(V)$ response, such that the superinsulating hyperactivated behaviour of the resistance observed in sufficiently large samples changes to a metal-like behaviour in sufficiently small systems with $L \lesssim d_s$. This size-dependence corresponds exactly to the transition from the confinement regime at large scale to the asymptotic free regime inside the "electric mesons".

To gain more insight about this transition, let us focus on the interaction energy $U(r)$ between charges separated by a distance $r$, derived from the compact QED model of superinsulation (we henceforth restore physical units) is

$$U(r) = \sigma(T)r - \frac{c\hbar\pi}{24r} + a\left[\ln\left(\frac{\lambda_{el}}{r_0}\right) - K_0\left(\frac{r}{\lambda_{el}}\right)\right],$$

(27)

where the second term is the so-called Lüscher term [34] and the third term, containing the MacDonald function $K_0$, is the screened 2D Coulomb potential, reducing to $a \ln(r/r_0)$ for $r \ll \lambda_{el}$ while decaying exponentially at $r \gg \lambda_{el}$, with $r_0 \approx$ the superconducting coherence length. For $r > d \simeq r_0$, the Lüscher term is negligible, so that $U(r_0) \simeq 0$. Near the SIT, the strength of the Coulomb potential becomes [8]

$$a = (4e^2/2\pi\varepsilon_0\varepsilon d)(f(\kappa)/g).$$

(28)

The exact form of $f(\kappa)$ is given in [8] and is not relevant here.

When samples are very big, with their dimension $L$ such that $L \gg d_s$, charges are confined and we expect the usual hyperactivated behaviour of the resistance as a function of temperature of superinsulators. However, for samples with dimensions in the range $\lambda_{el} < L < d_s$ Cooper pairs sufficiently far apart feel neither the string tension, since the string is loose on these scales, nor the Coulomb interaction, which is screened on the scale $\lambda_{el}$. We expect thus to observe a transition from hyperactivated resistance behaviour to a metallic saturation at the lowest temperatures when the sample size is decreased. This is exactly what has been observed in a NbTiN superinsulating film by varying the bridge length on which the external voltage is applied [12], as shown in Figure 5. For large bridge lengths the film displays hyperactivated resistances, for the smallest bridge length 0.2 mm, however, we see metallic saturated behaviour at low temperatures. The crossover from hyperactivation to metallic behaviour should take place around a bridge length $L \approx d_s$. The typical string size can be estimated from experimental data as follows. The energy $k_B T_{\mathrm{dec}}$ is the energy necessary to break up the string by raising the temperature. So it is a measure of $\sqrt{\sigma}$ and, therefore,

$$d_s \approx \frac{\hbar v}{k_B T_{\mathrm{dec}}} , \tag{29}$$

where we have reinstated physical units with $v = (1/\sqrt{\varepsilon})c$. Using the experimentally determined deconfinement temperature $T_{\mathrm{dec}} \approx 400~mK$ and the known dielectric constant of NbTiN near the SIT [6], $\epsilon \approx 800$, one can obtain an estimate $d_s \approx 0.13$ mm in excellent quantitative agreement with the observation of the metallic crossover. This is the first direct experimental evidence of asymptotic freedom.

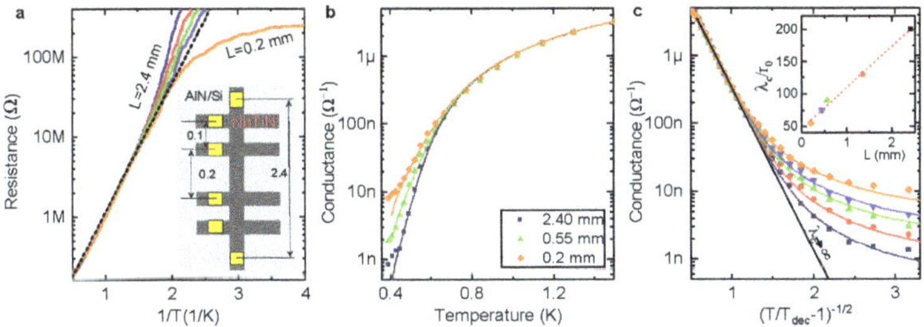

**Figure 5.** Sheet resistance of a NbTiN superinsulating film as a function of the effective sample size (bridge length) (**a**) Logarithmic plot of sheet resistance $R_\square$ vs. inverse temperature $1/T$ for bridges of various length $L$. The dashed straight line shows the Arrhenius behaviour $R \propto \exp(1/T)$. Inset: experimental setup. The Si substrate with AlN buffer layer is shown with light gray and the Hall bridge of NbTiN is dark grey. The square gold contacts are given in yellow. All lateral sizes are given in millimetres. (**b**) Same data as in (**a**) but replotted in terms of the conductance $G = 1/R_\square$ vs. $T$ in log-line scale. The dotted lines are fits using a two dimensional Coulomb gas model that generalises the Berezinskii-Kosterlitz-Thouless (BKT) formula for the conductance $G \propto \exp[-(T/T_{\mathrm{dec}} - 1)^{1/2}]$ by incorporating a self-consistent solution of the effects of electrostatic screening, where the screening length $\lambda_c$ and $T_{\mathrm{dec}}$ enter as fitting parameters. For all bridges the deconfinement temperature is $T_{dec} \approx 400$ mK. (**c**) Same data as in (**b**) but for temperature renormalized as $(T/T_{\mathrm{dec}} - 1)^{1/2}$. The solid line corresponds to the case of an infinite electrostatic screening length $\lambda_c \to \infty$.

## 4. (3+1) Dimensions

In this section, we will generalise our theory to the (3+1)-dimensional case. We will use in what follow a relativistic notation. The relevant degrees of freedom, Cooper pairs, and Josephson vortices, can acquire topological ABC phases when one is encircling the other. However, vortices are now one-dimensional extended objects and their world-surfaces are described by the two-index antisymmetric tensor $m_{\mu\nu}$ (we will use for the moment a continuous notation). Due to this the generalisation of the Chern-Simons representations of the linking number will include a Kalb-Ramond antisymmetric tensor fields $b_{\mu\nu}$ [35] which couples to the vortex current giving the BF action, generalising (3):

$$\mathcal{L} = \frac{1}{4\pi} b_{\mu\nu} \epsilon^{\mu\nu\alpha\beta} \partial_\alpha a_\beta + a_\mu j^\mu + \frac{1}{2} b_{\mu\nu} m^{\mu\nu} , \tag{30}$$

where $j^\mu$ and $m^{\mu\nu}$ are the charge and vortex currents, respectively. Although the field strength associated to $a_\mu$ is, as usual, $f_{\mu\nu} = \partial_\mu a_\nu - \partial_\nu a_\mu$, in (3+1) dimensions its the dual field strength is a 2-tensor,

$$\tilde{f}^{\mu\nu} = \frac{1}{2} \epsilon^{\mu\nu\alpha\beta} f_{\alpha\beta} = \epsilon^{\mu\nu\alpha\beta} \partial_\alpha a_\beta . \tag{31}$$

The field strength associated with the tensor field $b_{\mu\nu}$ is a 3-tensor

$$h_{\mu\nu\alpha} = \partial_\mu b_{\nu\alpha} + \partial_\nu b_{\alpha\mu} + \partial_\alpha b_{\mu\nu} , \tag{32}$$

and its dual field strength is, thus, a vector:

$$h^\mu = \frac{1}{6} \epsilon^{\mu\nu\alpha\beta} h_{\nu\alpha\beta} = \frac{1}{2} \epsilon^{\mu\nu\alpha\beta} \partial_\nu b_{\alpha\beta} . \tag{33}$$

These field strengths $f_{\mu\nu}$ and $h_{\mu\nu\alpha}$ can be used to add dynamics to the purely topological BF term (30),

$$\mathcal{L} = \frac{1}{12\Lambda^2} h_{\mu\nu\alpha} h^{\mu\nu\alpha} + \frac{1}{4\pi} b_{\mu\nu} \epsilon^{\mu\nu\alpha\beta} \partial_\alpha a_\beta - \frac{1}{4f^2} f_{\mu\nu} f^{\mu\nu} , \tag{34}$$

where $f$ is a dimensionless coupling and $\Lambda$ has canonical dimension [1/length]. This action was introduced as a field theory for a condensed matter system in [2]. The BF model is topological, since it is metric-independent. In addition to the usual gauge transformations $a_\mu \to a_\mu + \partial_\mu \zeta$, (34), is also invariant under gauge transformations of the second kind,

$$b_{\mu\nu} \to b_{\mu\nu} + \partial_\mu \lambda_\nu - \partial_\nu \lambda_\mu . \tag{35}$$

When using the BF term to model the emergent behaviour of condensed matter systems, one identifies the topologically conserved charge current $j^\mu$ and vortex current $m^{\mu\nu}$ as

$$
\begin{aligned}
j^\mu &= \tfrac{1}{2\pi} h^\mu = \tfrac{1}{4\pi} \epsilon^{\mu\nu\alpha\beta} \partial_\nu b_{\alpha\beta} , \\
m^{\mu\nu} &= \tfrac{1}{2\pi} \tilde{f}^{\mu\nu} = \tfrac{1}{2\pi} \epsilon^{\mu\nu\alpha\beta} \partial_\alpha a_\beta ,
\end{aligned}
\tag{36}
$$

with Cooper pairs measured in integer units of $2e$ and vortices in integer units of $2\pi/2e = \pi/e$.

To formulate the gauge-invariant lattice $BF$-term, we follow [2] and introduce the lattice $BF$ operators

$$
\begin{aligned}
k_{\mu\nu\rho} &\equiv S_\mu \epsilon_{\mu\alpha\nu\rho} d_\alpha , \\
\hat{k}_{\mu\nu\rho} &\equiv \epsilon_{\mu\nu\alpha\rho} \hat{d}_\alpha \hat{S}_\rho ,
\end{aligned}
\tag{37}
$$

The two lattice $BF$ operators are interchanged (no minus sign) upon summation by parts on the lattice and are gauge invariant so that:

$$
\begin{aligned}
k_{\mu\nu\rho} d_\nu &= k_{\mu\nu\rho} d_\rho = \hat{d}_\mu k_{\mu\nu\rho} = 0 , \\
\hat{k}_{\mu\nu\rho} d_\rho &= \hat{d}_\mu \hat{k}_{\mu\nu\rho} = \hat{d}_\nu \hat{k}_{\mu\nu\rho} = 0 ,
\end{aligned}
\tag{38}
$$

and satisfy the equations

$$\hat{k}_{\mu\nu\rho}k_{\rho\lambda\omega} = -\left(\delta_{\mu\lambda}\delta_{\nu\omega} - \delta_{\mu\omega}\delta_{\nu\lambda}\right)\nabla^2$$

$$+\left(\delta_{\mu\lambda}d_{\nu}\hat{d}_{\omega} - \delta_{\nu\lambda}d_{\mu}\hat{d}_{\omega}\right) + \left(\delta_{\nu\omega}d_{\mu}\hat{d}_{\lambda} - \delta_{\mu\omega}d_{\nu}\hat{d}_{\lambda}\right), \tag{39}$$

$$\hat{k}_{\mu\nu\rho}k_{\rho\nu\omega} = k_{\mu\nu\rho}\hat{k}_{\rho\nu\omega} = 2\left(\delta_{\mu\omega}\nabla^2 - d_{\mu}\hat{d}_{\omega}\right),$$

where $\nabla^2 = \hat{d}_{\mu}d_{\mu}$ is the lattice Laplacian. We use the notation $\Delta_{\mu}$ and $\hat{\Delta}_{\mu}$ for the forward and backwards finite difference operators.

As in the (2+1)-dimensional case, in the Euclidean lattice formulation, $Q_{\mu}$ and $M_{\mu\nu}$ becomes integer link and plaquette variables $Q_{\mu}$ and $M_{\mu\nu}$. In 4 Euclidean dimensions they describes the Euclidean world-lines of point charges and Euclidean world-surfaces of vortices. In materials, the velocity of light will be $v = 1/\sqrt{\varepsilon\mu} < 1$ by defining the Euclidean time lattice spacing as $\ell_0 = \ell/v$, where $\varepsilon$ is the electric permittivity and $\mu$ is the magnetic permeability we incorporate this velocity by rescaling all time derivatives, currents, and zero-components of gauge fields by the factor $1/v$. As a consequence, both gauge fields acquire a dispersion relation $E = \sqrt{m^2v^4 + v^2\mathbf{p}^2}$ with the topological mass given by $m = f\Lambda/2\pi v$, and we thus obtain the lattice action:

$$S = \sum_x \frac{\ell^4}{4f^2} f_{\mu\nu} f_{\mu\nu} + i\frac{\ell^4}{4\pi} a_{\mu}k_{\mu\alpha\beta}b_{\alpha\beta} + \frac{\ell^4}{12\Lambda^2} h_{\mu\nu\alpha}h_{\mu\nu\alpha} + i\ell a_{\mu}Q_{\mu} + i\ell^2\frac{1}{2}b_{\mu\nu}M_{\mu\nu}. \tag{40}$$

The dimensionless parameter $f = O(e)$ encodes the effective Coulomb interaction strength in the material, $\Lambda$ is the magnetic scale, $\Lambda = O(1/\lambda_{\rm L})$, where $\lambda_{\rm L}$ is the London penetration depth of the superconducting granules.

To find the topological action for monopoles, we start from Equation (40) and integrate out fictitious gauge fields $a_{\mu}$ and $b_{\mu\nu}$

$$S_{\rm top} = \sum_x \frac{f^2}{2\ell^2}\, Q_{\mu}\frac{\delta_{\mu\nu}}{(mv)^2 - \nabla^2}Q_{\nu} + \frac{g^2}{8}\, M_{\mu\nu}\frac{\delta_{\mu\alpha}\delta_{\nu\beta} - \delta_{\mu\beta}\delta_{\nu\alpha}}{(mv)^2 - \nabla^2}M_{\alpha\beta}$$

$$+ i\frac{\pi(mv)^2}{2\ell}\, Q_{\mu}\frac{k_{\mu\alpha\beta}}{\nabla^2((mv)^2 - \nabla^2)}M_{\alpha\beta}. \tag{41}$$

The last term can be neglected since it represents the Aharonov-Bohm phases of charged particles around vortices of width $\lambda_L$. In fact we consider scales much larger than $\lambda_L$, the denominator in (41) reduces to $(mv)^2\nabla^2$ and this last term becomes $(i2\pi - \text{integer})$, reflecting the absence of Aharonov-Bohm phases between charges $ne$ and magnetic fluxes $2\pi/ne$.

Gauge invariance requires closed vortex loops. The presence of magnetic monopoles at the endpoints of open vortices will break the gauge symmetry of the second kind (35) and the longitudinal components of the tensor gauge field $b_{\mu\nu}$ will become usual vector gauge fields for the magnetic monopoles. What is the effect of this gauge breaking term? Monopoles will experience the same type of Coulomb interaction experienced by charges, but this interaction is subdominant with respect to the linear tension created by the vortices between a monopole–antimonopole pair. We can thus neglect it for the determination of the phase structure and admit open vortices with magnetic monopoles at the endpoints.

The important consequence of the topological interactions is that they induce self-energies in form of the mass of Cooper pairs and tension for vortices between magnetic monopoles. These self-energies are encoded in the short-range kernels in the action (41), which we approximate by a constant. World-lines and world-surfaces are thus assigned energies, that are nothing else that their Euclidean actions in the present statistical field theory setting, proportional to their length $N$ and area $A$ which we measure in numbers of links and plaquettes,

$$S_N = 2\pi(mv\ell)G \frac{f}{\Lambda\ell} Q^2 N \,,$$

$$S_A = 2\pi(mv\ell)G \frac{\Lambda\ell}{f} M^2 A \,. \tag{42}$$

Here $Q$ and $M$ are the integer quantum numbers carried by the two kinds of topological excitations and $G = O(G(mv\ell))$, where $G(mv\ell)$ is the diagonal element of the lattice kernel $G(x-y)$ representing the inverse of the operator $\ell^2\left((mv)^2 - \nabla^2\right)$. As in the (2+1)-dimensional case, to construct the free energy, we need to estimate the entropy of link strings and plaquette surfaces. The entropy is, for string, proportional to their length $\mu_N N$, and for surfaces proportional to their area [36] $\mu_A A$. Both coefficients $\mu$ are non-universal: for strings $\mu_N \simeq \ln(7)$ since at each step the non-backtracking string can choose among 7 possible directions on how to continue, while, for surfaces, $\mu_A$ does not have such a simple interpretation but can be estimated numerically. The total free energy that we obtain is:

$$F = 2\pi(mv\ell)G\left[\left(\frac{f}{\Lambda\ell}Q^2 - \frac{1}{\eta_Q}\right)N + \left(\frac{\Lambda\ell}{f}M^2 - \frac{1}{\eta_M}\right)A\right]\,,$$

where we have defined

$$\eta_Q = \frac{2\pi(mv\ell)G}{\mu_N}\,, \qquad \eta_M = \frac{2\pi(mv\ell)G}{\mu_A}\,. \tag{43}$$

When the self-energy dominates, large string and surface configurations are suppressed in the partition function and Cooper pairs or vortices are gapped excitations, suppressed by their large action. On the contrary, when the entropy dominates large string and surface configurations are favoured in the "free energy" (effective action) and they condense. The phase in which long world-lines of Cooper pairs condense is a superconducting phase characterised by a charge Bose condensate. The phase in which a Bose condensate of magnetic monopoles forms, instead, is a superinsulator.

The formation of larger world-surface implies that the strings binding monopoles and antimonopoles into neutral pairs become loose on distance scales $\gg 1/vm$. This implies that magnetic monopoles at the endpoints of the loose vortices become deconfined and Bose condense.

The combined energy-entropy balance equations are best viewed as defining the interior of an ellipse on a 2D integer lattice of electric and magnetic quantum numbers,

$$\frac{Q^2}{r_Q^2} + \frac{M^2}{r_M^2} < 1\,, \tag{44}$$

where the semi-axes are given by

$$r_Q^2 = \frac{\ell\Lambda}{f}\frac{1}{\eta_Q} = \frac{\ell\Lambda}{f}\sqrt{\frac{\mu_N}{\mu_A}}\frac{1}{\eta}\,,$$

$$r_M^2 = \frac{f}{\ell\Lambda}\frac{1}{\eta_M} = \frac{f}{\ell\Lambda}\sqrt{\frac{\mu_A}{\mu_N}}\frac{1}{\eta}\,, \tag{45}$$

with

$$\eta = \sqrt{\eta_Q\eta_M} = 2\pi(mv\ell)G/\sqrt{\mu_N\mu_A}\,. \tag{46}$$

In (3+1)-dimensional case, however, only configurations with $\{0,M\}$ or $\{Q,0\}$ have to be considered, and configurations with $Q \neq 0$ and $M \neq 0$ must be excluded since the two types of excitations are different. The phase diagram is found by establishing which integer charges lie within the ellipse when the semi-axes are varied. We thus obtain a phase diagram that is essentially as in (2+1) dimensions:

$$
\begin{aligned}
\eta < 1 \rightarrow &\begin{cases} g < 1 \text{, charge Bose condensate,} \\ g > 1 \text{, monopole Bose condensate,} \end{cases} \\
\eta > 1 \rightarrow &\begin{cases} g < \frac{1}{\eta} \text{, charge Bose condensate,} \\ \frac{1}{\eta} < g < \eta \text{, bosonic insulator,} \\ g > \eta \text{, monopole Bose condensate,} \end{cases}
\end{aligned}
\tag{47}
$$

with the tuning parameter $g$ given in this case by:

$$
g = \frac{f}{\ell \Lambda} \sqrt{\frac{\mu_{\mathrm{N}}}{\mu_{\mathrm{A}}}} .
\tag{48}
$$

## 5. (3+1) Dimensions Superinsulating Phase

To derive the effective action for a superinsulator in (3+1) dimensions we follow exactly the same steps as in the (2+1)-dimensional case and we add the minimal coupling of the charge current $j^\mu$ to the electromagnetic field:

$$
\mathcal{L} \rightarrow \mathcal{L} + i \sum_x \ell^4 A_\mu j_\mu = \mathcal{L} + i \sum_x \ell^4 \frac{1}{4\pi} A_\mu k_{\mu\alpha\beta} b_{\alpha\beta} ,
\tag{49}
$$

and we compute its effective action by integrating over the fictitious gauge fields $a_\mu$ and $b_{\mu\nu}$. Using summation by parts, however, the above coupling amounts only to a shift

$$
M_{\mu\nu} \rightarrow M_{\mu\nu} + \frac{1}{2\pi} \ell^2 \hat{k}_{\mu\nu\alpha} A_\alpha ,
\tag{50}
$$

in (40). The electromagnetic response $S_{\mathrm{eff}}(A_\mu)$ is then obtained by integrating over the fictitious gauge fields and setting $Q_\mu = 0$:

$$
e^{-S_{\mathrm{eff}}(A_\mu)} = \sum_{M_{\mu\nu}} e^{-\frac{1}{8f^2} \sum_{x,\mu,\nu} \left( \tilde{F}_{\mu\nu} - 2\pi M_{\mu\nu} \right)^2} .
\tag{51}
$$

Equation (51) is the action of Polyakov's compact QED in (3+1) dimensions.

To prove linear confinement of charges we introduce two external probe charges $\pm q_{\mathrm{ext}}$ and compute the expectation value for the corresponding Wilson loop operator W(C), where C is a closed loop, now in 4D Euclidean space-time:

$$
\langle W(C) \rangle = \frac{1}{Z_{A_\mu, M_{\mu\nu}}} \sum_{\{M_{\mu\nu}\}} \int_{-\pi}^{+\pi} \mathcal{D}A_\mu \, e^{-\frac{1}{8f^2} \sum_{x,\mu,\nu} \left( \tilde{F}_{\mu\nu} - 2\pi M_{\mu\nu} \right)^2} e^{i q_{\mathrm{ext}} \sum_C l_\mu A_\mu} ,
\tag{52}
$$

where $l_\mu = 1$ on the links forming the closed loop C and $l_\mu = 0$ everywhere else. We can now use the lattice Stoke's theorem and, for small values of the coupling $f$, the saddle-point approximation to rewrite Equation (52) as:

$$
\langle W(C) \rangle = \frac{1}{Z_{A_\mu, M_{\mu\nu}}} \sum_{\{M_{\mu\nu}\}} \int_{-\pi}^{+\pi} \mathcal{D}A_\mu \, e^{-\frac{1}{8f^2} \sum_x \left( \tilde{F}_{\mu\nu} - 2\pi M_{\mu\nu} \right)^2} e^{i \frac{q_{\mathrm{ext}}}{2} \sum_S S_{\mu\nu} \left( \tilde{F}_{\mu\nu} - 2\pi M_{\mu\nu} \right)} ,
\tag{53}
$$

where the quantities $S_{\mu\nu}$ are unit surface elements perpendicular (in 4D) to the plaquettes forming the surface S encircled by the loop C and vanish on all other plaquettes. We have also multiplied the Wilson loop operator by 1 in the form $\exp(-i\pi q_{\mathrm{ext}} \sum_x S_{\mu\nu} M_{\mu\nu})$.

At this point, we can simply repeat the computation of Polyakov [10] which shows an area law behaviour for the expectation value of the Wilson loop:

$$\langle W(C) \rangle = \mathrm{e}^{-\sigma A} \tag{54}$$

where $A$ is the area of the surface $S$ enclosed by the loop $C$ and the string tension is given by

$$\sigma = \frac{32f}{\pi\sqrt{\varepsilon\mu}} \frac{1}{\ell^2} \exp\left( -\frac{\pi G(0)}{8f^2} \right) , \tag{55}$$

where $G(0) = 0.155$ is the value of the 4D lattice Coulomb potential at coinciding points. The monopole condensate, thus, generates a string binding together charges and preventing charge transport in systems of a sufficient size. A magnetic monopole condensate is a 3D superinsulator, characterized by an infinite resistance at finite temperatures [2,8,14]. The critical value of the effective Coulomb interaction strength for the transition to the superinsulating phase is $f_{\mathrm{crit}} = O(\ell/\lambda_L)$.

These results shows that the string confinement mechanism of superinsulation allows to generalise the concept of a superinsulator to (3+1) dimensions. The SIT has, however, been experimentally found only in (2+1) dimensions. What will be the experimental hallmark of superinsulation in (3+1) dimensions and, at the same time, unequivocally discriminate between the 3d and 2d superinsulators, exposing the linear nature of the underlying confinement? We can gain insight on this problem by looking at the finite temperature behaviour and the deconfinement transition at which string confinement of Cooper pairs ceases to exist. At a critical temperature $T_{\mathrm{dc}}$ the linear tension of the string turns to zero and the superinsulator transforms into a conventional insulator. In [37], we have shown that the confining string theory description of superinsulation leads to a deconfinement criticality that depends on the space dimension. In fact the critical behaviour is embodied by the behaviour of the (dimensionless) correlation length that is proportional to the inverse of the square root of the string tension near the critical temperature. In (2+1) dimensions when approaching the deconfinement transition from below the correlation length at the transition diverges according to the law

$$\xi_{\pm} \propto \exp\left[ \frac{b_{\pm}}{\sqrt{|T/T_{\mathrm{c}} - 1|}} \right] , \tag{56}$$

reproducing thus the BKT [17–19] criticality, typical of the 2D XY model, criticality that was predicted for compact QED in (2+1) dimensions by Svetitsky and Yaffe [38].This behaviour has been experimentally observed in [6]. In (3+1) dimensions, instead, we predicted in [37] that the finite-temperature confinement–deconfinement transition is in the Vogel-Fulcher-Tamman class [39], a quasi-2D behaviour in which the correlation length at the transition diverges according to the law

$$\xi_{\mathrm{corr}} \propto e^{\frac{\xi}{|T-T_{\mathrm{cr}}|}} . \tag{57}$$

This criticality differs from the one of the 2D XY model only by the power in the exponent. This critical behaviour has been detected in InO films [40], in which the thickness is much larger than the superconducting coherence length. While it seems premature to view this result as a conclusive evidence, yet one can view it as a possible indication of linear confinement in 3d superinsulators.

## 6. Conclusions

Even after decades of intense research the problem of quark confinement has not yet been completely understood. One of the most promising ways to explain confinement is that confinement of colour is produced by dual superconductivity [1,7,10]: the chromoelectric field produced by quark–antiquark pairs is constrained by the dual Meissner effect into Abrikosov flux tubes in the same way as magnetic field is confined in usual superconductors of type II. This produces an energy proportional to the distance of the pairs, $E = \sigma R$, with $\sigma$ the string tension, leading to confinement. Magnetic monopoles, however, have never been observed as elementary particles. In this review, we have shown that they exist as emergent excitations in superconducting films exhibiting the SIT as instantons, where they can form a plasma, and as particles in 3D materials, where they can form a Bose condensate. Monopoles give thus rise to a new state of matter, the superinsulator, in which electric fields are squeezed into flux tubes by the dual Meissner effect leading to linear confinement of charges. Superinsulators realize thus a single-colour version of quantum chromodynamics (QCD) with Cooper pairs playing the role of quarks. Due to the Abelian nature of QED, although in strong coupling, for superinsulators it is possible to derive analytically the linear confinement by electric strings. In QCD, instead, this is possible only through numerical computations. Superinsulators are, thus, a toy model for exploring and testing the fundamental implications of confinement by monopoles and asymptotic safety via desktop experiments on superconductors.

**Author Contributions:** M.C.D. and C.A.T. the authors equally contributed to this work. Both authors have read and agreed to the published version of the manuscript.

**Funding:** This research received no external funding.

**Conflicts of Interest:** The authors declare no conflict of interest.

# References


1. Hooft, G.T. On the phase transition towards permanent quark confinement. *Nucl. Phys.* **1978**, *B138*, 1–25. [CrossRef]
2. Diamantini, M.C.; Sodano, P.; Trugenberger, C.A. Gauge theories of Josephson junction arrays. *Nucl. Phys. B* **1996**, *474*, 641–677. [CrossRef]
3. Sambandamurthy, G.; Engels, L.W.; Johansson, A.; Peled, E.; Shahar, D. Experimental Evidence for a Collective Insulating State in Two-Dimensional Superconductors. *Phys. Rev. Lett.* **2005**, *94*, 017003. [CrossRef]
4. Vinokur, V.M.; Baturina, T.I.; Fistul, M.V.; Mironov, A.Y.; Baklanov, M.R.; Strunk, C. Superinsulator and quantum synchronization. *Nature* **2008**, *452*, 613–615. [CrossRef] [PubMed]
5. Baturina, T.I.; Vinokur, V.M. Superinsulator–Superconductor duality in two dimensions. *Ann. Phys.* **2013**, *331*, 236–257. [CrossRef]
6. Mironov, A.Y.; Silevitch, D.M.; Proslier, T.; Postolova, S.V.; Burdastyh, M.V.; Gutakovskii, A.K.; Rosenbaum, T.F.; Vinokur, V.M.; Baturina, T.I. Charge Berezinskii-Kosterlitz-Thouless transition in superconducting NbTiN films. *Sci. Rep.* **2018**, *8*, 4082. [CrossRef]
7. Mandelstam, S. Vortices and quark confinement in non-Abelian gauge theories. *Phys. Rep.* **1976**, *23*, 245–249. [CrossRef]
8. Diamantini, M.C.; Trugenberger, C.A.; Vinokur, V.M. Confinement and asymptotic freedom with Cooper pairs. *Commun. Phys.* **2018**, *1*, 1–7. [CrossRef]
9. Polyakov, A.M. Compact gauge fields and the infrared catastrophe. *Phys. Lett.* **1975**, *59*, 82–84. [CrossRef]
10. Polyakov, A.M. *Gauge Fields and Strings*; Harwood Academic Publisher: Chur, Switzerland, 1987.
11. Milton, K.A. Theoretical and experimental status of magnetic monopoles. *Rep. Prog. Phys.* **2006**, *69*, 1637–1712. [CrossRef]
12. Diamantini, M.C.; Gammaitoni, L.; Strunk, C.; Postolova, S.V.; Mironov, A.Y.; Trugenberger, C.A.; Vinokur, V.M. Direct probe of the interior of an electric pion in a Cooper pair superinsulator. *Nat. Commun. Phys.* **2020**, *3*, 142. [CrossRef]
13. Diamantini, M.C.; Gammaitoni, L.; Trugenberger, C.A.; Vinokur, V.M. The superconductor-superinsulator transition: S-duality and the QCD on the desktop. *J. Supercond. Novel Magn.* **2019**, *51*, 32–47. [CrossRef]
14. Diamantini, M.C.; Trugenberger, C.A.; Vinokur, V.M. Quantum magnetic monopole condensate. *Nature Commun. Phys.* **2021**, *4*, 25.
15. Caselle, M.; Panero, M.; Vadacchino, D. Width of the flux tube in compact U(1) gauge theory in three dimensions. *J. High Energy Phys.* **2016**, *2*, 180. [CrossRef]
16. Kleinert, H.; Nogueira, F.S.; Sudbo, A. Kosterlitz-Thouless-like deconfinement mechanism in the (2+1)-dimensional Abelian Higgs model. *Nucl. Phys.* **2003**, *B666*, 361–395. [CrossRef]
17. Berezinskii, V.L. Destruction of long-range order in one-dimensional and two-dimensional systems having a continuous symmetry group I. Classical systems. *Sov. Phys. JETP* **1970**, *32*, 493–500.
18. Kosterlitz, J.M.; Thouless, D.J. Long range order and metastability in two dimensional solids and superfluids. (Application of dislocation theory). *J. Phys. C Solid State Phys.* **1972**, *5*, L124 [CrossRef]
19. Kosterlitz, J.M.; Thouless, D.J. Ordering, metastability and phase transitions in two-dimensioal systems. *J. Phys. C Solid State Phys.* **1973**, *6*, 1181–1203. [CrossRef]
20. Lu, Y.-M.; Vishwanath, A. Theory and Classification of interacting integer topological phases in two dimensions: A Chern-Simons approach. *Phys. Rev. B* **2012**, *86*, 125119. [CrossRef]
21. Wang, C.; Senthil, T. Boson topological insulators: A window into highly entangled quantum phases. *Phys. Rev. B* **2013**, *87*, 235122. [CrossRef]

22. Chen, X.; Gu, Z.-C.; Liu, Z.-X.; Wen, X.-G. Symmetry protected topological orders and the group cohomology of their symmetry group. *Phys. Rev. B* **2013**, *87*, 155114. [CrossRef]
23. Jackiw, R.; Templeton, S. How super-renormalizable interactions cure infrared divergences. *Phys. Rev.* **1981**, *D23*, 2291. [CrossRef]
24. Deser, S.; Jackiw, R.; Templeton, S. Three-dimensional massive gauge theories. *Phys. Rev. Lett.* **1982**, *48*, 975. [CrossRef]
25. Deser, S.; Jackiw, R.; Templeton, S. Topologically massive gauge theories. *Ann. Phys. (N.Y.)* **1982**, *140*, 372–411. [CrossRef]
26. Dunne, G.; Jackiw, R.; Trugenberger, C.A. Topological (Chern-Simons) quantum mechanics. *Phys. Rev.* **1990**, *D41*, 661–666. [CrossRef]
27. Dunne, G.; Jackiw, R.; Trugenberger, C.A. Chern-Simons theory in the Schrödinger representation. *Ann. Phys.* **1989**, *194*, 197–223. [CrossRef]
28. Cardy, J.L.; Rabinovici, E. Phase structure of Z(p) models in presence of the theta parameter. *Nucl. Phys.* **1982**, *B205*, 1–16. [CrossRef]
29. Banks, T.; Myerson, R.; Kogut, J. Phase Transitions in Abelian Lattice Gauge Theories. *Nucl. Phys.* **1977**, *B129*, 493–510. [CrossRef]
30. Diamantini, M.C.; Mironov, A.Y.; Postolova, S.V.; Liu, X.; Hao, Z.; Silevitch, D.M.; Kopelevich, Y.; Kim, P.; Trugenberger, C.A.; Vinokur, V.M. Bosonic topological intermediate state in the superconductor-insulator transition. *Phys. Lett.* **2020**, *A384*, 126570. [CrossRef]
31. Trugenberger, C.A.; Diamantini, M.C.; Poccia, N.; Nogueira, F.S.; Vinokur, V.M. Magnetic monopoles and superinsulation in Josephson junction arrays. *Quantum Rep.* **2020**, *2*, 388–399. [CrossRef]
32. Zinn-Justin, J. *Quantum Field Theory and Critical Phenomena*; Clarendon Press: Oxford, UK, 1989.
33. Greensite, J. *An Introduction to the Confinement Problem*; Springer: Berlin, Germany, 2011.
34. Lüscher, M. Symmetry-breaking aspects of the roughening transition in gauge theories. *Nucl. Phys. B* **1981**, *180*, 317–329. [CrossRef]
35. Kalb, M.; Ramond, P. Classical direct interstring action. *Phys. Rev.* **1974**, *D9*, 2273–2284. [CrossRef]
36. Nelson, D.; Piran, T.; Weinberg, S. *Statistical Mechanics of Membranes and Surfaces*; World Scientific: Singapore, 2004.
37. Diamantini, M.C.; Gammaitoni, L.; Trugenberger, C.A.; Vinokur, V.M. Vogel-Fulcher-Tamman criticality of 3D superinsulators. *Sci. Rep.* **2018**, *8*, 15718. [CrossRef] [PubMed]
38. Svetitsky, B.; Yaffe, L.G. Critical behavior at finite temperature confinement transitions. *Nucl. Phys.* **1982**, *B210*, 423–447. [CrossRef]
39. Anderson, P.W. Lectures on Amorphous Systems. In *Les Houches, Session XXXI*; Balian, R., Maynard, R., Toulouse, G., Eds.; World Scientific, North Holland: Amsterdam, The Netherlands, 1978.
40. Ovadia, M.; Kalok, D.; Tamir, I.; Mitra, S.; Sacépé, B.; Shahar, D. Evidence for a finite-temperature insulator. *Sci. Rep.* **2015**, *5*, 13503. [CrossRef] [PubMed]

*Article*

# Local Correlation among the Chiral Condensate, Monopoles, and Color Magnetic Fields in Abelian Projected QCD

Hideo Suganuma [1,*] and Hiroki Ohata [2]

[1] Department of Physics, Kyoto University, Kitashirakawaoiwake, Sakyo, Kyoto 606-8502, Japan
[2] Yukawa Institute for Theoretical Physics, Kyoto University, Sakyo, Kyoto 606-8502, Japan;
hiroki.ohata@yukawa.kyoto-u.ac.jp
* Correspondence: suganuma@scphys.kyoto-u.ac.jp

**Abstract:** Using the lattice gauge field theory, we study the relation among the local chiral condensate, monopoles, and color magnetic fields in quantum chromodynamics (QCD). First, we investigate idealized Abelian gauge systems of (1) a static monopole–antimonopole pair and (2) a magnetic flux without monopoles, on a four-dimensional Euclidean lattice. In these systems, we calculate the local chiral condensate on quasi-massless fermions coupled to the Abelian gauge field, and find that the chiral condensate is localized in the vicinity of the magnetic field. Second, using SU(3) lattice QCD Monte Carlo calculations, we investigate Abelian projected QCD in the maximally Abelian gauge, and find clear correlation of distribution similarity among the local chiral condensate, monopoles, and color magnetic fields in the Abelianized gauge configuration. As a statistical indicator, we measure the correlation coefficient $r$, and find a strong positive correlation of $r \simeq 0.8$ between the local chiral condensate and an Euclidean color-magnetic quantity $\mathcal{F}$ in Abelian projected QCD. The correlation is also investigated for the deconfined phase in thermal QCD. As an interesting conjecture, like magnetic catalysis, the chiral condensate is locally enhanced by the strong color-magnetic field around the monopoles in QCD.

**Keywords:** QCD; chiral symmetry; monopole; lattice QCD; spontaneous symmetry breaking; Abelian projection; magnetic catalysis

**Citation:** Suganuma, H.; Ohata, H. Local Correlation among the Chiral Condensate, Monopoles, and Color Magnetic Fields in Abelian Projected QCD. *Universe* **2021**, *7*, 318. https://doi.org/10.3390/universe7090318

Academic Editor: Dmitri Antonov

Received: 17 August 2021
Accepted: 25 August 2021
Published: 28 August 2021

**Publisher's Note:** MDPI stays neutral with regard to jurisdictional claims in published maps and institutional affiliations.

## 1. Introduction

Quantum chromodynamics (QCD) is an SU($N_c$) gauge theory to describe the strong interaction, and has presented many interesting subjects full of variety and difficult problems in physics. Actually, in spite of the simple form of the QCD action, this miracle theory creates hundreds of hadrons and leads to various interesting non-perturbative phenomena, such as color confinement and dynamical chiral-symmetry breaking [1].

This magic is due to the strong coupling of QCD in the low-energy region, and this strong-coupling nature drastically changes the vacuum structure itself. Therefore, a perturbative technique is no more workable and analytical treatment of QCD is fairly difficult in the strong-coupling region. As a reliable standard technique, lattice QCD Monte Carlo simulations have been applied to analyze non-perturbative QCD [2,3].

Among the non-perturbative properties of QCD, spontaneous chiral-symmetry breaking is particularly important in our real world. Indeed, chiral symmetry breaking drastically influences the vacuum structure and gives a non-trivial vacuum expectation value of the chiral condensate $\langle \bar{q}q \rangle$, which plays the role of an order parameter. Additionally, it is considered that chiral symmetry breaking leads to dynamical quark-mass generation [1,4], and creates most of the matter mass of our Universe, apart from the dark matter, because only small masses of u, d, current quarks, and electrons are Higgs-origin in atoms [5] and their contribution to the nucleon mass is estimated to be small [6]. In addition, chiral symmetry breaking inevitably accompanies light pions of the Nambu–Goldstone bosons, and their small mass gives range of the nuclear force.

In non-perturbative QCD, color confinement is also one of the most important phenomena in physics, and presents an extremely difficult mathematical problem. Experiments for hadron spectra and lattice QCD studies for various inter-quark potentials [7–10] show that the quark confining force is basically characterized by a universal physical quantity of the string tension $\sigma \simeq 0.89$ GeV/fm. This universal string tension is physically explained by one-dimensional squeezing of the color electric flux, i.e., the color flux-tube formation in hadrons, as is also indicated by lattice QCD for both mesons [3] and baryons [11]. As for the relation between color confinement and chiral symmetry breaking, it is not yet clarified directly from QCD. Although almost coincidence between deconfinement and chiral-restoration temperatures [12] suggests their close correlation, a lattice QCD analysis using the Dirac-mode expansion based on the Banks–Casher relation [13] indicates some independence of these phenomena in QCD [14,15].

For the quark confinement mechanism, Nambu [16], 't Hooft [17], and Mandelstam [18] proposed the dual superconductivity scenario, paying attention to analogy with the Abrikosov vortex in the superconductivity, where Cooper-pair condensation leads to the Meissner effect, and the magnetic flux is excluded or squeezed like a one-dimensional tube as the Abrikosov vortex. If the QCD vacuum can be regarded as the dual version of the superconductor, the electric-type color flux is squeezed between (anti)quarks in hadrons, and quark confinement can be physically explained by the dual Meissner effect. Because of the electromagnetic duality, the dual Meissner effect inevitably needs condensation of magnetic objects, i.e., color magnetic monopoles, which correspond to the dual version of the electric Cooper-pair bosonic field.

In the dual-superconductor picture for the QCD vacuum, however, there are two large gaps with QCD.

1. Although QCD is a non-Abelian gauge theory, the dual-superconductor picture is based on an Abelian gauge theory subject to the Maxwell-type equations including magnetic currents, where electromagnetic duality is manifest;
2. Although QCD includes only color electric variables, i.e., quarks and gluons, as the elementary degrees of freedom, the dual-superconductor picture requires condensation of color magnetic monopoles as a key concept.

Historically, to bridge between QCD and the dual-superconductivity, 't Hooft proposed Abelian gauge fixing [19], partial gauge fixing which only remains Abelian gauge degrees of freedom in QCD. By Abelian gauge fixing, QCD reduces into an Abelian gauge theory, where off-diagonal gluons behave as charged matter fields similar to $W_\mu^\pm$ in the Weinberg–Salam model and give the color electric current $j_\mu$ in terms of the residual Abelian gauge symmetry. As a remarkable fact in the Abelian gauge, color-magnetic monopoles appear as topological objects corresponding to the non-trivial homotopy group $\Pi_2(\mathrm{SU}(N_c)/\mathrm{U}(1)^{N_c-1}) = \mathbf{Z}_\infty^{N_c-1}$ in a similar manner to appearance of 't Hooft–Polyakov monopoles [20] in the SU(2) non-Abelian Higgs theory. Thus, in the Abelian gauge, QCD is reduced into an Abelian gauge theory, including both electric current $j_\mu$ and magnetic current $k_\mu$, which is expected to give a theoretical basis of the dual-superconductor picture for the confinement mechanism, although off-diagonal gluons remain as charged matter fields.

From the viewpoint of Abelianzation of QCD, the maximally Abelian (MA) gauge [21] is an interesting special Abelian gauge. In the MA gauge, off-diagonal gluons have a large effective mass of about 1 GeV in both SU(2) and SU(3) lattice QCD [22–24], so that off-diagonal gluons become infrared inactive, and only the Abelian gluon is relevant at distances larger than about 0.2 fm. Additionally, monopole condensation is suggested from appearance of long entangled monopole worldlines [21,25] and the magnetic screening in lattice QCD [26,27].

In this way, by taking the MA gauge, the QCD vacuum can be regarded as an Abelian dual superconductor at a large scale, and color magnetic monopoles seem to capture essence of non-perturbative QCD. Note, however, that, even without gauge fixing, there is an evidence of monopole condensation in non-Abelian gauge theories [27], and, therefore, it might be possible to define infrared-relevant monopoles in QCD and to construct the dual superconductor system in more general manner. In fact, MA gauge fixing gives a concrete way to extract infrared-relevant Abelian gauge manifold and monopoles from QCD.

In the context of the dual superconductor picture, close correlation between monopoles and chiral symmetry breaking was pointed out in the dual Ginzburg–Landau theory [28], in SU(2) lattice QCD in the MA gauge [29,30], and in SU(3) lattice QCD [31,32]. Since most of the pioneering lattice studies were done in SU(2) QCD or on a small lattice as $8^3 \times 4$, we recently investigated SU(3) QCD with a large volume, and find a clear correlation between monopoles and the chiral condensate in SU(3) lattice QCD in the MA gauge [33].

In this paper, as a continuation of Ref. [33], we proceed the lattice works for the relation between chiral symmetry breaking and color magnetic objects including monopoles. In particular, as a new point of this paper, we quantitatively study correlation of the local chiral condensate with color magnetic fields using the lattice gauge theory.

The organization of this paper is as follows. In Section 2, we review the MA gauge and Abelianization of QCD in SU(3) lattice formalism. In Section 3, we prepare magnetic objects in Abelian projected QCD. In Section 4, we consider the local chiral condensate and chiral symmetry breaking in Abelian gauge systems. In Section 5, we present idealized Abelian gauge systems of a static monopole–antimonopole pair on a lattice, and investigate the relation of the local chiral condensate with the magnetic objects. In Section 6, we perform SU(3) lattice QCD Monte Carlo calculations and study the relation among monopoles, magnetic fields, and the local chiral condensate in Abelian projected QCD in the MA gauge. Section 7 is devoted for summary and conclusion.

## 2. Maximally Abelian Gauge and Abelianization of QCD

To begin with, we briefly review the lattice formalism for maximally Abelian (MA) gauge fixing and Abelianization in QCD.

Continuum QCD is described with the quark field $q(x)$, the gluon field $A_\mu(x) \in$ su($N_c$) and the QCD gauge coupling $g$. In SU($N_c$) lattice QCD [3], the gluon field is described as the SU($N_c$) link variable $U_\mu(s) \equiv \exp\left(iagA_\mu(s)\right) \in$ SU($N_c$) on four-dimensional Euclidean lattices with the spacing $a$ and the volume $V = L_x L_y L_z L_t$.

Using the Cartan subalgebra $\vec{H} \equiv (T_3, T_8)$ in SU(3), MA gauge fixing is defined so as to maximize

$$R_{\mathrm{MA}}[U_\mu(s)] \equiv \sum_s \sum_{\mu=1}^4 \mathrm{tr}\left(U_\mu^\dagger(s)\vec{H}U_\mu(s)\vec{H}\right) = \sum_s \sum_{\mu=1}^4 \left(1 - \frac{1}{2}\sum_{i \neq j}|U_\mu(s)_{ij}|^2\right) \tag{1}$$

by the SU(3) gauge transformation, and, therefore, this gauge fixing strongly suppresses all the off-diagonal fluctuation of the SU(3) gauge field. In the MA gauge, the SU(3) gauge group is partially fixed remaining its maximal torus subgroup $U(1)_3 \times U(1)_8$ with the global Weyl (color permutation) symmetry [34], and QCD is reduced to an Abelian gauge theory.

From the SU(3) link variable $U_\mu^{\mathrm{MA}}(s) \in$ SU(3) in the MA gauge, we extract the Abelian link variable

$$u_\mu(s) = e^{i\vec{\theta}_\mu(s)\cdot\vec{H}} = \mathrm{diag}\left(e^{i\theta_\mu^1(s)}, e^{i\theta_\mu^2(s)}, e^{i\theta_\mu^3(s)}\right) \in U(1)_3 \times U(1)_8 \subset \mathrm{SU}(3) \tag{2}$$

by maximizing the overlap

$$R_{\mathrm{Abel}} \equiv \frac{1}{3}\mathrm{Re}\,\mathrm{tr}\left\{U_\mu^{\mathrm{MA}}(s)u_\mu^\dagger(s)\right\} \in \left[-\frac{1}{2}, 1\right]. \tag{3}$$

Note that the distance between $u_\mu(s)$ and $U_\mu^{\mathrm{MA}}(s)$ becomes the smallest in the SU(3) manifold, and there is a constraint $\sum_{i=1}^3 \theta_\mu^i(s) = 0 \,(\mathrm{mod}\, 2\pi)$ reflecting the uni-determinant of $u_\mu(s)$. Here, $\theta_\mu^i(s)$ $(i = 1, 2, 3)$ is taken to be the principal value of $-\pi \leq \theta_\mu^i(s) < \pi$.

The Abelian projection is defined by the simple replacement of the SU(3) link variable $U_\mu(s)$ by the Abelian link variable $u_\mu(s)$ for each gauge configuration, that is, $O[U_\mu(s)] \rightarrow$

$O[u_\mu(s)]$ for QCD operators. Abelian projected QCD is thus extracted from SU(3) QCD. The case of $\langle O[U_\mu(s)]\rangle \simeq \langle O[u_\mu(s)]\rangle$ is called "Abelian dominance" for the operator $O$ [35].

As a remarkable fact, Abelian dominance of quark confinement is shown in both SU(2) [36] and SU(3) lattice QCD [37–39]. Additionally, Abelian dominance of chiral symmetry breaking is observed in SU(2) [29–31] and SU(3) lattice QCD [33].

### 3. Magnetic Objects in Abelian Projected QCD

In this section, we prepare magnetic objects in Abelian projected QCD in four-dimensional Euclidean space-time.

*3.1. Monopoles in Abelian Projected QCD*

In this subsection, we define monopoles in Abelian projected QCD in lattice formalism [40]. Like the ordinary SU(3) plaquette, the Abelian plaquette variable is defined as

$$
\begin{aligned}
u_{\mu\nu}(s) &\equiv u_\mu(s)u_\nu(s+\hat{\mu})u_\mu^\dagger(s+\hat{\nu})u_\nu^\dagger(s) = e^{i\vec{\theta}_{\mu\nu}(s)\cdot\vec{H}} \\
&= \mathrm{diag}(e^{i\theta_{\mu\nu}^1(s)}, e^{i\theta_{\mu\nu}^2(s)}, e^{i\theta_{\mu\nu}^3(s)}) \in \mathrm{U}(1)^2 \subset \mathrm{SU}(3),
\end{aligned}
\tag{4}
$$

where $\hat{\mu}$ is the $\mu$-directed unit vector in the lattice unit. The Abelian field strength $\theta_{\mu\nu}^i(s)$ ($i = 1, 2, 3$) is the principal value of the exponent in $u_{\mu\nu}(s)$, and is defined as

$$
\begin{aligned}
&\partial_\mu\theta_\nu^i(s) - \partial_\nu\theta_\mu^i(s) = \theta_{\mu\nu}^i(s) - 2\pi n_{\mu\nu}^i(s), \\
&-\pi \leq \theta_{\mu\nu}^i(s) < \pi, \quad n_{\mu\nu}^i(s) \in \mathbb{Z},
\end{aligned}
\tag{5}
$$

with the forward derivative $\partial_\mu$. Note that $\theta_{\mu\nu}^i(s)$ is $\mathrm{U}(1)^2$ gauge invariant and corresponds to the regular Abelian field strength in the continuum limit of $a \to 0$, while $n_{\mu\nu}^i(s)$ corresponds to the singular gauge-variant Dirac string [40].

The electric current $j_\mu^i$ and the monopole current $k_\mu^i$ are defined from the Abelian field strength $\theta_{\mu\nu}^i$ as

$$
\begin{aligned}
j_\nu^i(s) &\equiv \partial_\mu'\theta_{\mu\nu}^i(s), \tag{6} \\
k_\nu^i(s) &\equiv \partial_\mu\tilde{\theta}_{\mu\nu}^i(s)/2\pi = \partial_\mu\tilde{n}_{\mu\nu}^i(s) \in \mathbb{Z}, \tag{7}
\end{aligned}
$$

where $\partial_\mu'$ is the backward derivative and $\tilde{\theta}_{\mu\nu}$ is the dual tensor of $\tilde{\theta}_{\mu\nu} \equiv \frac{1}{2}\epsilon_{\mu\nu\alpha\beta}\theta_{\alpha\beta}$. Both electric and monopole currents are $\mathrm{U}(1)^2$ gauge invariant, according to $\mathrm{U}(1)^2$ gauge invariance of $\theta_{\mu\nu}^i(s)$. In the lattice formalism, $k_\mu^i(s)$ is located at the dual lattice $L_{\mathrm{dual}}^4$ of $s^\alpha + \frac{1}{2}$ with flowing in $\mu$ direction [41].

In this way, Abelian projected QCD includes both electric current $j_\mu^i$ and monopole current $k_\mu^i$. Remarkably, lattice QCD shows monopole dominance, i.e., dominant role of monopoles for quark confinement in the MA gauge [42]. Additionally, lattice QCD shows monopole dominance for chiral symmetry breaking, that is, monopoles in the MA gauge crucially contribute to spontaneous chiral-symmetry breaking in both SU(2) [29,31] and SU(3) lattice QCD [33].

In the lattice formalism, the monopole current $k_\mu^i$ appears on the dual lattice $L_{\mathrm{dual}}^4$ of $s^\alpha + \frac{1}{2}$, and, therefore, we define the local monopole density

$$
\rho_{\mathrm{L}}(s) \equiv \frac{1}{3\cdot 2^4}\sum_{i=1}^{3}\sum_{s'\in P(s)}\sum_{\mu=1}^{4}\left|k_\mu^i(s')\right|,
\tag{8}
$$

where $P(s)$ denotes the dual lattices in the vicinity of $s$, i.e., $P(s) = \{s' \in L_{\mathrm{dual}}^4 \big| |s - s'| = 1\}$. Note here that the distance between the site $s$ and its closest dual site $s'$ is $|s - s'| = \sqrt{\sum_1^4(\frac{1}{2})^2} = 1$ in the four-dimensional Euclidean space-time.

*3.2. General Argument for Magnetic Instability and Magnetic Objects in the QCD Vacuum*

In QCD in the MA gauge, color magnetic monopoles generally appear, and play an important role in non-perturbative properties, which might looks curious, since the original QCD action does not have monopoles.

However, some active roles of magnetic objects would be natural in QCD, because QCD itself has color magnetic instability, and spontaneous generation of color magnetic fields generally takes place, as Savvidy first pointed out in 1977 [43,44].

In fact, in the QCD vacuum in the Minkowski space-time, the gluon condensate $\langle G_{\mu\nu}^a G_{\mu\nu}^a \rangle$ takes a large positive value, which physically means that the QCD vacuum is filled with color magnetic fields. Since the gluon condensate is expressed with color magnetic fields $\vec{H}_a$ and color electric fields $\vec{E}_a$ as

$$\langle G_{\mu\nu}^a G_{\mu\nu}^a \rangle = 2(\langle \vec{H}_a^2 \rangle - \langle \vec{E}_a^2 \rangle) > 0, \tag{9}$$

its large positivity means inevitable significant generation of color magnetic fields. Thus, some superior role of magnetic objects is expected instead of electric objects in the Minkowski QCD vacuum.

In the Euclidean space-time, because of the space-time $SO(4)$ symmetry, the roles of magnetic and electric fields become similar. Actually, the gluon condensate is written as $G_{\mu\nu}^a G_{\mu\nu}^a = 2(\vec{H}_a^2 + \vec{E}_a^2)$, where the electromagnetic duality is manifest. Then, in Euclidean QCD, the electric field often behaves as a magnetic field, and, therefore, we regard the Euclidean electric field as a sort of the magnetic field in this paper.

*3.3. Lorentz Invariant Quantities in Abelian Projected QCD*

In Abelian projected QCD, there are two Lorentz invariant quantities $\mathcal{F}$ and $\mathcal{G}$ in the Euclidean space-time:

$$\mathcal{F} \equiv \frac{1}{3} \sum_{i=1}^{3} \frac{1}{4} F_i^{\mu\nu} F_i^{\mu\nu} = \frac{1}{3} \sum_{i=1}^{3} \frac{1}{2} (\vec{H}_i^2 + \vec{E}_i^2), \tag{10}$$

$$\mathcal{G} \equiv \frac{1}{3} \sum_{i=1}^{3} \frac{1}{4} F_i^{\mu\nu} \tilde{F}_i^{\mu\nu} = \frac{1}{3} \sum_{i=1}^{3} \vec{H}_i \cdot \vec{E}_i, \tag{11}$$

with the color magnetic field $(\vec{H}_i)_j \equiv \frac{1}{2} \epsilon_{jkl} F_i^{kl}$ and the color electric field $(\vec{E}_i)_j \equiv F_i^{j4}$. These quantities are also invariant under the residual U(1)$^2$ gauge transformation and global Weyl transformation [34], i.e., permutation of the color index, in the MA gauge.

Here, $\mathcal{F}$ is parity-even and expresses total magnitude of magnetic fields in the Euclidean space-time, since the electric field behaves as a magnetic field there. In this paper, we simply call $\mathcal{F}$ "magnetic quantity" in Euclidean gauge theories. Note that $\mathcal{G}$ is parity-odd and is just the Abelian projected quantity of the topological charge density on instantons in QCD, which might relate to chiral symmetry breaking.

In the lattice formalism, the field strength tensor is a plaquette variable spanning at $s$, $s + \hat{\mu}, s + \hat{\nu}$, and $s + \hat{\mu} + \hat{\nu}$, so that we define the Abelian field strength $F_{\mu\nu}^i(s)$ as the local average of clover-type four plaquettes,

$$a^2 g F_{\mu\nu}^i(s) \equiv \frac{1}{4} \left( \theta_{\mu\nu}^i(s) + \theta_{\mu\nu}^i(s + \hat{\mu}) + \theta_{\mu\nu}^i(s + \hat{\nu}) + \theta_{\mu\nu}^i(s + \hat{\mu} + \hat{\nu}) \right), \tag{12}$$

and consider $\mathcal{F}$ and $\mathcal{G}$ as local quantities in each Abelian gauge configuration.

## 4. Local Chiral Condensate and Chiral Symmetry Breaking in Gauge Theories

In this section, we consider the chiral condensate and chiral symmetry breaking in the gauge theory in terms of the quark propagator.

### 4.1. Local Chiral Condensate in Lattice QCD

In this subsection, we briefly review the local chiral condensate in lattice formalism. The local chiral condensate can be calculated with the quark propagator for each gauge configuration $U = \{U_\mu(s)\}$ generated with the Monte Carlo method.

As the lattice fermion, we here adopt the Kogut–Susskind (KS) fermion [3]. For the KS fermion, the Dirac operator $\gamma_\mu D_\mu$ is expressed by $\eta_\mu D_\mu$ with the staggered phase $\eta_\mu(s) \equiv (-1)^{s_1+\cdots+s_{\mu-1}}$ $(\mu \geq 2)$ with $\eta_1(s) \equiv 1$. The KS Dirac operator is expressed as

$$(\eta_\mu D_\mu)_{ss'} = \frac{1}{2}\sum_{\mu=1}^{4}\sum_{\pm} \pm\eta_\mu(s)U_{\pm\mu}(s)\delta_{s\pm\hat{\mu},s'} \tag{13}$$

with $U_{-\mu}(s) \equiv U_\mu^\dagger(s - \hat{\mu})$, and the KS Dirac eigenvalue equation takes the form of

$$\frac{1}{2}\sum_{\mu=1}^{4}\sum_{\pm} \pm\eta_\mu(s)U_{\pm\mu}(s)\chi_n(s\pm\hat{\mu}) = i\lambda_n\chi_n(s). \tag{14}$$

Here, the quark field $q^\alpha(s)$ is described by a spinless Grassmann variable $\chi(s)$ [3], and the chiral condensate per flavor is evaluated as $\langle\bar{q}q\rangle = \langle\bar{\chi}\chi\rangle/4$ in the continuum limit.

The local chiral condensate can be calculated using the quark propagator of the KS fermion with a small quark mass $m$. The chiral-limit value is estimated by the chiral extrapolation of $m \to 0$. As a technical caution, the chiral and continuum limits do not commute for the KS fermion at the quenched level, although this problem would be absent in full QCD [45].

For the gauge configuration $U = \{U_\mu(s)\}$, the Euclidean KS fermion propagator is given by

$$G_U^{ij}(x,y) \equiv \langle\chi^i(x)\bar{\chi}^j(y)\rangle_U = \langle x,i|\left(\frac{1}{\eta_\mu D_\mu[U] + m}\right)|y,j\rangle \tag{15}$$

with the color index $i$ and $j$. This propagator is numerically obtained by solving the large-scale linear equation with a point source. The local chiral condensate for the gauge configuration $\{U_\mu(s)\}$ is expressed with the propagator as

$$\langle\bar{\chi}(x)\chi(x)\rangle_U = -\text{Tr}\,G_U(x,x). \tag{16}$$

Here, we consider the net chiral condensate by subtracting the contribution from the trivial vacuum $U = 1$ as

$$\langle\bar{\chi}\chi(x)\rangle_U \equiv \langle\bar{\chi}(x)\chi(x)\rangle_U - \langle\bar{\chi}\chi\rangle_{U=1}, \tag{17}$$

where the subtraction term is exactly zero in the chiral limit $m = 0$. The global chiral condensate is obtained by taking its average over the space-time $x$ and the gauge ensembles $U_1, U_2, ..., U_N$,

$$\langle\bar{\chi}\chi\rangle \equiv \sum_{x,i}\langle\bar{\chi}\chi(x)\rangle_{U_i}\Big/\sum_{x,i}1. \tag{18}$$

### 4.2. Chiral Symmetry Breaking in Abelian Gauge System

In this subsection, we analytically investigate relation between chiral symmetry breaking and the field strength in Euclidean Abelian gauge systems. For the simple argument, we consider Euclidean U(1) gauge systems with quasi-massless Dirac fermions coupled to the U(1) gauge field, although it is straightforward to generalize this argument to Abelian projected QCD with U(1)$^2$ gauge symmetry.

In the U(1) gauge system, the chiral condensate is proportional to the functional trace of the fermion propagator,

$$I \equiv \mathrm{Tr}\frac{1}{\slashed{D}+m} = -m\mathrm{Tr}\frac{1}{D^2 - m^2 + \frac{g}{2}\sigma \cdot F}, \tag{19}$$

with the covariant derivative $D_\mu \equiv \partial_\mu + igA_\mu$, the field strength $F_{\mu\nu}$, $\sigma \cdot F \equiv \sigma_{\mu\nu}F_{\mu\nu}$ and $\sigma_{\mu\nu} \equiv \frac{i}{2}[\gamma_\mu, \gamma_\nu]$. Note that $D^2 - m^2$ is a negative-definite operator, and all of its eigenvalues are negative. Since the trace of any odd-number product of $\gamma$-matrices is zero, we find

$$I = -m\mathrm{Tr}\frac{D^2 - m^2 - \frac{g}{2}\sigma \cdot F}{(D^2 - m^2)^2 - 2g^2(\mathcal{F} - \gamma_5\mathcal{G}) - \frac{g}{2}[D^2, \sigma \cdot F]}, \tag{20}$$

with

$$\mathcal{F} \equiv \frac{1}{4}F_{\mu\nu}F_{\mu\nu} = \frac{1}{2}(\vec{H}^2 + \vec{E}^2), \quad \mathcal{G} \equiv \frac{1}{4}F_{\mu\nu}\tilde{F}_{\mu\nu} = \vec{H} \cdot \vec{E}. \tag{21}$$

Because of the overall factor $m$ in $I$, $I$ goes to zero in the chiral limit $m \to 0$, unless the denominator becomes zero in this limit.

Since the operator $(D^2 - m^2)^2$ in the denominator is positive definite, to realize the zero denominator in $I$ in Equation (20), we need a significant negative contribution from the other three terms including $\mathcal{F}$, $\mathcal{G}$, or $[D^2, \sigma \cdot F]$. For instance, in the absence of the field strength, i.e., $F_{\mu\nu} \equiv 0$, one finds near $m \simeq 0$

$$I_{F_{\mu\nu}\equiv0} = m\mathrm{Tr}\frac{1}{p^2 + m^2} = m\gamma \int \frac{d^4p}{(2\pi)^4}\frac{1}{p^2 + m^2} = m\frac{\gamma}{16\pi^2}\int^{\Lambda^2}dp^2\frac{p^2}{p^2 + m^2} \simeq m\frac{\gamma\Lambda^2}{16\pi^2} \tag{22}$$

with the UV cut-off $\Lambda$ and the degeneracy $\gamma$. According to the positive denominator in the integrand, $I_{F_{\mu\nu}\equiv0}$ has no IR singularity to cancel $m$ of the numerator, and, therefore, $I_{F_{\mu\nu}\equiv0}$ goes to zero in the chiral limit of $m \to 0$.

To cancel $m$ in the numerator of $I$, we need a significantly large amount of the field strength so as to present zero mode in the denominator of $I$ and to keep $I$ non-zero in the chiral limit. Note here that $\mathcal{F}(\geq 0)$ always gives a negative (non-positive) contribution in the denominator of $I$, while the contribution from $\mathcal{G}$ or $[D^2, \sigma \cdot F]$ can be positive and negative. In fact, the magnetic quantity $\mathcal{F}$ can give the zero mode in the denominator of $I$, even without the contribution from $\mathcal{G}$ and $[D^2, \sigma \cdot F]$. In contrast, in Euclidean Abelian gauge systems, $\mathcal{F} \equiv \frac{1}{4}F_{\mu\nu}^2 = 0$ means $F_{\mu\nu} = 0$, and then $\mathcal{G} = [D^2, \sigma \cdot F] = 0$.

To conclude, the magnetic quantity $\mathcal{F}$ is expected to be significantly important to realize chiral symmetry breaking in Euclidean Abelian gauge theories, although, in some cases, the contribution from $\mathcal{G}$ and $[D^2, \sigma \cdot F]$ can assist the realization of chiral symmetry breaking.

In a special case of constant $F_{\mu\nu}$, one finds $[D^2, \sigma \cdot F] = 0$ for the Abelian system, and obtains

$$I = -m\mathrm{Tr}\frac{(D^2 - m^2)[(D^2 - m^2)^2 - 2g^2\mathcal{F}]}{[(D^2 - m^2)^2 - 2g^2\mathcal{F}]^2 - 4g^2\mathcal{G}^2}, \tag{23}$$

because of $\mathrm{tr}\gamma_5 = \mathrm{tr}\sigma_{\mu\nu} = \mathrm{tr}\gamma_5\sigma_{\mu\nu} = 0$. For more special case of a constant magnetic field, there occurs the Landau-level quantization, and the spatial degrees of freedom perpendicular to the magnetic field is frozen in the lowest Landau level. This infrared effective low-dimensionalization of the charged spinor dynamics induces chiral symmetry breaking in the chiral limit [46–48], which is known as magnetic catalysis [49].

## 5. Abelian Gauge System with a Static Monopole–Antimonopole Pair on a Lattice

In the QCD vacuum, complicated monopole world-lines generally emerge in the MA gauge [21,25], and, therefore, it is difficult to clarify the primary correlation with the chiral condensate among the magnetic objects, such as monopoles, $\mathcal{F}$ and $\mathcal{G}$.

In this section, to seek for the primary correlation with the chiral condensate, we create idealized Abelian gauge system with a monopole–antimonopole pair on a lattice, and investigate the relation among the local chiral condensate, monoples, and magnetic fields. Additionally, we consider a magnetic flux system without monopoles.

For simplicity, we here consider U(1) lattice gauge systems described by U(1) link variables

$$u_\mu(s) = e^{i\theta_\mu(s)} \in \mathrm{U}(1), \tag{24}$$

and quasi-massless Dirac fermions coupled to U(1) gauge fields with the coupling $g = 1$.

### 5.1. Static Monopole–Antimonopole Pair Systems

To begin with, we deal with an idealized Abelian gauge system of a static monopole–antimonopole pair on a periodic lattice of the four-dimensional Euclidean space-time.

In the three-dimensional space $\mathbf{R}^3$, let us consider a static monopole–antimonopole pair with the distance of $l$ in $z$-direction. To realize such a lattice gauge system, we set the Abelian link-variable $u_\mu(s)$ to be

$$u_x(s) = u_y(s + \hat{x}) = u_x^\dagger(s + \hat{y}) = u_y^\dagger(s) = i \quad \text{for} \quad s_x = s_y = 0, \ 1 \le s_z \le l, \tag{25}$$

otherwise $u_\mu(s) = 1$.

Figure 1 shows the building-block plaquette to realize a static monopole–antimonopole pair on the lattice. Here, only the red link-variables take a non-trivial value of $i$.

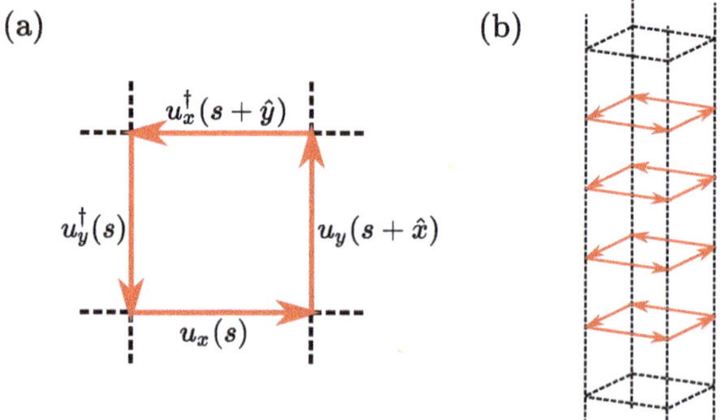



**Figure 1.** The building-block plaquette to realize a static monopole–antimonopole pair on the lattice in (**a**) the $x$-$y$ plane and (**b**) spatial $\mathbf{R}^3$ for $s = (0, 0, s_z, s_t)$ with $1 \le s_z \le l$. Only the red link-variables take a non-trivial value of $i$. The all-red plaquette induces the singular Dirac string at its center on the dual lattice. A physical magnetic field is also created in the neighboring plaquette $u_{xy}(s)$ including only one red link.

As for the phase variable $\theta_\mu(s)$, which corresponds to the Abelian gluon, one finds

$$\theta_x(s) = \theta_y(s + \hat{x}) = -\theta_x(s + \hat{y}) = -\theta_y(s) = \frac{\pi}{2} \quad \text{for} \quad s_x = s_y = 0, \ 1 \le s_z \le l, \tag{26}$$

otherwise $\theta_\mu(s) = 0$. For the all-red plaquette with $s_x = s_y = 0$ and $1 \le s_z \le l$, one gets

$$\partial_x \theta_y(s) - \partial_y \theta_x(s) = \theta_x(s) + \theta_y(s + \hat{x}) - \theta_x(s + \hat{y}) - \theta_y(s) = 2\pi, \tag{27}$$

which leads to the Dirac string of $n_{xy}(s) = -1$ and zero field strength $\theta_{xy}(s) = 0$, because of the definition of the field strength $\theta_{\mu\nu}(s)$ and the Dirac string $n_{\mu\nu}(s)$,

$$\partial_\mu \theta_\nu(s) - \partial_\nu \theta_\mu(s) = \theta_{\mu\nu}(s) - 2\pi n_{\mu\nu}(s), \quad -\pi \leq \theta_{\mu\nu}(s) < \pi, \quad n_{\mu\nu}(s) \in \mathbb{Z}. \tag{28}$$

Thus, the all-red plaquette induces the singular Dirac string at its center on the dual lattice. In fact, for the idealized system in Figure 1b, a Dirac string appears inside the all-red plaquette.

At the terminal of the Dirac string, a monopole or an anti-monopole appears on the dual lattice, as shown in Figure 2. Actually, the three-dimensional spatial cube including only one all-red plaquette has a static (anti)monopole at its center (on the dual lattice), because only one $n_{kl}(s)$ has non-zero value of $\pm 1$ among the six independent plaquettes composing the cube,

$$
\begin{aligned}
k_4(s) &= \partial_j \tilde{n}_{j4}(s) = \frac{1}{2}\epsilon_{jkl}\partial_j n_{kl}(s) = \frac{1}{2}\epsilon_{jkl}\{n_{kl}(s+\hat{j}) - n_{kl}(s)\} \\
&= n_{xy}(s+\hat{z}) - n_{xy}(s) + n_{yz}(s+\hat{x}) - n_{yz}(s) + n_{zx}(s+\hat{y}) - n_{zx}(s) \\
&= \pm 1.
\end{aligned}
\tag{29}
$$

Thus, this idealized system includes a static monopole at $(\frac{1}{2}, \frac{1}{2}, \frac{1}{2})$ and a static anti-monopole at $(\frac{1}{2}, \frac{1}{2}, l + \frac{1}{2})$ in spatial $\mathbf{R}^3$.

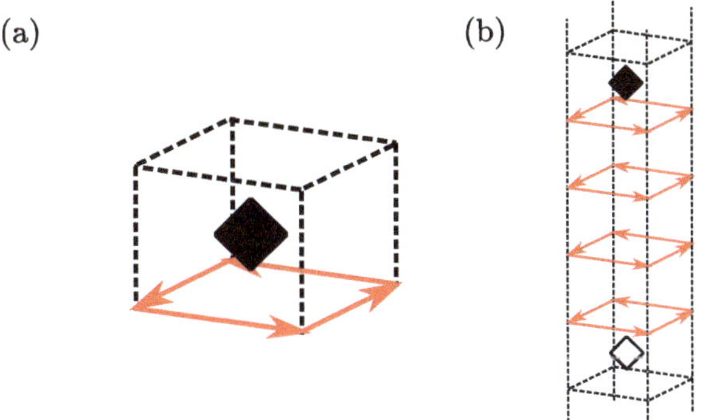

**Figure 2.** The link-variables to realize a static monopole–antimonopole pair on the lattice in spatial $\mathbf{R}^3$. Only the red link-variables take a non-trivial value of $i$. (**a**) The cube including only one all-red plaquette induces a magnetic monopole at its inside on the dual lattice. (**b**) A monopole (black diamond) and an anti-monopole (white diamond) appear at the two terminals of the red plaquette tower.

This monopole and anti-monopole system has also physical magnetic flux around the line segment connecting the monopole pair. In fact, a physical magnetic field is created in the neighboring plaquette $u_{xy}(s)$ of the all-red plaquette in Figure 1. In this idealized system, only the plaquette $u_{xy}(s)$ including one red link takes a non-trivial value as

$$u_{xy}(s) = -i = e^{-i\pi/2} \quad \text{in case with one nontrivial link,} \tag{30}$$

otherwise $u_{\mu\nu}(s) = 1$. Note here that, by gauge transformation, the location of the Dirac string is generally changed, but the physical field strength is never changed.

Figure 3 shows the local chiral condensate $\langle \bar{\chi}\chi(s)\rangle_u$ and the magnetic quantity $\mathcal{F} \equiv \frac{1}{4}F_{\mu\nu}^2 = \frac{1}{2}\vec{H}^2$ for $l = 4$ in the three dimensional space $\mathbf{R}^3$. In this demonstration, the quark mass is taken to be $m = 0.01$ in the lattice unit.

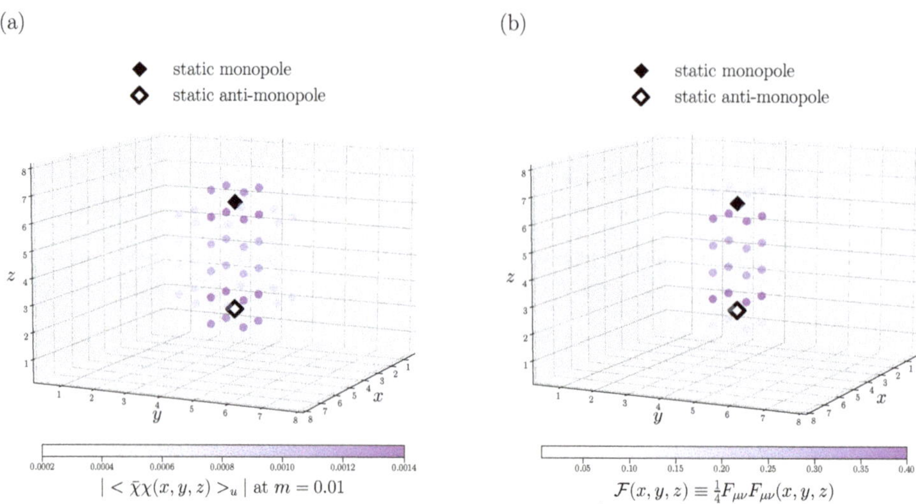



**Figure 3.** An idealized Abelian gauge system of a static monopole and anti-monopole pair with the distance of $l = 4$. In the three dimensional space $\mathbf{R}^3$, the value is visualized with the color graduation for (**a**) the local chiral condensate and (**b**) the magnetic quantity $\mathcal{F}$.

For this idealized static system, there actually appears a magnetic field $\vec{H}$, i.e., non-zero flux of $\mathcal{F} = \frac{1}{2}\vec{H}^2 > 0$, in space between the monopole and the anti-monopole, and the local chiral condensate takes a significant value in the vicinity of the magnetic field. In contrast, one finds $\mathcal{G} = \vec{H} \cdot \vec{E} = 0$ everywhere, since only spatial plaquettes take a non-trivial value and $\vec{E} = \vec{0}$. Thus, in this system, it is likely that the magnetic field stemming from monopoles has the primary correlation with the local chiral condensate.

### 5.2. Static Magnetic Flux System

Next, let us investigate a static magnetic flux system without monopoles. Owing to the spatial periodicity, the special case of $l = L_z$ in the static monopole–antimonopole system has no (anti)monopoles, because of the magnetic-charge cancellation. In this special case of $l = L_z$, there only exists a physical static magnetic flux along $z$-direction.

Figure 4 shows the local chiral condensate $\langle \bar{\chi}\chi(s) \rangle_u$ and the magnetic quantity $\mathcal{F} \equiv \frac{1}{4}F_{\mu\nu}^2 = \frac{1}{2}\vec{H}^2$ for $l = L_z$ in spatial $\mathbf{R}^3$, taking the quark mass of $m = 0.01$ in the lattice unit.

Again, the local chiral condensate takes a significant value in the vicinity of the magnetic field $\vec{H}$, i.e., non-zero flux of $\mathcal{F} = \frac{1}{2}\vec{H}^2 > 0$, even without (anti)monopoles. Note also that this system has $\mathcal{G} = \vec{H} \cdot \vec{E} = 0$ everywhere, because of $\vec{E} = \vec{0}$. Therefore, in this idealized system, we conclude that the magnetic field or $\mathcal{F}$ has the primary correlation with the local chiral condensate.

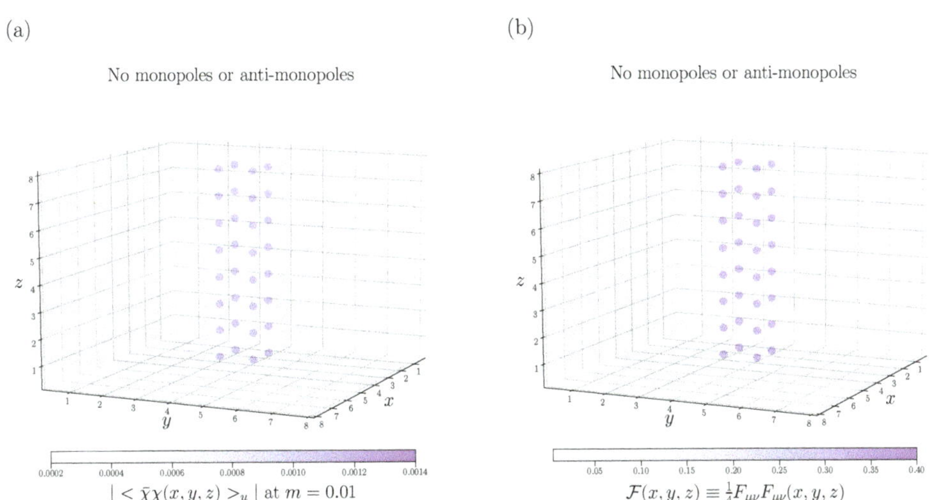



**Figure 4.** An idealized Abelian gauge system of a static magnetic flux without monopoles. In the three dimensional space $\mathbf{R}^3$, the value is visualized with the color graduation for (**a**) the local chiral condensate and (**b**) the magnetic quantity $\mathcal{F}$.

## 6. Lattice QCD Study for Local Chiral Condensate, Monopoles, and Magnetic Fields

In our previous study with lattice QCD, we observed a strong correlation between the local chiral condensate and monopoles in Abelian projected QCD [33]. As a possible reason of this correlation, we conjectured that the strong magnetic field around monopoles is responsible to chiral symmetry breaking in QCD, similarly to the magnetic catalysis [46–49].

In this section, using lattice QCD Monte Carlo simulations, we investigate the relation among the local chiral condensate, monopoles, and magnetic fields in Abelian projected QCD. In this paper, the SU(3) lattice QCD simulation is performed using the standard plaquette action at the quenched level. In each space-time direction, we impose the periodic boundary condition for link variables, and the anti-periodic boundary condition for quarks in order to describe also thermal QCD.

For the numerical Monte Carlo calculation, we basically adopt the lattice parameter of $\beta \equiv 2N_c/g^2 = 6.0$ and the size $V = 24^4$. The lattice spacing $a \simeq 0.1$ fm is obtained from the string tension $\sigma = 0.89$ GeV/fm [37]. Additionally, we adopt $\beta = 6.0$ and $V = 24^3 \times 6$ for the high-temperature deconfined phase at $T \simeq 330$ MeV above the critical temperature.

Using the pseudo-heat-bath algorithm, we generate 100 and 200 gauge configurations for $V = 24^4$ and $24^3 \times 6$, respectively. All the gauge configurations are taken every 500 sweeps after thermalization of 5000 sweeps. MA gauge fixing is performed with the stopping criterion that the deviation $\Delta R_{\mathrm{MA}}/(4V)$ becomes smaller than $10^{-5}$ in 100 iterations. For the calculation of the local chiral condensate, we use the quark propagator of the KS fermion with the quark mass of $m = 0.01, 0.015, 0.02$ in the lattice unit, Here, the quark mass is taken to be finite, since the chiral and continuum limits do not commute for the KS fermion at the quenched level [45]. The jackknife method is used for statistical error estimates.

For each lattice gauge configuration of Abelian projected QCD in the MA gauge, we calculate the local monopole density $\rho_L(s)$, the local chiral condensate, and the Lorentz invariants $\mathcal{F}$ and $\mathcal{G}$, defined in Section 3.

### 6.1. Distribution Similarity between Local Chiral Condensate and Magnetic Variables

To begin with, we pick up a gauge configuration generated in lattice QCD on $V = 24^4$ at $\beta = 6.0$, and investigate correlation between the local chiral condensate and magnetic variables.

Figure 5 shows the local chiral condensate $\langle \bar{\chi}\chi(s) \rangle_u$ with the quark mass of $m = 0.02$, the local monopole density $\rho_{\rm L}(s)$, and the Lorentz invariants $\mathcal{F}(s)$ and $|\mathcal{G}(s)|$, respectively, as well as the monopole location in the space $\mathbf{R}^3$ at a time slice, for a typical gauge configuration of Abelian projected QCD.

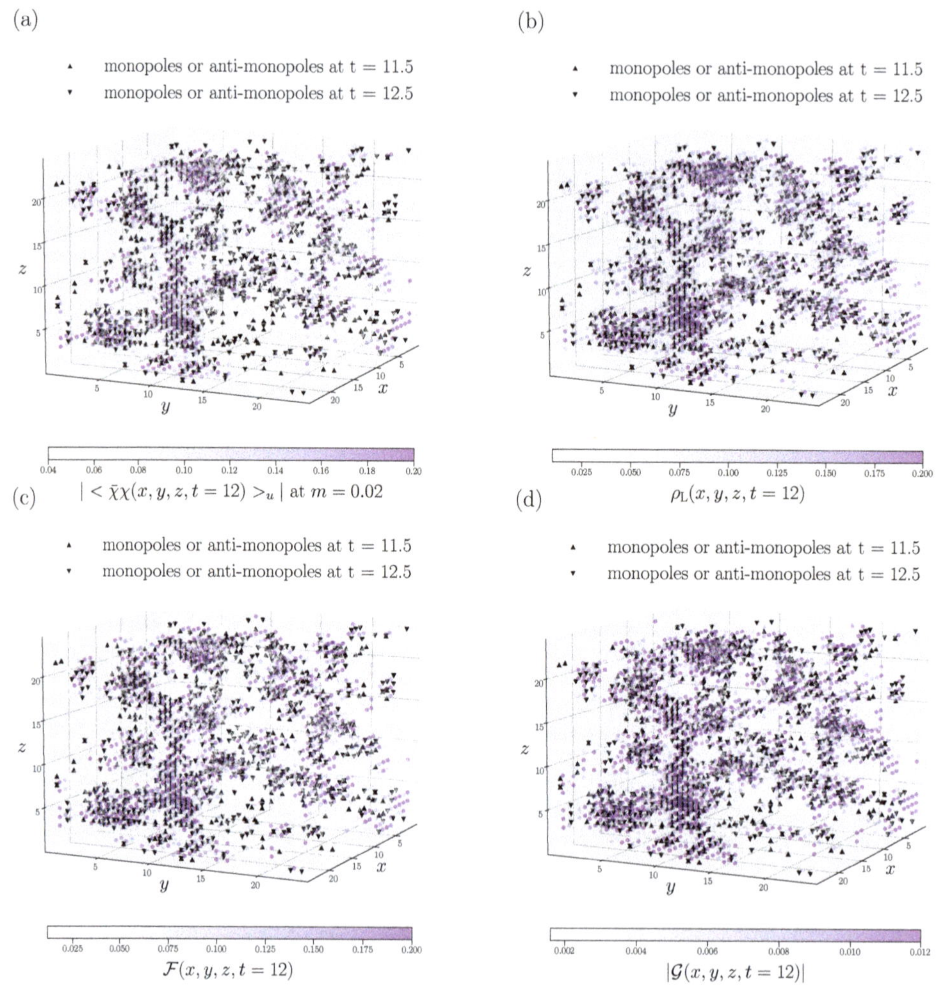



**Figure 5.** Lattice QCD results for (**a**) the local chiral condensate $\langle \bar{\chi}\chi(s) \rangle_u$ with the quark mass of $m = 0.02$, (**b**) the local monopole density $\rho_{\rm L}(s)$, and the Lorentz invariants (**c**) $\mathcal{F}(s)$ and (**d**) $|\mathcal{G}(s)|$ in spatial $\mathbf{R}^3$ at a time slice, for a typical gauge configuration of Abelian projected QCD. The value is visualized with the color graduation. Monopoles at $t = 11.5$ and $12.5$ are plotted with upper and lower triangles, respectively.

From Figure 5a, one finds that the local chiral condensate tends to take a large value near the monopole location [33]. Since monopoles appear on the dual lattice, we show the local monopole density $\rho_{\rm L}(s)$, as the average on closest dual sites. Of course, $\rho_{\rm L}(s)$ takes a large value near the monopole. The distribution of the the local monopole density resembles that of the local chiral condensate, as was pointed out in Ref. [33]. Figure 5c,d show the Lorentz invariants $\mathcal{F}$ and $\mathcal{G}$, respectively. As a new result in this paper, we find that the distributions of $\mathcal{F}$ and $\mathcal{G}$ also resemble that of the local chiral condensate.

The close relation of monopoles with $\mathcal{F}$ and $\mathcal{G}$ might be understood, since the field strength tensor relates to monopoles as $\partial_\mu \tilde{F}^i_{\mu\nu} = k^i_\nu$. Roughly speaking, the monopole can be a kind of source of $\mathcal{F}$ and $\mathcal{G}$. In contrast, their similarity with the local chiral condensate is fairly non-trivial.

In any case, we find clear correlation of distribution similarity among the local chiral condensate, the local monopole density, and the Lorentz invariants $\mathcal{F}$ and $\mathcal{G}$ in Abelian projected QCD in the MA gauge.

### 6.2. Correlation Coefficients between Local Chiral Condensate and Magnetic Variables

In this subsection, we quantify the similarity between the local chiral condensate $\langle \bar{\chi}\chi(s) \rangle_u$ and magnetic variables, i.e., $\rho_L(s)$, $\mathcal{F}(s)$ and $\mathcal{G}(s)$, defined in Section 3. To this end, we use all the generated 100 gauge configurations in lattice QCD on $V = 24^4$ at $\beta = 6.0$, and calculate the local chiral condensate at $2^4$ distant space-time points for each gauge configuration, resulting 1600 data points at each quark mass.

Figure 6 shows the scatter plot between the local chiral condensate $\langle \bar{\chi}\chi(s) \rangle_u$ and magnetic variables, i.e., the local monopole density $\rho_L(s)$, Lorentz invariants $\mathcal{F}(s)$ and $|\mathcal{G}(s)|$, respectively, using 100 gauge configurations of Abelian projected QCD in the MA gauge, with the quark mass of $m = 0.01, 0.015, 0.02$ in the lattice unit. In Figure 6, positive correlation is qualitatively found between the local chiral condensate and the magnetic variables, $\rho_L$, $\mathcal{F}$ and $|\mathcal{G}|$, respectively.

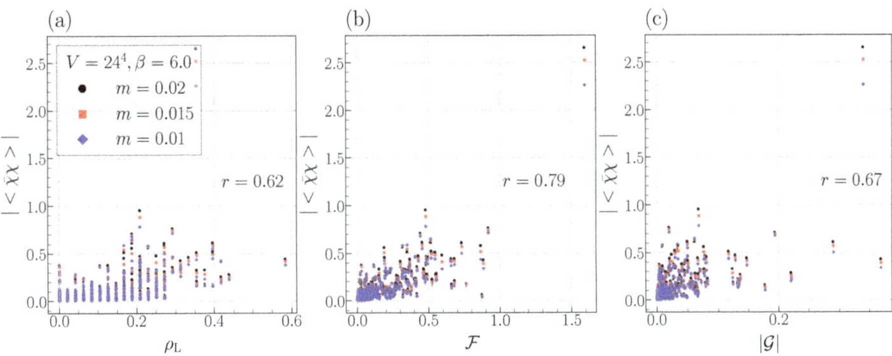


**Figure 6.** The scatter plot between the local chiral condensate $\langle \bar{\chi}\chi(s) \rangle_u$ and (**a**) the local monopole density $\rho_L(s)$, (**b**) $\mathcal{F}(s)$ and (**c**) $|\mathcal{G}(s)|$, using 100 Abelianized gauge configurations in SU(3) lattice QCD with $\beta = 6.0$ and $V = 24^4$ at each quark mass $m$.

Next, we consider a quantitative analysis using correlation coefficients between the local chiral condensate and the magnetic variables, as a statistical indicator of correlation. In general, for arbitrary two statistical ensembles $\{A_i\}$ and $\{B_i\}$, their correlation coefficient $r$ is defined as

$$ r \equiv \frac{\langle (A - \langle A \rangle)(B - \langle B \rangle) \rangle}{\sigma_A \sigma_B}, \tag{31} $$

using the average notation $\langle\ \rangle$ and the standard deviation $\sigma_A \equiv \sqrt{\langle (A - \langle A \rangle)^2 \rangle}$ and $\sigma_B \equiv \sqrt{\langle (B - \langle B \rangle)^2 \rangle}$. Here, $r = 1$ means perfect positive linear correlation, and $r \gtrsim 0.7$ indicates strong positive linear correlation.

We measure correlation coefficients between the local chiral condensate $|\langle \bar{\chi}\chi(s) \rangle_u|$ and three magnetic variables, $\rho_L(s)^\alpha$, $\mathcal{F}(s)^\alpha$ and $|\mathcal{G}(s)|^\alpha$, at various exponent $\alpha$, using 100 gauge configurations of Abelian projected QCD in SU(3) lattice QCD at $\beta = 6.0$ on $V = 24^4$, for the quark mass $m = 0.01$ in the lattice unit. Table 1 shows the result for the correlation coefficients.

**Table 1.** Correlation coefficients between the local chiral condensate $|\langle \bar{\chi}\chi(s)\rangle_u|$ and three magnetic variables, $\rho_{\mathrm{L}}(s)^\alpha$, $\mathcal{F}(s)^\alpha$, and $|\mathcal{G}(s)|^\alpha$ at various $\alpha$, using 100 gauge configurations of Abelian projected QCD in SU(3) lattice QCD at $\beta = 6.0$ on $V = 24^4$, for the quark mass $m = 0.01$ in the lattice unit.

| Lattice | $\alpha$ | $\rho_{\mathrm{L}}^\alpha$ | $\mathcal{F}^\alpha$ | $|\mathcal{G}|^\alpha$ |
|---|---|---|---|---|
| $V = 24^4$, $\beta = 6.0$ | 0.25 | 0.47 | 0.63 | 0.60 |
| | 0.5 | 0.55 | 0.71 | 0.67 |
| | 1 | 0.62 | 0.79 | 0.67 |
| | 1.5 | 0.63 | 0.81 | 0.60 |
| | 2 | 0.60 | 0.80 | 0.55 |

Quantitatively, the magnetic quantity $\mathcal{F}$ has the strongest correlation with the chiral condensate rather than $\rho_{\mathrm{L}}$ and $\mathcal{G}$. As a conclusion of this paper, we find a strong positive correlation of $r \simeq 0.8$ between the local chiral condensate $|\langle \bar{\chi}\chi(s)\rangle_u|$ and the magnetic quantity $\mathcal{F}(s)$ in the confined vacuum of Abelian projected QCD.

*6.3. High-Temperature Deconfined Phase*

Finally, we also investigate a high-temperature deconfined phase in lattice QCD on $V = 24^3 \times 6$ at $\beta = 6.0$, where the temperature is $T \simeq 330$ MeV above the critical temperature. We generate 200 gauge configurations, and calculate the local chiral condensate at $2^3$ distant space points at a time slice for each gauge configuration, resulting 1600 data points at each quark mass.

Figure 7 shows the scatter plot between the local chiral condensate $|\langle \bar{\chi}\chi(s)\rangle_u|$ and magnetic variables, i.e., the local monopole density $\rho_{\mathrm{L}}(s)$, and Lorentz invariants $\mathcal{F}(s)$ and $|\mathcal{G}(s)|$, respectively, using 200 gauge configurations of Abelian projected QCD in the MA gauge, with the quark mass of $m = 0.01, 0.015, 0.02$ in the lattice unit.

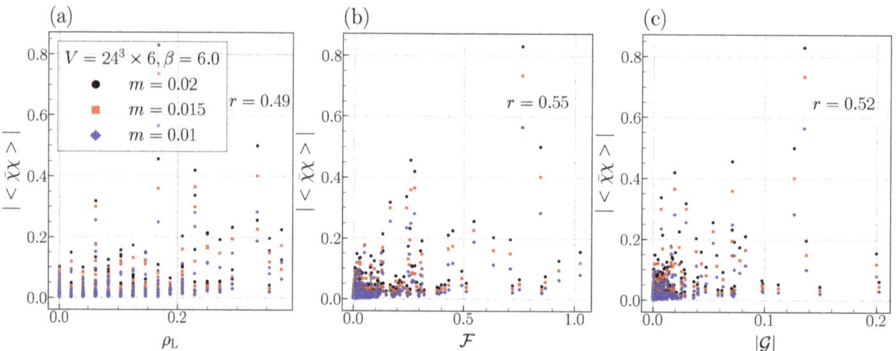



**Figure 7.** Result of the high-temperature deconfined phase for the scatter plot between the local chiral condensate $\langle \bar{\chi}\chi(s)\rangle_u$ and (**a**) $\rho_{\mathrm{L}}(s)$, (**b**) $\mathcal{F}(s)$ and (**c**) $|\mathcal{G}(s)|$, using 200 Abelianized gauge configurations in SU(3) lattice QCD with $\beta = 6.0$ and $V = 24^3 \times 6$ at each quark mass $m$.

We show in Table 2 correlation coefficients between the local chiral condensate $|\langle \bar{\chi}\chi(s)\rangle_u|$ and three magnetic variables, $\rho_{\mathrm{L}}(s)^\alpha$, $\mathcal{F}(s)^\alpha$ and $|\mathcal{G}(s)|^\alpha$, at various exponent $\alpha$, using 200 gauge configurations of Abelian projected QCD in SU(3) lattice QCD at $\beta = 6.0$ on $V = 24^3 \times 6$, for the quark mass $m = 0.01$ in the lattice unit.

From Figure 7 and Table 2, all the correlations between the local chiral condensate and the three magnetic variables, $\rho_{\mathrm{L}}$, $\mathcal{F}$ and $\mathcal{G}$, become weaker in the deconfined phase, where the chiral condensate itself goes to zero in the chiral limit.

**Table 2.** Correlation coefficients in the deconfined phase between the local chiral condensate $|\langle \bar{\chi}\chi(s)\rangle_u|$ and three magnetic variables, $\rho_L(s)^\alpha$, $\mathcal{F}(s)^\alpha$ and $|\mathcal{G}(s)|^\alpha$ at various $\alpha$ in Abelian projected QCD of SU(3) lattice QCD at $\beta = 6.0$ on $V = 24^3 \times 6$ for $m = 0.01$.

| Lattice | $\alpha$ | $\rho_L^\alpha$ | $\mathcal{F}^\alpha$ | $|\mathcal{G}|^\alpha$ |
|---|---|---|---|---|
| $V = 24^3 \times 6$, $\beta = 6.0$ | 0.25 | 0.37 | 0.49 | 0.49 |
| | 0.5 | 0.43 | 0.55 | 0.55 |
| | 1 | 0.49 | 0.55 | 0.52 |
| | 1.5 | 0.50 | 0.51 | 0.45 |
| | 2 | 0.49 | 0.46 | 0.38 |

## 7. Summary and Conclusions

We have studied the relation among the local chiral condensate, monopoles, and magnetic fields, using the lattice gauge theory, as a continuation of Ref. [33].

First, we have created idealized Abelian gauge systems of (1) a static monopole–antimonopole pair, and (2) a magnetic flux without monopoles, on a four-dimensional Euclidean lattice. In these systems, we have calculated the local chiral condensate on quasi-massless fermions coupled to the Abelian gauge field, and have found that the chiral condensate is localized in the vicinity of the magnetic field.

Second, performing SU(3) lattice QCD Monte Carlo simulations, we have investigated Abelian projected QCD in the maximally Abelian gauge, and have found clear correlation of distribution similarity among the local chiral condensate, color monopoles, and color magnetic fields in the Abelianized gauge configuration.

As a statistical indicator, we have measured the correlation coefficient $r$, and have found a strong positive correlation of $r \simeq 0.8$ between the local chiral condensate and the Euclidean color-magnetic quantity $\mathcal{F}$.

We have also examined the local correlation in the deconfined phase of thermal QCD, and have found that the correlation between the local chiral condensate and magnetic variables becomes weaker.

Thus, in this paper, we have observed a strong correlation between the local chiral condensate and magnetic fields in both idealized Abelian gauge systems and Abelian projected QCD. From these results, we conjecture that the chiral condensate is locally enhanced by the strong color-magnetic field around the monopoles in Abelian projected QCD, like magnetic catalysis.

Note, however, that this correlation does not necessarily mean that chiral symmetry breaking is caused by the non-uniform magnetic field. To realize spontaneous chiral-symmetry breaking, as was discussed in Section 4.2, we need some zero mode in the denominator of $I$ in the chiral limit. In the context of the dual superconductor picture, this might be realized by condensation of monopoles, as was suggested in the dual Ginzburg-Landau theory [28].

To conclude, once chiral symmetry is spontaneously broken, the local chiral condensate is expected to have a strong correlation with the color magnetic field.

**Author Contributions:** Conceptualization, H.S.; methodology, H.S.; software, H.O.; validation, H.O.; formal analysis, H.S. and H.O.; investigation, H.O.; resources, H.O.; data curation, H.O.; writing—original draft preparation, H.S. and H.O.; writing—review and editing, H.S. and H.O.; visualization, H.O.; supervision, H.S.; project administration, H.S.; funding acquisition, H.S. and H.O. All authors have read and agreed to the published version of the manuscript.

**Funding:** H.S. was supported in part by the Grants-in-Aid for Scientific Research [19K03869] from Japan Society for the Promotion of Science. H.O. was supported by a Grant-in-Aid for JSPS Fellows (Grant No. 21J20089).

**Data Availability Statement:** The basic data of this study are available from the authors.

**Acknowledgments:** Most of numerical calculations have been performed on OCTOPUS, at the Cybermedia Center, Osaka University. We have used PETSc to solve linear equations for the Dirac operator [50–52].

**Conflicts of Interest:** The authors declare no conflict of interest.

## References


1. Nambu, Y.; Jona-Lasinio, G. Dynamical Model of Elementary Particles Based on an Analogy with Superconductivity. 1. *Phys. Rev.* **1961**, *122*, 345–358. [CrossRef]
2. Creutz, M. Monte Carlo Study of Quantized SU(2) Gauge Theory. *Phys. Rev. D* **1980**, *21*, 2308–2315. [CrossRef]
3. Rothe, H.J. *Lattice Gauge Theories*, 4th ed.; World Scientific: Singapore, 2012.
4. Higashijima, K. Dynamical Chiral Symmetry Breaking. *Phys. Rev. D* **1984**, *29*, 1228. [CrossRef]
5. Particle Data Group; Zyla, P.A.; Barnett, R.M.; Beringer, J.; Dahl, O.; Dwyer, D.A.; Groom, D.E.; Lin, C.J.; Lugovsky, K.S.; Pianori, E. Review of Particle Physics. *Prog. Theor. Exp. Phys. (PTEP)* **2020**, *2020*, 083C01. [CrossRef]
6. Ji, X.D. Breakup of hadron masses and energy—Momentum tensor of QCD. *Phys. Rev. D* **1995**, *52*, 271–281. [CrossRef]
7. Takahashi, T.T.; Matsufuru, H.; Nemoto, Y.; Suganuma, H. The Three quark potential in the SU(3) lattice QCD. *Phys. Rev. Lett.* **2001**, *86*, 18–21. [CrossRef]
8. Takahashi, T.T.; Suganuma, H.; Nemoto, Y.; Matsufuru, H. Detailed analysis of the three quark potential in SU(3) lattice QCD. *Phys. Rev. D* **2002**, *65*, 114509. [CrossRef]
9. Okiharu, F.; Suganuma, H.; Takahashi, T.T. Detailed analysis of the tetraquark potential and flip-flop in SU(3) lattice QCD. *Phys. Rev. D* **2005**, *72*, 014505. [CrossRef]
10. Okiharu, F.; Suganuma, H.; Takahashi, T.T. First study for the pentaquark potential in SU(3) lattice QCD. *Phys. Rev. Lett.* **2005**, *94*, 192001. [CrossRef]
11. Bornyakov, V.; Ichie, H.; Mori, Y.; Pleiter, D.; Polikarpov, M.; Schierholz, G.; Streuer, T.; Stuben, H.; Suzuki, T. Baryonic Flux in Quenched and Two Flavor Dynamical QCD. *Phys. Rev. D* **2004**, *70*, 054506. [CrossRef]
12. Karsch, F. Lattice QCD at high temperature and density. *Lect. Notes Phys.* **2002**, *583*, 209–249._6. [CrossRef]
13. Banks, T.; Casher, A. Chiral Symmetry Breaking in Confining Theories. *Nucl. Phys. B* **1980**, *169*, 103–125. [CrossRef]
14. Doi, T.M.; Suganuma, H.; Iritani, T. Relation between Confinement and Chiral Symmetry Breaking in Temporally Odd-number Lattice QCD. *Phys. Rev. D* **2014**, *90*, 094505. [CrossRef]
15. Suganuma, H.; Doi, T.M.; Iritani, T. Analytical Formulae of the Polyakov and Wilson Loops with Dirac Eigenmodes in Lattice QCD. *Prog. Theor. Exp. Phys. (PTEP)* **2016**, *2016*, 013B06. [CrossRef]
16. Nambu, Y. Strings, Monopoles and Gauge Fields. *Phys. Rev. D* **1974**, *10*, 4262. [CrossRef]
17. 't Hooft, G. *Gauge Theories with Unified Weak, Electromagnetic, and Strong Interactions*; High-Energy Particle Physics; Editorice Compositori: Bologna, Italy, 1975.
18. Mandelstam, S. Vortices and Quark Confinement in Nonabelian Gauge Theories. *Phys. Rep.* **1976**, *23*, 245–249. [CrossRef]
19. 't Hooft, G. Topology of the Gauge Condition and New Confinement Phases in Nonabelian Gauge Theories. *Nucl. Phys. B* **1981**, *190*, 455–478. [CrossRef]
20. Shnir, Y. Magnetic monopoles. *Phys. Part. Nucl. Lett.* **2011**, *8*, 749–754. [CrossRef]
21. Kronfeld, A.S.; Laursen, M.; Schierholz, G.; Wiese, U. Monopole Condensation and Color Confinement. *Phys. Lett. B* **1987**, *198*, 516–520. [CrossRef]
22. Amemiya, K.; Suganuma, H. Off-diagonal Gluon Mass Generation and Infrared Abelian Dominance in the Maximally Abelian Gauge in Lattice QCD. *Phys. Rev. D* **1999**, *60*, 114509. [CrossRef]
23. Gongyo, S.; Iritani, T.; Suganuma, H. Off-diagonal Gluon Mass Generation and Infrared Abelian Dominance in Maximally Abelian Gauge in SU(3) Lattice QCD. *Phys. Rev. D* **2012**, *86*, 094018. [CrossRef]
24. Gongyo, S.; Suganuma, H. Gluon Propagators in Maximally Abelian Gauge in SU(3) Lattice QCD. *Phys. Rev. D* **2013**, *87*, 074506. [CrossRef]
25. Kronfeld, A.S.; Schierholz, G.; Wiese, U. Topology and Dynamics of the Confinement Mechanism. *Nucl. Phys. B* **1987**, *293*, 461–478. [CrossRef]
26. Suganuma, H.; Amemiya, K.; Tanaka, A.; Ichie, H. Quark confinement physics from quantum chromodynamics. *Nucl. Phys. A* **2000**, *670*, 40–47. [CrossRef]
27. Hoelbling, C.; Rebbi, C.; Rubakov, V.A. Free energy of an SU(2) monopole - anti-monopole pair. *Phys. Rev. D* **2001**, *63*, 034506. [CrossRef]
28. Suganuma, H.; Sasaki, S.; Toki, H. Color Confinement, Quark Pair Creation and Dynamical Chiral Symmetry Breaking in the Dual Ginzburg-Landau Theory. *Nucl. Phys. B* **1995**, *435*, 207–240. [CrossRef]
29. Miyamura, O. Chiral Symmetry Breaking in Gauge Fields dominated by Monopoles on SU(2) Lattices. *Phys. Lett. B* **1995**, *353*, 91–95. [CrossRef]
30. Woloshyn, R. Chiral Symmetry Breaking in Abelian Projected SU(2) Lattice Gauge Theory. *Phys. Rev. D* **1995**, *51*, 6411–6416. [CrossRef] [PubMed]
31. Lee, F.X.; Woloshyn, R.; Trottier, H.D. Abelian Dominance of Chiral Symmetry Breaking in Lattice QCD. *Phys. Rev. D* **1996**, *53*, 1532–1536. [CrossRef]

32. Thurner, S.; Feurstein, M.; Markum, H. Instantons and Monopoles are locally correlated with the Chiral Condensate. *Phys. Rev. D* **1997**, *56*, 4039–4042. [CrossRef]

33. Ohata, H.; Suganuma, H. Clear correlation between monopoles and the chiral condensate in SU(3) QCD. *Phys. Rev. D* **2021**, *103*, 054505. [CrossRef]

34. Ichie, H.; Suganuma, H. Maximally Abelian gauge and the gauge invariance condition. *Phys. Rev. D* **1999**, *60*, 077501. [CrossRef]

35. Ezawa, Z.; Iwazaki, A. Abelian Dominance and Quark Confinement in Yang-Mills Theories. *Phys. Rev. D* **1982**, *25*, 2681. [CrossRef]

36. Suzuki, T.; Yotsuyanagi, I. A Possible Evidence for Abelian Dominance in Quark Confinement. *Phys. Rev. D* **1990**, *42*, 4257–4260. [CrossRef]

37. Sakumichi, N.; Suganuma, H. Perfect Abelian Dominance of Quark Confinement in SU(3) QCD. *Phys. Rev. D* **2014**, *90*, 111501. [CrossRef]

38. Sakumichi, N.; Suganuma, H. Three-quark potential and Abelian dominance of confinement in SU(3) QCD. *Phys. Rev. D* **2015**, *92*, 034511. [CrossRef]

39. Ohata, H.; Suganuma, H. Gluonic-excitation energies and Abelian dominance in SU(3) QCD. *Phys. Rev. D* **2020**, *102*, 014512. [CrossRef]

40. DeGrand, T.A.; Toussaint, D. Topological Excitations and Monte Carlo Simulation of Abelian Gauge Theory. *Phys. Rev. D* **1980**, *22*, 2478–2489. [CrossRef]

41. Ichie, H.; Suganuma, H. Monopoles and Gluon Fields in QCD in the Maximally Abelian Gauge. *Nucl. Phys. B* **2000**, *574*, 70–106. [CrossRef]

42. Stack, J.D.; Neiman, S.D.; Wensley, R.J. String Tension from Monopoles in SU(2) Lattice Gauge Theory. *Phys. Rev. D* **1994**, *50*, 3399–3405. [CrossRef] [PubMed]

43. Savvidy, G.K. Infrared Instability of the Vacuum State of Gauge Theories and Asymptotic Freedom. *Phys. Lett. B* **1977**, *71*, 133–134. [CrossRef]

44. Nielsen, N.K.; Olesen, P. An Unstable Yang-Mills Field Mode. *Nucl. Phys. B* **1978**, *144*, 376–396. [CrossRef]

45. Bernard, C. Order of the chiral and continuum limits in staggered chiral perturbation theory. *Phys. Rev. D* **2005**, *71*, 094020. [CrossRef]

46. Suganuma, H.; Tatsumi, T. On the Behavior of Symmetry and Phase Transitions in a Strong Electromagnetic Field. *Ann. Phys.* **1991**, *208*, 470–508. [CrossRef]

47. Suganuma, H.; Tatsumi, T. Chiral symmetry and quark - anti-quark pair creation in a strong color electromagnetic field. *Prog. Theor. Phys.* **1993**, *90*, 379–404. [CrossRef]

48. Klevansky, S. The Nambu-Jona-Lasinio Model of Quantum Chromodynamics. *Rev. Mod. Phys.* **1992**, *64*, 649–708. [CrossRef]

49. Gusynin, V.; Miransky, V.; Shovkovy, I. Dimensional Reduction and Catalysis of Dynamical Symmetry Breaking by a Magnetic Field. *Nucl. Phys. B* **1996**, *462*, 249–290. [CrossRef]

50. Balay, S.; Abhyankar, S.; Adams, M.F.; Brown, J.; Brune, P.; Buschelman, K.; Dalcin, L.; Dener, A.; Eijkhout, V.; Gropp, W.D.; et al. PETSc Web Page. 2019. Available online: https://www.mcs.anl.gov/petsc (accessed on 5 September 2020).

51. Balay, S.; Abhyankar, S.; Adams, M.F.; Brown, J.; Brune, P.; Buschelman, K.; Dalcin, L.; Dener, A.; Eijkhout, V.; Gropp, W.D.; et al. *PETSc Users Manual*; Technical Report ANL-95/11—Revision 3.14; Argonne National Laboratory: Lemont, IL, USA, 2020.

52. Balay, S.; Gropp, W.D.; McInnes, L.C.; Smith, B.F. Efficient Management of Parallelism in Object Oriented Numerical Software Libraries. In *Modern Software Tools in Scientific Computing*; Arge, E., Bruaset, A.M., Langtangen, H.P., Eds.; Birkhäuser Press: Boston, MA, USA, 1997; pp. 163–202.

 universe

*Review*

# Localization of Dirac Fermions in Finite-Temperature Gauge Theory

**Matteo Giordano [1,*] and Tamás G. Kovács [1,2]**

1   Institute for Theoretical Physics, ELTE Eötvös Loránd University, Pázmány Péter Sétány 1/A, H-1117 Budapest, Hungary
2   Institute for Nuclear Research, Bem Tér 18/c, H-4026 Debrecen, Hungary
*   Correspondence: giordano@bodri.elte.hu

**Abstract:** It is by now well established that Dirac fermions coupled to non-Abelian gauge theories can undergo an Anderson-type localization transition. This transition affects eigenmodes in the lowest part of the Dirac spectrum, the ones most relevant to the low-energy physics of these models. Here we review several aspects of this phenomenon, mostly using the tools of lattice gauge theory. In particular, we discuss how the transition is related to the finite-temperature transitions leading to the deconfinement of fermions, as well as to the restoration of chiral symmetry that is spontaneously broken at low temperature. Other topics we touch upon are the universality of the transition, and its connection to topological excitations (instantons) of the gauge field and the associated fermionic zero modes. While the main focus is on Quantum Chromodynamics, we also discuss how the localization transition appears in other related models with different fermionic contents (including the quenched approximation), gauge groups, and in different space-time dimensions. Finally, we offer some speculations about the physical relevance of the localization transition in these models.

**Keywords:** localization; QCD; lattice gauge theory; finite temperature

**Citation:** Giordano, M.; Kovács, T.G. Localization of Dirac Fermions in Finite-Temperature Gauge Theory. *Universe* **2021**, *7*, 194. https://doi.org/10.3390/universe7060194

Academic Editor: Dmitri Antonov

Received: 28 April 2021
Accepted: 4 June 2021
Published: 8 June 2021

**Publisher's Note:** MDPI stays neutral with regard to jurisdictional claims in published maps and institutional affiliations.



## 1. Introduction

Quantum Chromodynamics (QCD) is currently our best microscopic description of strong interactions. As is well known, QCD is a gauge theory with gauge group SU(3), coupling six "flavors" of *quarks*, which are spin-$\frac{1}{2}$ Dirac fermions transforming in the fundamental representation of the group, to the eight spin-1 gauge bosons (known as *gluons*) associated with the local SU(3) symmetry. Despite their apparently simple form, the interactions of quarks and gluons, as dictated by the gauge principle and encoded in the Dirac operator, give rise to a wide variety of phenomena. Most notably, the low-energy properties of strongly interacting matter are largely determined by the phenomena of confinement and chiral symmetry breaking (see, e.g., Refs. [1–4]). At zero temperature, quarks and gluons are in fact confined within hadrons by a linearly rising potential, up to distances where a quark-antiquark pair can be created out of the vacuum. Furthermore, there is an approximate chiral symmetry associated with the lightest quarks, which in the limit of exactly massless quarks is broken spontaneously. This determines most of the properties of light hadrons, once the effects of the explicit breaking by the light quark masses is taken into account.

Confinement and chiral symmetry breaking persist also at nonzero but low temperature and densities. It is well established that, at vanishing chemical potential, QCD undergoes a finite-temperature transition to a deconfined, chirally restored phase (*quark-gluon plasma*), around $T_c \approx 155$ MeV [5,6]. This transition is a rapid but analytic crossover [7], with both confining and chiral properties of the theory changing dramatically in a relatively narrow interval of temperatures.

In particular, the confining properties are determined by the fate of an approximate $\mathbb{Z}_3$ center symmetry, i.e., a symmetry under gauge transformations which are periodic in time up to an element of the group center. At low temperature, center symmetry is only explicitly and mildly broken by the presence of quarks; at higher temperatures, instead, center symmetry is strongly broken spontaneously by the ordering of the Polyakov loop, i.e., the holonomy of the gauge field along a straight path winding around the temporal direction. In the "quenched" limit of infinitely heavy quarks, QCD reduces to SU(3) pure gauge theory, where center symmetry is an exact symmetry of the action, realized at low temperature *à la* Wigner-Weyl in the Hilbert space of the system, while spontaneously broken at high temperatures. Around the same temperature at which the Polyakov loop starts becoming ordered in QCD, corresponding roughly to the spontaneous breaking of center symmetry, also the chiral properties of the system change radically. The approximate order parameter of chiral symmetry, i.e., the chiral condensate $\langle \bar{\psi}\psi \rangle$, decreases rapidly around $T_c$, corresponding to the disappearance of the effects of spontaneous breaking in the chiral limit. The net effect is an effective restoration of chiral symmetry, up to the explicit breaking due to the quark masses.

While the existence and the nature of the finite-temperature transition are by now well established, the mechanisms of confinement and chiral symmetry breaking, and similarly of deconfinement and chiral symmetry restoration, are not fully understood yet; nor is the apparently close relation between these two phenomena that are in principle completely unrelated. An important role in the breaking and restoration of chiral symmetry is played by the topological properties of the gauge field configurations.

It might be possible to understand the formation of a chiral condensate in the low temperature phase of QCD in terms of instantons and the associated zero modes [8–15]. It is well known that there is an exact zero mode of the Dirac operator associated with an isolated instanton or anti-instanton, and with their finite-temperature versions known as calorons [16–23]. Typical gauge field configurations can be interpreted as a more or less dense medium of instantons and anti-instantons. For a sufficiently high density of topological objects, the associated zero modes will strongly mix and form a finite band of near-zero Dirac modes, which in turn gives rise to a finite condensate via the Banks-Casher relation [24]. At higher temperatures the density of topological objects decreases, and so do the density of near-zero Dirac modes and the chiral condensate, until the symmetry is effectively restored. This is the "disordered medium scenario" for chiral symmetry breaking. The mechanism of confinement is understood less clearly, and various proposals have been put forth: we invite the interested reader to consult Refs. [1,2].

Relatively recently, a third phenomenon has been found to take place in correspondence with deconfinement and chiral symmetry restoration in QCD, namely the localization of the low-lying eigenmodes of the Dirac operator [25–40]. Localization is a widely studied subject in condensed matter physics, since Anderson's work on the absence of diffusion in random lattice system [41]. In his seminal paper, Anderson showed how the presence of disorder causes the spatial localization of energy eigenmodes. For electrons in a disordered medium, such as a conductor with impurities, localized modes appear at the band edge, beyond a "mobility edge" separating extended and localized modes. As the amount of impurities/disorder increases, the mobility edge moves towards the band center, eventually leading to all modes becoming localized, and to the conducting sample turning into an insulator. It is outside the scope of this paper (and frankly quite a Herculean task) to provide an exhaustive account of the developments in the theory of Anderson localization, and we refer the interested reader to the reviews [42–46].

It has been shown that a similar localization phenomenon takes place for the low-lying modes of the Dirac operator in QCD above the pseudocritical temperature [27,28,30,33,34,39,40]: up to a critical point in the spectrum, i.e., the analogue of the "mobility edge", low modes are spatially localized on the scale of the inverse temperature [33,39]. The mobility edge depends on the temperature $T$, and its extrapolation towards the confined phase vanishes at a temperature compatible with $T_c$ [33,40]. At the mobility edge, a second-order phase transition takes place in the spectrum, which has been shown to be a genuine Anderson transition [34–37].

A disorder-driven transition, such as the Anderson transition in the Dirac spectrum, obviously needs a source of disorder. This was first identified in the local fluctuations of the topological charge density, treated as a dilute ensemble of pseudoparticles (calorons) [25–28]. As already mentioned above, these topological objects individually support localized zero modes of the Dirac operator; since they overlap, the corresponding modes mix and shift away from zero, but for a dilute ensemble this effect is small and modes remain localized and near zero. While evidence was produced supporting the connection between localization and topological objects, it turned out that not all localized modes could be explained this way [31], and at least another source of disorder was needed. This was identified in the fluctuations of the Polyakov loop [31], a hypothesis supported by numerical results [31,39,40] and by the critical properties found at the mobility edge [34–37]. This led to the so-called "sea/islands picture" of localization, proposed in Ref. [31] and further elaborated in Refs. [47–49]: Dirac eigenmodes tend to localize on "islands" of Polyakov loop fluctuations away from its ordered value, which form an extended "sea" in the deconfined phase. The sea/islands picture requires only the existence of a phase with ordered Polyakov loop in order for localization to appear, and so leads one to expect the localization of the low Dirac modes in a generic gauge theory with a deconfinement transition. This has been verified in a variety of models [32,49–56], including ones without topology, thus providing further support to the sea/islands picture. This also clearly suggests a strong connection between localization of the low Dirac modes and deconfinement.

The relation between localization and chiral symmetry restoration has received less attention, mostly because of the intrinsic difficulty of studying gauge theories with massless fermions, where chiral symmetry is exact. It is, however, clearly established that localization of the low modes is accompanied by evident changes in the spectral density at the low end of the spectrum. In Refs. [51,54,57,58] it was observed that in the quenched theory a peak of near-zero modes of topological origin forms, followed by a spectral range with low mode density. A similar peak of localized modes was observed in Ref. [38] in the presence of dynamical fermions. (The presence of this peak was discussed before in Ref. [59], and has been studied recently in Refs. [60,61], although the localization properties of the eigenmodes are not studied there.) It is shown in Ref. [58] that the near-zero peak can be explained in terms of a dilute instanton/anti-instanton gas and the associated zero modes. Recently, it has been proposed that the presence of a finite density of near-zero localized modes in the chiral limit can lead to the disappearance of the finite-temperature massless excitations predicted by the finite-temperature version of Goldstone's theorem [62].

While the presence of localization at high temperature, and its connection with the finite-temperature transition in QCD and in other gauge theories are by now fairly well established, the physical meaning of localization of Dirac modes, and a detailed understanding of the aforementioned connection with deconfinement and chiral symmetry restoration, have proved to be quite elusive. The hope is that a better understanding of localization can shed light on the mechanisms of confinement and chiral symmetry breaking, and on the finite-temperature deconfining and chirally restoring transition.

In this review we will discuss developments in the study of the localization of Dirac modes in finite-temperature gauge theories. While the main focus is on QCD, we will discuss also other models, showing in particular that the connection between localization and deconfinement is a general phenomenon. As this review is aimed both at the particle physics and condensed matter communities (and expected to disappoint them both), we provide brief introductions to the subjects of finite-temperature QCD and Anderson localization in an attempt to bridge the gaps. Older results were already reviewed in Refs. [36,63].

Localization in gauge theories also occurs in a few other contexts, albeit those correspond to physical situations very different from the one discussed in the present review. For this reason here we do not discuss them in detail, and only mention them for completeness. A non-exhaustive list of references includes the following papers. For results concerning the relation between the localization properties of the low Dirac modes and the topological structure of the vacuum in gauge theories, we refer to Refs. [64–68] and references therein. The role played by localization in the Aoki phase of quenched QCD with Wilson fermions, especially concerning the fate of Goldstone modes, is studied in Refs. [69,70]. Localization properties of the eigenmodes of the covariant Laplacian in Yang-Mills theories are studied in Refs. [71,72].

The plan of this paper is as follows. In Section 2 we review finite-temperature QCD and related issues. In Section 3 we review the topic of localization and Anderson transitions in some generality. In Section 4 we discuss localization in QCD at finite temperature. The disordered medium scenario and the sea/islands picture, providing mechanisms for localization, are discussed in Section 5. Localization in gauge theories other than QCD is discussed in Section 6. Finally, in Section 7 we draw our conclusions and show some prospects for the future.

## 2. QCD at Finite Temperature

In strongly interacting systems such as QCD localization takes place as the systems cross from the hadronic to the high-temperature quark-gluon plasma state. To put localization in QCD in the proper context, in the present section we summarize some basic facts about this finite temperature transition. For an introduction to gauge theories at finite temperature we refer the reader to the literature (see, e.g., Refs. [73,74]).

In an extended sense, QCD is a gauge theory with $N_f$ flavors of quarks transforming in the fundamental representation of the gauge group $SU(N_c)$ and interacting via the corresponding gauge field, the excitations of which are the gluons. In a stricter sense, in QCD $N_f$ is fixed to six, the number of known quark flavors, and $N_c = 3$. At finite temperature, the theory is formally defined by the Euclidean partition function

$$Z_{QCD} = \int [dA] \, e^{-S_{YM}[A]} \prod_f \det\left( \slashed{D}[A] + m_f \right), \tag{1}$$

where the product runs over the quark flavors with masses $m_f$, $A$ is the gauge field, $S_{YM}$ is the Euclidean Yang-Mills action and

$$\slashed{D}[A] = \sum_{\mu=1}^{4} \gamma_\mu(\partial_\mu + ig A_\mu) \tag{2}$$

is the Euclidean Dirac operator, with $\gamma_\mu$ the Euclidean, Hermitean gamma matrices and $g$ the coupling constant. The temporal direction is compactified to a circle of size equal to the inverse temperature, and periodic boundary conditions in the temporal direction are imposed on the gauge fields. In this form of the partition function the quark fields, appearing in the action quadratically, have been explicitly integrated out, resulting in the quark determinant. The Dirac operator is an anti-Hermitean operator with purely imaginary spectrum, which is furthermore symmetric about zero thanks to the property $\{\gamma_5, \slashed{D}\} = 0$.

It is instructive to consider the theory as a function of its parameters, that in reality are fixed by the observed properties of hadrons. The only such parameters of QCD are the quark masses.[1] In particular, the low-energy properties of light hadrons are completely determined by the masses of the lightest quarks, the $u$ and $d$ quark, and to some extent the heavier $s$ quark. The other three known quark flavors are so heavy that they have little influence on the low energy physics.

Quark masses are also important parameters from a theoretical point of view, because they crucially influence some symmetries of the system. Even though in nature these are only approximate symmetries, considering them helps to better understand the finite temperature transition of QCD. In an imaginary world with two massless quark flavors, i.e., when $m_u = m_d = 0$, QCD would have an exact $SU(2)_V \times SU(2)_A \times U(1)_A$ chiral symmetry. Here $SU(2)_V$ is a rotation in the two-dimensional $(u, d)$ flavor-space that acts identically on all the Dirac components. In contrast, the flavor non-singlet axial symmetry $SU(2)_A$ not only mixes the two flavors, but also transforms the left and right Dirac components with opposite phases. Finally, the $U(1)_A$ flavor-singlet axial symmetry acts trivially in flavor space and rotates the left and right Dirac components with opposite phases. Even though these are all symmetries of the classical Lagrangian, after quantization the $U(1)_A$ part of the symmetry is anomalously broken. Furthermore, at zero temperature the $SU(2)_A$ axial symmetry is spontaneously broken. The emerging three Goldstone bosons are the analogues of the pions, and the order parameter of the symmetry breaking is the light quark condensate $\langle \bar{\psi}\psi \rangle$. Finally, the vector part of the symmetry $SU(2)_V$ remains intact even for finite, but equal quark masses.

In reality, the nonzero and non-equal masses of the $u$ and $d$ quarks explicitly break these symmetries; however, the spontaneous and anomalous breaking inherited from the massless theory both turn out to be much stronger than this explicit breaking. In fact, in an imaginary world with zero $u$ and $d$ quark masses, the low-energy properties of the light hadrons would be much the same as they are in the real world. The only important exceptions would be the pions, which in that case would be exact Goldstone bosons with zero mass.

If one imagines changing the quark masses, the other interesting limit is the one in which quarks are much heavier than in reality. In particular, in the limit of infinitely heavy quarks the quark determinant in the path integral completely decouples and the back-reaction of the quarks on the gauge field disappears. This is the so-called quenched theory. In this limit QCD has a different exact symmetry, the symmetry group being the center of the gauge group, in the case at hand $\mathbb{Z}_3$. The symmetry transformation in question is a gauge transformation that is singular along a spacelike hypersurface, and its singularity is characterized by an element of the center $\mathbb{Z}_3$. Recalling that the system is finite in the temporal direction with periodic boundary conditions for the gauge field, this symmetry transformation is a gauge transformation that is not periodic in the temporal direction (hence singular), and it multiplies by the same $\mathbb{Z}_3$ center element all the holonomies (gauge parallel transporters) going around the system in the temporal direction. The holonomies wrapping around the system in the temporal direction along a straight path are also called Polyakov loops.

Gauge invariant local gluonic quantities are defined in terms of holonomies around small loops, and those never wrap around the system. These types of loops cross the hypersurface where the gauge transformation is singular the same number of times in both directions. As a result, the $\mathbb{Z}_3$ factors along such a loop always cancel, and gauge invariant local quantities are invariant with respect to the $\mathbb{Z}_3$ center transformations. In contrast, fermionic quantities are affected, since such a singular gauge transformation essentially introduces an extra $\mathbb{Z}_3$ twist for the temporal boundary condition of the fermions through the covariant derivative in the Dirac operator. The boundary condition affects the spectrum of the Dirac operator and also its determinant that appears in the path integral. At low temperatures where the correlation length is much smaller than the temporal extent of the system, and Polyakov loops fluctuate locally with little correlation, the temporal boundary condition has only a small impact on the Dirac determinant. Consequently, the quarks only mildly break the $\mathbb{Z}_3$ symmetry. However, at high temperature, where the correlation length becomes comparable to or larger than the temporal size of the system, and the Polyakov loops tend to align with each other, this picture changes drastically.

To understand exactly how that happens, let us first recall that in finite temperature quantum field theory the temporal boundary condition for fermions is antiperiodic. For free massless fermions this implies a gap in the spectrum of the Dirac operator equal to the first Matsubara frequency. If by a singular gauge transformation (as defined above) we introduce an additional $\mathbb{Z}_3$ twist in the boundary condition then the gap will decrease, because the twist $\pi$ corresponding to the antiperiodic boundary condition will decrease to $\pi \pm 2\pi/3 = \pm\pi/3 \ (\mathrm{mod}\ 2\pi)$. In the interacting theory there is no gap in the spectrum, but through this mechanism the low end of the spectrum is much denser when the spatially averaged Polyakov loop is in the complex center sectors (i.e., close to one of the complex center elements) than when it is in the real one (i.e., close to the identity), where the effective twist comes only from the antiperiodic boundary condition. As a result, the fermion determinant, that disfavors larger low-mode density, strongly favors the real Polyakov loop sector, and for finite quark mass this is the only sector that contributes to the path integral. This is how fermions explicitly break the $\mathbb{Z}_3$ center symmetry and select the real Polyakov loop sector out of the three sectors that would be equivalent in their absence. This mechanism is at work also at low temperature, but much less effective there since the average Polyakov loop fluctuates around zero.[2]

Now going back to the quenched theory, at zero and low temperature, the exact $\mathbb{Z}_3$ symmetry of its Lagrangian remains intact, while above a critical temperature this symmetry is spontaneously broken and its order parameter, the trace of the Polyakov loop, develops a nonzero expectation value. In fact, in the quenched theory, the logarithm of the expectation value of the Polyakov loop is proportional to the gauge field energy it costs to insert an infinitely heavy static quark in the system. In this way, the vanishing of the Polyakov loop in the low-temperature phase shows that no free quarks can exist there, so quarks are confined into hadrons. In contrast, the nonzero expectation value of the Polyakov loop in the high temperature phase implies that quarks are not confined there.

The nature of the finite-temperature transition in extended QCD is governed by the chiral and the $\mathbb{Z}_3$ symmetries, which—as we have already seen—depend on the quark masses. In the quenched limit (infinite quark mass) lattice simulations have shown the transition to be weakly first order [78,79], and this behavior persists for large enough, but finite quark masses. If the quarks become lighter, the transition weakens and for intermediate quark masses there is a wide region where it is only a crossover. In particular, the light-quark masses in nature fall in this range [7]. For even smaller quark masses, the transition is again expected to become a true phase transition, but its order depends on the number of light quark flavors. For two light flavors (and physical strange quark mass) it is expected to be second order, whereas for three light flavors a first order phase transition is anticipated. However, the presence of these phase transitions, previously predicted based on an epsilon expansion [80] (see also [81] for the role played by the $U(1)_A$ anomaly), have not yet been confirmed by lattice simulations, because simulations close to the chiral limit are technically challenging.

Most of the results discussed in this review are based on numerical calculations on the lattice. Lattice field theory is a nonperturbative approach to the quantization of quantum field theories, based on the discretization of the relevant path integrals that define the theory in the path-integral approach. We provide here only a very brief introduction to this subject, referring the interested reader to the extensive literature (e.g., the books [73,74,82–84]). In the lattice approach to gauge theories devised by Wilson [85], the SU(3) gauge fields of QCD are replaced by unitary SU(3) matrices (*link variables*) associated with the links of a finite hypercubic lattice. In continuum language, these correspond to the parallel transporters of the gauge fields along the paths connecting neighbouring lattice points. After a suitable discretization of the gauge action, the relevant path integrals are obtained by integrating over the gauge fields, which in practice means integrating the link variables over the group manifold with the invariant (Haar) group measure. The desired, continuum field theory is obtained (if this is possible) by properly tuning the parameters in the action, so that the correlation length of the system in lattice units diverges, and the system "forgets" about the underlying lattice. For pure gauge theories, the only available parameter is the lattice inverse gauge coupling (usually denoted by $\beta$), which ceases to be a freely adjustable parameter and turns instead into a measure of the lattice spacing.[3]

The approach outlined above is easily generalized to other gauge theories based on different gauge groups, by simply replacing the SU(3) link variables and the corresponding Haar measure with elements of the relevant gauge group and the corresponding Haar measure. The inclusion of fermions instead is not straightforward, especially for what concerns the implementation of chiral symmetry. Nonetheless, there are several viable discretization of the Dirac operator, which are expected to all lead to the same results in the continuum limit. Since they appear below in Section 4, we mention Wilson fermions, staggered fermions (possibly rooted), domain wall fermions, overlap fermions, and twisted mass fermions (see Ref. [83,84] and references therein for details). We finally mention that several improvement schemes exist that bring the system closer to the continuum limit, i.e., that reduce the effects due to the finiteness of the lattice spacing. Such schemes exist both for the gauge action and for the fermionic determinant (see Ref. [83,84] and references therein for details).

### 2.1. Finite-Temperature Transition, Dirac Spectrum, and Localization—An Overview

We have seen that in the two extreme cases, the quenched limit and the chiral limit, two different symmetries, the $\mathbb{Z}_3$ center symmetry and the $SU(2)_A$ axial symmetry govern the transition. The respective order parameters, the Polyakov loop and the quark condensate, signal spontaneous breaking of the symmetry in the high temperature phase for the $\mathbb{Z}_3$ symmetry and in the low temperature phase for the $SU(2)_A$ axial symmetry. In nature, both symmetries are only approximate, the transition is a crossover and the order parameters have only inflection points in the crossover region. It also follows that in real QCD there is no sharply defined transition temperature. In contrast, regardless of the quark mass, the localization transition, i.e., the appearance of the first localized modes at the low edge of the Dirac spectrum, occurs at a sharply defined critical temperature. Moreover—as anticipated in the Introduction, and as we will see below in Sections 4 and 6—the localization transition occurs in the temperature range of the deconfining and chiral crossover in the case of real QCD, and exactly at the deconfining temperature in the quenched limit. This suggests that there might be a connection between the thermodynamic (chiral and deconfining) transitions on the one hand, and the localization transition on the other hand.

Besides the coincidence of their respective critical or pseudocritical temperatures, these phenomena are also connected through the degrees of freedom playing the most important role in their respective dynamics. When the system crosses into the high temperature phase, the spectral density of the Dirac operator around zero drops considerably, exactly vanishing in the chiral limit. This is how the chiral symmetry, spontaneously broken at low temperature, is restored above the transition. Indeed, through the Banks-Casher relation [24], the order parameter of chiral symmetry breaking, the quark condensate, is proportional in the chiral limit to the spectral density at zero, and generally strongly sensitive to the low end of the spectrum. Lower spectral density also means that eigenmodes close to each other in the spectrum are less likely to be mixed by fluctuations of the gauge field, which might lead to localization at the low end of the Dirac spectrum.

The spectral density, however, is not the only important parameter that influences localization. In the quark mass regions numerically explored so far, where the transition is either a crossover (near and below the physical values of the light quark masses) or a true phase transition governed by (approximate) center symmetry (heavy quark limit), the spectral density does not immediately drop to zero at the (pseudo)critical temperature. In particular, in the quenched limit just above the transition a narrow but tall spike at zero appears in the spectral density (see Figure 1). This is due to near-zero modes associated with a dilute gas of calorons and anticalorons, local fluctuations of the topological charge [58]. Even though the spectrum is dense in the spike, eigenmodes there are localized [86]. A similar peak of near-zero modes is also found for physical, near-physical, and below-physical light-quark masses [38,59–61]. For near-physical masses these modes are found to be localized [38], and most likely this persists as the mass is decreased.

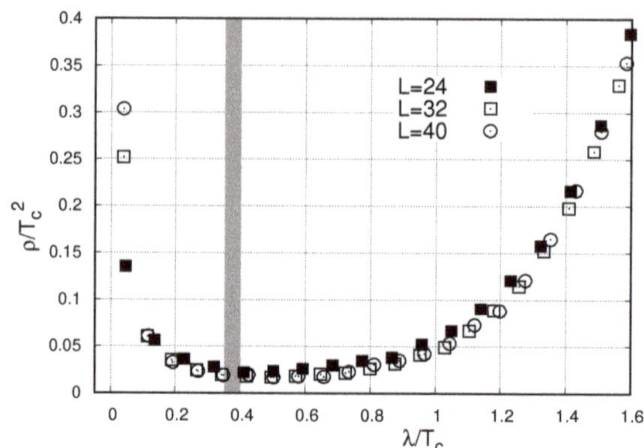

**Figure 1.** The spectral density, Equation (9) (here normalized by the volume), of the overlap Dirac operator in quenched QCD (i.e., SU(3) pure gauge theory), just above the phase transition at $T = 1.045T_c$. The grey band indicates the point separating the lowest, topological modes from the rest (its width equals the corresponding uncertainty). From Ref. [58]. (Figure adapted from R.Á. Vig and T.G. Kovács, arXiv:2101.01498 (2021), and used under a CC-BY 4.0 license (https://creativecommons.org/licenses/by/4.0)).

Recently, the spike in the spectral density received another interpretation. It was argued that it signals the appearance of a new, previously undiscovered "phase" of QCD, intermediate between the low-temperature confined and the high-temperature deconfined phase [87]. In a more recent paper the same authors studied a newly defined infrared dimension $d_{IR}$ of the eigenmodes in the low end of the spectrum of the chirally symmetric overlap Dirac operator. They concluded that the exact zero modes have $d_{IR} = 3$, and in the spectral peak $d_{IR}$ changes rapidly but smoothly from 2 to 1 as one moves up in the spectrum [88]. This behavior persists up to the bulk of the spectrum, where the spectral density, together with the infrared dimension $d_{IR}$ of the modes starts to increase again. This nontrivial change in the infrared dimension all happens in the region where based on the spectral statistics and the scaling of the participation ratio with the volume, the eigenmodes are thought to be localized. It would be interesting to further investigate how $d_{IR}$ relates to the usual fractal dimension $D_2$ (see Equation (6)), and what kind of spatial structure in the eigenmodes gives rise to this nontrivial behavior. This could also depend on the chiral and locality properties of the particular discretization of the Dirac operator.

Topological fluctuations and the localization of the eigenmodes are both intimately related to fluctuations of the Polyakov loop, the order parameter of the quenched transition. The spatial localization of low Dirac eigenmodes is found to strongly correlate with local fluctuations of the Polyakov loop away from its symmetry-breaking equilibrium value [31,39,40]. This gives rise to the sea/islands picture of localization that we will discuss in Section 5.2 of the present paper in more details. Localization on calorons and localization on Polyakov loop fluctuations are, however, not mutually exclusive, as calorons always contain large fluctuations of the Polyakov loop. In fact, within a caloron, the Polyakov loop wraps around the gauge group in a topologically nontrivial way. The connection between calorons and Polyakov loop fluctuations is also shown by the strong correlation between the Polyakov loop and the topological susceptibility that can be observed close to the transition in the high temperature phase (see Figure 2).

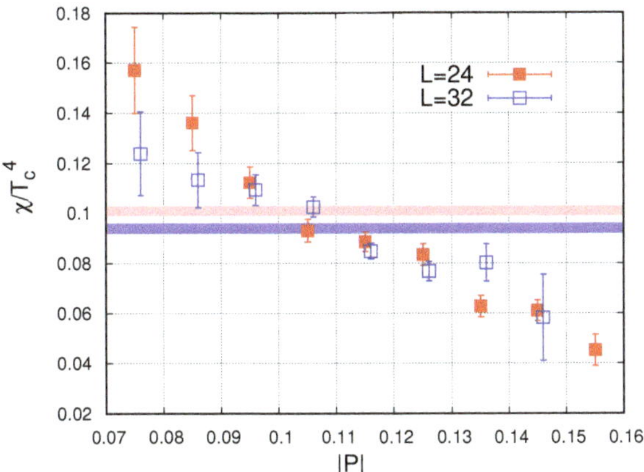

**Figure 2.** The dependence of the topological susceptibility on the value of the spatially averaged Polyakov loop in quenched lattice simulations. The susceptibility was computed by dividing the configurations in sets according to the spatially averaged Polyakov loop. The averages for the whole ensembles (in two volumes) are shown by the horizontal bands, the widths indicating the uncertainties. From Ref. [58]. (Figure adapted from R.Á. Vig and T.G. Kovács, arXiv:2101.01498 (2021), and used under a CC-BY 4.0 license (https://creativecommons.org/licenses/by/4.0)).

The finite temperature transition of QCD is a result of the interplay of all those mechanisms that we just discussed, involving the Polyakov loop, the topological fluctuations and the spectral density of the Dirac operator around zero. Since localization is intimately related to all these aspects, it might hold the key to a better intuitive understanding of the dynamics of the transition.

### 3. Localization and Anderson Transitions

In a classic paper [41], Anderson showed that a sufficiently large amount of disorder in a lattice system prevents quantum-mechanical diffusion. Working in the one-particle tight-binding approximation, and mimicking the effect of disorder by supplementing the tight-binding Hamiltonian with a random potential on the lattice sites, Anderson showed that all the eigenfunctions of the system are localized for sufficiently strong disorder (i.e., for a sufficiently broad distribution for the random potential).

A practical example of this situation is a "dirty" crystal where some of the lattice atoms are replaced by impurities. Anderson's results imply that all the electron eigenstates become localized for a sufficiently large concentration of impurities. This prevents electron diffusion and the associated transport phenomena; in particular, the d.c. conductivity at zero temperature vanishes [89,90]. Localization then provides a possible mechanism for a disorder-induced metal-insulator transition (MIT).

Anderson's original arguments were later scrutinized and clarified by several authors [91–97]. Since then, the topic of disorder-induced localization, or *Anderson localization*, has been extensively studied in the condensed matter community, and it is impossible for us to provide here a comprehensive survey, or even do justice to the related literature. In this section we limit ourselves to a short review of the main aspects of Anderson localization, especially those relevant to gauge theories, discussed in the next Section. We invite the interested reader to consult the reviews [42–46].

*3.1. The Anderson Model*

In its simplest form, the (orthogonal) Anderson model Hamiltonian reads

$$H^{\mathrm{AM}}_{\vec{x},\vec{y}} = \varepsilon_{\vec{x}}\delta_{\vec{x},\vec{y}} + \sum_{\mu=1}^{3}\left(\delta_{\vec{x}+\hat{\mu},\vec{y}} + \delta_{\vec{x}-\hat{\mu},\vec{y}}\right), \tag{3}$$

where $\vec{x}, \vec{y}$ label the sites of a simple cubic lattice with lattice vectors $\hat{\mu}$, $\mu = 1, 2, 3$, and $\varepsilon_{\vec{x}}$ is a random on-site potential, with uniform probability distribution in the interval $[-\frac{W}{2}, \frac{W}{2}]$. The lattice spacing and the hopping energy are set to 1 for simplicity. The width $W$ of the distribution is a measure of the amount of disorder in the system, with $W = 0$ corresponding to a perfectly pure crystal. In this case, for a lattice of side $L$ with periodic boundary conditions the eigenstates of $H^{AM}$ are plane waves with wave vectors $\vec{p} = \frac{2\pi\vec{k}}{L}$, with $k_\mu = 0, 1, \ldots, L - 1$. However, as soon as even a small amount of disorder is put into the system, i.e., $W \neq 0$, the eigenmodes $\psi(\vec{x})$ at the band edge become exponentially localized, i.e., $|\psi(\vec{x})|^2 \sim e^{-|\vec{x}-\vec{x}_0|/\xi}$ for $E$ beyond critical energies $\pm E_c(W)$ called "mobility edges" [90] (see Figure 3). As the amount of disorder $W$ in the system increases, the mobility edge moves towards the band center. Eventually, for $W$ larger than a critical disorder, $W_c$, all the modes become localized. If Equation (3) describes the conduction band of an electron in some "dirty" crystalline system, for large enough $W$ the Fermi energy will lie in the localized part of the band; d.c. transport then takes place through hopping of electrons from one localized state to another, which has an exponentially small probability of happening, and in the limit of infinite size leads to the absence of charge transport. As the amount of impurities increases past the critical value, the system then undergoes a metal-to-insulator transition.

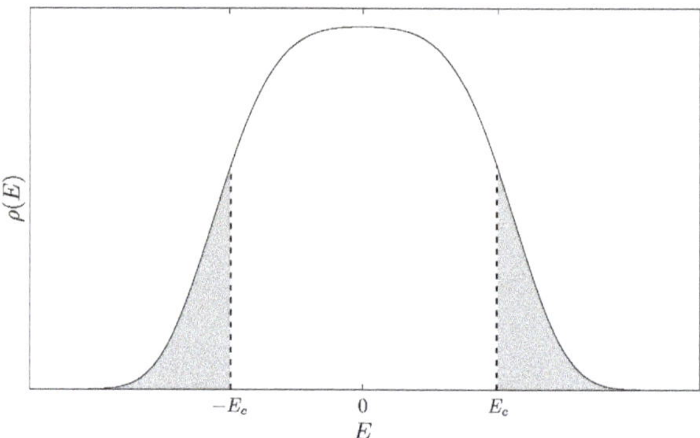

**Figure 3.** Sketch of the density of states $\rho$ (see Equation (9)) as a function of energy $E$ in the Anderson model, Equation (3). Localized modes are present in the shaded region beyond the mobility edges $\pm E_c$.

### 3.2. Anderson Transitions

When the energy of the modes crosses the mobility edge $E_c(W)$ at fixed $W$, or equivalently when the disorder in the system crosses the energy-dependent critical disorder $W_c(E)$ at fixed mode energy $E$, the nature of the eigenmodes of $H^{AM}$ changes from delocalized to localized. As argued in Ref. [98], in three dimensions the associated transition is a second-order phase transition (*Anderson transition*), with divergent correlation length $\xi(E) \sim |E - E_c|^{-\nu}$ or $\xi(W) \sim |W - W_c|^{-\nu}$, where the same exponent $\nu$ is expected.

This prediction is based on the so-called *scaling theory of localization* (see Ref. [43] for an introduction): first proposed in Ref. [98] based on previous ideas exposed in Refs. [42,99–101], it was later put on a firmer basis through a field-theoretical description of disordered systems and Anderson transitions [102–104] (see Ref. [45] for a full list of references). The basic idea is that the change in the conductance $G(L)$ of the system[4] as its size $L$ is increased is controlled only by the localized or delocalized nature of the energy eigenmodes, which in turn is measured by the conductance itself, as a proxy for the disorder in the system. This implies a scaling behavior of the conductance, $\frac{d \ln G(L)}{d \ln L} = \beta(G(L))$. Using the asymptotics of the $\beta$ function obtained from localized or delocalized modes is then enough to show that in three dimensions there is an unstable fixed point (in the renormalization-group sense), and so a mobility edge in the energy spectrum and a phase transition at some critical amount of disorder (see Figure 4). In one dimension no Anderson transition is expected as all modes are localized in the presence of disorder [105,106], while the situation in two dimensions is more complicated (see below).

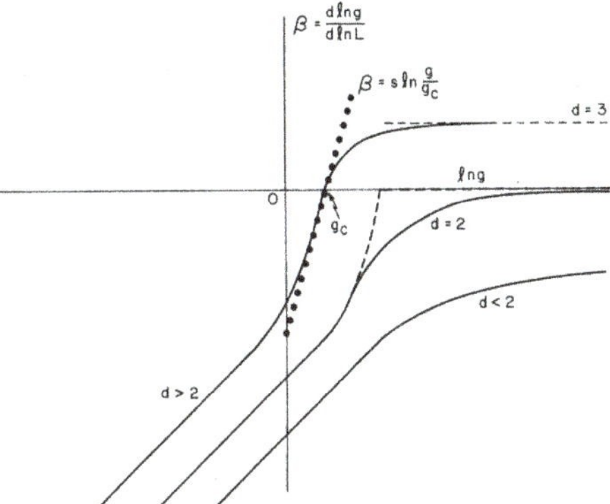


**Figure 4.** The scaling function $\beta(g)$ against the dimensionless conductance $g = \frac{2\hbar G}{e^2}$ in various dimensions. From Ref. [98]. (Reprinted figure with permission from E. Abrahams, P.W. Anderson, D.C. Licciardello, and T.V. Ramakrishnan, Phys. Rev. Lett. 42, 673 (1979). Copyright (1979) by the American Physical Society).

The Anderson model that we just described, Equation (3), is but the simplest disordered Hamiltonian in three dimensions, and can be generalized in various ways. One can consider different probability distributions for the on-site disorder (e.g., the Lloyd model [107]), add off-diagonal disorder by making also the hopping terms random [108–110], increase the range of interaction by adding more hopping terms (e.g., Ref. [111]), and so on. However, according to the general theory of the renormalization group (see, e.g., Ref. [112]), critical properties at a second-order phase transition are shared by systems in the same *universality class*, determined only by general properties such as the dimensionality and the symmetries of the system.[5]

From a technical point of view, the Anderson model is a model of (sparse) random matrices. The properties of these models are the subject of Random Matrix Theory (RMT) [113–115]. For random systems, the relevant symmetry classification has been provided by Dyson [116], later extended by Verbaarschot [117], and completed by Altland and Zirnbauer [118–121] (see also Refs. [45,122]). The main *symmetry classes*, relevant to the models discussed in this review, are determined by the existence (or not) of an antiunitary symmetry operator $T$ ("time reversal") commuting with the Hamiltonian, $[T, H] = 0$, and further specified by whether $T^2 = 1$ or $T^2 = -1$. If $T$ exists and $T^2 = 1$, the system is in the *orthogonal class* (O): this is the case for the model in Equation (3). If $T$ does not exist, the system is in the *unitary class* (U). Perhaps the simplest example of a system in this class is the so-called unitary Anderson model (UAM),

$$H_{\vec{x},\vec{y}}^{\mathrm{UAM}} = \varepsilon_{\vec{x}}\delta_{\vec{x},\vec{y}} + \sum_{\mu=1}^{3}(\delta_{\vec{x}+\hat{\mu},\vec{y}} + \delta_{\vec{x}-\hat{\mu},\vec{y}})e^{i\phi_{\vec{x},\vec{y}}}\,, \qquad \phi_{\vec{y},\vec{x}} = -\phi_{\vec{x},\vec{y}}\,, \tag{4}$$

which includes also off-diagonal disorder in the form of random phases $\phi_{\vec{x},\vec{y}}$ in the hopping terms, mimicking the presence of a random magnetic field. Finally, if $T$ exists and $T^2 = -1$, the system is in the *symplectic class* (S). This classification is complete as far as the statistical properties in the bulk of the spectrum are concerned.

A refined classification is needed if one wants to discuss statistical spectral properties near the origin. In this case, one has to consider whether also a "particle-hole" symmetry exists, realized in terms of an antiunitary operator $C$ obeying $\{C, H\} = 0$, and if so whether

$C^2 = \pm 1$. This gives rise to nine different combinations. The eight combinations obtained when at least $T$ or $C$ exists correspond to eight different symmetry classes. If both $T$ and $C$ exist, it automatically follows that a unitary operator $\Gamma = TC$ exists, anticommuting with the Hamiltonian, $\{\Gamma, H\} = 0$, and satisfying $\Gamma^2 = 1$. However, a $\Gamma$ satisfying this property can exist also if $T$ and $C$ are both absent. In this case there are two further symmetry classes, corresponding to whether such a $\Gamma$ exists or not, for a total of ten. The classification is summarized in Table 1. In particular, if $\Gamma$ exists and commutes with $T$ (if this also exists), the system belongs to one of the chiral classes (chO, chU, and chS). Examples of systems of this type are provided by certain lattice models with random hopping terms, and no on-site potential, on bipartite lattices.

**Table 1.** Symmetry classes of random matrix ensembles. Entries corresponding to time-reversal ($T$) and particle-hole ($C$) symmetry indicate whether the symmetry is absent (0) or, if present, what is its square ($\pm$); entries corresponding to chiral symmetry ($\Gamma$) indicate whether it is absent or present (0 or 1).

| $T$ | $C$ | $\Gamma$ | Class | |
|---|---|---|---|---|
| | | | Wigner-Dyson classes | |
| 0 | 0 | 0 | A | (unitary) |
| + | 0 | 0 | AI | (orthogonal) |
| − | 0 | 0 | AII | (symplectic) |
| | | | chiral classes | |
| 0 | 0 | 1 | AIII | (chiral unitary) |
| + | + | 1 | BDI | (chiral orthogonal) |
| − | − | 1 | CII | (chiral symplectic) |
| | | | Bogoliubov-de Gennes classes | |
| 0 | − | 0 | C | |
| + | − | 1 | CI | |
| 0 | + | 0 | D | |
| − | + | 1 | DIII | |

*3.3. Detecting Localization: Eigenmode Observables*

A convenient way to study the localization properties of the eigenmodes of a random lattice Hamiltonian and how they change along the spectrum is by means of the *inverse participation ratios* (IPRs),[6]

$$\mathrm{IPR}_q \equiv \sum_x |\psi(x)|^{2q} , \tag{5}$$

where it is assumed that eigenmodes obey the usual normalization condition, i.e., $\sum_x |\psi(x)|^2 = 1$, and $x$ now labels the sites of the relevant lattice, assumed finite and of volume $V = L^d$, with $L$ the linear size and $d$ the dimensionality. Unless specified otherwise, in the following both the term and the notation IPR, without subscript, will be used to refer specifically to the case $q = 2$. For modes extended throughout the whole system, one has qualitatively $|\psi_{\mathrm{ext}}(x)|^2 \sim 1/V$, and so $\mathrm{IPR}_q \sim V^{1-q}$: after averaging over the possible realizations of disorder, which will be denoted with $\langle \ldots \rangle$, and taking the large-volume limit, one has then $\langle \mathrm{IPR}_q \rangle \to 0$ as $V \to \infty$ (for $q > 1$). For modes localized in a region of volume $V_0$ one has $|\psi_{\mathrm{loc}}(x)|^2 \sim 1/V_0$ inside the localization region and negligible outside, and so $\mathrm{IPR}_q \sim V_0^{1-q}$: one has then $\langle \mathrm{IPR}_q \rangle \to$ const. as $V \to \infty$. At the mobility edge, instead, the scaling of $\mathrm{IPR}_q$ with the volume depends on $q$ in a highly nontrivial way. One has in general

$$\mathrm{IPR}_q \sim L^{-(q-1)D_q} , \tag{6}$$

where $D_q = d$ for extended modes and $D_q = 0$ for localized modes, while at criticality $D_q$ is not a constant. This reflects the multifractal nature of eigenmodes at $E_c$ [124,125], and leads to define a set of multifractal exponents characterizing the critical behavior at the Anderson transition (see Ref. [45]).

Closely related to the IPR is the *participation ratio*,

$$\mathrm{PR} \equiv \frac{1}{V}\mathrm{IPR}^{-1} \qquad \left(= \frac{1}{V}\mathrm{IPR}_2^{-1}\right), \tag{7}$$

which measures the fraction of the system effectively occupied by the mode. For localized modes one has in the infinite volume limit $\langle \mathrm{PR}\rangle \to 0$, while for delocalized modes extended throughout the system one finds $\langle \mathrm{PR}\rangle \to$ a nonzero constant. Another equivalent way to measure the localization properties is to use the mode "size", i.e., $V \cdot \mathrm{PR} = \mathrm{IPR}^{-1}$, which as $V \to \infty$ (after averaging over the disorder) remains constant for localized modes and diverges for delocalized modes. For systems with nontrivial spin and/or internal degrees of freedom, the eigenvectors $\psi_{\alpha,c}(x)$ possess extra spin ($\alpha$) and/or internal indices ($c$). In these cases it is convenient to employ a definition of the IPR which is invariant under spacetime and internal (unitary) rotations, i.e.,

$$\mathrm{IPR} = \sum_x \left(\sum_{\alpha,c} |\psi_{\alpha,c}(x)|^2\right)^2 = \sum_x \left(\psi(x)^\dagger \psi(x)\right)^2, \tag{8}$$

where the normalization condition $\sum_x \psi(x)^\dagger \psi(x) = 1$ is understood. For example, for eigenmodes of the continuum, Wilson, or overlap Dirac operators, $\alpha = 1,\ldots,4$ is the Dirac index, and $c = 1,\ldots,N_c$ is the gauge group ("color") index; for eigenmodes of the staggered operator $\alpha$ is absent but $c$ is present.

### 3.4. Detecting Localization: Eigenvalue Observables

Another useful tool to detect localization are the statistical properties of the eigenvalues $\lambda_i$ of a random lattice Hamiltonian, which are closely related to the localization properties of its eigenvectors [126]. Localized modes are in fact expected to be sensitive only to local fluctuations in the disorder, and so the corresponding eigenvalues are expected to fluctuate independently. More precisely, after removal of non-universal, model-dependent features by means of the so-called *unfolding* procedure [113] (see below), the unfolded eigenvalues corresponding to localized modes should obey Poisson statistics. Delocalized modes, on the other hand, are expected to be mixed easily by fluctuations in the disorder, and so the corresponding unfolded spectrum should behave like that of a dense random matrix, and display the statistics of the Gaussian ensemble of Random Matrix Theory [113–115] in the appropriate symmetry class.

Unfolding is a monotonic mapping of the eigenvalues $\lambda_i$ that makes the spectral density equal to 1 throughout the spectrum. The spectral density is defined as

$$\rho(\lambda) \equiv \left\langle \sum_i \delta(\lambda - \lambda_i)\right\rangle. \tag{9}$$

The unfolded eigenvalues $x_i$ are given by

$$\lambda_i \to x_i = \int^{\lambda_i} d\lambda\, \rho(\lambda), \tag{10}$$

and it is easy to see that they have unit density, $\bar{\rho}(x) = \frac{d\lambda}{dx}\rho(\lambda) = 1$. For random matrix models with dense matrices, the bulk statistical properties of the unfolded spectrum are expected to be universal (i.e., to not depend on the details of the model) and uniform throughout the spectrum. This has been proved rigorously for a large class of matrix ensembles (see Refs. [127,128] and references therein). One can then determine these properties in the exactly solvable Gaussian ensembles in the various symmetry classes (orthogonal, unitary, symplectic) [113]. In particular, the probability distribution of the unfolded spacings $s_i = x_{i+1} - x_i$, or *unfolded level spacing distribution* (ULSD), $p_{\mathrm{ULSD}}(s)$, can be obtained exactly, although not in closed form. A good approximation for the ULSD is provided by the so-called *Wigner surmise* in the appropriate symmetry class [114] (see Figure 5),

$$p_{\mathrm{WS}}^{(\beta)}(s) = a_\beta s^\beta e^{-b_\beta s^2} , \tag{11}$$

where $\beta$ is the *Dyson index* of the Gaussian ensemble, and one has for the various symmetry classes[7]

$$
\begin{aligned}
\text{orthogonal}: \quad & \beta = 1, \quad & a_1 = \frac{\pi}{2}, \quad & b_1 = \frac{\pi}{4}, \\
\text{unitary}: \quad & \beta = 2, \quad & a_2 = \frac{32}{\pi^2}, \quad & b_2 = \frac{4}{\pi}, \\
\text{symplectic}: \quad & \beta = 4, \quad & a_4 = \frac{262144}{729\pi^3}, \quad & b_4 = \frac{64}{9\pi}.
\end{aligned}
\tag{12}
$$

For chiral classes, the same ULSD is found as for the corresponding non-chiral classes.

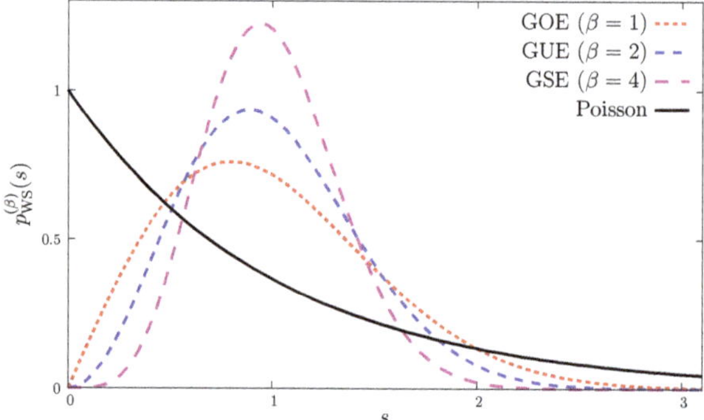

**Figure 5.** Wigner surmise for the various symmetry classes. The exponential distribution for Poisson statistics is also shown.

For independent eigenvalues, obeying Poisson statistics, the ULSD is the exponential distribution,

$$p_{\mathrm{Poisson}}(s) = e^{-s} . \tag{13}$$

Both for localized and delocalized modes, exact analytical results are then available for the statistical properties of the unfolded spectrum, and the transition from one type of modes to the other can be easily monitored across the spectrum. This allows in particular to identify the mobility edge, where a different, critical statistics is expected instead of Poisson or RMT statistics [129,130]. Various families of random matrix models have been developed to describe the critical statistics [131–138].

We mention in passing an alternative approach to the study of universal statistical properties of the spectrum, based on the use of the ratio of consecutive level spacings [139]. Since this ratio is independent of the local spectral density, this approach has the advantage of not requiring any unfolding, and has been shown to provide more precise results than those obtained from the unfolded spectrum in a variety of many-body systems (see Ref. [140] and references therein).

*3.5. Finite-Size Scaling at the Anderson Transition*

The mobility edge and the correlation-length critical exponent $\nu$ can be obtained by means of a finite-size scaling study [130] of the average $\mathcal{O}_L(\lambda)$ of suitable observables built out of the unfolded spectrum, and computed locally in the spectrum, for lattices of linear size $L$. Local statistics are defined formally as

$$\mathcal{O}_L(\lambda) \equiv \frac{1}{\rho(\lambda)} \langle \sum_i \delta(\lambda - \lambda_i)\mathcal{O}(\lambda_i)\rangle \equiv \langle \mathcal{O}\rangle_\lambda, \tag{14}$$

where $\mathcal{O}(\lambda_i)$ is some function of the eigenvalues (e.g., the level spacing $\Delta\lambda_i = \lambda_{i+1} - \lambda_i$, or the unfolded level spacing $s_i = x_{i+1}(\lambda_{i+1}) - x_i(\lambda_i)$ and its powers), and the average is over the ensemble. The last equality sets an alternative notation for local averages.

In practice, for a finite sample of disorder realizations obtained numerically, local statistics are computed by dividing the spectrum in small bins and averaging over modes within those, as well as over the sample. Unfolding is done by first fitting some smooth $\rho_{\text{average}}(\lambda)$ to the numerical data and then applying Equation (10). Alternatively, one can sort the eigenvalues of the sample by magnitude and replace them by their rank divided by the number of disorder realizations. If one is interested only in unfolded level spacings, one can divide $\Delta\lambda_i$ by the average level spacing $\langle\Delta\lambda\rangle_\lambda$ in the relevant spectral region, $s_i = \frac{\Delta\lambda_i}{\langle\Delta\lambda\rangle_\lambda}$ (notice that in the infinite-volume, infinite-statistics limit one has $\langle\Delta\lambda\rangle_\lambda = 1/\rho(\lambda)$).

For a finite-size scaling study, convenient observables are obtained from the unfolded level spacing distribution, $p_{\text{ULSD}}(s)$, defined above. Commonly used are the second moment, $\langle s^2\rangle = \int_0^\infty ds\, p_{\text{ULSD}}(s)s^2$, and the integrated probability distribution $I_{s_0} = \int_0^{s_0} ds\, p_{\text{ULSD}}(s)$. As the system size grows, $\mathcal{O}_L(\lambda)$ tends to its value for Poisson statistics in spectral regions where modes are localized, and to its value for (the appropriate) RMT statistics in spectral regions where modes are delocalized. Near the mobility edge $\lambda_c$, renormalization-group arguments and the one-parameter scaling hypothesis [98] imply that $\mathcal{O}_L(\lambda)$ depends on $\lambda$ and $L$ only through the combination $\xi(\lambda)/L$, where $\xi$ is the correlation length,[8] that diverges at $\lambda_c$ like $\xi(\lambda) \sim |\lambda - \lambda_c|^{-\nu}$. Since $\mathcal{O}_L(\lambda)$ is analytic in $\lambda$ for finite $L$, it must then take the form $\mathcal{O}_L(\lambda) = f((\lambda - \lambda_c)L^{1/\nu})$. Corrections to one-parameter scaling due to irrelevant operators can also be included, and the corresponding critical exponents be measured [141] (see Ref. [142] for an introduction). The goodness of one-parameter scaling can be visualized by means of the so-called "shape analysis" [143], obtained by plotting one spectral observable against another. If the scaling hypothesis is correct, only $\xi/L$ should determine the statistical properties of the spectrum, and so points corresponding to different $\lambda$ and system sizes should all lie on a single curve, corresponding to a path in the space of probability distributions connecting RMT and Poisson going through the critical statistics.[9] Thanks to the persistence of a remnant of multifractality near the mobility edge [144], one can apply similar finite-size scaling techniques also to the study of eigenmodes near criticality, in order to obtain the multifractal exponents [145], as well as the correlation-length exponent $\nu$ [146].

### 3.6. Anderson Transitions in Specific Models: Analytic Predictions and Numerical Results

Critical properties at the Anderson transition have been extensively studied by means of numerical simulations in the case of the conventional symmetry classes (O, U, and S), see Refs. [45,142,147,148] and references therein. According to the scaling theory of localization [98], a second-order Anderson transition is expected in all the conventional classes in three dimensions. The existence of these Anderson transitions has been confirmed numerically, and measurements of the correlation length critical exponent $\nu$ have shown that the three classes belong to different universality classes [141,148–152] (see Table 2). The expected nontrivial multifractal structure has also been found [145,146,148,151,152]. Universality has been explicitly demonstrated for the orthogonal class using different disorder distributions [141], and for the unitary class using different Hamiltonians [148,152].

**Table 2.** Correlation-length critical exponent for Anderson transitions in the conventional symmetry classes.

| Symmetry Class | Method | $\nu$ | Reference |
|---|---|---|---|
| orthogonal | localization length of quasi-1d bar | $1.57^{+0.02}_{-0.02}$ | [141] |
| | multifractal finite-size scaling | $1.590^{+0.012}_{-0.011}$ | [151] |
| | multifractal finite-size scaling | $1.595^{+0.014}_{-0.013}$ | [148] |
| unitary | localization length of quasi-1d bar | $1.43^{+0.04}_{-0.04}$ | [149] |
| | multifractal finite-size scaling | $1.437^{+0.011}_{-0.011}$ | [148] |
| | multifractal finite-size scaling | $1.446^{+0.006}_{-0.006}$ | [152] |
| symplectic | localization length of quasi-1d bar | $1.375^{+0.016}_{-0.016}$ | [150] |
| | multifractal finite-size scaling | $1.383^{+0.029}_{-0.024}$ | [148] |

In two dimensions, the predictions of the scaling theory of localization depend strongly on the details of the model. Absence of an Anderson transition is predicted in the orthogonal Anderson model, where all modes are expected to be localized for nonzero disorder, while an Anderson transition is predicted in the symplectic case (Ando model) [153,154]. The inclusion of topological effects in the field-theoretical description of disordered systems led one to expect an Anderson transition also in the theory of the integer Quantum Hall Effect [155], which belongs to the unitary class (see Ref. [156] for a review). While numerical evidence qualitatively supported this idea [157], significant quantitative discrepancies between different microscopic models were observed (see Ref. [158] for a summary), in contrast with the expected universality of the transition. A better understanding of the field theory describing the critical point was obtained only recently, in terms of a conformal field theory deformed only by marginal perturbations, that emerge from the spontaneous breaking of the replica (super)symmetry of the relevant nonlinear sigma model [159]. This proposal is quantitatively supported by numerical results, and can explain the apparent numerical discrepancies [158]. A transition between localized and delocalized modes was observed in the two-dimensional unitary Anderson model [160]. This transition is a disorder-induced transition of topological (Berezinskiĭ-Kosterlitz-Thouless [161–163]) type, with exponentially divergent correlation length, $\log \xi \sim |\lambda - \lambda_c|^{-1/2}$, in contrast to the usual second-order transition. For the unitary Anderson model there are conflicting theoretical predictions: while perturbative contributions lead to all states being localized (see, e.g., Refs. [45,154]), the inclusion of nonperturbative terms can possibly lead to the presence of an Anderson transition (see references cited in Ref. [160]). A similar transition of topological type was also observed in a model for disordered graphene with strong long-range impurities [164], belonging to the orthogonal class in two dimensions.

For our purposes, it is important to discuss the effect of off-diagonal disorder on localization. In the orthogonal class, theoretical arguments [109,110] suggest that off-diagonal disorder alone cannot localize modes at the band center; only increasing the on-site disorder leads eventually to localization of all the modes. This is confirmed by numerical results in three dimensions for the orthogonal Anderson model with random hopping [165–168]. The mobility edges $\pm E_c$ separating extended and localized modes move towards the band center as the on-site disorder $W$ is increased, and all modes are localized for $W > W_c$. In the absence of diagonal disorder ($W = 0$) for bipartite lattices, this model belongs to the chiral orthogonal class. The critical exponent $\nu$ characterizing the Anderson transition at $E_c \neq 0$ when $W = 0$ is found to be in agreement with that of the (non-chiral) orthogonal class, as well as with the one characterizing the transition at $E = 0$ as the critical on-site disorder $W_c$ is reached [166,167].

The origin $E = 0$ is singled out when the system has chiral symmetry (which is always the case when the lattice is bipartite and only off-diagonal disorder is present). In two dimensions, theoretical arguments predict critical behavior of modes at $E = 0$ (i.e., modes are extended but not fully delocalized) [111,169–175], while all other modes are localized. Numerical results indicate that modes are indeed critical at $E = 0$ [174,176–181], but an Anderson transition to localized modes can also appear [174,182]. It has been argued that such an Anderson transition can be present due to non-perturbative, topological effects [183]. In three-dimensional models with chiral symmetry, Anderson transitions at the origin ($E_c = 0$) are expected to show critical properties differing from those of the correponding conventional class (and from those at $E_c \neq 0$). Ref. [184] studied the Anderson transition at $E \sim 0$ in a chiral unitary model with purely off-diagonal disorder, finding multifractal exponents differing from those of the corresponding non-chiral class. In Refs. [185,186] the Anderson transition at the origin was studied in two-band models with on-site disorder in the chiral orthogonal and chiral unitary classes (as well as in other non-conventional symmetry classes), finding correlation length critical exponents differing from those of the corresponding non-chiral classes (and not entirely universal).

The critical properties of Anderson transitions at $E_c \neq 0$ in systems with chiral symmetry are instead not expected to differ from those in the corresponding conventional classes. Ref. [187] provides evidence of localization near the band center in a chiral unitary model mimicking fermions in a background of correlated spins with antiferromagnetic coupling in three dimensions, with the same critical properties as the non-chiral class; in two dimensions states near the band center seem instead to remain extended. Anticipating the results discussed in the following Sections, Refs. [27,34,35,37,51,53] provide examples of models in the chiral unitary class, both in three and two dimensions, displaying Anderson transitions at finite energy, and showing the same critical properties at the mobility edge as the corresponding non-chiral classes.

## 4. Localization and Deconfinement in QCD at Finite Temperature

The study of localization in QCD was initially motivated by the idea that the spontaneous breaking of chiral symmetry could have a similar origin as conductivity in a disordered medium [8–15]. The basic idea of the disordered medium scenario is that the near-zero modes responsible for the breaking of chiral symmetry originate from the mixing of the zero modes associated with overlapping instantons (or, more precisely, calorons at finite temperature). At finite temperature these zero modes are exponentially localized on the scale of the inverse temperature. If instantons/calorons overlap sufficiently, mixing of the corresponding zero modes will transform the zero eigenvalues into a near-zero band of levels, and lead to delocalized eigenmodes [10].[10] This is analogous, for example, to the Anderson-Mott insulator-metal transition[11] driven by the impurity concentration in doped semiconductors (see, e.g., Ref. [189]). Starting from the chirally symmetric phase and decreasing $T$, the overlap of calorons increases and eventually leads to a finite density of near-zero, delocalized modes. This leads to expect a localization-delocalization transition in the near-zero region.

As we will see below, this scenario is most likely only a part of the story, and overlooks the important role played by deconfinement in localizing the low Dirac modes. Instead of sticking to the disordered medium scenario, we prefer to adopt a more general point of view, looking at the Dirac operator as a random matrix, ignoring initially any relation with topological objects and deconfinement. After a few introductory remarks on this approach, we review the available results regarding localization and Anderson transitions in various lattice approximations for QCD, following a chronological order. Some of the results discussed here deal with the pure gauge theory, sometimes for $N_c = 2$, and are included in this section mostly for historical reasons. A summary of the results for QCD proper and organized by topic is provided at the end of the section.

### 4.1. The Dirac Operator as a Random Matrix

The Dirac operator in the background of fluctuating gauge fields can be intepreted as a sparse random matrix, and the properties of its eigenvalues and eigenvectors can be studied with the machinery discussed in the previous Section. For the continuum anti-Hermitian Dirac operator, $-i\slashed{D}$ can be formally treated as the Hamiltonian of a disordered system, with disorder provided by the gauge fields. If an Anderson transition is present in its spectrum, its critical properties are expected to be determined by the symmetry class of the Dirac operator and by the dimensionality of the space-time over which it is defined.

Concerning the symmetry class, the four-dimensional Dirac operator for fundamental fermions in $SU(N_c)$ theories belongs to the chiral unitary class for $N_c > 2$, and to the chiral orthogonal class for $N_c = 2$.[12] The spectral correlations of Dirac eigenvalues in QCD ($N_c = 3$) are then expected to display GUE-type bulk statistics,[13] as long as the corresponding eigenvectors are delocalized.[14] If localized modes are present, they are expected to obey Poisson statistics, regardless of the symmetry class.

The discretization of the Dirac operator on a lattice is known to be tricky due to the doubling problem (see Refs. [73,74,82–84]), and in some cases its chiral properties are changed (see Ref. [188] and Ref. [115], Section 5.2.1). For staggered fermions [196–198] a remnant of the continuum chiral symmetry preserves the chiral nature of the symmetry class, and the staggered Dirac operator belongs to the chiral unitary class, as the continuum operator, for $N_c \geq 3$. For $N_c = 2$ the symmetry class is instead changed to the chiral symplectic one.[15] Overlap fermions [199–202] possess an exact lattice chiral symmetry, and belong to the same symmetry class as their continuum counterpart. More precisely, since the overlap operator is not anti-Hermitean, this is true for its anti-Hermitean part. It is then understood, unless specified otherwise, that the imaginary part of the overlap eigenvalues is considered in the following. For the low modes this is an adequate approximation, that becomes exact in the continuum limit. Moreover, since the unfolded spectrum is unaffected by any monotonic mapping, the statistical properties of the low modes are unchanged if one uses other types of projection on the imaginary axis (i.e., eigenvalue magnitude, stereographic projection).[16]

Concerning the dimensionality of the problem, in finite-temperature field theory the temporal size of the system is fixed (in physical units) in the thermodynamic limit, and only the size of the $d$ spatial directions is sent to infinity in the thermodynamic limit. For $d + 1$ spacetime dimensions, the dimensionality of the disordered system described by the Dirac operator in a gauge-field background is then equal to $d$, while the temporal direction can be technically seen as an internal degree of freedom.

*4.2. Numerical Results on the Lattice*

The disordered medium scenario was investigated in Ref. [25] by means of numerical simulations of quenched QCD on the lattice on both sides of the finite-temperature transition. They used a single spatial volume, employed the staggered discretization of the Dirac operator, and studied the rotation- and gauge-invariant version of the IPR of an eigenmode $\psi$, Equation (8). They observed that in the physical $\mathbb{Z}_3$ sector (real Polyakov loop sector) the IPR of the lowest modes was considerably larger above the transition than below the transition (see Figure 6). Moreover, above $T_c$ it was larger in the real sector than in the complex Polyakov loop sectors, where it does not change much across the transition. Sensitivity to the Polyakov loop sector is equivalent to sensitivity to the temporal boundary conditions, and shows that the low modes cannot be localized in the temporal direction on a scale much shorter than the temporal size.[17] This suggests the presence of a localization transition in the physical sector, with spatially localized low modes at high temperature. Evidence for some of the localized modes being related to calorons was also provided. More evidence for localization of the low modes in the real sector appeared in Ref. [26], where more volumes and a chirally improved discretization of the Dirac operator were used. The volume scaling of the IPR of the low modes in the real sector was found to be in qualitative agreement with that expected for localized modes.

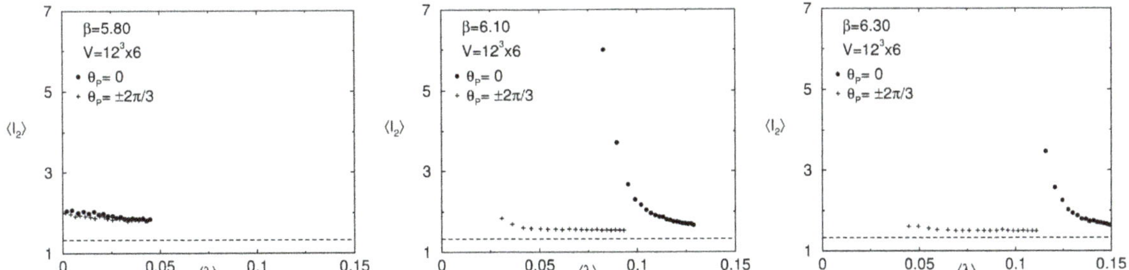

**Figure 6.** IPR of the low staggered modes on quenched configurations below (**left**), slightly above (**center**), and well above (**right**) the deconfinement transition for real ($\theta_P = 0$) and complex ($\theta_P = \pm\frac{2}{3}$) Polyakov loop sectors. From Ref. [25]. (Reprinted figure with permission from M. Göckeler, P.E.L. Rakow, A. Schäfer, W. Söldner, and T. Wettig, Phys. Rev. Lett. 87, 042001 (2001). Copyright (2001) by the American Physical Society).

The disordered medium scenario was investigated further by García-García and Osborn in Refs. [27,28]. In Ref. [27] they considered an Instanton Liquid Model (ILM) for the QCD vacuum (see Ref. [204]), and studied the behavior of the instantonic zero modes. Changing the temperature, and so the spatial extension of the zero modes, they observed the appearance of a mobility edge near the origin, both in the quenched approximation and in the presence of fermions. In the quenched case, the multifractal properties of the near-zero modes at the transition were found to be consistent with those of the 3d unitary Anderson transition (see Refs. [45,148,152]). With two massless flavors, the mobility edge appears at the same temperature where the chiral condensate shows a drop (see Figure 7, left). Although the thermodynamic limit was not studied, this was taken as an indication that localization of the low modes coincides with the chiral transition.

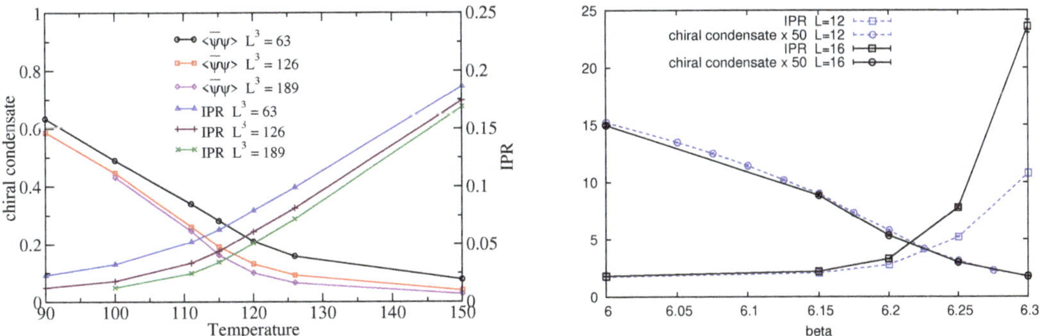

**Figure 7. Left**: IPR of the lowest Dirac mode and chiral condensate in an ILM model for QCD with two massless fermions for various system sizes (in fm$^3$). From Ref. [27]. (Reprinted from Nucl. Phys. A, 770, A.M. García-García and J.C. Osborn, "Chiral phase transition and Anderson localization in the Instanton Liquid Model for QCD", Pages 141–161, Copyright (2006), with permission from Elsevier). **Right**: IPR (here times the volume $V = L^3$) of the low Dirac modes and chiral condensate in 2+1 flavor QCD with staggered fermions. From Ref. [28]. (Reprinted figure with permission from A.M. García-García and J.C. Osborn, Phys. Rev. D 75, 034503 (2007). Copyright (2007) by the American Physical Society).

A test of the disordered medium scenario in a more realistic context was presented in Ref. [28]. There the localization properties of the near-zero modes were studied on the lattice in quenched QCD, i.e., pure gauge SU(3) theory, and "unquenched" QCD, i.e., with 2+1 flavors of dynamical quarks of relatively large masses, leading to heavier-than-physical pions [205,206]. The one-loop Symanzik improved gauge action was used, and the Asqtad-improved [207–210] staggered discretization was employed for the lattice Dirac operator. In both cases, they found indications of critical (i.e., volume-independent) spectral statistics (from the second moment of the ULSD) at a temperature $T_c^{\mathrm{loc}}$, where also IPR $\cdot$ $V$ starts increasing with the volume (for the unquenched case see Figure 7, right). In the quenched case, indications of a vanishing spectral density and of an increase of the Polyakov loop are found at a similar temperature $T_c^{\mathrm{dec}/\chi}$, identified with the deconfinement temperature (in the physical $\mathbb{Z}_3$ sector). In the unquenched case, $T_c^{\mathrm{loc}}$ is close to the crossover temperature $T_c^{\chi}$ obtained from the chiral susceptibility. Although the use of small lattices does not allow a full quantitative assessment, these indications suggest that an Anderson transition takes place near the origin of the spectrum as the system crosses over from the low-temperature to the high-temperature phase, with the low-lying Dirac modes turning from delocalized to localized, and the formation of a mobility edge that separates them from delocalized modes in the bulk of the spectrum.

Studies of localization in QCD-like settings includes also the case of two flavors of dynamical staggered fermions [29], and that of quenched two-color QCD (i.e., pure gauge SU(2) theory) analyzing overlap [30,31] and staggered [32] spectra. In the two-flavor three-color case Ref. [29] found that IPR $\cdot$ $V$ of the low modes was volume-independent below the transition temperature $T_c^{2f}$, but above that it scaled with the volume in a manner compatible with localization (see Figure 8, left). Ref. [30] shows evidence of absence of correlations in the low-lying overlap spectrum, which is typical of localized modes, at $T = 2.6T_c^{\mathrm{SU}(2)}$, where $T_c^{\mathrm{SU}(2)}$ is the deconfinement temperature of the pure gauge SU(2) theory. Ref. [32] shows clear evidence of localization of the low-lying staggered modes, and of the presence of a mobility edge separating them from delocalized bulk modes, again at $T = 2.6T_c^{\mathrm{SU}(2)}$. This is obtained by studying how (i) the scaling with the lattice spatial volume of the spatial "size" $(\mathrm{IPR}^{-1}/N_t)^{\frac{1}{3}}$ of the eigenmodes, and (ii) the ULSD of the corresponding eigenvalues change along the spectrum. The spatial extension of the low modes is volume-independent, while higher up in the spectrum it is seen to increase with the lattice size (see Figure 8, right). Looking at the ULSD in different spectral regions, it is observed that it matches the exponential distribution of Poisson statistics for the lowest modes, changing towards the symplectic Wigner surmise[18] as one moves towards the bulk. Finally, in Ref. [31] the transition in the overlap spectrum from localized to delocalized modes is studied via the ULSD at $T = 2.6T_c^{\mathrm{SU}(2)}$. A clear change from the exponential to the orthogonal Wigner surmise is observed.[19] Moreover, assuming that there are no strong interactions among instantons and anti-instantons, it is argued that the instanton density is too low to match the density of localized modes at this temperature. Indications of correlations between localized modes and local fluctuations of the Polyakov loop away from its ordered value (i.e., 1, in the physical sector) are also reported (see Section 5.2 for a detailed discussion).

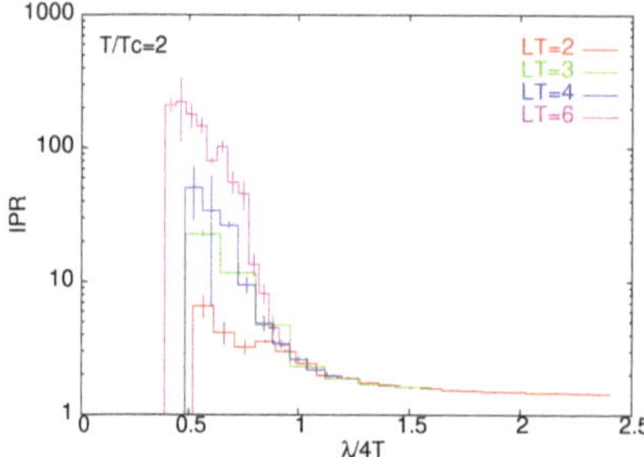
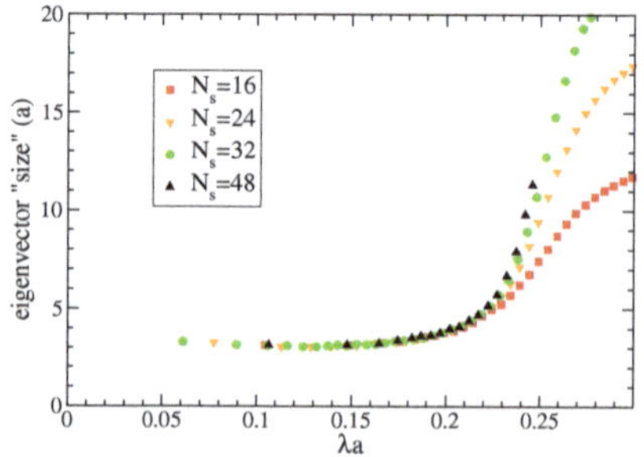

**Figure 8. Left**: IPR (here times the volume $V = L^3$) of low staggered modes for various volumes at $T = 2T_c^{2f}$ in QCD with two flavors of dynamical staggered fermions. From Ref. [29]. (Reprinted figure with permission from R.V. Gavai, S. Gupta, and R. Lacaze, Phys. Rev. D 77, 114506 (2008). Copyright (2008) by the American Physical Society). **Right**: Mode size for low staggered modes in the background of quenched SU(2) configurations at $T = 2.6T_c^{\mathrm{SU}(2)}$. From Ref. [32]. (Reprinted figure with permission from T.G. Kovács and F. Pittler, Phys. Rev. Lett. 105, 192001 (2010). Copyright (2010) by the American Physical Society).

A comprehensive study of localization in the high-temperature phase of real-world QCD appeared in Ref. [33], using a tree-level Symanzik improved gauge action and a two-level stout smeared [211] staggered fermion action for 2+1 quark flavors with physical mass [212]. Several volumes, aspect ratios and lattice spacings were used, covering the temperature range $1.7T_c < T < 5T_c$ (here $T_c = 155$ MeV [5]) with lattices of linear size $2\,\text{fm} \leq L \leq 6\,\text{fm}$. Localized modes were observed at the low end of the spectrum (see Figure 9, left). A temperature-dependent mobility edge $\lambda_c(T)$ separating low-lying, localized modes from delocalized bulk modes was found in the whole temperature range, studying how the spectral statistics change along the spectrum from Poisson to unitary RMT type. More precisely, $\lambda_c$ was estimated as the inflection point of the variance of the ULSD, $\langle s^2 \rangle_\lambda - \langle s \rangle_\lambda^2 = \langle s^2 \rangle_\lambda - 1$, computed locally in the spectrum. As $T$ increases, $\lambda_c(T)$ increases as well. Extrapolation to the continuum is studied at $T = 400$ MeV. The mobility edge is expected to renormalize like a quark mass, and the ratio $\lambda_c/m_{\text{ud}}$ is indeed shown to be independent of the lattice spacing within numerical uncertainties. The localization length $l \equiv a\langle \text{IPR}^{-\frac{1}{4}} \rangle$ of the low modes is also shown to extrapolate to a finite continuum limit, and $lT$ is found to be between 0.7 and 0.9 for all the lattice ensembles. A second-order polynomial fit to the RG-invariant quantity $\lambda_c(T)/m_{\text{ud}}$ shows that it extrapolates to zero at $T_c^{\text{loc}} = 170$ MeV (see Figure 9, right), which is within the temperature range where the system undergoes a crossover from the low-temperature phase to the high-temperature phase [5,6]. This is consistent with localization of the low modes appearing as the system changes from confined and chirally broken to deconfined and chirally restored. The density of localized modes (number of modes per unit spatial volume) is seen to increase with $T$ (see Figure 13 below).

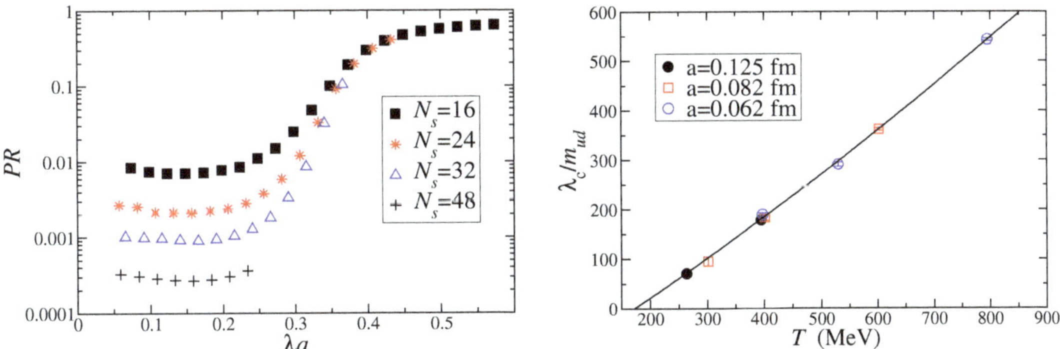

**Figure 9.** Average PR of the staggered Dirac modes as a function of the eigenvalue $\lambda$ for various volumes at $T = 394$ MeV, $a = 0.125$ fm (**left**) and renormalized mobility edge $\lambda_c/m_{\text{ud}}$ as a function of $T$ (**right**) in 2+1 flavor QCD with staggered fermions and physical quark masses. From Ref. [33]. (Reprinted figures with permission from T.G. Kovács and F. Pittler, Phys. Rev. D 86, 114515 (2012). Copyright (2012) by the American Physical Society).

The critical behavior of the eigenmodes at the mobility edge was studied in Refs. [34,35,37]. All these references use the same setup as Ref. [33] with $N_t = 4$ and $a = 0.125$ fm, corresponding to $T = 2.6T_c$. In Ref. [34] it was established, by means of a finite size scaling analysis of the integrated ULSD $I_{s_0}$, that the transition from localized to delocalized modes at the mobility edge is indeed an Anderson transition (see Figure 10, left). The critical exponent was found to be $\nu = 1.43(6)$, in agreement with the one obtained for the 3d unitary Anderson model [149] (see Table 2). In Ref. [35] the critical eigenvalue statistics at the mobility edge was studied in terms of the one-parameter family of deformed random matrix ensembles of Refs. [136,137]. The critical statistics was shown to be indeed volume-independent, and well described by a deformed random matrix ensemble, with deformation parameter consistent with the one found for the 3d unitary Anderson model [148,149,152]. Finally, in Ref. [37] the critical exponent $\nu$ and the multifractal exponents where studied using the finite size scaling techniques for the eigenmode density developed in Refs. [145,146]. All exponents were found to be in agreement with those of the 3d unitary Anderson model [148,152] (see Figure 10, right).

While mostly focussed on the properties of the spectrum, Ref. [38] briefly discussed the localization properties of the low Dirac modes in QCD near the crossover temperature. The spectrum of the overlap operator was studied in the background of gauge configurations generated with tree-level improved Symanzik gauge action and 2+1 flavors of highly improved staggered quarks (HISQ) [213] with near-physical quark masses ($m_l/m_s = 1/20$, $m_\pi = 160\,\text{MeV}$). Evidence was found of a small peak of localized near-zero modes at $T = 1.2T_c$ and $T = 1.5T_c$ (here $T_c = 154\,\text{MeV}$). Localization was inferred from the smallness of the PR; the volume scaling was not discussed. It was suggested that near-zero modes in the peak correspond to an approximate superposition of the exact zero modes associated with instanton–anti-instanton pairs. Comparison of the PR of zero and near-zero modes shows however that only a fraction of near-zero modes is compatible with this interpretation. The large fluctuations of the PR of the near-zero modes suggests instead a large variability of the number of topological lumps participating in the superposition.

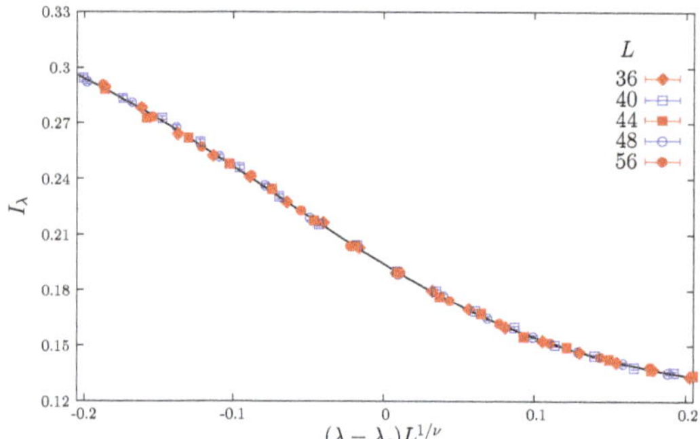
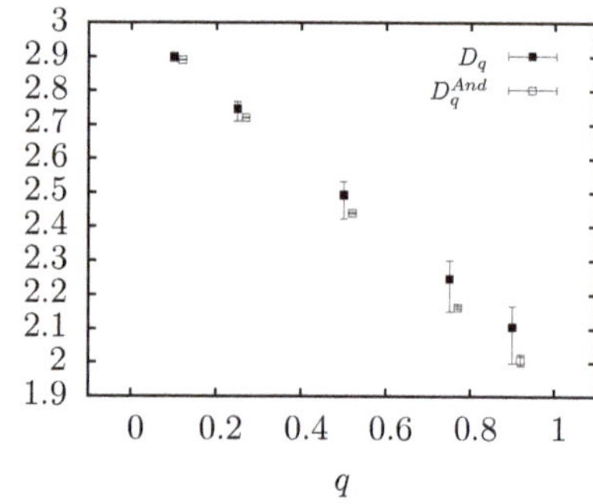

**Figure 10. Left**: scaling of $I_{s_0}$ near the mobility edge in 2+1 QCD with staggered quarks, at $T = 394\,\text{MeV}$ and $a = 0.125\,\text{fm}$. The linear size $L$ of the lattice is expressed in lattice units. Here $\lambda_c \simeq 0.3364$, $\nu \simeq 1.43$, and the critical value is $I_{s0}|_{\lambda_c} \simeq 0.194$. From Ref. [34]. (Reprinted figure with permission from M. Giordano, T.G. Kovács and F. Pittler, Phys. Rev. Lett. 112, 102002 (2014). Copyright (2014) by the American Physical Society). **Right**: multifractal exponents $D_q$, governing the scaling of $\text{IPR}_q \sim L^{-D_q(q-1)}$ at large linear size $L$ at the mobility edge, in 2+1 QCD with staggered quarks ($T = 394\,\text{MeV}$, $a = 0.125\,\text{fm}$) and in the 3d unitary Anderson model. From Ref. [37]. (Reprinted figure with permission from L. Ujfalusi, M. Giordano, F. Pittler, T.G. Kovács and I. Varga, Phys. Rev. D 92, 094513 (2015). Copyright (2015) by the American Physical Society).

Localization in two-flavor QCD was studied in Ref. [39] using tree-level improved Symanzik gauge action and dynamical Möbius domain-wall fermions [214–216] with stout smearing, and looking at the spectrum of the Hermitian operator $\gamma_5 D$, with $D$ the four-dimensional effective Dirac operator of the five-dimensional domain-wall fermion. This operator is in the chiral unitary class. The temperature range was $0.9T_c \leq T \leq 1.9T_c$ with $T_c \simeq 175\,\text{MeV}$ the deconfinement temperature estimated from the average Polyakov loop. A range of bare quark masses, two spatial volumes and two temporal extensions (in lattice units) were used. Above $T_c$, the scaling of the PR of the eigenmodes shows that the lowest modes are localized, while moving up in the spectrum modes become delocalized (see Figure 11, left). The size $v \equiv V \cdot \text{PR}$ of the low modes increases along the spectrum and decreases with $T$ (see Figure 11, right). The localization length $l \equiv v^{\frac{1}{4}}$ of the lowest (nonzero) mode was shown to be of the order of the inverse temperature, $lT \sim 1.3$. The ULSD computed locally in the spectrum was seen to change from Poisson-type to RMT-type in the unitary class as one moves from the lowest modes towards the bulk. By contrast, below $T_c$ RMT statistics was observed everywhere in the spectrum. Changing boundary conditions to periodic in time, low modes were found to be delocalized with

RMT statistics also above $T_c$. A clear correlation between the spatial density $\psi^\dagger\psi(x)$ of the low modes and the local fluctuations of the Polyakov loop $P(\vec{x})$ away from its ordered value was observed, favoring sites with $\mathrm{Re}\,\mathrm{tr}P(\vec{x})$ close to $-1$, and becoming stronger as $T$ increases (see Figure 12 below). Correlation with action ($s(x)$) and topological charge ($q(x)$) densities was also observed, with localized modes favoring sites with large $s$ and $q$, in particular "(anti)self-dual" sites where $|q|/s \sim 1$. The overlap of the left- and right-chirality components of the modes was seen to be the smallest for the lowest modes, and to increase as one moves towards the bulk; it was also seen to increase with temperature, and showed little dependence on the bare quark mass.

In Ref. [40] (see also Ref. [217]), localization was studied in 2+1+1 flavor QCD with physical strange and charm masses but heavy pions ($m_\pi \simeq 370\,\mathrm{MeV}$), looking at the (stereographically projected) spectrum of the overlap operator in the background of configurations generated with Iwasaki gauge action and dynamical twisted-mass Wilson fermions [218–220]. Above $T_c \simeq 188\,\mathrm{MeV}$, a very small PR is found for the lowest modes, which increases to around 0.8 in the bulk. The position of the mobility edge was estimated as the inflection point of the PR in the spectrum. As a function of $T$, $\lambda_c(T)$ appears to be linear, and its extrapolation vanishes at a temperature compatible with $T_c$. A strong anticorrelation of the localized modes with $\mathrm{Re}\,\mathrm{tr}P(\vec{x})$ was observed.

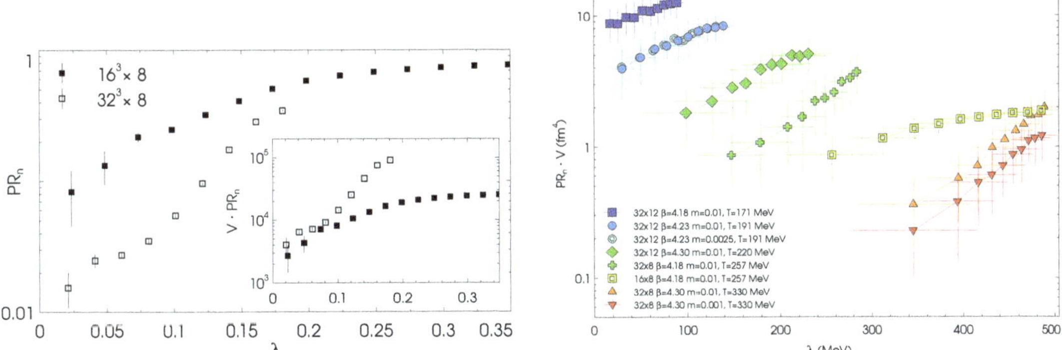

**Figure 11.** Scaling of the PR (and of $V \cdot \mathrm{PR}$ in the inset) for $\beta = 4.18$, $N_t = 8$ ($T = 257\,\mathrm{MeV} \simeq 1.5T_c$) and bare mass $m = 0.01$ (**left**), and dependence on $T$ and $m$ of the PR of the 10 lowest modes (**right**) in two-flavor QCD with Möbius domain-wall fermions. From Ref. [39]. (Figures adapted from G. Cossu and S. Hashimoto, J. High Energy Phys. 06, 56 (2016), and used under a CC-BY 4.0 license (https://creativecommons.org/licenses/by/4.0)).

### 4.3. Summary

Let us summarize the results discussed in this section, restricting to QCD "proper", i.e., gauge group SU(3) and dynamical quarks. QCD-like models and other gauge theories are discussed in detail below in Section 6.

- *Low Dirac modes are localized in lattice QCD in the high-temperature phase.* More precisely, a large amount of evidence indicates that the low Dirac modes are localized in lattice QCD, for temperatures above the finite-temperature transition, for more or less physical quark content and masses, and different fermion discretizations [28,33,38–40]. The available evidence suggests that localization is not a lattice artifact and survives the continuum limit: both the localization length and the renormalized mobility edge seem in fact to possess a continuum limit. Evidence is, however, limited to a single study, and at a single temperature [33].

- *Localization appears approximately at the transition.* As the transition is only a crossover, this statement can only be of qualitative nature. In all the cases discussed above [28,33,39,40], localization appears somewhere in the range of temperatures where the crossover takes place.

- *The localization length is of the order of the inverse temperature* [33,39].
- *An Anderson transition takes place in the Dirac spectrum in the high-temperature phase.* More precisely, a mobility edge separating localized and delocalized modes in the spectrum is observed on the lattice [33–35,37,39,40]. For staggered fermions it has been shown that a genuine Anderson transition takes place at the mobility edge [34,35,37].
- *Localized modes correlate with local fluctuations in the confining and topological properties of the configurations.* More precisely, the spatial position of localized modes shows correlations with the local fluctuations of the Polyakov loop away from order [39,40], as well as with positive fluctuations of the action density and of the magnitude of the topological charge density, especially at (anti)self-dual points [39].

We now list a few remarks.

- As disordered systems, almost all the models discussed in this section are in the 3d chiral unitary class.[20] The appearance of localized modes at the band center contrasts with the delocalized nature of the band center in the 3d chiral orthogonal Anderson model [166,167]. On the other hand, it agrees with what was found in the 3d chiral unitary Anderson model [184], and in the Anderson model with correlated disorder of Ref. [187] in the same class.
- The results of Refs. [34,35,37] indicate that a genuine second-order Anderson transition is present in the staggered Dirac spectrum in high-temperature QCD, in the universality class of the 3d unitary Anderson model. Since QCD is in the 3d chiral unitary class, this suggests that the Anderson transition at nonzero eigenvalue for the 3d chiral and non-chiral unitary classes belong to the same universality class. This is not surprising, as chiral symmetry is not expected to play an important role in the bulk of the spectrum, but only near the origin, around which the spectrum is symmetric precisely due to chiral symmetry. Further support to the lack of any differences in the transition of the chiral and non-chiral model is given by the findings of Ref. [27] concerning the multifractal exponents in the ILM model for QCD, and by the critical statistics found in the Anderson model with correlated disorder of Ref. [187]. A different critical behavior is found instead when the Anderson transition is at the origin in 3d chiral models [184–186].
- In the ILM model of Ref. [27], both in the quenched and unquenched cases, a second mobility edge was observed higher up in the spectrum, moving towards the high end as the temperature is decreased. While this part of the spectrum is not representative of real QCD, as the model neglects nonzero modes at the outset, it is nonetheless possible that a similar localization mechanism at the high end of the spectrum applies in QCD as well.[21]
- It is now clear that the Dirac spectral density does not vanish in the deconfined phase of pure gauge SU(3) theory, if one uses sufficiently fine lattices, or lattice discretizations of the Dirac operator with good chiral properties; instead, a peak is formed near the origin (see Refs. [51,57,59]). A sort of "chiral transition" still takes place at deconfinement, where the peak structure appears.
- The disordered medium scenario requires that the densities of instantons and of localized modes match in the high-temperature phase. As observed in Ref. [31] and, in the pure gauge case, in Ref. [86] (see Section 6), the instanton density, obtained assuming an ideal (non-interacting) instanton gas approximation, is lower than the density of localized modes (number of modes per unit spatial volume). Moreover, the latter is seen to increase with $T$ [33], while the instanton density decreases. This indicates that topology can only partially explain the localization of the low Dirac modes.
- An alternative interpretation of localization in terms of topological objects was proposed in Ref. [39]. The authors suggest that localized modes favor regions where $L$-type (Kaluza-Klein) monopole-antimonopole pairs are located. These are one of the types of monopole constituents inside calorons [20]. This interpretation is supported

by the correlation with Polyakov-loop fluctuations, action and topological density, and chirality. A direct identification of monopoles or a quantitative estimate of their density is, however, unavailable.

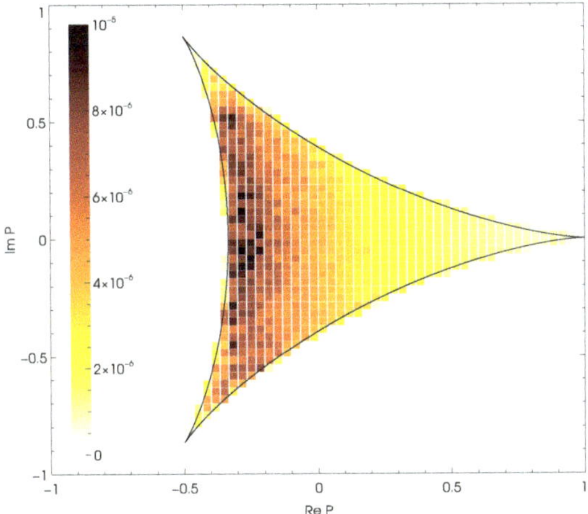

**Figure 12.** Density plot of the average local norm $\psi^\dagger\psi$ of low Dirac modes in the Polyakov loop plane (here $\mathrm{P} = \frac{1}{3}\mathrm{tr}\,P$) in the high-temperature phase. From Ref. [39]. (Figure adapted from G. Cossu and S. Hashimoto, J. High Energy Phys. 06, 56 (2016), and used under a CC-BY 4.0 license (https://creativecommons.org/licenses/by/4.0)).

## 5. Mechanisms for Localization

In this section, we discuss in some detail the two mechanisms, or more precisely the two sources of disorder, proposed so far to explain localization of the low Dirac modes in QCD: the disordered medium scenario, based on topology fluctuations; and the sea/islands picture, based on fluctuations of the Polyakov loop. We have already briefly discussed the disordered medium scenario in the previous section; here we discuss it again, both to keep this section self-contained, and to give more details.

### 5.1. The Disordered Medium Scenario

As is well known, the continuum Dirac operator in the background of a gauge configuration of topological charge $Q$ has $n_\pm$ exact zero modes of definite chirality $\pm 1$, with $Q = n_+ - n_-$ (index theorem). In particular, for instantons (resp. anti-instantons) of topological charge 1 (resp. $-1$) one finds an exact zero mode of positive (resp. negative) chirality. The same holds for the finite-temperature generalization of instantons known as calorons [16–23].[22] The zero modes supported by instantons (i.e., at $T = 0$) are algebraically localized, decaying like $1/R^3$ with the distance $R$ from the instanton center. The zero modes supported by calorons are instead fully delocalized in the temporal direction, and exponentially localized in the spatial directions, decaying like $e^{-rT}$, with $r$ the spatial distance from the caloron center and $T$ the temperature of the system. For a dilute ensemble of these objects, their associated zero modes are not exact Dirac eigenmodes any longer, due to the fact that instantons/calorons overlap. The low-lying Dirac eigenmodes are instead linear combinations of these "unperturbed" zero modes,[23] obtained by diagonalizing the "perturbed" Dirac operator, which in the zero-mode basis reads (see, e.g., Ref. [204])

$$i\slashed{D} = \begin{pmatrix} \mathbf{0} & T_{IA} \\ T_{AI} & \mathbf{0} \end{pmatrix}, \tag{15}$$

with $T_{IA}$ and $T_{AI} = T_{IA}^\dagger$ the matrices of the overlap integrals of $i\slashed{D}$ between the unperturbed zero modes associated with an instanton–anti-instanton pair.[24] For a dilute ensemble the total topological charge $Q$ is expected to be simply equal to the sum of the individual charges. Out of all the unperturbed zero modes, $Q$ are preserved by topology despite mixing,[25] while the remaining ones are not protected by topology and spread around $\lambda = 0$ forming a band. The extent of this spreading and the resulting density of near-zero Dirac modes for typical gauge configurations are dynamical issues, determined by the typical density and size of topological objects. In the quenched case, a finite spectral density of near-zero modes is expected to survive as long as a non-negligible density of topological objects supports them. In the presence of dynamical fermions, the fermionic determinant tends to suppress configurations with a higher density of near-zero modes, and so suppresses topological excitations, with respect to the quenched case, but a finite spectral density is still possible. In any case, the details of the dynamics, including especially the temperature and the fermion masses, determine whether a nonzero density of near-zero modes is formed, i.e., loosely speaking, whether chiral symmetry is spontaneously broken.

For sufficiently low temperature, and not too many quark flavors, the density of topological objects and the effect of mixing become strong enough to overcome the repulsive effect of the fermionic determinant, and chiral symmetry breaks spontaneously through the formation of a nonzero density of near-zero modes. This is the disordered medium scenario for chiral symmetry breaking [8–15]. The mixing of the unperturbed modes is also expected to spread them out in space, over topological objects that overlap non-negligibly with their original location. If the typical spatial distance $n^{-\frac{1}{3}}$ between topological objects is large compared to the typical spatial range $1/T$ of the corresponding unperturbed zero modes, one expects the resulting perturbed near-zero modes to remain localized on a few objects only. Here $n = \frac{\mathcal{N}_{\text{top}}}{V}$ is the spatial density of $\mathcal{N}_{\text{top}}$ calorons and anti-calorons in a finite spatial volume $V$. At high temperatures both density and range are small, and near-zero modes are expected to be localized. As the temperature decreases, both density and range increase, with more and more topological objects overlapping, and near-zero modes are expected to eventually delocalize over the whole system [10]. It is reasonable to expect that delocalization will take place around the same temperature as chiral symmetry breaking (in the loose sense explained above). It should be clear, however, that finite spectral density and delocalization of modes near the origin are not automatically linked.[26]

According to the scenario above, near-zero localized modes should be associated with local lumps of topological charge. It is worth noting that (anti)calorons in $SU(N_c)$ gauge theory are made up of $N_c$ (anti)monopole constituents [20], and that when these are well separated the associated zero mode is localized on a single constituent; which one depends on the holonomy (Polyakov loop) of the gauge field at asymptotic distance from the core [222,223]. For typical high-temperature ordered configurations with Polyakov loop in the trivial sector, the relevant constituent is the type-$L$ monopole, which also has the largest action and topological charge densities, as well as the smallest size (see Ref. [221]). This further characterizes the favorable locations for modes according to the disordered medium scenario.

Refs. [25,26,38,39,86] provide evidence that *some* of the modes are indeed localized on topological objects. In particular, Ref. [39] shows that the locations favored by some of the localized low modes have all the features of $L$-type monopoles and antimonopoles: large action and topological charge densities, near (anti)self-duality, and near degeneracy of two eigenvalues of the untraced Polyakov loop (see Figure 12).[27] In Refs. [58,86] it is shown that for pure gauge $SU(3)$ theory the distribution of the number of near-zero modes in the peak of the spectral density near zero (see Figure 1) is consistent with the distribution of a dilute gas of topological objects. This suggests that the peak of near-zero modes indeed originates from the zero modes associated with topological objects. This provides further evidence supporting the disordered medium scenario as a viable mechanism for localization, and its close relation with spontaneous chiral symmetry breaking.

If localization were entirely due to mixing topological would-be zero modes, then the density of localized modes would be equal to that of the topological objects (calorons).[28] Since above the transition the density of calorons decreases sharply with increasing temperature, we would expect the same behavior of the density of localized modes. However, as shown in Ref. [33], the density of localized modes actually increases with temperature (see Figure 13). In the pure gauge case, no more than half of the localized modes seem to be of topological origin for temperatures as low as $1.03 T_c$ [86] and as the temperature increases, this fraction rapidly decreases. In Figure 14 we show the temperature dependence of the fraction of localized modes that can be associated with near-zero modes of topological origin [51,86]. We show results obtained with the overlap Dirac operator, and with the staggered Dirac operator for three different values of the lattice spacing. All the results are consistent and show that with increasing temperature a rapidly decreasing fraction of the localized modes are of topological origin. We can conclude that while topology-related localized modes may suffice to explain the near-zero peak, they cannot explain all the remaining localized modes found in a typical high-temperature gauge configuration, which therefore require a different supplementary mechanism.

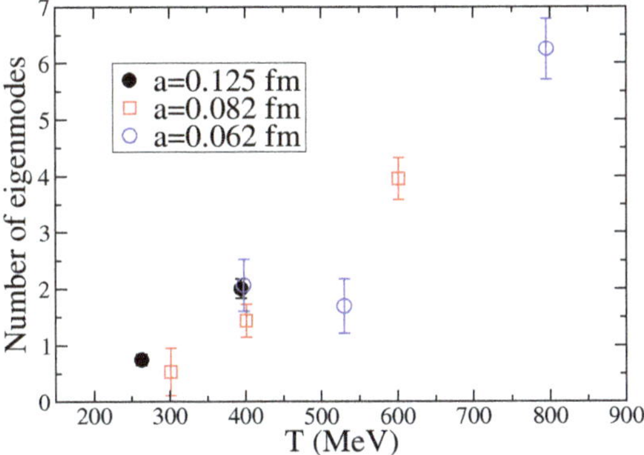


**Figure 13.** Density of localized modes (modes per cubic fermi) in high-temperature QCD. From Ref. [33]. (Reprinted figure with permission from T.G. Kovács and F. Pittler, Phys. Rev. D 86, 114515 (2012). Copyright (2012) by the American Physical Society).

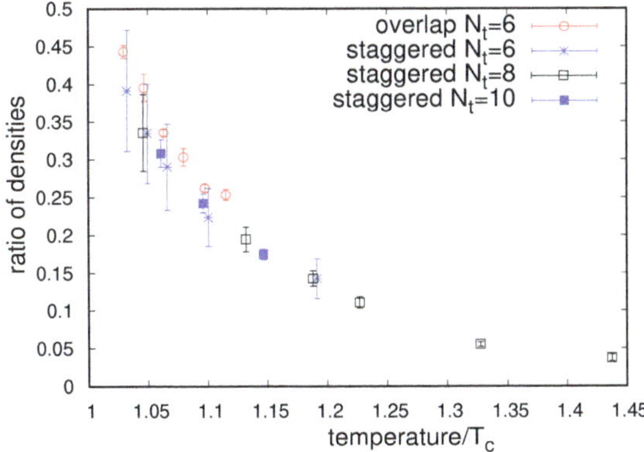


**Figure 14.** The fraction of localized modes in the quenched theory that are topology-related near-zero modes. The figure shows data calculated with the overlap Dirac operator on $N_t = 6$ lattices, as well as with the staggered Dirac operator using three different lattice spacings corresponding to $N_t = 6, 8$ and 10. From Ref. [86]. (Figure adapted from T.G. Kovács and R.Á. Vig, PoS LATTICE2018, 258 (2019), and used under a CC-BY-NC-ND 4.0 license (https://creativecommons.org/licenses/by-nc-nd/4.0)).

### 5.2. The Sea/Islands Picture

An alternative mechanism has been proposed in Ref. [31], and further elaborated in Refs. [47–49]. The basic observation is that the eigenvalues of the untraced Polyakov loop at a spatial site $\vec{x}$ effectively change the temporal boundary condition felt locally by the quark eigenfunctions $\psi(t,\vec{x})$. Indeed, working for simplicity in the continuum, if one fixes the gauge to the temporal gauge $A_4(t,\vec{x}) = 0$, the eigenvalue problem reduces to

$$\slashed{D}\psi = (\partial_4 \gamma_4 + \slashed{D}_{(3)})\psi = i\lambda\psi\,, \qquad \slashed{D}_{(3)} = \sum_{j=1}^{3}\gamma_j(\partial_j + igA_j)\,, \tag{16}$$

while the effect of a nontrivial (untraced) Polyakov loop $P(\vec{x})$ is to change the temporal boundary condition from antiperiodic to

$$\psi(1/T,\vec{x}) = -P(\vec{x})\psi(0,\vec{x})\,. \tag{17}$$

Equations (16) and (17) define the eigenvalue problem in temporal gauge. Clearly, the effective local boundary condition Equation (17) affects the contribution of site $\vec{x}$ to the Dirac eigenvalue, $i\lambda = (\psi, \slashed{D}\psi)$.

To gain some insight on the effects of the Polyakov loop, it is useful to study a family of configurations for which the eigenvalue problem can be solved exactly, namely those with $A_\mu(t,\vec{x}) = 0$ everywhere, and with a constant but nontrivial Polyakov loop. This can always be diagonalized by means of a global gauge transformation, so without loss of generality we can take $P(\vec{x}) = \mathrm{diag}(e^{i\phi_1}, e^{i\phi_2}, e^{i\phi_3})$, with $e^{i(\phi_1+\phi_2+\phi_3)} = 1$. On these configurations the eigenfunctions of $-\slashed{D}^2$ are plane waves,

$$\psi_c^{(a,k,\vec{p})}(t,\vec{x}) = \delta_{ca}e^{i(\omega_{ak}t+\vec{p}\cdot\vec{x})}\,, \tag{18}$$

where $c$ is the color index,[29] with temporal frequency (*effective Matsubara frequency*) given by

$$\omega_{ak} = T[(2k+1)\pi + \phi_a]\,, \quad k \in \mathbb{Z}\,, \tag{19}$$

and corresponding eigenvalues

$$\lambda_{ak}(\vec{p})^2 = \omega_{ak}^2 + \vec{p}^{\,2}\,. \tag{20}$$

Restricting without loss of generality to $\phi_{1,2} \in (-\pi,\pi]$, $\phi_1 + \phi_2 + \phi_3 = 0$, the lowest positive Dirac eigenvalue is seen to be

$$\lambda_{\min} = T(\pi - \max_a |\phi_a|)\,, \tag{21}$$

i.e., it decreseas monotonically and symmetrically as one moves away from $\phi_a = 0\ \forall a$, and vanishes when at least one of the Polyakov loop eigenvalues equals $-1$.

While the configuration discussed above is obviously unrealistic, the result Equation (21) allows understanding qualitatively which sites will be favored by a low Dirac eigenmode when the Polyakov loop configuration is mostly ordered near $P(\vec{x}) \approx \mathbf{1}$, with "islands" of fluctuations in the "sea" of ordered Polyakov loops, as it happens at high temperature. One can in fact interpret Equation (21), now with $\vec{x}$-dependent phases $\phi_a = \phi_a(\vec{x})$, as a sort of three-dimensional local potential for the quarks, to which one should add the appropriate "hopping terms" originating from the spatial dependence of the Polyakov loops, as well as from the spatial components of the gauge potential. From this point of view, fluctuations of the Polyakov loop away from order provide regions of lower potential that can "trap" the quarks.

More precisely, at high temperatures $\phi_a(\vec{x}) \approx 0$ in an extended region, and neglecting in a first approximation the effect of the islands and of hopping, one finds fully delocalized modes. In the same approximation one finds a spectral gap, with the corresponding (positive) eigenvalues starting at $T\pi$, i.e., the usual lowest (fermionic) Matsubara frequency. The presence of islands and the effect of the interactions are expected to reduce this gap, but the lowest eigenvalue that can be reached by a delocalized mode is expected to remain separated from the origin. On the other hand, localizing on an island of fluctuations can be "energetically" more favorable, and bring the corresponding eigenvalues inside the gap, as long as the gain in potential energy achieved by avoiding the sea of ordered Polyakov loops is sufficiently larger than the price paid for localization in terms of spatial momenta. This leads to expect the following scenario, at least when the islands are sufficiently distant from each other: a region of low spectral density, or *pseudogap*, opens between the origin and some point $\lambda_c$ in the spectrum, or *mobility edge*, above which modes are extended throughout the whole space; modes in the pseudogap can exist only if they are localized on energetically convenient islands of fluctuations.

The scenario described above, which has been dubbed the sea/islands picture of localization, shows a clear similarity with the Anderson-type models of condensed matter physics, and the terminology has been chosen precisely to reflect this similarity. From the point of view of random Hamiltonians, the local fluctuations of the Polyakov loop provide a three-dimensional source of on-site disorder. This would naturally explain the fact that the critical behavior found at the mobility edge in QCD is the same as that of the three-dimensional unitary Anderson model.[30] There are, however, important differences with the simple unitary Anderson model of Equation (4). In that case, localization starts from the band edges and moves towards the band center as the amount of disorder, as measured by the width of its probability distribution, is increased. In QCD, while localization may as well be present at the band edges (cf. the results of Ref. [27] in the ILM model and Ref. [56] in $\mathbb{Z}_2$ gauge theory), it is its appearance directly at the band center that characterizes the deconfined phase. Moreover, the actual source of disorder are the eigenvalues of the Polyakov loops, which are complex numbers lying on the unit circle, and so the magnitude of the disorder is actually bounded.

Another important difference, which is relevant also to the problem of spontaneous chiral symmetry breaking, is the different structure of the "free" Hamiltonian associated with the two cases in the absence of fluctuations. For the Anderson model this is simply $H_{\text{free}}^{(\text{AM})} = \sum_{j=1}^{3} T_j + T_j^{\dagger}$ with $T_j$ the translation operator in direction $j$, while for the Dirac operator one has (for a naive lattice discretization) $\slashed{D}_{\text{free}} = \sum_{\mu=1}^{4} \gamma_\mu (T_\mu - T_\mu^{\dagger})$. Here periodic boundary conditions are understood in the spatial directions; the effective boundary conditions Equation (17) are assumed for the temporal direction. While the spectrum of $H_{\text{free}}^{(\text{AM})}$, $E(\vec{p}) = \sum_j \cos p_j$, with $p_j = \frac{2\pi k_j}{L}$, is dense near the origin, the presence of the gamma matrices in $\slashed{D}_{\text{free}}$ leads to $\lambda_{ak}(\vec{p}) = \sqrt{\omega_{ak}^2 + \vec{p}^2}$. Even in the case $\omega_{ak} = 0$, in which there is no sharp spectral gap, the spectral density near the origin is low and vanishes at $\lambda = 0$.

An important aspect of this scenario is that deconfinement is naturally associated with the two effects that lead to localization of the low modes in high-temperature QCD. The first such effect is of course the formation of a sea of ordered Polyakov loops close to the identity, which can cause the opening of a spectral pseudogap and so make modes that localize on the islands of fluctuations stable against delocalization, as explained above. However, the appearance of the pseudogap requires also a second effect due to the ordering of Polyakov loops at deconfinement, namely the increased correlation between gauge fields on different time slices. The discussion of this effect requires a more detailed description of the Dirac operator in the language of Anderson models, in what can be called the "Dirac-Anderson approach". Here we sketch the discussion in the continuum in the temporal gauge; a more detailed and mathematically more precise analysis is presented in Ref. [48] for staggered fermions on the lattice.

Due to its compactness, the temporal direction can be treated as an internal degree of freedom, in particular by expanding the quark eigenfunctions on a complete basis of plane waves $e^{i\omega_{ak}t}$ obeying the appropriate effective boundary conditions, Equation (17). Here $\omega_{ak}$ are the effective Matsubara frequencies of Equation (19), now $\vec{x}$-dependent, which provide a random on-site potential of the form $\omega_{ak}(\vec{x})\gamma_4$. For every color $a$ with associated

Polyakov-loop phase $\phi_a(\vec{x})$, in correspondence to each wave number $k$ there is a different branch of the on-site potential, and so a different associated three-dimensional Anderson-type model, built by adding the on-site disorder to the spatial part of the Dirac operator (projected on the $a, k$ subspace). We will refer to each of these models as a Dirac-Anderson model. The full Dirac operator is obtained by putting the various Dirac-Anderson models together, and by including their coupling induced by the hopping terms (i.e., the spatial part of the operator). The strength of the coupling among the Dirac-Anderson models turns out to be inversely related to the correlation of the gauge fields on different time slices. At low temperatures this correlation is small, the Dirac-Anderson models are strongly coupled, and the internal degree of freedom is effectively one more direction in which the modes can extend, thus facilitating their delocalization. This is in agreement with the effectively four-dimensional nature of QCD in the low temperature phase. As a matter of fact, the pseudogap does not open at low temperatures, where the spectral density is finite near zero, and this can only happen if the various Dirac-Anderson models do mix with each other (see the discussion about the free Dirac operator). In the absence of a pseudogap, localization of a mode is generally unstable against mixing with modes of similar energies. At high temperature, instead, the Polyakov loops become ordered inducing stronger correlations among different time slices, and the Dirac-Anderson models decouple making the problem effectively three-dimensional. In particular, the pseudogap is now expected to appear: it would be present for exactly decoupled Dirac-Anderson models (see again the discussion about the free Dirac operator), and their limited mixing is not sufficient to close it. Localized modes near the band center can then be supported by Polyakov loop fluctuations, as discussed above. As shown in Ref. [48] in a toy model where ordering of the Polyakov loop and correlation of the time slices can be varied independently, both effects are required for localization to appear at the band center.

An important aspect of the sea/islands picture is that it is not incompatible with the growing density of localized modes observed in QCD. Differently from the case of topological charge, Polyakov-loop fluctuations are not quantized. As $T$ grows in the deconfined phase, the volume $V_{\text{fluct}}$ occupied by Polyakov-loop fluctuations is expected to decrease, $V_{\text{fluct}} \sim T^{-c_1}$, as the Polyakov loop becomes more and more ordered. On the other hand, the typical size $V_0$ of the islands of fluctuations is also expected to decrease, $V_0 \sim T^{-c_2}$. The number of localized modes is expected to be directly related to the number of islands, $V_{\text{fluct}}/V_0 \sim T^{c_2-c_1}$, and whether this number increases or decreases with temperature depends on the details of the dynamics. For example, while increasing in QCD [33] up to $T \sim 5T_c$, it is seen to decrease in 2+1-dimensional SU(3) gauge theory above $T \sim 1.1 \div 1.2T_c$ [53].

Perhaps the most appealing feature of the sea/islands picture is its simplicity: all that it needs to work is the ordering of the Polyakov loop. This leads immediately to expect that localized modes will appear at the low end of the spectrum whenever an ordering transition takes place, independently of details such as the gauge group and its representation, fermionic content, nontrivial topological features, dimensionality,[31] and so on. This is discussed in the next section.

## 6. Localization in Other Gauge Theories

In this section, we discuss localization of the low Dirac modes in gauge theories other than QCD. Some of the references have been already discussed in Section 4 in connection with QCD, where they where treated as approximations. Here they are briefly discussed again, focussing more on the differences than on the similarities with QCD.

The main motivation in studying more general gauge theories is to investigate further the extent of the connection between localization on one side, and deconfinement and chiral restoration on the other. In particular, studying localization in models with genuine deconfining and/or chirally restoring phase transitions allows one to investigate this connection in a more clear-cut setting than in QCD, where it is somewhat blurred by the crossover nature of the transition. Studying more general gauge theories also allows one to test the sea/islands picture discussed in the previous section, and its generic prediction of localization of low modes in the high-temperature, "ordered" phase.

Genuine deconfining phase transitions are found in pure gauge $SU(N_c)$ theory. In 3+1 dimensions the transition is second order for $N_c = 2$ and first order for $N_c \geq 3$, while in 2+1 dimensions it is second order for $N_c = 2, 3$ and first order for $N_c \geq 4$ (see, e.g., Refs. [78,79,224,225]). As already mentioned in Section 4, localized low Dirac modes have been found in pure gauge SU(2) theory in 3+1 dimensions, both with the overlap [30,31] and with the staggered [31,32] Dirac operator (see Figure 8, right), above the deconfinement temperature $T_c$. (Further details can be found in Section 4 and will not be repeated here.) No sign of localized modes was found instead below $T_c$. From the random-matrix point of view, the SU(2) case differs from $SU(N_c \geq 3)$ as the symmetry class is the symplectic instead of the unitary one. This is reflected in the different behavior of the unfolded spectrum in the bulk, which agrees with the symplectic Wigner surmise, Equations (11) and (12). A detailed study of the Anderson transition was not pursued.

Results for pure gauge SU(3) have been presented in Refs. [51,54,86] for the 3+1-dimensional case, and in Ref. [53] for the 2+1-dimensional case. Localized low Dirac modes are found in both cases in the deconfined phase. In 3+1 dimensions the temperature dependence of the mobility edge $\lambda_c$ was studied using the Wilson gauge action both with staggered [51] and overlap [54] fermions (using in this case the magnitude of the eigenvalues), smearing the gauge fields with two steps of stout smearing [211] in the staggered case, and two steps of hex smearing [226] in the overlap case. The integrated ULSD computed locally in the spectrum, $I_{s_0}(\lambda)$, was used to determine $\lambda_c$ as the point where $I_{s_0}$ takes its critical value $I_{s_0}^{(c)}$ [34], i.e., $I_{s_0}(\lambda_c) = I_{s_0}^{(c)}$. Here use was made of the universality of the critical properties of the Anderson transition, which should be shared by QCD and pure gauge SU(3) theory, as they are both in the 3d unitary class. This was confirmed by the volume-independence of the resulting $\lambda_c$. For both discretizations, $\lambda_c$ is seen to extrapolate to zero at a temperature which agrees with the deconfinement temperature (see Refs. [78,79] and references therein) within numerical errors (see Figure 15).

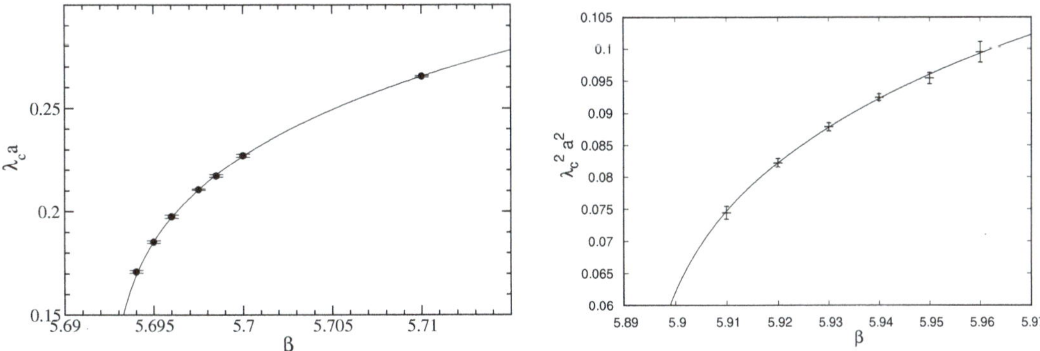

**Figure 15.** Mobility edge in pure gauge SU(3) theory in the staggered (**left**) and overlap (**right**) Dirac spectrum. From Refs. [51,54]. (Figures adapted from T.G. Kovács and R.Á. Vig, Phys. Rev. D 97, 014502 (2018) (left), and from R.Á. Vig and T.G. Kovács, Phys. Rev. D 101, 094511 (2020) (right), and used under a CC-BY 4.0 license (https://creativecommons.org/licenses/by/4.0)).

The 2+1-dimensional case was studied in Ref. [53] using the Wilson gauge action and the staggered discretization (without smearing). Universality arguments lead to expect that the Anderson transition is of BKT type with exponentially divergent correlation length, as found in Ref. [160] for the 2d unitary Anderson model. The results of Ref. [53] support this scenario. In particular, spectral statistics are critical, i.e., volume independent for all $\lambda$ above $\lambda_c$, as expected for a BKT-type Anderson transition [227], see Figure 16, left. The mobility edge was determined by means of a finite size scaling study, and found to extrapolate to zero at a temperature compatible with the deconfinement temperature [225] (although with much larger numerical uncertainty), see Figure 16, right. In the confined phase no localization was found, but low modes were seen to display a nontrivial fractal dimension $D_2 < 2$ (see Equation (6)).

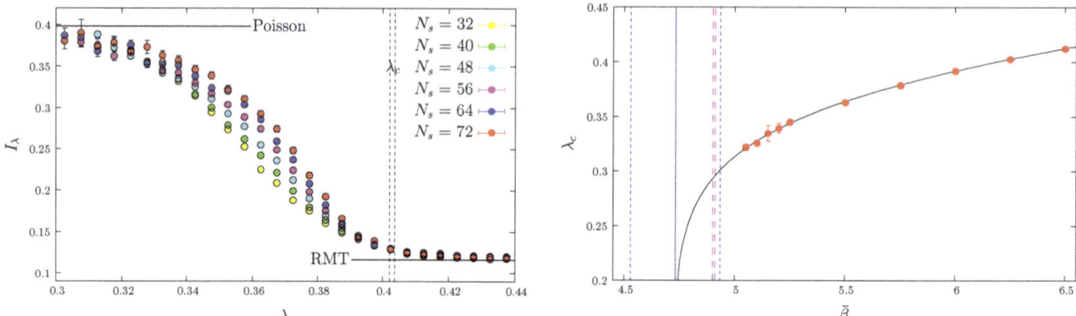


**Figure 16.** Spectral statistics $I_{s_0}$ along the spectrum for various volumes at coupling $\bar{\beta} = 6.25$ (**left**), and mobility edge as a function of $\bar{\beta}$ (**right**) for the staggered operator in 2+1-dimensional pure gauge SU(3) theory. (Here $\bar{\beta} = \beta/3$ with $\beta$ the Wilson action coupling.) In the left panel, Poisson and RMT predictions and the position of the mobility edge are also shown. In the right panel, a power-law fit (black solid line) to $\lambda_c$, the position (blue solid line) and error band (blue dashed lines) of $\bar{\beta}_{\text{loc}}$ at which $\lambda_c$ extrapolates to zero, and the error band of the critical $\bar{\beta}_c$ (magenta dashed lines) are also shown. From Ref. [53]. (Figures adapted from M. Giordano, J. High Energy Phys. 05, 204 (2019), and used under a CC-BY 4.0 license (https://creativecommons.org/licenses/by/4.0)).

Localization of Dirac modes was studied in $\mathbb{Z}_2$ pure gauge theory in 2+1 dimensions in Ref. [56], probed with unimproved staggered fermions. This model has the simplest gauge group, and the lowest dimensionality in which a deconfining transition is found. Studying the fractal dimension $D_2$, it was shown that low modes are localized ($D_2 = 0$) in the high-temperature, deconfined phase of the theory in the positive center sector (i.e., positive spatially averaged Polyakov loop), while they are delocalized (with $D_2 < 2$) in the low-temperature, confined phase, and in the high temperature phase in the negative center sector (i.e., negative spatially averaged Polyakov loop). Localized modes are also found at the high end of the spectrum, independently of the phase and of the center sector. Significant correlation between localized modes and both Polyakov loops and clusters of negative plaquettes was observed.

While a genuine phase transition is expected for SU(3) gauge group in the presence of $N_f = 3$, light enough dynamical fermions [80], so far a critical point has been observed only on coarse lattices, and disappears in the continuum limit [228–230]. Although only a toy model for QCD, the SU(3) theory with $N_f = 3$ flavors of unimproved staggered fermions on $N_t = 4$ lattices is nonetheless a well-defined statistical model with a genuine first order transition, affecting both its chiral and confining properties, despite the absence of exact chiral and center symmetries. More precisely, as the coupling $\beta$ crosses the critical value $\beta_c$, the chiral condensate jumps downwards to a much smaller but still finite value; and the average Polyakov loop jumps upwards from its small but nonzero value to a considerably larger value. Evidence of localization of the low staggered Dirac modes was reported in Ref. [50] for bare fermion mass $m = 0.01$, below the critical value $m_c = 0.0259$ [230], where genuine first-order phase transitions are present. A mobility edge was shown to be present for $\beta > \beta_c$: it increases with $\beta$, and extrapolates to zero close to $\beta_c$. The lowest mode was also seen to turn from delocalized to localized at a coupling $\beta_{\text{loc}}$ compatible with $\beta_c$, i.e., in correspondence with the finite-temperature transition (see Figure 17, left).

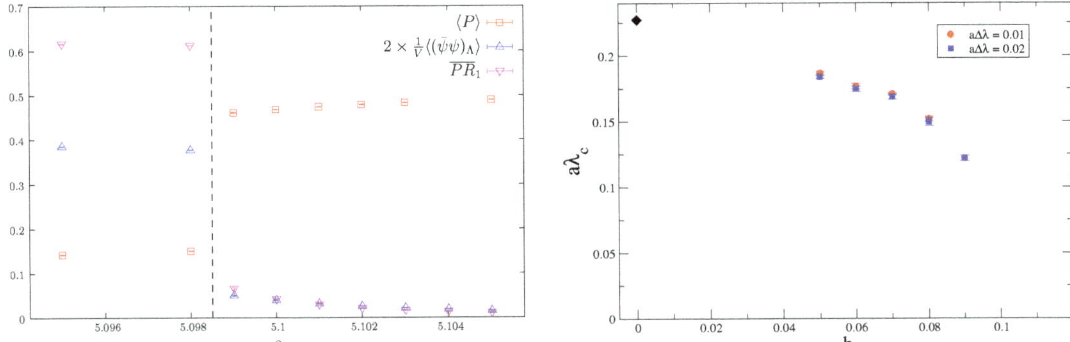

**Figure 17. Left**: average Polyakov loop (red squares), chiral condensate (upward blue triangles) and average PR of the lowest staggered mode (downward magenta triangle) for $N_f = 3$ unimproved staggered fermions of bare mass $m = 0.01$ on $N_t = 4$ lattices. The critical coupling $\beta_c = 5.0985$ is also shown. From Ref. [50]. (Figure adapted from M. Giordano, S. D. Katz, T. G. Kovács, and F. Pittler, J. High Energy Phys. 02, 055 (2017), and used under a CC-BY 4.0 license (https://creativecommons.org/licenses/by/4.0)). **Right**: mobility edge as a function of the deformation parameter $h$ in trace-deformed SU(3) gauge theory in the high-temperature deconfined phase ($\beta = 6.0$). Here the critical deformation parameter for reconfinement is $h_c = 0.1$. The black diamond is the $h = 0$ result of Ref. [51]. From Ref. [55]. (Figure adapted from C. Bonati, M. Cardinali, M. D'Elia, M. Giordano, and F. Mazziotti, Phys. Rev. D 103, 034506 (2021), and used under a CC-BY 4.0 license (https://creativecommons.org/licenses/by/4.0)).

The relation between localization and deconfinement was tested at a different deconfinement phase transition in trace-deformed [231,232] pure gauge SU(3) theory at finite temperature in Ref. [55]. In this model a deformation term $\Delta S = h \sum_{\vec{x}} |\text{tr} P(\vec{x})|^2$ is added to the action, which (for $h > 0$) tends to locally suppress a nonzero trace for the Polyakov loop $P(\vec{x})$. For temperatures above the deconfinement temperature, $\Delta S$ pushes the theory towards a "reconfined" phase where $\text{tr} P(\vec{x}) \sim 0$. This happens when the deformation parameter $h$ crosses a (temperature dependent) critical value $h_c$. Ref. [55] studied the spectrum of the two-stout smeared staggered spectrum at $\beta = 6.0$ on $N_t = 6$ lattices for various volumes and deformation parameters. Results showed that localized modes are present for $h < h_c$, but disappear as the system crosses over into the reconfined phase. The mobility edge was determined by comparing the fractal dimension of the modes with its value at criticality [148]. While monotonically decreasing with $h$, it is not clear whether it vanishes continuously at $h_c$ or jumps to zero discontinuously (see Figure 17, right).

Finally, the connection between localization and ordering of the background configuration was studied in spin models in Refs. [47–49,52]. Ref. [47] used a simple 3d Hamiltonian in the orthogonal class with on-site disorder provided by the spins of a continuous-spin Ising-type model. In the ordered phase of the spin model, localization was observed for the low modes, with a mobility edge separating them from higher modes, and critical behavior compatible with that of the 3d orthogonal Anderson model. Refs. [48,49] dealt with the Dirac-Anderson form of the staggered operator (see Section 5.2), so in the 3d chiral unitary class, in the case $N_t = 2$ in the background of Polyakov loops constructed from a spin model. Localized low modes are observed in the ordered phase of the model [48], appearing at the critical temperature [49]. Ref. [52] reports on the $CP^3$ model in 1+1 and 2+1 dimensions. While in 1+1 dimensions localized modes are found in both phases of the model, as expected in one spatial dimension, in the 2+1 case localized modes are found only in the ordered phase. This model belongs to the 2d chiral unitary class.

## 7. Conclusions and Outlook

The presence of localized modes in the spectrum of the Dirac operator in the high-temperature phase of gauge theories is by now well established. Numerical studies on the lattice have shown that above the transition temperature the low-lying Dirac modes are spatially localized on the scale of the inverse temperature, in QCD and QCD-like theories, as well as in several pure gauge theories and related models in 3+1 and 2+1 dimensions [25–40,47–56,86].

The physical significance of localization has so far remained quite elusive. First of all we should emphasize an important difference between localization in electron systems and localization in QCD. In the former case the mobility edge in the spectrum can be "accessed" by tuning a suitable control parameter, such as an electric field or the density of electrons. As the Fermi energy crosses the mobility edge, the system undergoes a genuine phase transition, with the zero-temperature conductivity changing non-analytically. In contrast, the mobility edge in the QCD Dirac spectrum cannot be directly connected to a thermodynamic transition. This is because in that case, in general, there is no control parameter that can be adjusted to make the system sensitive to just the eigenmodes at the mobility edge in the spectrum. The only exception is when the mobility edge is at zero, which happens only at the critical temperature of localization. If at the same time the quark masses are set to zero, the system becomes most sensitive to the lowest Dirac eigenmodes, the ones closest to zero. Thus, only in this double limit when the temperature tends to the critical temperature of localization and the quark mass to zero can one possibly directly connect the localization transition to a genuine thermodynamic phase transition. Unfortunately, this limit is out of the reach of present day lattice simulations and we have no numerical evidence of what happens there.

On the other hand, some progress has been made to understand the physical significance of localization in QCD. A clear connection with deconfinement has emerged: in all the models investigated so far, localization of the low modes shows up when the system transitions from the confined, low-temperature phase to the deconfined, high-temperature phase [28,33,49–51,53–56]. Convincing evidence has been presented for the crucial role played by the ordering of the Polyakov loop and by its fluctuations in the formation of a mobility edge in the Dirac spectrum, separating low-lying, localized modes from the delocalized bulk modes [31,39,40,48,56]. As the Polyakov loop is the (approximate, in the case of QCD) order parameter for confinement, the observed connection between localization and deconfinement has a dynamical explanation, further backed by a viable mechanism (the sea/islands picture [31,47–49], see Section 5.2) relating the two phenomena. This raises the hope that further studies can lead to a better understanding of confinement, and possibly to the uncovering of the mechanism behind this remarkable property of gauge theories. In this context, it would be interesting to further elucidate the relation between localization and center symmetry, since so far only models with nontrivial gauge group center have been investigated.

Localization could also help in explaining the close relation observed between deconfinement and restoration of chiral symmetry. These two phenomena in fact take place at the same temperature, or in a relatively narrow interval of temperatures, where also localized low Dirac modes appear. Localization could then provide the key to understanding this relation between in principle unrelated phenomena. Unfortunately, while the connection between deconfinement and localization can be easily studied in a clear-cut situation by investigating pure gauge theories with a genuine deconfining phase transition, studying the connection between chiral symmetry restoration and localization by means of numerical lattice simulations faces the considerable difficulties involved in taking the chiral limit. Studies of this type would be of great interest, especially in the light of the possible role played by localized modes in suppressing the finite-temperature Goldstone excitations, suggested in Ref. [62]. A particularly interesting case would be that of adjoint massless fermions, for which both chiral and center symmetries are exact, and an intermediate, deconfined but chirally broken phase was observed on the lattice for two flavors in Ref. [233]. This suggests that a nonzero density of near-zero, localized modes is present in this phase, and that no Goldstone excitation is present.

Nonetheless, even in theories such as QCD where chiral symmetry is only approximate, the study of the relation between localized modes and topological fluctuations of the gauge fields sheds indirectly some light on the interplay of chiral symmetry and localization. Indeed, a peak of near-zero [59–61] localized [38] modes of topological origin appears around the QCD pseudocritical temperature. These modes originate most likely from the mixing of the localized zero modes associated with isolated instantons and anti-instantons, which in the high-temperature phase form a dilute gas of topological excitations (the disordered medium scenario [8–15], see Section 5.1). In contrast, in the low temperature phase these excitations form a dense medium, and the mixing of the associated zero modes leads to a band of near-zero delocalized modes giving rise to a nonzero spectral density near the origin, and so a large increase of the chiral condensate.

An interesting observation is that in the quenched limit of QCD this peak of near-zero modes can be accurately described in terms of a non-interacting gas of topological objects [58,86]. The absence (or near absence) of interactions could be related to why the associated zero modes do not mix efficiently, thus remaining localized and failing to spread in a near-zero band of eigenvalues. On the other hand, as this occurs only in the high-temperature phase, it is natural to expect that deconfinement is responsible for the radical change in the behavior of topological excitations. This aspect surely deserves more attention.

In this review, we summarized what is known (to us, at least) about localization of Dirac modes in the deconfined phase of gauge theories, and highlighted the connections between localized modes, ordering of the Polyakov loop, density of low modes, and topological objects. We hope that this will motivate further investigations of the interplay of confinement, chiral symmetry, topology, and localization in finite-temperature gauge theories.

**Author Contributions:** Writing—original draft preparation, M.G. and T.G.K; writing—review and editing, M.G. and T.G.K. Both authors have read and agreed to the published version of the manuscript.

**Funding:** M.G. was partially supported by the NKFIH grant KKP-126769.

**Acknowledgments:** We thank R. Shindou and M.R. Zirnbauer for useful correspondence.

**Conflicts of Interest:** The authors declare no conflict of interest.

## Notes

1.  In principle, there is also the gauge coupling, but it turns out not to be a freely adjustable parameter, instead it runs with the energy scale. See Ref. [75], ch. 18.
2.  This type of argument first appeared in Ref. [76] to explain the difference in the chiral condensate observed in the various center sectors of quenched lattice QCD in the deconfined phase [77].
3.  As the lattice spacing corresponds to the inverse of the largest energy attainable on the lattice, this indicates that the gauge coupling runs with the energy scale, see footnote 1, and Ref. [82], ch. 13.
4.  The *conductance* $G(L)$ for a $d$-dimensional (hyper)cubic sample of linear size $L$ equals $G(L) = \sigma L^{d-2}$ where $\sigma$ is the *conductivity* of the system.
5.  The dimensionality and the symmetry class do not always determine uniquely the universality class of the Anderson transition: see Ref. [45].
6.  The participation ratio (see below) was introduced in Ref. [123]; its inverse as a measure of localization is discussed in Ref. [42].
7.  Notice that by construction one has for a generic random matrix ensemble $\int_0^\infty ds\, p_{\mathrm{ULSD}}(s) = \int_0^\infty ds\, p_{\mathrm{ULSD}}(s)s = 1$. This follows from the fact that the average spacing equals the inverse of the spectral density, which is 1 for the unfolded spectrum.
8.  On the insulator side of the transition, $\zeta$ can be identified with the localization length, while on the metallic side it can be related to the conductivity [44].
9.  Shape analysis in QCD is discussed in Refs. [35,36].
10. The possibility of localization taking place in QCD was mentioned in Ref. [188].
11. *Mott transitions* are MITs driven by electron-electron interactions, in contrast to the disorder-driven Anderson transition. In Anderson-Mott transitions both interactions and disorder play an important role.

12    The detailed symmetry classification of the Dirac operator in various dimensions is provided in Refs. [117] (four dimensions), Refs. [190–192] (three dimensions), and Ref. [193] (two dimensions).

13    Bulk statistics are not affected by the chiral symmetry, and should not be confused with the microscopic statistics near $\lambda = 0$, which, in contrast, are affected if a nonzero density of modes is present. In the chirally broken phase, where $\Sigma = \frac{1}{V}\pi\rho(0) \neq 0$, with $V$ the volume, the statistical properties of the microscopic spectrum $z_i \equiv \lambda_i \Sigma \frac{V}{T}$ near $\lambda = 0$ are described by the microscopic correlations of the chGUE. See Ref. [115] for a detailed review.

14    RMT is expected to govern correlations up to some characteristic separation scale between eigenvalues ("Thouless energy"), both for microscopic and bulk statistics [13,14,194]. For the role played by fluctuations in the ensemble in determining this scale in the case of bulk statistics, see Ref. [195].

15    The symmetry class of the staggered operator is actually independent of the spacetime dimension.

16    We note in passing that for adjoint fermions in four dimensions the relevant class is the chiral symplectic class for all $N_c$ in the continuum, and on the lattice with overlap fermions; for staggered fermions it is instead the chiral orthogonal class for all $N_c$ [115,203], independently of the dimension.

17    From this observation, Ref. [25] concluded that modes are actually extended in the temporal direction. This is actually not necessary: localization in the temporal direction on a scale comparable with the temporal size is sufficient for modes to be sensitive to the boundary conditions.

18    For the fundamental representation of the gauge group SU(2) the staggered operator is in the symplectic class due to the property $\sigma_2 U \sigma_2 = U^*$ of SU(2) matrices $U$ (see Ref. [188] and Ref. [115], Section 5.2.1).

19    The overlap operator is in the same symmetry class as the corresponding continuum operator, so the orthogonal class for fundamental fermions and gauge group SU(2) (see Ref. [188] and Ref. [115], Section 5.2.1).

20    The only exception is the SU(2) theory studied in Refs. [30–32], which is in the 3d chiral orthogonal or chiral symplectic class depending on the fermion discretization, see footnotes 18 and 19.

21    Localized modes at the high end of the staggered Dirac spectrum have also been found in 2+1-dimensional $\mathbb{Z}_2$ gauge theory [56], see Section 6.

22    For a review of instantons and calorons we refer the reader to Refs. [10,204,221].

23    In a first approximation, the nonzero unperturbed modes associated with topological objects can be neglected.

24    Overlap integrals vanish for a pair of instantons or anti-instantons due the definite (and equal) chirality of the zero modes.

25    While the index theorem requires only $Q = n_+ - n_-$, it is expected that only zero modes of one chirality appear in typical gauge configurations.

26    For example, modes are localized at the band center in the Anderson model above the critical disorder, but with finite spectral density; and in the near-zero spike found right above $T_c$ in QCD and pure gauge SU(3) theory.

27    The claim of Ref. [39] is actually stronger: localized low modes do localize on $L - \bar{L}$ monopole-antimonopole pairs. We believe that this claim is not fully supported by the available evidence. On the one hand, while both selfdual and anti-selfdual points are clearly favored by localized modes, there is no clear evidence that these modes localize where selfdual and anti-selfdual points are spatially close. On the other hand, $L$-type anti(monopoles) are located at sites where a pair of the eigenvalues $(e^{i\phi_1}, e^{i\phi_2}, e^{-i(\phi_1+\phi_2)})$ of the untraced Polyakov loop is nearly degenerate and close to $-1$ (fluctuations of the degenerate pair around $-1$ correspond to fluctuations of the Polyakov loop at spatial infinity around 1), and while these sites are among the favorable localization points, sites with $\mathrm{Re}\,\mathrm{tr}P = -1$ but without eigenvalue degeneracy are at least equally (if not more) favorable, see Figure 12.

28    This observation applies also to the case in which the relevant objects are the $L$-type monopoles and antimonopoles, independently of them being part of calorons, as suggested in Ref. [39].

29    Since $\slashed{D}^2$ is trivial in Dirac space in this case, the Dirac index is omitted.

30    Notice that the mobility edge is generally far from the near-zero zone where localized modes are of topological origin [86].

31    Dimensionality should not matter as long as both deconfinement and localization are allowed. For example, no Anderson transition should be found in 1+1-dimensional gauge theories at finite temperature: no deconfinement transition is present there, and all modes are expected to be localized in one spatial dimension.

## References


1. Greensite, J. The Confinement problem in lattice gauge theory. *Prog. Part. Nucl. Phys.* **2003**, *51*, 1. [CrossRef]
2. Greensite, J. *An Introduction to the Confinement Problem*; Lecture Notes in Physics; Springer: Berlin/Heidelberg, Germany, 2011; Volume 821. [CrossRef]
3. Chandrasekharan, S.; Wiese, U.J. An Introduction to chiral symmetry on the lattice. *Prog. Part. Nucl. Phys.* **2004**, *53*, 373. [CrossRef]

4.  Faber, M.; Höllwieser, R. Chiral symmetry breaking on the lattice. *Prog. Part. Nucl. Phys.* **2017**, *97*, 312. [CrossRef]

5.  Borsányi, S.; Fodor, Z.; Hoelbling, C.; Katz, S.D.; Krieg, S.; Ratti, C.; Szabó, K.K. Is there still any $T_c$ mystery in lattice QCD? Results with physical masses in the continuum limit III. *J. High Energy Phys.* **2010**, *9*, 73. [CrossRef]

6.  Bazavov, A.; Brambilla, N.; Ding, H.T.; Petreczky, P.; Schadler, H.P.; Vairo, A.; Weber, J.H. Polyakov loop in 2+1 flavor QCD from low to high temperatures. *Phys. Rev. D* **2016**, *93*, 114502. [CrossRef]

7.  Aoki, Y.; Endrődi, G.; Fodor, Z.; Katz, S.D.; Szabó, K.K. The Order of the quantum chromodynamics transition predicted by the standard model of particle physics. *Nature* **2006**, *443*, 675. [CrossRef]

8.  Diakonov, D.; Petrov, V.Y. Chiral condensate in the instanton vacuum. *Phys. Lett. B* **1984**, *147*, 351. [CrossRef]

9.  Diakonov, D.; Petrov, V.Y. A Theory of Light Quarks in the Instanton Vacuum. *Nucl. Phys. B* **1986**, *272*, 457. [CrossRef]

10.  Diakonov, D. Chiral symmetry breaking by instantons. *Proc. Int. Sch. Phys. Fermi* **1996**, *130*, 397. [CrossRef]

11.  Smilga, A.V. Vacuum fields in the Schwinger model. *Phys. Rev. D* **1992**, *46*, 5598. [CrossRef]

12.  Janik, R.A.; Nowak, M.A.; Papp, G.; Zahed, I. Chiral disorder in QCD. *Phys. Rev. Lett.* **1998**, *81*, 264. [CrossRef]

13.  Osborn, J.C.; Verbaarschot, J.J.M. Thouless energy and correlations of QCD Dirac eigenvalues. *Phys. Rev. Lett.* **1998**, *81*, 268. [CrossRef]

14.  Osborn, J.C.; Verbaarschot, J.J.M. Thouless energy and correlations of QCD Dirac eigenvalues. *Nucl. Phys. B* **1998**, *525*, 738. [CrossRef]

15.  García-García, A.M.; Osborn, J.C. The QCD vacuum as a disordered medium: A Simplified model for the QCD Dirac operator. *Phys. Rev. Lett.* **2004**, *93*, 132002. [CrossRef] [PubMed]

16.  Harrington, B.J.; Shepard, H.K. Periodic Euclidean Solutions and the Finite Temperature Yang-Mills Gas. *Phys. Rev. D* **1978**, *17*, 2122. [CrossRef]

17.  Harrington, B.J.; Shepard, H.K. Thermodynamics of the Yang-Mills Gas. *Phys. Rev. D* **1978**, *18*, 2990. [CrossRef]

18.  Kraan, T.C.; van Baal, P. Exact T duality between calorons and Taub - NUT spaces. *Phys. Lett. B* **1998**, *428*, 268. [CrossRef]

19.  Kraan, T.C.; van Baal, P. Periodic instantons with nontrivial holonomy. *Nucl. Phys. B* **1998**, *533*, 627. [CrossRef]

20.  Kraan, T.C.; van Baal, P. Monopole constituents inside SU($n$) calorons. *Phys. Lett. B* **1998**, *435*, 389. [CrossRef]

21.  Lee, K.M.; Yi, P. Monopoles and instantons on partially compactified D-branes. *Phys. Rev. D* **1997**, *56*, 3711. [CrossRef]

22.  Lee, K.M. Instantons and magnetic monopoles on $\mathbb{R}^3 \times S^1$ with arbitrary simple gauge groups. *Phys. Lett. B* **1998**, *426*, 323. [CrossRef]

23.  Lee, K.M.; Lu, C.h. SU(2) calorons and magnetic monopoles. *Phys. Rev. D* **1998**, *58*, 025011. [CrossRef]

24.  Banks, T.; Casher, A. Chiral symmetry breaking in confining theories. *Nucl. Phys. B* **1980**, *169*, 103. [CrossRef]

25.  Göckeler, M.; Rakow, P.E.L.; Schäfer, A.; Söldner, W.; Wettig, T. Calorons and localization of quark eigenvectors in lattice QCD. *Phys. Rev. Lett.* **2001**, *87*, 042001. [CrossRef]

26.  Gattringer, C.; Göckeler, M.; Rakow, P.E.L.; Schaefer, S.; Schäfer, A. A Comprehensive picture of topological excitations in finite temperature lattice QCD. *Nucl. Phys. B* **2001**, *618*, 205. [CrossRef]

27.  García-García, A.M.; Osborn, J.C. Chiral phase transition and Anderson localization in the Instanton Liquid Model for QCD. *Nucl. Phys. A* **2006**, *770*, 141. [CrossRef]

28.  García-García, A.M.; Osborn, J.C. Chiral phase transition in lattice QCD as a metal-insulator transition. *Phys. Rev. D* **2007**, *75*, 034503. [CrossRef]

29.  Gavai, R.V.; Gupta, S.; Lacaze, R. Eigenvalues and Eigenvectors of the Staggered Dirac Operator at Finite Temperature. *Phys. Rev. D* **2008**, *77*, 114506. [CrossRef]

30.  Kovács, T.G. Absence of correlations in the QCD Dirac spectrum at high temperature. *Phys. Rev. Lett.* **2010**, *104*, 031601. [CrossRef] [PubMed]

31.  Bruckmann, F.; Kovács, T.G.; Schierenberg, S. Anderson localization through Polyakov loops: Lattice evidence and random matrix model. *Phys. Rev. D* **2011**, *84*, 034505. [CrossRef]

32.  Kovács, T.G.; Pittler, F. Anderson Localization in Quark-Gluon Plasma. *Phys. Rev. Lett.* **2010**, *105*, 192001. [CrossRef]

33.  Kovács, T.G.; Pittler, F. Poisson to Random Matrix Transition in the QCD Dirac Spectrum. *Phys. Rev. D* **2012**, *86*, 114515. [CrossRef]

34.  Giordano, M.; Kovács, T.G.; Pittler, F. Universality and the QCD Anderson Transition. *Phys. Rev. Lett.* **2014**, *112*, 102002. [CrossRef] [PubMed]

35.  Nishigaki, S.M.; Giordano, M.; Kovács, T.G.; Pittler, F. Critical statistics at the mobility edge of QCD Dirac spectra. *PoS* **2014**, *LATTICE2013*, 018. [CrossRef]

36.  Giordano, M.; Kovács, T.G.; Pittler, F. Anderson localization in QCD-like theories. *Int. J. Mod. Phys. A* **2014**, *29*, 1445005. [CrossRef]

37.  Ujfalusi, L.; Giordano, M.; Pittler, F.; Kovács, T.G.; Varga, I. Anderson transition and multifractals in the spectrum of the Dirac operator of Quantum Chromodynamics at high temperature. *Phys. Rev. D* **2015**, *92*, 094513. [CrossRef]

38.  Dick, V.; Karsch, F.; Laermann, E.; Mukherjee, S.; Sharma, S. Microscopic origin of $U_A(1)$ symmetry violation in the high temperature phase of QCD. *Phys. Rev. D* **2015**, *91*, 094504. [CrossRef]

39.  Cossu, G.; Hashimoto, S. Anderson Localization in high temperature QCD: Background configuration properties and Dirac eigenmodes. *J. High Energy Phys.* **2016**, *6*, 56. [CrossRef]

40. Holicki, L.; Ilgenfritz, E.M.; von Smekal, L. The Anderson transition in QCD with $N_f = 2 + 1 + 1$ twisted mass quarks: Overlap analysis. *PoS* **2018**, *LATTICE2018*, 180. [CrossRef]
41. Anderson, P.W. Absence of Diffusion in Certain Random Lattices. *Phys. Rev.* **1958**, *109*, 1492. [CrossRef]
42. Thouless, D.J. Electrons in disordered systems and the theory of localization. *Phys. Rep.* **1974**, *13*, 93. [CrossRef]
43. Lee, P.A.; Ramakrishnan, T.V. Disordered electronic systems. *Rev. Mod. Phys.* **1985**, *57*, 287. [CrossRef]
44. Kramer, B.; MacKinnon, A. Localization: Theory and experiment. *Rep. Prog. Phys.* **1993**, *56*, 1469. [CrossRef]
45. Evers, F.; Mirlin, A.D. Anderson transitions. *Rev. Mod. Phys.* **2008**, *80*, 1355. [CrossRef]
46. Abrahams, E. (Ed.) *50 Years of Anderson Localization*; World Scientific: Singapore, 2010. [CrossRef]
47. Giordano, M.; Kovács, T.G.; Pittler, F. An Ising-Anderson model of localisation in high-temperature QCD. *J. High Energy Phys.* **2015**, *4*, 112. [CrossRef]
48. Giordano, M.; Kovács, T.G.; Pittler, F. An Anderson-like model of the QCD chiral transition. *J. High Energy Phys.* **2016**, *6*, 7. [CrossRef]
49. Giordano, M.; Kovács, T.G.; Pittler, F. Localization and chiral properties near the ordering transition of an Anderson-like toy model for QCD. *Phys. Rev. D* **2017**, *95*, 074503. [CrossRef]
50. Giordano, M.; Katz, S.D.; Kovács, T.G.; Pittler, F. Deconfinement, chiral transition and localisation in a QCD-like model. *J. High Energy Phys.* **2017**, *2*, 55. [CrossRef]
51. Kovács, T.G.; Vig, R.Á. Localization transition in SU(3) gauge theory. *Phys. Rev. D* **2018**, *97*, 014502. [CrossRef]
52. Bruckmann, F.; Wellnhofer, J. Anderson localization in sigma models. *EPJ Web Conf.* **2018**, *175*, 07005. [CrossRef]
53. Giordano, M. Localisation in 2+1 dimensional SU(3) pure gauge theory at finite temperature. *J. High Energy Phys.* **2019**, *05*, 204. [CrossRef]
54. Vig, R.Á.; Kovács, T.G. Localization with overlap fermions. *Phys. Rev. D* **2020**, *101*, 094511. [CrossRef]
55. Bonati, C.; Cardinali, M.; D'Elia, M.; Giordano, M.; Mazziotti, F. Reconfinement, localization and thermal monopoles in $SU(3)$ trace-deformed Yang-Mills theory. *Phys. Rev. D* **2021**, *103*, 034506. [CrossRef]
56. Baranka, G.; Giordano, M. Localisation of Dirac modes in finite-temperature $\mathbb{Z}_2$ gauge theory on the lattice. *arXiv* **2021**, arXiv:2104.03779.
57. Edwards, R.G.; Heller, U.M.; Kiskis, J.E.; Narayanan, R. Chiral condensate in the deconfined phase of quenched gauge theories. *Phys. Rev. D* **2000**, *61*, 074504. [CrossRef]
58. Vig, R.Á.; Kovács, T.G. Ideal topological gas in the high temperature phase of SU(3) gauge theory. *arXiv* **2021**, arXiv:2101.01498.
59. Alexandru, A.; Horváth, I. Phases of SU(3) Gauge Theories with Fundamental Quarks via Dirac Spectral Density. *Phys. Rev. D* **2015**, *92*, 045038. [CrossRef]
60. Ding, H.T.; Li, S.T.; Mukherjee, S.; Tomiya, A.; Wang, X.D.; Zhang, Y. Correlated Dirac eigenvalues and axial anomaly in chiral symmetric QCD. *Phys. Rev. Lett.* **2021**, *126*, 082001. [CrossRef] [PubMed]
61. Kaczmarek, O.; Mazur, L.; Sharma, S. Eigenvalue spectra of QCD and the fate of $U_A(1)$ breaking towards the chiral limit. *arXiv* **2021**, arXiv:2102.06136.
62. Giordano, M. Localised Dirac eigenmodes, chiral symmetry breaking and Goldstone's theorem. *arXiv* **2020**, arXiv:2009.00486.
63. Giordano, M. Localisation, chiral symmetry and confinement in QCD and related theories. *PoS* **2019**, *Confinement2018*, 045. [CrossRef]
64. Zakharov, V.I. Matter of resolution: From quasiclassics to fine tuning. In *Sense of Beauty in Physics: A Volume in Honour of Adriano Di Giacomo*; D'Elia, M., Konishi, K., Meggiolaro, E., Rossi, P., Eds.; Edizioni Plus srl: Pisa, Italy, 2006.
65. de Forcrand, P. Localization properties of fermions and bosons. *AIP Conf. Proc.* **2007**, *892*, 29. [CrossRef]
66. Ilgenfritz, E.M.; Koller, K.; Koma, Y.; Schierholz, G.; Streuer, T.; Weinberg, V. Exploring the structure of the quenched QCD vacuum with overlap fermions. *Phys. Rev. D* **2007**, *76*, 034506. [CrossRef]
67. Hollwieser, R.; Faber, M.; Heller, U.M. Intersections of thick Center Vortices, Dirac Eigenmodes and Fractional Topological Charge in SU(2) Lattice Gauge Theory. *J. High Energy Phys.* **2011**, *6*, 52. [CrossRef]
68. Ilgenfritz, E.M.; Martemyanov, B.V.; Müller-Preussker, M. Topology near the transition temperature in lattice gluodynamics analyzed by low lying modes of the overlap Dirac operator. *Phys. Rev. D* **2014**, *89*, 054503. [CrossRef]
69. Golterman, M.; Shamir, Y. Localization in lattice QCD. *Phys. Rev. D* **2003**, *68*, 074501. [CrossRef]
70. Golterman, M.; Shamir, Y.; Svetitsky, B. Localization properties of lattice fermions with plaquette and improved gauge actions. *Phys. Rev. D* **2005**, *72*, 034501. [CrossRef]
71. Greensite, J.; Olejník, S.; Polikarpov, M.; Syritsyn, S.; Zakharov, V. Localized eigenmodes of covariant Laplacians in the Yang-Mills vacuum. *Phys. Rev. D* **2005**, *71*, 114507. [CrossRef]
72. Greensite, J.; Kovalenko, A.V.; Olejník, S.; Polikarpov, M.I.; Syritsyn, S.N.; Zakharov, V.I. Peculiarities in the spectrum of the adjoint scalar kinetic operator in Yang-Mills theory. *Phys. Rev. D* **2006**, *74*, 094507. [CrossRef]
73. Rothe, H.J. *Lattice Gauge Theories: An Introduction*; World Scientific: Singapore, 2012; Volume 82, pp. 1–606.
74. Montvay, I.; Münster, G. *Quantum Fields on a Lattice*; Cambridge Monographs on Mathematical Physics; Cambridge University Press: Cambridge, UK, 1997. [CrossRef]
75. Weinberg, S. *The Quantum Theory of Fields. Vol. 2: Modern Applications*; Cambridge University Press: Cambridge, UK, 2013. [CrossRef]

76. Stephanov, M.A. Chiral symmetry at finite T, the phase of the Polyakov loop and the spectrum of the Dirac operator. *Phys. Lett. B* **1996**, *375*, 249. [CrossRef]
77. Chandrasekharan, S.; Christ, N.H. Dirac spectrum, axial anomaly and the QCD chiral phase transition. *Nucl. Phys. B Proc. Suppl.* **1996**, *47*, 527. [CrossRef]
78. Francis, A.; Kaczmarek, O.; Laine, M.; Neuhaus, T.; Ohno, H. Critical point and scale setting in SU(3) plasma: An update. *Phys. Rev. D* **2015**, *91*, 096002. [CrossRef]
79. Lucini, B.; Teper, M.; Wenger, U. The High temperature phase transition in SU(N) gauge theories. *J. High Energy Phys.* **2004**, *01*, 061. [CrossRef]
80. Pisarski, R.D.; Wilczek, F. Remarks on the Chiral Phase Transition in Chromodynamics. *Phys. Rev. D* **1984**, *29*, 338. [CrossRef]
81. Pelissetto, A.; Vicari, E. Relevance of the axial anomaly at the finite-temperature chiral transition in QCD. *Phys. Rev. D* **2013**, *88*, 105018. [CrossRef]
82. Creutz, M. *Quarks, Gluons and Lattices*; Cambridge Monographs on Mathematical Physics; Cambridge University Press: Cambridge, UK, 1985.
83. DeGrand, T.; DeTar, C. *Lattice Methods for Quantum Chromodynamics*; World Scientific: Singapore, 2006. [CrossRef]
84. Gattringer, C.; Lang, C.B. *Quantum Chromodynamics on the Lattice*; Springer: Berlin/Heidelberg, Germany, 2010; Volume 788. [CrossRef]
85. Wilson, K.G. Confinement of Quarks. *Phys. Rev. D* **1974**, *10*, 2445. [CrossRef]
86. Kovács, T.G.; Vig, R.Á. Localization and topology in high temperature QCD. *PoS* **2019**, *LATTICE2018*, 258. [CrossRef]
87. Alexandru, A.; Horváth, I. Possible New Phase of Thermal QCD. *Phys. Rev. D* **2019**, *100*, 094507. [CrossRef]
88. Alexandru, A.; Horváth, I. Unusual Features of QCD Low-Energy Modes in IR Phase. *arXiv* **2021**, arXiv:2103.05607.
89. Halperin, B.I. Properties of a particle in a one-dimensional random potential. *Adv. Chem. Phys.* **1967**, *13*, 123. [CrossRef]
90. Mott, N.F. Electrons in disordered structures. *Adv. Phys.* **1967**, *16*, 49. [CrossRef]
91. Ziman, J.M. Localization of electrons in ordered and disordered systems II. Bound bands. *J. Phys. C Solid State Phys.* **1969**, *2*, 1230. [CrossRef]
92. Anderson, P.W. The Fermi Glass: Theory and Experiment. In *A Career in Theoretical Physics*; World Scientific: Singapore, 2005; pp. 353–359.
93. Thouless, D.J. Anderson's theory of localized states. *J. Phys. C Solid State Phys.* **1970**, *3*, 1559. [CrossRef]
94. Mott, N. Conduction in non-crystalline systems IV. Anderson localization in a disordered lattice. *Phil. Mag.* **1970**, *22*, 7. [CrossRef]
95. Economou, E.; Cohen, M. Localization in disordered materials: Existence of mobility edges. *Phys. Rev. Lett.* **1970**, *25*, 1445. [CrossRef]
96. Economou, E.; Cohen, M. Existence of mobility edges in Anderson's model for random lattices. *Phys. Rev. B* **1972**, *5*, 2931. [CrossRef]
97. Abou-Chacra, R.; Anderson, P.W.; Thouless, D.J. A selfconsistent theory of localization. *J. Phys. C Solid State Phys.* **1973**, *6*, 1734. [CrossRef]
98. Abrahams, E.; Anderson, P.W.; Licciardello, D.C.; Ramakrishnan, T.V. Scaling Theory of Localization: Absence of Quantum Diffusion in Two Dimensions. *Phys. Rev. Lett.* **1979**, *42*, 673. [CrossRef]
99. Edwards, J.T.; Thouless, D.J. Numerical studies of localization in disordered systems. *J. Phys. C Solid State Phys.* **1972**, *5*, 807. [CrossRef]
100. Licciardello, D.C.; Thouless, D.J. Conductivity and mobility edges for two-dimensional disordered systems. *J. Phys. C Solid State Phys.* **1975**, *8*, 4157. [CrossRef]
101. Wegner, F.J. Electrons in disordered systems. Scaling near the mobility edge. *Z. Physik B* **1976**, *25*, 327. [CrossRef]
102. Wegner, F. The mobility edge problem: continuous symmetry and a conjecture. *Z. Physik B* **1979**, *35*, 207. [CrossRef]
103. Efetov, K.B.; Larkin, A.I.; Kheml'nitskiĭ, D.E. Interaction of diffusion modes in the theory of localization. *Sov. Phys. JETP* **1980**, *52*, 568.
104. Efetov, K.B. Supersymmetry and theory of disordered metals. *Adv. Phys.* **1983**, *32*, 53. [CrossRef]
105. Mott, N.F.; Twose, W.D. The theory of impurity conduction. *Adv. Phys.* **1961**, *10*, 107. [CrossRef]
106. Borland, R.E. The nature of the electronic states in disordered one-dimensional systems. *Proc. R. Soc. Lond. A* **1963**, *274*, 529. [CrossRef]
107. Lloyd, P. Exactly solvable model of electronic states in a three-dimensional disordered Hamiltonian: Non-existence of localized states. *J. Phys. C Solid State Phys.* **1969**, *2*, 1717. [CrossRef]
108. Theodorou, G.; Cohen, M.H. Extended states in a one-dimensional system with off-diagonal disorder. *Phys. Rev. B* **1976**, *13*, 4597. [CrossRef]
109. Antoniou, P.D.; Economou, E.N. Absence of Anderson's transition in random lattices with off-diagonal disorder. *Phys. Rev. B* **1977**, *16*, 3768. [CrossRef]
110. Economou, E.N.; Antoniou, P.D. Localization and off-diagonal disorder. *Solid State Commun.* **1977**, *21*, 285. [CrossRef]
111. Inui, M.; Trugman, S.A.; Abrahams, E. Unusual properties of midband states in systems with off-diagonal disorder. *Phys. Rev. B* **1994**, *49*, 3190. [CrossRef] [PubMed]
112. Cardy, J. *Scaling and Renormalization in Statistical Physics*; Cambridge Lecture Notes in Physics; Cambridge University Press: Cambridge, UK, 1996. [CrossRef]

113. Mehta, M.L. *Random Matrices*, 3rd ed.; Pure and Applied Mathematics; Academic Press: Cambridge, MA, USA, 2004; Volume 142.
114. Guhr, T.; Müller-Groeling, A.; Weidenmüller, H.A. Random matrix theories in quantum physics: Common concepts. *Phys. Rept.* **1998**, *299*, 189. [CrossRef]
115. Verbaarschot, J.J.M.; Wettig, T. Random matrix theory and chiral symmetry in QCD. *Ann. Rev. Nucl. Part. Sci.* **2000**, *50*, 343. [CrossRef]
116. Dyson, F.J. The threefold way. Algebraic structure of symmetry groups and ensembles in quantum mechanics. *J. Math. Phys.* **1962**, *3*, 1199. [CrossRef]
117. Verbaarschot, J.J.M. The Spectrum of the QCD Dirac operator and chiral random matrix theory: The Threefold way. *Phys. Rev. Lett.* **1994**, *72*, 2531. [CrossRef] [PubMed]
118. Altland, A.; Zirnbauer, M.R. Random Matrix Theory of a Chaotic Andreev Quantum Dot. *Phys. Rev. Lett.* **1996**, *76*, 3420. [CrossRef] [PubMed]
119. Altland, A.; Zirnbauer, M.R. Nonstandard symmetry classes in mesoscopic normal-superconducting hybrid structures. *Phys. Rev. B* **1997**, *55*, 1142. [CrossRef]
120. Zirnbauer, M.R. Riemannian symmetric superspaces and their origin in random-matrix theory. *J. Math. Phys.* **1996**, *37*, 4986. [CrossRef]
121. Zirnbauer, M.R. Symmetry Classes. In *The Oxford Handbook of Random Matrix Theory*; Akemann, G., Baik, J., Di Francesco, P., Eds.; Oxford University Press: Oxford, UK, 2015; Chapter 3. [CrossRef]
122. Chiu, C.K.; Teo, J.C.; Schnyder, A.P.; Ryu, S. Classification of topological quantum matter with symmetries. *Rev. Mod. Phys.* **2016**, *88*, 035005. [CrossRef]
123. Bell, R.J.; Dean, P. Atomic vibrations in vitreous silica. *Disc. Faraday Soc.* **1970**, *50*, 55. [CrossRef]
124. Wegner, F. Inverse participation ratio in 2+ $\varepsilon$ dimensions. *Z. Physik B* **1980**, *36*, 209. [CrossRef]
125. Castellani, C.; Peliti, L. Multifractal wavefunction at the localisation threshold. *J. Phys. A Math. Gen.* **1986**, *19*, L429. [CrossRef]
126. Al'tshuler, B.L.; Shklovskii, B.I. Repulsion of energy levels and conductivity of small metal samples. *Sov. Phys. JETP* **1986**, *64*, 127.
127. Erdős, L.; Yau, H.T.; Yin, J. Bulk universality for generalized Wigner matrices. *Probab. Theory Relat. Fields* **2012**, *154*, 341. [CrossRef]
128. Tao, T.; Vu, V. Random matrices: Universality of local eigenvalue statistics. *Acta Math.* **2011**, *206*, 127. [CrossRef]
129. Al'tshuler, B.; Zharekeshev, I.K.; Kotochigova, S.A.; Shklovskii, B. Repulsion between energy levels and the metal-insulator transition. *Zh. Eksp. Teor. Fiz* **1988**, *94*, 343.
130. Shklovskii, B.I.; Shapiro, B.; Sears, B.R.; Lambrianides, P.; Shore, H.B. Statistics of spectra of disordered systems near the metal-insulator transition. *Phys. Rev. B* **1993**, *47*, 11487. [CrossRef]
131. Muttalib, K.A.; Chen, Y.; Ismail, M.E.H.; Nicopoulos, V.N. New family of unitary random matrices. *Phys. Rev. Lett.* **1993**, *71*, 471. [CrossRef] [PubMed]
132. Moshe, M.; Neuberger, H.; Shapiro, B. A generalized ensemble of random matrices. *Phys. Rev. Lett.* **1994**, *73*, 1497. [CrossRef]
133. Mirlin, A.D.; Fyodorov, Y.V.; Dittes, F.M.; Quezada, J.; Seligman, T.H. Transition from localized to extended eigenstates in the ensemble of power-law random banded matrices. *Phys. Rev. E* **1996**, *54*, 3221. [CrossRef]
134. Canali, C.M. Model for a random-matrix description of the energy-level statistics of disordered systems at the Anderson transition. *Phys. Rev. B* **1996**, *53*, 3713. [CrossRef]
135. Kravtsov, V.E.; Muttalib, K.A. New Class of Random Matrix Ensembles with Multifractal Eigenvectors. *Phys. Rev. Lett.* **1997**, *79*, 1913. [CrossRef]
136. Nishigaki, S.M. Level spacing distribution of critical random matrix ensembles. *Phys. Rev. E* **1998**, *58*, R6915. [CrossRef]
137. Nishigaki, S.M. Level spacings at the metal-insulator transition in the Anderson Hamiltonians and multifractal random matrix ensembles. *Phys. Rev. E* **1999**, *59*, 2853. [CrossRef]
138. García-García, A.M.; Verbaarschot, J.J.M. Chiral random matrix model for critical statistics. *Nucl. Phys. B* **2000**, *586*, 668. [CrossRef]
139. Oganesyan, V.; Huse, D.A. Localization of interacting fermions at high temperature. *Phys. Rev. B* **2007**, *75*, 155111. [CrossRef]
140. Atas, Y.Y.; Bogomolny, E.; Giraud, O.; Roux, G. Distribution of the Ratio of Consecutive Level Spacings in Random Matrix Ensembles. *Phys. Rev. Lett.* **2013**, *110*, 084101. [CrossRef]
141. Slevin, K.; Ohtsuki, T. Corrections to Scaling at the Anderson Transition. *Phys. Rev. Lett.* **1999**, *82*, 382. [CrossRef]
142. Kramer, B.; MacKinnon, A.; Ohtsuki, T.; Slevin, K. Finite size scaling analysis of the Anderson transition. *Int. J. Mod. Phys. B* **2010**, *24*, 1841. [CrossRef]
143. Varga, I.; Hofstetter, E.; Schreiber, M.; Pipek, J. Shape analysis of the level-spacing distribution around the metal-insulator transition in the three-dimensional Anderson model. *Phys. Rev. B* **1995**, *52*, 7783. [CrossRef]
144. Cuevas, E.; Kravtsov, V.E. Two-eigenfunction correlation in a multifractal metal and insulator. *Phys. Rev. B* **2007**, *76*, 235119. [CrossRef]
145. Rodríguez, A.; Vasquez, L.J.; Römer, R.A. Multifractal Analysis with the Probability Density Function at the Three-Dimensional Anderson Transition. *Phys. Rev. Lett.* **2009**, *102*, 106406. [CrossRef]
146. Rodríguez, A.; Vasquez, L.J.; Slevin, K.; Römer, R.A. Critical Parameters from a Generalized Multifractal Analysis at the Anderson Transition. *Phys. Rev. Lett.* **2010**, *105*, 046403. [CrossRef] [PubMed]

147. Römer, R.A.; Schreiber, M. Numerical investigations of scaling at the Anderson transition. In *Anderson Localization and Its Ramifications*; Brandes, T., Kettemann, S., Eds.; Springer: Berlin/Heidelberg, Germany, 2003; pp. 3–19. [CrossRef]
148. Ujfalusi, L.; Varga, I. Finite-size scaling and multifractality at the Anderson transition for the three Wigner-Dyson symmetry classes in three dimensions. *Phys. Rev. B* **2015**, *91*, 184206. [CrossRef]
149. Slevin, K.; Ohtsuki, T. The Anderson transition: time reversal symmetry and universality. *Phys. Rev. Lett.* **1997**, *78*, 4083. [CrossRef]
150. Asada, Y.; Slevin, K.; Ohtsuki, T. Anderson transition in the three dimensional symplectic universality class. *J. Phys. Soc. Jpn.* **2005**, *74*, 238. [CrossRef]
151. Rodríguez, A.; Vasquez, L.J.; Slevin, K.; Römer, R.A. Multifractal finite-size scaling and universality at the Anderson transition. *Phys. Rev. B* **2011**, *84*, 134209. [CrossRef]
152. Lindinger, J.; Rodríguez, A. Multifractal finite-size scaling at the Anderson transition in the unitary symmetry class. *Phys. Rev. B* **2017**, *96*, 134202. [CrossRef]
153. Hikami, S.; Larkin, A.I.; Nagaoka, Y. Spin-Orbit Interaction and Magnetoresistance in the Two Dimensional Random System. *Prog. Theor. Phys.* **1980**, *63*, 707. [CrossRef]
154. Wegner, F. Four-loop-order $\beta$-function of nonlinear $\sigma$-models in symmetric spaces. *Nucl. Phys. B* **1989**, *316*, 663. [CrossRef]
155. Pruisken, A.M.M. On localization in the theory of the quantized Hall effect: A two-dimensional realization of the $\theta$-vacuum. *Nucl. Phys. B* **1984**, *235*, 277. [CrossRef]
156. Huckestein, B. Scaling theory of the integer quantum Hall effect. *Rev. Mod. Phys.* **1995**, *67*, 357. [CrossRef]
157. Slevin, K.; Ohtsuki, T. Critical exponent for the quantum Hall transition. *Phys. Rev. B* **2009**, *80*, 041304. [CrossRef]
158. Dresselhaus, E.J.; Sbierski, B.; Gruzberg, I.A. Numerical evidence for marginal scaling at the integer quantum Hall transition. *arXiv* **2021**, arXiv:2101.01716.
159. Zirnbauer, M.R. The integer quantum Hall plateau transition is a current algebra after all. *Nucl. Phys. B* **2019**, *941*, 458. [CrossRef]
160. Xie, X.C.; Wang, X.; Liu, D. Kosterlitz-Thouless-type metal-insulator transition of a 2D electron gas in a random magnetic field. *Phys. Rev. Lett.* **1998**, *80*, 3563. [CrossRef]
161. Berezinskiĭ, V.L. Destruction of long range order in one-dimensional and two-dimensional systems having a continuous symmetry group. I. Classical systems. *Sov. Phys. JETP* **1971**, *32*, 493.
162. Berezinskiĭ, V.L. Destruction of long-range order in one-dimensional and two-dimensional systems possessing a continuous symmetry group. II. Quantum systems. *Sov. Phys. JETP* **1971**, *34*, 610.
163. Kosterlitz, J.M.; Thouless, D.J. Ordering, metastability and phase transitions in two-dimensional systems. *J. Phys. C Solid State Phys.* **1973**, *6*, 1181. [CrossRef]
164. Zhang, Y.Y.; Hu, J.; Bernevig, B.A.; Wang, X.R.; Xie, X.C.; Liu, W.M. Localization and the Kosterlitz-Thouless transition in disordered graphene. *Phys. Rev. Lett.* **2009**, *102*, 106401. [CrossRef] [PubMed]
165. Weaire, D.; Srivastava, V. Numerical results for Anderson localisation in the presence of off-diagonal disorder. *Solid State Commun.* **1977**, *23*, 863. [CrossRef]
166. Cain, P.; Römer, R.A.; Schreiber, M. Phase diagram of the three-dimensional Anderson model of localization with random hopping. *Ann. Phys. (Leipzig)* **1999**, *8*, 507.
167. Biswas, P.; Cain, P.; Römer, R.A.; Schreiber, M. Off-Diagonal Disorder in the Anderson Model of Localization. *Phys. Stat. Sol.* **2000**, *218*, 205. [CrossRef]
168. Evangelou, S.N.; Katsanos, D.E. Spectral statistics in chiral-orthogonal disordered systems. *J. Phys. A* **2003**, *36*, 3237. [CrossRef]
169. Wegner, F.J. Disordered system with $n$ orbitals per site: $n = \infty$ limit. *Phys. Rev. B* **1979**, *19*, 783. [CrossRef]
170. Oppermann, R.; Wegner, F. Disordered system with $n$ orbitals per site: $1/n$ expansion. *Z. Physik B* **1979**, *34*, 327. [CrossRef]
171. Gade, R.; Wegner, F. The $n = 0$ replica limit of U($n$) and U($n$)SO($n$) models. *Nucl. Phys. B* **1991**, *360*, 213. [CrossRef]
172. Gade, R. Anderson localization for sublattice models. *Nucl. Phys. B* **1993**, *398*, 499. [CrossRef]
173. Fabrizio, M.; Castellani, C. Anderson localization in bipartite lattices. *Nucl. Phys. B* **2000**, *583*, 542. [CrossRef]
174. Motrunich, O.; Damle, K.; Huse, D.A. Particle-hole symmetric localization in two dimensions. *Phys. Rev. B* **2002**, *65*, 064206. [CrossRef]
175. Mudry, C.; Ryu, S.; Furusaki, A. Density of states for the $\pi$-flux state with bipartite real random hopping only: A weak disorder approach. *Phys. Rev. B* **2003**, *67*, 064202. [CrossRef]
176. Soukoulis, C.; Webman, I.; Grest, G.S.; Economou, E.N. Study of electronic states with off-diagonal disorder in two dimensions. *Phys. Rev. B* **1982**, *26*, 1838. [CrossRef]
177. Eilmes, A.; Römer, R.A.; Schreiber, M. The two-dimensional Anderson model of localization with random hopping. *Eur. Phys. J. B* **1998**, *1*, 29. [CrossRef]
178. Xiong, S.J.; Evangelou, S.N. Power-law localization in two and three dimensions with off-diagonal disorder. *Phys. Rev. B* **2001**, *64*, 113107. [CrossRef]
179. Markoš, P.; Schweitzer, L. Critical conductance of two-dimensional chiral systems with random magnetic flux. *Phys. Rev. B* **2007**, *76*, 115318. [CrossRef]
180. Schweitzer, L.; Markoš, P. Scaling at chiral quantum critical points in two dimensions. *Phys. Rev. B* **2012**, *85*, 195424. [CrossRef]
181. Markoš, P.; Schweitzer, L. Logarithmic scaling of Lyapunov exponents in disordered chiral two-dimensional lattices. *Phys. Rev. B* **2010**, *81*, 205432. [CrossRef]

182. Bocquet, M.; Chalker, J.T. Network models for localization problems belonging to the chiral symmetry classes. *Phys. Rev. B* **2003**, *67*, 054204. [CrossRef]
183. König, E.J.; Ostrovsky, P.M.; Protopopov, I.V.; Mirlin, A.D. Metal-insulator transition in two-dimensional random fermion systems of chiral symmetry classes. *Phys. Rev. B* **2012**, *85*, 195130. [CrossRef]
184. García-García, A.M.; Cuevas, E. Anderson transition in systems with chiral symmetry. *Phys. Rev. B* **2006**, *74*, 113101. [CrossRef]
185. Luo, X.; Xu, B.; Ohtsuki, T.; Shindou, R. Critical behavior of Anderson transitions in three-dimensional orthogonal classes with particle-hole symmetries. *Phys. Rev. B* **2020**, *101*. [CrossRef]
186. Wang, T.; Ohtsuki, T.; Shindou, R. Universality classes of the Anderson transition in three-dimensional symmetry classes AIII, BDI, C, D and CI. *arXiv* **2021**, arXiv:2105.02500.
187. Takaishi, T.; Sakakibara, K.; Ichinose, I.; Matsui, T. Localization and delocalization of fermions in a background of correlated spins. *Phys. Rev. B* **2018**, *98*, 184204. [CrossRef]
188. Halász, Á.M.; Verbaarschot, J.J.M. Universal fluctuations in spectra of the lattice Dirac operator. *Phys. Rev. Lett.* **1995**, *74*, 3920. [CrossRef]
189. Zvyagin, I., Charge Transport via Delocalized States in Disordered Materials. In *Charge Transport in Disordered Solids with Applications in Electronics*; John Wiley & Sons, Ltd.: Hoboken, NJ, USA, 2006; Chapter 1, pp. 1–47. [CrossRef]
190. Verbaarschot, J.J.M.; Zahed, I. Random matrix theory and QCD in three-dimensions. *Phys. Rev. Lett.* **1994**, *73*, 2288. [CrossRef] [PubMed]
191. Magnea, U. The Orthogonal ensemble of random matrices and QCD in three-dimensions. *Phys. Rev. D* **2000**, *61*, 056005. [CrossRef]
192. Magnea, U. Three-dimensional QCD in the adjoint representation and random matrix theory. *Phys. Rev. D* **2000**, *62*, 016005. [CrossRef]
193. Kieburg, M.; Verbaarschot, J.J.M.; Zafeiropoulos, S. Dirac spectra of two-dimensional QCD-like theories. *Phys. Rev. D* **2014**, *90*, 085013. [CrossRef]
194. Verbaarschot, J.J.M. Universal scaling of the valence quark mass dependence of the chiral condensate. *Phys. Lett. B* **1996**, *368*, 137. [CrossRef]
195. Guhr, T.; Ma, J.Z.; Meyer, S.; Wilke, T. Statistical analysis and the equivalent of a Thouless energy in lattice QCD Dirac spectra. *Phys. Rev. D* **1999**, *59*, 054501. [CrossRef]
196. Kogut, J.B.; Susskind, L. Hamiltonian Formulation of Wilson's Lattice Gauge Theories. *Phys. Rev. D* **1975**, *11*, 395. [CrossRef]
197. Banks, T.; Susskind, L.; Kogut, J.B. Strong Coupling Calculations of Lattice Gauge Theories: (1+1)-Dimensional Exercises. *Phys. Rev. D* **1976**, *13*, 1043. [CrossRef]
198. Susskind, L. Lattice Fermions. *Phys. Rev. D* **1977**, *16*, 3031. [CrossRef]
199. Narayanan, R.; Neuberger, H. Chiral determinant as an overlap of two vacua. *Nucl. Phys.* **1994**, *B412*, 574. [CrossRef]
200. Narayanan, R.; Neuberger, H. Chiral fermions on the lattice. *Phys. Rev. Lett.* **1993**, *71*, 3251. [CrossRef]
201. Neuberger, H. Exactly massless quarks on the lattice. *Phys. Lett. B* **1998**, *417*, 141. [CrossRef]
202. Neuberger, H. More about exactly massless quarks on the lattice. *Phys. Lett. B* **1998**, *427*, 353. [CrossRef]
203. Bruckmann, F.; Keppeler, S.; Panero, M.; Wettig, T. Polyakov loops and spectral properties of the staggered Dirac operator. *Phys. Rev. D* **2008**, *78*, 034503. [CrossRef]
204. Schäfer, T.; Shuryak, E.V. Instantons in QCD. *Rev. Mod. Phys.* **1998**, *70*, 323. [CrossRef]
205. Bernard, C.; Burch, T.; DeTar, C.; Gottlieb, S.; Heller, U.M.; Hetrick, J.E.; Levkova, L.; Maresca, F.; Renner, D.B.; Sugar, R.; et al. The equation of state for QCD with 2+1 flavors of quarks. *PoS* **2006**, *LAT2005*, 156. [CrossRef]
206. Aubin, C.; Bernard, C.; DeTar, C.; Osborn, J.; Gottlieb, S.; Gregory, E.B.; Toussaint, D.; Heller, U.M.; Hetrick, J.E.; Sugar, R. Light hadrons with improved staggered quarks: Approaching the continuum limit. *Phys. Rev. D* **2004**, *70*, 094505. [CrossRef]
207. Orginos, K.; Toussaint, D. Testing improved actions for dynamical Kogut-Susskind quarks. *Phys. Rev. D* **1999**, *59*, 014501. [CrossRef]
208. Toussaint, D.; Orginos, K. Tests of improved Kogut-Susskind fermion actions. *Nucl. Phys. B Proc. Suppl.* **1999**, *73*, 909. [CrossRef]
209. Lepage, P. Perturbative improvement for lattice QCD: An Update. *Nucl. Phys. B Proc. Suppl.* **1998**, *60*, 267. [CrossRef]
210. Lepage, G.P. Flavor symmetry restoration and Symanzik improvement for staggered quarks. *Phys. Rev. D* **1999**, *59*, 074502. [CrossRef]
211. Morningstar, C.; Peardon, M.J. Analytic smearing of SU(3) link variables in lattice QCD. *Phys. Rev. D* **2004**, *69*, 054501. [CrossRef]
212. Aoki, Y.; Fodor, Z.; Katz, S.D.; Szabó, K.K. The Equation of state in lattice QCD: With physical quark masses towards the continuum limit. *J. High Energy Phys.* **2006**, *1*, 89. [CrossRef]
213. Follana, E.; Mason, Q.; Davies, C.; Hornbostel, K.; Lepage, G.P.; Shigemitsu, J.; Trottier, H.; Wong, K. Highly improved staggered quarks on the lattice, with applications to charm physics. *Phys. Rev. D* **2007**, *75*, 054502. [CrossRef]
214. Brower, R.C.; Neff, H.; Orginos, K. Möbius fermions: Improved domain wall chiral fermions. *Nucl. Phys. B Proc. Suppl.* **2005**, *140*, 686. [CrossRef]
215. Brower, R.C.; Neff, H.; Orginos, K. Möbius fermions. *Nucl. Phys. B Proc. Suppl.* **2006**, *153*, 191. [CrossRef]
216. Brower, R.C.; Neff, H.; Orginos, K. The Möbius domain wall fermion algorithm. *Comput. Phys. Commun.* **2017**, *220*, 1. [CrossRef]
217. Holicki, L. Quark Localization and the Anderson Transition in Lattice Quantum Chromodynamics. Ph.D. Thesis, Giessen University, Giessen, Germany, 2019.

218. Burger, F.; Hotzel, G.; Müller-Preussker, M.; Ilgenfritz, E.M.; Lombardo, M.P. Towards thermodynamics with $N_f = 2 + 1 + 1$ twisted mass quarks. *PoS* **2013**, *LATTICE2013*, 153. [CrossRef]
219. Burger, F.; Ilgenfritz, E.M.; Lombardo, M.P.; Müller-Preussker, M.; Trunin, A. Towards the quark–gluon plasma Equation of State with dynamical strange and charm quarks. *J. Phys. Conf. Ser.* **2016**, *668*, 012092. [CrossRef]
220. Burger, F.; Ilgenfritz, E.M.; Lombardo, M.P.; Müller-Preussker, M.; Trunin, A. Topology (and axion's properties) from lattice QCD with a dynamical charm. *Nucl. Phys. A* **2017**, *967*, 880. [CrossRef]
221. Diakonov, D. Topology and confinement. *Nucl. Phys. B Proc. Suppl.* **2009**, *195*, 5. [CrossRef]
222. García Pérez, M.; González-Arroyo, A.; Pena, C.; van Baal, P. Weyl-Dirac zero mode for calorons. *Phys. Rev. D* **1999**, *60*, 031901. [CrossRef]
223. Chernodub, M.N.; Kraan, T.C.; van Baal, P. Exact fermion zero mode for the new calorons. *Nucl. Phys. B Proc. Suppl.* **2000**, *83*, 556. [CrossRef]
224. Lucini, B.; Teper, M.; Wenger, U. Properties of the deconfining phase transition in SU(N) gauge theories. *J. High Energy Phys.* **2005**, *2*, 33. [CrossRef]
225. Liddle, J.; Teper, M. The Deconfining phase transition in D=2+1 SU(N) gauge theories. *arXiv* **2008**, arXiv:0803.2128.
226. Capitani, S.; Dürr, S.; Hoelbling, C. Rationale for UV-filtered clover fermions. *J. High Energy Phys.* **2006**, *11*, 028. [CrossRef]
227. Barber, M.N. Finite-size scaling. In *Phase Transitions and Critical Phenomena*; Domb, C., Lebowitz, J.L., Eds.; Academic Press: London, UK, 1983; Volume 8, pp. 146–266.
228. Karsch, F.; Laermann, E.; Schmidt, C. The Chiral critical point in three-flavor QCD. *Phys. Lett. B* **2001**, *520*, 41. [CrossRef]
229. de Forcrand, P.; Philipsen, O. The QCD phase diagram for three degenerate flavors and small baryon density. *Nucl. Phys. B* **2003**, *673*, 170. [CrossRef]
230. de Forcrand, P.; Philipsen, O. The Chiral critical point of $N_f = 3$ QCD at finite density to the order $(\mu/T)^4$. *J. High Energy Phys.* **2008**, *11*, 012. [CrossRef]
231. Myers, J.C.; Ogilvie, M.C. New phases of SU(3) and SU(4) at finite temperature. *Phys. Rev. D* **2008**, *77*, 125030. [CrossRef]
232. Ünsal, M.; Yaffe, L.G. Center-stabilized Yang-Mills theory: Confinement and large N volume independence. *Phys. Rev. D* **2008**, *78*, 065035. [CrossRef]
233. Karsch, F.; Lütgemeier, M. Deconfinement and chiral symmetry restoration in an SU(3) gauge theory with adjoint fermions. *Nucl. Phys. B* **1999**, *550*, 449. [CrossRef]

 universe



*Article*

# Influence of Fermions on Vortices in SU(2)-QCD

**Zeinab Dehghan [1], Sedigheh Deldar [1], Manfried Faber [2,*], Rudolf Golubich [2] and Roman Höllwieser [3]**

[1] Department of Physics, University of Tehran, Tehran 1439955961, Iran; zeinab.dehghan@ut.ac.ir (Z.D.); sdeldar@ut.ac.ir (S.D.)

[2] Atominstitut, Technische Universität Wien, 1040 Wien, Austria; rudolf.golubich@gmail.com

[3] Department of Physics, Bergische Universität Wuppertal, 42119 Wuppertal, Germany; roman.hoellwieser@gmail.com

* Correspondence: faber@kph.tuwien.ac.at

**Abstract:** Gauge fields control the dynamics of fermions, and, in addition, a back reaction of fermions on the gauge field is expected. This back reaction is investigated within the vortex picture of the QCD vacuum. We show that the center vortex model reproduces the string tension of the full theory also in the presence of fermionic fields.

**Keywords:** quantum chromodynamics; confinement; center vortex model; vacuum structure

**PACS:** 11.15.Ha; 12.38.Gc

**Citation:** Dehghan, Z.; Deldar, S.; Faber, M.; Golubich, R.; Höllwieser, R. Influence of Fermions on Vortices in SU(2)-QCD. *Universe* **2021**, *7*, 130. https://doi.org/10.3390/universe7050130

Academic Editor: Dmitri Antonov

Received: 2 April 2021
Accepted: 1 May 2021
Published: 4 May 2021

**Publisher's Note:** MDPI stays neutral with regard to jurisdictional claims in published maps and institutional affiliations.

## 1. Introduction

The QCD vacuum is highly nontrivial and has magnetic properties, as we have known since Savvidy's article [1]. The QCD vacuum should explain the non-perturbative properties of QCD, including confinement [2] and chiral symmetry breaking [3]. Lattice QCD puts the means at our disposal to answer the question about the important degrees of freedom of this non-perturbative vacuum. In the center vortex picture [4–6], the QCD vacuum is seen as a condensate of closed quantized magnetic flux tubes. These flux tubes have random shapes and evolve in time and therefore form closed surfaces in the dual space. They may expand and shrink, fuse and split and percolate in the confinement phase in all space-time directions and pierce Wilson loops randomly. Thus, Wilson loops asymptotically follow an exponential decay with the area. This is the area law of Wilson loops, which allows attributing the string tension to center vortices. The finite temperature phase transition is characterized by a loss of center symmetry and correspondingly by a loss of percolation in time direction. Therefore, vortices get static and only spatial Wilson loops keep showing the area law behavior.

Color electric charges are sources of electric flux according to Gauss's law. The electric flux between opposite color charges does not like to penetrate this magnetic "medium" of center vortices and shrinks to the well-known electric flux tube. On the other hand, the magnetic flux does not like to enter the electric string. Since fermions carry color charges, their dynamics is controlled by the gauge field. The presence of a fermion condensate is expected to suppress the quantized magnetic flux lines, and as a result the gluon condensate and therefore the string tension are reduced. Since, as usual, the lattice spacing is determined via the string tension, taking into account dynamical fermions leads to a decrease of the lattice spacing. In this article, we show a careful investigation of the string tension within the vortex picture of the QCD vacuum.

SU(2) and SU(3) QCD have equivalent non-perturbative properties. In a first study, we restrict our analysis to the simpler case of SU(2)-QCD. The most important difference between SU(2) and SU(3) QCD is the order of the finite temperature phase transition for a pure gluonic Lagrangian. There is a natural explanation for this difference from the structure of SU(2) and SU(3) vortices. There is only one non-trivial center element in the group SU(2) and therefore one type of center vortices, whereas there are two non-trivial center elements for SU(3) and two types of vortices, allowing two vortices of the same type to fuse to the other type. This leads to a more stable structure of the net of vortices for SU(3) and to a first order phase transition, whereas in SU(2) the transition is of second order.

We investigate the fermionic back reaction on the gluonic degrees of freedom in SU(2) QCD. Visualizing the distribution of center vortices, this back reaction can be easily observed (see Figure 1). One can clearly see that dynamical fermions decrease the percolation of vortices. It is difficult to draw a closed surface in four dimensions. Therefore, we restrict ourselves to the three dimensional diagram of a time-slice and indicate the continuation to other slices by line stubs.

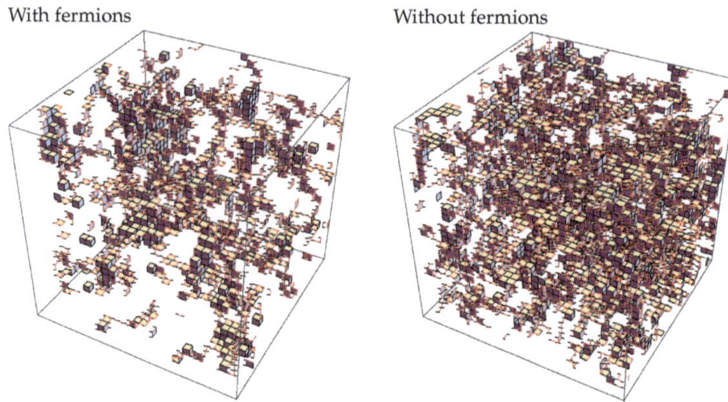



**Figure 1.** The closed vortex surface is visualized by showing the dual P-plaquettes of three-dimensional lattice slices. Stubs of red lines indicate plaquettes that are not fully part of the lattice slice shown. We clearly see that with fermions (**left**) an overall smaller amount of P-plaquettes is observed compared with the pure gluonic case (**right**). In both cases, one big vortex cluster dominates.

We want to quantify the effect in more detail. We are especially interested in the center vortex model [4–6] and its sensitivity to the fermionic back reaction. We also analyze the influence of fermionic fields on the geometric structure of the center vortex surface. This work compares four different estimates of the string tension, with and without fermions, in the full theory and in the vortex picture:

- via the potential calculated from the center degrees of freedom only, in pure gluonic ensembles;
- via the potential calculated from the center degrees of freedom only, in the presence of fermionic fields;
- via the potential in the full theory, in pure gluonic ensembles; and
- via the potential in the full theory, in the presence of fermionic fields.

With this comparison, we study the sensitivity of the center vortex model to the fermionic back reaction.

Our work is based on the QCD path integral which defines the vacuum to vacuum transition amplitude. In lattice QCD, we usually evaluate this amplitude on a lattice periodic in Euclidean time. Inserting a complete set of eigenstates of QCD with the quantum numbers of the vacuum in this amplitude results in an exponential decay of the eigenstates with the physical time extent $aN_t$ of the lattice, where $a$ is the lattice spacing and $N_t$ is the number of lattice sites in time direction. The inverse of this time extent therefore acts as a temperature of the ensemble. In a Monte-Carlo simulation, the states are occupied with the corresponding Boltzmann factor. The higher is the excitation, the smaller is the Boltzmann factor and the more difficult is the measurement of its properties. Finally, the excited states are vanishing in the noise.

The potential, as well as the string tension, can be calculated using *Wilson loops*

$$W(R,T) = \text{Tr} \mathcal{P} e^{i g \int_{R \times T} A_\mu^a(x) t_a dx^\mu}.$$ (1)

A loop of size $R \times T$ in space-time represents the world-line of a quark-antiquark system at distance $R$ propagating in the QCD vacuum for a time $T$. On a Euclidean lattice in SU(2)-QCD, a path ordered loop is determined by the product of link variables $U_\mu(x) \in$ SU(2) along the loop. Inserting a complete set of eigenstates of the quark-antiquark system into the expectation value $\langle W(R,T) \rangle$, the contributions of the eigenstates decay exponentially with Euclidean time $T$. The expectation values of Wilson loops can therefore be expanded in a series of eigenstates of the quark-antiquark system

$$\langle W(R,T) \rangle = \sum_{i=0}^{\infty} c_i e^{-\varepsilon_i(R)T},$$ (2)

For large times, $\langle W(R,T) \rangle$ is dominated by the ground state energy $\varepsilon_0(R)$. The more precise we determine $\lim_{T \to \infty} \langle W(R,T) \rangle$, the better is the precision of the quark-antiquark potential $V(R) := \varepsilon_0(R)$. Since the energy of the quark-antiquark system increases with the distance $R$, it follows from the above discussion that for increasing $R$ the signal for $V(R)$ is vanishing soon in the noise. How we handle this noise and how center vortices are detected is explained in Section 2. We assume that the potential is dominated by a Coulombic part at small $R$ but rises linearly for large $R$,

$$\varepsilon_0(R) = V(R) = V_0 + \sigma R - \frac{\alpha}{R}.$$ (3)

We use $\langle W(R,T) \rangle$ to approximate $\varepsilon_0(R)$, denoted as *1-exp fit*. $V_0$ parameterizes the scale dependent self-energy of the quark-antiquark sources. Wilson loops extracted from the center degrees of freedom are dominated by the long-range fluctuations of the QCD vacuum, hence we describe the potential within these degrees of freedom by

$$V_{\text{CP}}(R) = v_0 + \sigma_{\text{CP}} R.$$ (4)

The aim of this article is to investigate whether we can understand the string tension and its modification in the presence of fermions in the vortex model of confinement. Further, we present and discuss conceptual improvements to the gauge fixing procedure, required for the center vortex detection.

For systems with dynamical fermions one would expect string breaking when the energy of the system rises above twice the pion mass, but string breaking has been detected only using mesonic channels (see [7]). The center vortex model explains the asymptotic behavior of Wilson loops. There are indications that center vortices are sensitive to string breaking [8,9], but a direct measurement is not possible. From the vortex structure, we do not find any indication for string breaking which could show up as disintegration of the percolating vortex.

## 2. Materials and Methods

This section starts with a description of the parameters of the lattice configurations, used for our analysis. Then, our method of detecting center vortices with some novel improvements is discussed. We explain how the information about the geometric structure of the vortex surface can be acquired by smoothing procedures and we end with a detailed explanation of our method to extract the potential from Wilson loops. In each subsection, we list the intermediate results.

### 2.1. Simulation Specifications

We study the configurations described in [10] for chemical potential $\mu = 0$ with $S_G$ defined by a tree level improved Symanzik gauge action [11,12]

$$S_G = \beta \left( c_0 \sum_{\square} (1 - \frac{1}{2}\mathrm{Tr}\,\square) + c_1 \sum_{\square\square} (1 - \frac{1}{2}\mathrm{Tr}\,\square\square) \right), \tag{5}$$

with coefficients $c_0 = 5/3$ and $c_1 = -1/12$. The first sum corresponds to the Wilson action with $\square$ indicating single unoriented plaquettes, while the second sum uses rectangular Wilson loops built of 6 links, symbolized by $\square\square$. The inverse coupling is defined as $\beta = \frac{4}{g^2}$ for SU(2).

For the fermionic degrees of freedom, staggered fermions are used with an action of the form

$$S_F = \sum_{x,y} \bar{\psi}_x M(m)_{x,y} \psi_y + \frac{\lambda}{2} \sum_x \left( \psi_x^T \tau_2 \psi_x + \bar{\psi}_x \tau_2 \bar{\psi}_x^T \right), \tag{6}$$

with $\tau_i$ being the Pauli matrices and

$$M(m)_{xy} = m\delta_{xy} + \frac{1}{2} \sum_{\nu=1}^{4} \eta_\nu(x) \left[ U_{x,\nu}\delta_{x+h_\nu,y} - U_{x-h_\nu,\nu}^\dagger \delta_{x-h_\nu,y} \right], \tag{7}$$

where $\bar{\psi}$, $\psi$ are staggered fermion fields, $a$ is the lattice spacing, $m$ is the bare quark mass, $U_{x,\nu}$ is a SU(2) element corresponding to a link at position $x$ is in direction and $\mu$ and $\eta_\nu(x)$ are the standard staggered phase factors: $\eta_1(x) = 1$, $\eta_\nu(x) = (-1)^{x_1+\cdots+x_{\nu-1}}, \nu = 2,3,4$. The total action is given by $S = S_G + S_F$. Integrating out the fermionic degrees of freedom, the partition function with $N_f = 2$ is given by

$$Z = \int DU\, e^{-S_G} \left( \det(M^\dagger M) + \lambda^2 \right)^{\frac{1}{4}}. \tag{8}$$

The properties of 1000 configurations of size $32^4$ with $\beta = 1.8$, quark mass parameter $m = 0.0075$ (corresponding to $m_\pi = 740(40)$ MeV with lattice spacing $a = 0.044$ fm), and $\lambda = 0.00075$ are compared to 1000 pure gluonic configurations at the same inverse coupling $\beta$. For both sets of 1000 configurations, we extract the potentials from all available Wilson loop data and compare them with the string tensions resulting from the two sets of $40 \times 100$ center projected configurations. In this way, we try to answer the question, if in the presence of dynamical fermions the center degrees of freedom determine the string tension of the gluonic flux tube in quark–antiquark systems.

### 2.2. Center Vortex Detection

Assuming that center excitations are the relevant degrees of freedom for confinement, we detect these center vortices within the lattice configurations. We first identify gauge matrices $\Omega(x) \in$ SU(2) at each site $x^\mu$ maximizing the functional

$$R_F = \sum_x \sum_\mu |\,\mathrm{Tr}[\acute{U}_\mu(x)]\,|^2 \quad \text{with} \quad \acute{U}_\mu(x) = \Omega(x + e_\mu)U_\mu(x)\Omega^\dagger(x). \tag{9}$$

After fixing the gauge, the link variables $\acute{U}_\mu(x)$ are projected on the center degrees of freedom, that is $\pm 1$ for SU(2), to neglect short range properties and keep only long-range effects

$$U_\mu(x) \to Z_\mu(x) \equiv \mathrm{sign}\,\mathrm{Tr}[U_\mu(x)]. \tag{10}$$

After performing the center projection, the center projected plaquettes resulting from the vortex detection are the products of four center elements. The projected plaquettes are non-trivial, known as P-plaquettes, $U_\square = -1$, if one or three links are non-trivial. In the four-dimensional lattice, a given link belongs to six plaquettes. On the dual lattice, the corresponding six plaquettes build the surface of a cube. Therefore, the duals of P-plaquettes form closed surfaces, dual P-vortices which correspond to the closed flux line evolving in time.

This procedure is the original DMCG [13] in which a gradient climb with *over relaxation* was used to maximize the gauge functional. From a few gauge copies only, produced in this way, the one with the highest value of the functional is usually chosen for further analysis. This method leads to promising results, but improvements at maximizing the gauge functional using *simulated annealing* have brought a flaw to light—the many local maxima of $R_F$ do not necessarily correspond to the same physics. Bornyakov et al. [14] showed that there exist local maxima of the gauge functional that underestimate the string tension. We have been able to resolve these problems for smaller lattices using improved version of the gauge fixing routines based on non-trivial center regions [15–17], but our implementation was not able to handle the big lattices used in this work. Taking a closer look at the problem at hand, we can look for a different approach. We now consider Creutz ratios to estimate the string tension

$$\sigma \approx \chi(R) = -\ln \frac{\langle W(R+1, R+1)\rangle \, \langle W(R, R)\rangle}{\langle W(R, R+1)\rangle \, \langle W(R+1, R)\rangle},$$ (11)

with Wilson loops $W(R, R)$ of size $R = T$. Some probability densities for the relation between the values of the gauge functional $R_F$ and the Creutz ratio $\chi(R)$ for individual configurations are shown in Figure 2. This determination is based on 40 configurations with 100 gauge copies for configurations with (left) and without (right) dynamical fermions. For Creutz ratios of small Wilson loops, we observe a nearly linear relation between the two quantities reflecting the finding of Bornyakov et al. [14]: there exist gauge copies of the configurations with maximal $R_F$ and very low $\sigma$. With increasing size of Wilson loops, this correlation weakens. Nevertheless, the request to maximize the gauge functional (9) fails.

Another observation is of high interest: extremely small and large values of the gauge functional are strongly suppressed in the probability densities. Instead of looking for higher local maxima of the gauge functional, we propose a different approach: "ensemble averaged maximal center gauge" (EaMCG). We produce many random gauge copies, approach the next local maximum by the gradient method and take the average of the ensemble. The idea is that not the best local maximum alone carries the physical meaning, but the average over all local maxima does: maxima with a higher value of the gauge functional result in a reduced string tension, but they are not dominating the ensemble. The same holds for lower valued maxima, possibly overestimating the string tension.

Taking again a look at Figure 2, it can be seen that the average values and the most probable values are in good agreement for small loops. This is shown in more detail in Figure 3 for Creutz ratios of different loop-sizes. The fact that differences increase with loop sizes can probably be explained by the lack of statistics for the Creutz ratios of single configurations. Until the values start to deviate from one another, there is a variation of 10% over the whole $R$-region. Despite the low statistics of a single configuration, the intermediate loop sizes already reproduce the asymptotic behavior and let us expect the possibility for a more precise determination. First averaging over Wilson loops and then calculating Creutz ratios gives much more stable results (see $\chi_W(R)$ in Figure 3). The final estimate of the string tensions in Sections 2.4 and 3 is based on the determination of the potential.

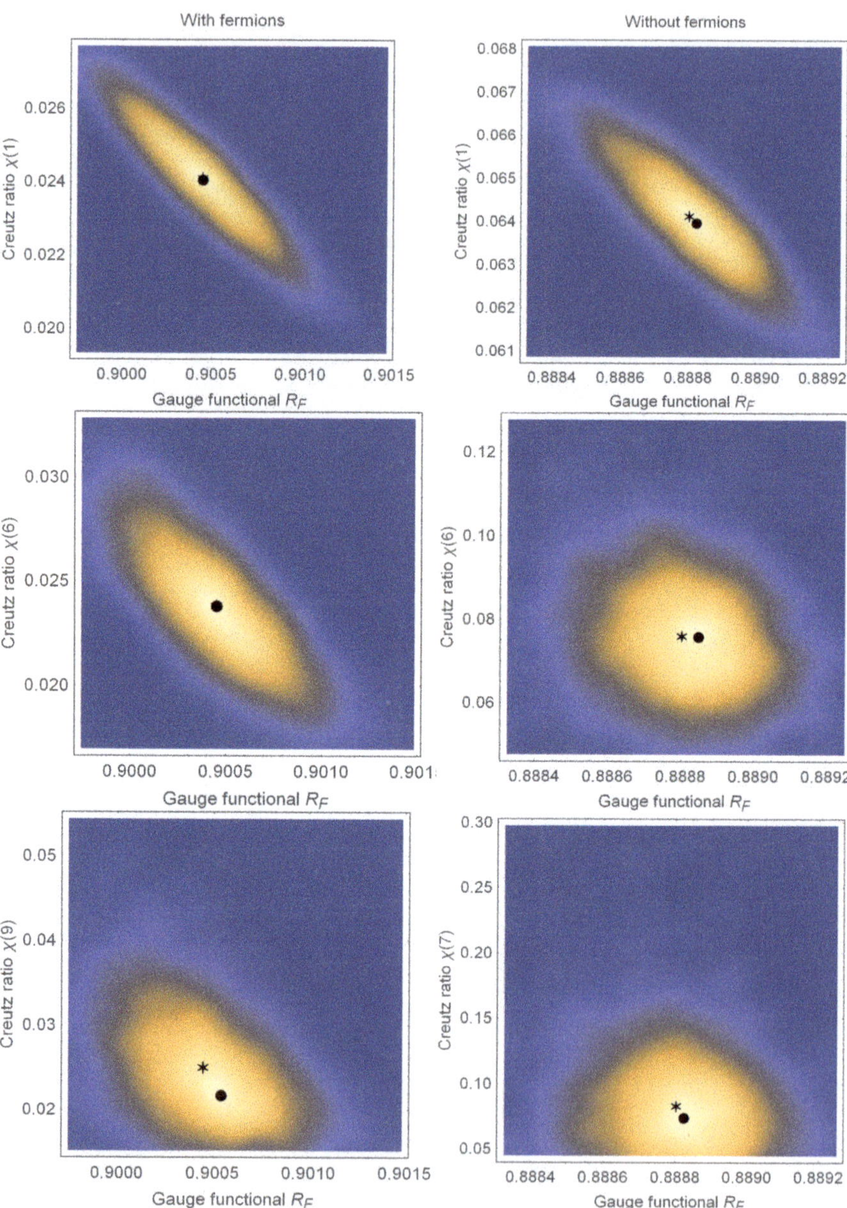

**Figure 2.** These probability densities specify the relation between the values of gauge functional and Creutz ratio for individual configurations. This determination is based on 40 configurations with 100 gauge copies for the configurations with dynamical fermions (**left**) and without (**right**). For Creutz ratios of small Wilson loops, we observe a nearly linear relation. With increasing size of the Wilson loops this correlation weakens. We marked the average values (star) and the most probable values (circle) of the distributions.

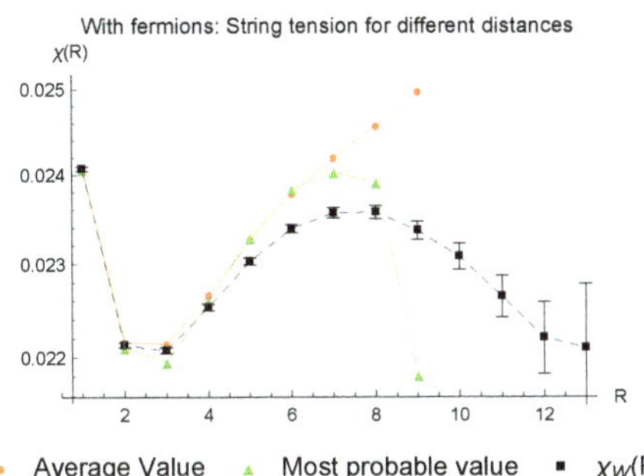

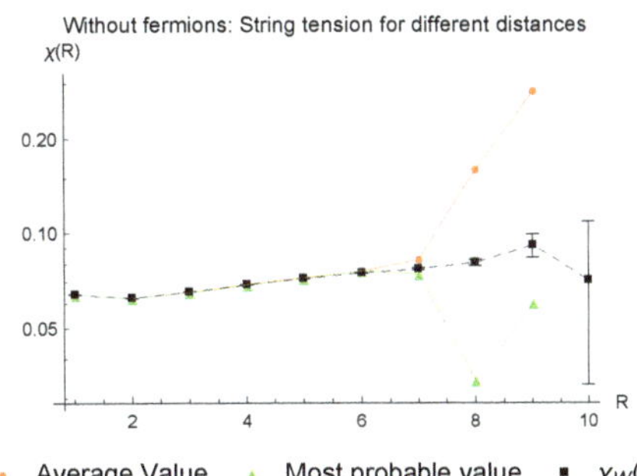

**Figure 3.** The average and most probable values of $\chi(R)$ are compared for simulations with and without fermions. This complements the probability densities of Figure 2. The increasing discrepancy between the two quantities with larger $R$ can probably be explained by the low precision of Creutz ratios of single configurations. Until the values start to deviate from one another, the variations of $\chi(R)$ are of the order of 10%. For comparison, we show also the more precise Creutz ratios $\chi_W(R)$ extracted from averages of Wilson loops.

Thus far, we have calculated the Creutz ratios for single configurations of the ensemble and have taken the average afterwards. The EaMCG itself does not average over Creutz ratios, but combines first the Wilson loops of all gauge copies and configurations. From this fact, it is possible to extract the quark anti-quark potential, which allows a more precise determination of the string tension from the center vortex model.

In the respective single configurations, we observe one percolating large cluster that is surrounded and traversed by small fluctuations. These result in an increased number of P-plaquettes that do not contribute to the string tension. Analyzing these distortions, we gain insight on the influence of fermions on the geometric structure of the vortex surface.

### 2.3. Smoothing the Vortex Surface

There exist several procedures for smoothing the vortex surface by removing distortions. These procedures are discussed in detail in [18]. They do not modify the long range effects of the configuration. To get information about the smoothness of the vortex surface with and without fermions, we use the smoothing steps depicted in Figure 4. The smoothing 0 is not depicted, which removes unit-cubes.

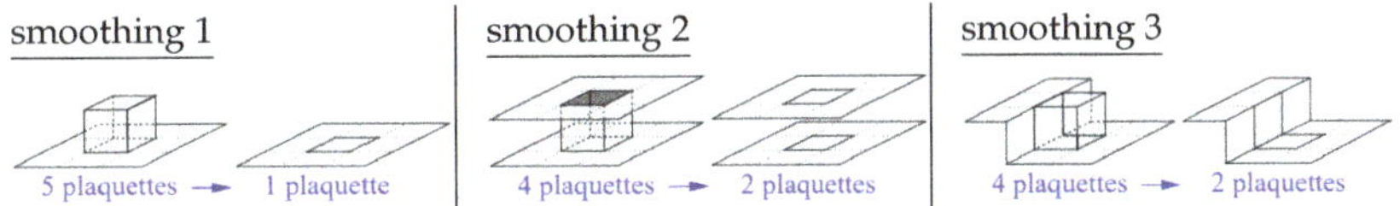

**Figure 4.** The effect of the smoothing procedures on the vortex surface is depicted, taken from ([19] Figure 5.8). We distinguish *warts* (**left**), *bottlenecks* (**middle**) and *stumbling blocks* (**right**). The *unit cubes* are not depicted, which are simply deleted.

The smoothing steps 1–3 cut out parts of the vortex surface and closes the emerging holes with a flat surface. In this way, short-range fluctuations of the vortex surface are suppressed. We first count the P-plaquettes without any smoothing performed, and then the loss of P-plaquettes for the respective smoothing steps is determined. The results are given in Table 1.

**Table 1.** Reduction of the total count of P-plaquettes for different smoothing procedures.

| P-plaquette Reduction | smoothing 0 | smoothing 1 | smoothing 2 | smoothing 3 |
|---|---|---|---|---|
| With fermions | 12.5% | 10.1% | 24% | 10.2% |
| Without fermions | 7% | 10.6% | 27.8% | 10.9% |

This quantifies the percentage of the respective structures depicted in Figure 4. When fermions are present, we clearly have a higher proportion of unit cubes and a lower proportion of bottlenecks than without fermions.

By restricting this analysis to the single percolating vortex cluster, we gain information about the long range excitations. The results are given in Table 2. The reduction in the proportion of bottlenecks is also seen here. The presence of fermions leads to a smoother surface of the percolating cluster.

**Table 2.** Reduction of P-plaquettes for the percolating vortex cluster for different smoothing procedures.

| Reduction within Cluster | smoothing 1 | smoothing 2 | smoothing 3 |
|---|---|---|---|
| With fermions | 8.6% | 24.5% | 8.8% |
| Without fermions | 9.6% | 28.1% | 10% |

*2.4. Potential Fits and Noise Handling*

When extracting the potential from Wilson loops, two effects have to be taken care of:

- for small areas, the loop averages are influenced by short range fluctuations; and
- with increasing area, the data suffer from statistical noise and soon the errors get larger than the signal.

An example for a 1-exponential fit to Wilson loops $\langle W(R,T)\rangle$ for given $R$ and $T \geq T_i$ (see Equation (2)) is shown in the left diagram of Figure 5. The dependence of this example on the initial $T = T_i$ is depicted in the right diagram.

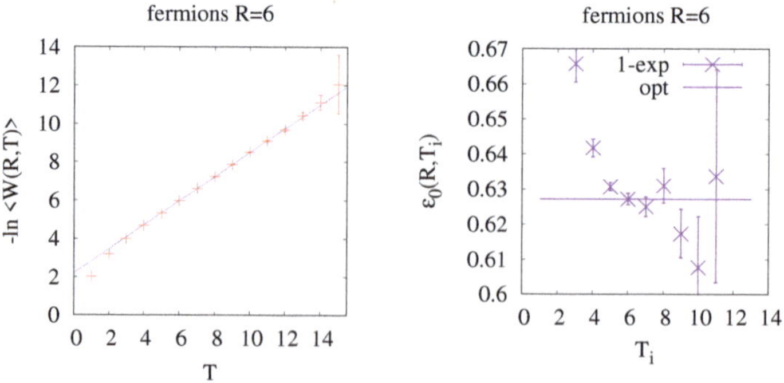


**Figure 5.** (**Left**) Example of the optimal 1-exponential fit of Wilson loops for given $R$. (**Right**) Dependence of $\varepsilon_0(R, T_i)$ on the fit region $T \geq T_i$. The line marks the fit for the optimal value for $T_i$.

At lower $T_i$, an increase of $T_i$ causes large changes of the fit parameters, but with growing $T_i$ these changes become smaller until a most stationary point is reached which may be hidden behind a strong increase of error bars. With the naked eye, one sees data that result in quite good fits, but finding analytic or numeric criteria for the choice of $T_i$ proves difficult. The smaller is the change of the values of the fit parameters, the smaller are the error bars of Wilson loops, and a rapid increase of the $p$-value of the fits often coincide, but this is not a general rule. Our criteria to choose $T_i$ is based on identifying the first local minimum of an error quantifier

$$\text{Err} := \frac{2}{3}\langle\Delta_{\delta i}\rangle + \frac{1}{3}\langle\Delta_{\text{err}}\rangle. \tag{12}$$

Here, $\langle\Delta_{\delta i}\rangle$ denotes the average change of the fit parameter $\varepsilon(R, T_{i\pm1})$ when decreasing or increasing $T_i$; and $\langle\Delta_{\text{err}}\rangle$ denotes the average over the error bars of $\varepsilon(R, T_{i-1})$, $\varepsilon(R, T_i)$, and $\varepsilon(R, T_{i+1})$. The weight factors are chosen to avoid the choice of occasionally nearly stationary regions with large error bars. For $R > 3$, we prevent any further increase of $T_i$, because with increasing $R$ the error bars start to grow earlier. The example in Figure 5 tries to convince that the selection of $T_i$ based on the error quantifier results in optimal fits under the boundary conditions of systematic deviations for low $T_i$ and increasing error bars for high $T_i$. Using this procedure, we determine the potential for the whole range of $R$-values, which allows extracting the slope of the potential at large values of $R$.

## 3. Summarized Results and Discussion

The fermionic back reaction on the string tension is clearly observed in the full theory as well as for EaMCG (Ensemble averaged Maximal Center Gauge), where the link variables of the gauge field are projected to $Z_2$. The potentials for the gluonic and fermionic configurations are depicted in Figure 6 and compare the full SU(2) theory with the $Z_2$ theory. The string tension was extracted by fitting the respective Equation (3) or (4) to the data describing the potential. The resulting parameters of these fits are given in Table 3. The relevant parameters to compare are $\sigma$ and $\sigma_{\text{CP}}$.

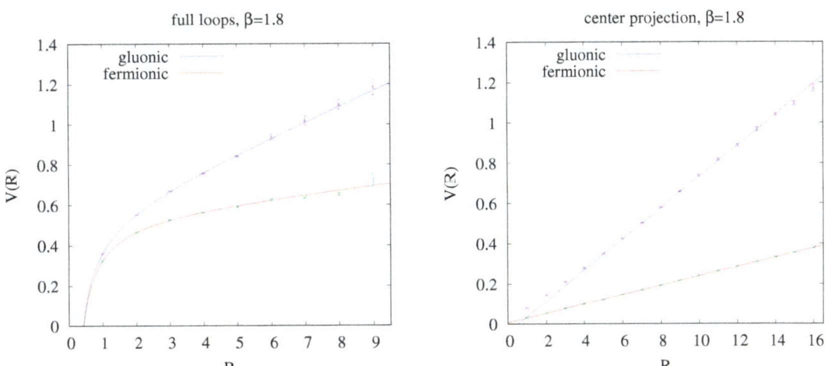


**Figure 6.** (**Left**) Potential $V(R)$ in lattice units between two sources in the fundamental representation. There is a large difference between the string tensions for pure gluonic configurations ("gluonic") and in the presence of one species of dynamical fermions. (**Right**) Potentials extracted from Wilson loops after ensemble averaged maximal center projection are depicted for pure gluonic configurations and for configurations with dynamical fermions. Due to the removal of short range fluctuations the potentials are in both cases almost linearly increasing with the lattice distance $R$. Data are fitted by linear functions. For gluonic (fermionic) configurations, only data with $R \geq 6(2)$ are fitted.

Without fermions, both estimates for $\sigma$ are compatible within errors to one another: In the full $SU(2)$, we observe $\sigma = 0.0756(12)$ compared to $\sigma_{\text{CP}} = 0.07691(13)$ in the $Z_2$ description. With fermions the full SU(2) theory results with $\sigma = 0.0199(9)$ a lower value than the $Z_2$ theory with $\sigma_{\text{CP}} = 0.02291(5)$. In all cases, we clearly observe that the presence of fermions reduces the string tension in lattice units: The back reaction is observed in the full SU(2) theory and also reproduced by the center vortices. The determination of the lattice spacing via the usual formula ($a = \sqrt{\chi}/2.23$ fm, corresponding to $\chi = (440\ \text{MeV})^2$) results in $0.123(1)$ fm for the gluonic configurations and $0.0633(15)$ fm for the fermionic configurations.

**Table 3.** The parameters for the fits according Equations (3) and (4) in Figure 6 allow a direct comparison of the respective string tensions. A strong suppression of the Coulomb part can be seen in the $Z_2$ theory.

| Theory | SU(2) | | | $Z_2$ | |
|---|---|---|---|---|---|
| **Parameter** | $V_0$ | $\sigma$ | $\alpha$ | $v_0$ | $\sigma_{\text{CP}}$ |
| gluonic | 0.5175(38) | 0.0756(12) | 0.2326(26) | −0.0366(8) | 0.07691(13) |
| fermionic | 0.5464(27) | 0.0199(9) | 0.2414(19) | 0.01027(13) | 0.02291(5) |

Concerning the geometric structure of the vortex surface, we observe that the presence of fermions increases the number of isolated short range fluctuations (see Table 1): without fermions, about 6.98% of the P-plaquettes are part of isolated unit cubes, whereas, with fermions, this proportion increases to 12.45%. The proportion of P-plaquettes belonging to bottlenecks is in total decreased from 27.81% to 24%. Fermions increase the amount of unit cubes, but decrease the amount of bottlenecks.

Restricting the analysis to the long-ranged cluster we observe a decrease of fluctuations, especially bottlenecks, when fermions are present (see Table 2): the proportion of P-plaquettes belonging to bottlenecks is reduced from 28.12% to 24.45%. All other fluctuations are only reduced by about 1%.

From this, we can conclude that the presence of fermions causes short range fluctuations to detach from the vortex surface, resulting in a more smooth vortex surface that is surrounded by an increased number of isolated short range fluctuations.

**Author Contributions:** Conceptualization, M.F. and R.H.; methodology, Z.D., M.F., R.H. and R.G.; validation, S.D.; investigation, Z.D. and R.G.; data curation, Z.D. and M.F.; writing—original draft preparation, Z.D. and R.G.; writing—review and editing, Z.D. and S.D.; visualization, R.G. and R.H. All authors have read and agreed to the published version of the manuscript.

**Funding:** This research received no external funding.

**Data Availability Statement:** The data presented in this study are available on request from the corresponding author.

**Acknowledgments:** We thank Aleksandr Nikolaev, Nikita Astrakhantsev and Andrey Kotov for their cooperation in the early stage of this investigation and Vitaly Bornyakov for important advice.

**Conflicts of Interest:** The authors declare no conflict of interest.

## References

1. Savvidy, G. Infrared Instability of the Vacuum State of Gauge Theories and Asymptotic Freedom. *Phys. Lett.* **1977**, *B71*, 133. [CrossRef]
2. Greensite, J. *An Introduction to the Confinement Problem*; Springer Nature: New York, NY, USA, 2020; p. 271.
3. Faber, M.; Höllwieser, R. Chiral symmetry breaking on the lattice. *Prog. Part. Nucl. Phys.* **2017**, *97*, 312–355. [CrossRef]
4. 't Hooft, G. On the phase transition towards permanent quark confinement. *Nucl. Phys. B* **1978**, *138*, 1–25. [CrossRef]
5. Cornwall, J.M. Quark confinement and vortices in massive gauge-invariant QCD. *Nucl. Phys. B* **1979**, *157*, 392–412. [CrossRef]
6. Del Debbio, L.; Faber, M.; Giedt, J.; Greensite, J.; Olejnik, S. Detection of center vortices in the lattice Yang-Mills vacuum. *Phys. Rev.* **1998**, *D58*, 094501. [CrossRef]
7. Bulava, J.; Hörz, B.; Knechtli, F.; Koch, V.; Moir, G.; Morningstar, C.; Peardon, M. String breaking by light and strange quarks in QCD. *Phys. Lett.* **2019**, *B793*, 493–498. [CrossRef]
8. Höllwieser, R.; Altarawneh, D. Center Vortices, Area Law and the Catenary Solution. *Int. J. Mod. Phys. A* **2015**, *30*, 1550207. [CrossRef]
9. Altarawneh, D.; Höllwieser, R.; Engelhardt, M. Confining Bond Rearrangement in the Random Center Vortex Model. *Phys. Rev. D* **2016**, *93*, 054007. [CrossRef]
10. Astrakhantsev, N.Y.; Bornyakov, V.G.; Braguta, V.V.; Ilgenfritz, E.M.; Kotov, A.Y.; Nikolaev, A.A.; Rothkopf, A. Lattice study of static quark-antiquark interactions in dense quark matter. *J. High Energy Phys.* **2019**, *2019*, 171. [CrossRef]

11. Weisz, P. Continuum limit improved lattice action for pure Yang-Mills theory (I). *Nucl. Phys. B* **1983**, *212*, 1–17. [CrossRef]
12. Curci, G.; Menotti, P.; Paffuti, G. Symanzik's improved lagrangian for lattice gauge theory. *Phys. Lett. B* **1983**, *130*, 205–208. [CrossRef]
13. Del Debbio, L.; Faber, M.; Greensite, J.; Olejnik, S. Center dominance, center vortices, and confinement. In Proceedings of the NATO Advanced Research Workshop on Theoretical Physics: New Developments in Quantum Field Theory, Zakopane, Poland, 14–20 June 1997; pp. 47–64.
14. Bornyakov, V.; Komarov, D.; Polikarpov, M. P-vortices and drama of Gribov copies. *Phys. Lett. B* **2001**, *497*, 151–158. [CrossRef]
15. Golubich, R.; Faber, M. The Road to Solving the Gribov Problem of the Center Vortex Model in Quantum Chromodynamics. *Acta Phys. Pol. B Proc. Suppl.* **2020**, *13*, 59–65. [CrossRef]
16. Golubich, R.; Faber, M. Center Regions as a Solution to the Gribov Problem of the Center Vortex Model. *Acta Phys. Pol. B Proc. Suppl.* **2021**, *14*, 87. [CrossRef]
17. Golubich, R.; Faber, M. Improving Center Vortex Detection by Usage of Center Regions as Guidance for the Direct Maximal Center Gauge. *Particles* **2019**, *2*, 30. [CrossRef]
18. Bertle, R.; Faber, M.; Greensite, J.; Olejník, S. The structure of projected center vortices in lattice gauge theory. *J. High Energy Phys.* **1999**, *64*, 019. [CrossRef]
19. Bertle, R. The Vortex Model in Lattice Quantum Chromo Dynamics. Ph.D. Thesis, Vienna University of Technology, Vienna, Austria, 2005.

*Article*

# A Possible Resolution to Troubles of SU(2) Center Vortex Detection in Smooth Lattice Configurations

**Rudolf Golubich * and Manfried Faber**

Atominstitut, Technische Universität Wien, 1040 Wien, Austria; faber@kph.tuwien.ac.at
* Correspondence: rudolf.golubich@gmail.com

**Abstract:** The *center vortex model* of quantum-chromodynamics can explain confinement and chiral symmetry breaking. We present a possible resolution for problems of the vortex detection in smooth configurations and discuss improvements for the detection of center vortices.

**Keywords:** quantum chromodynamics; confinement; center vortex model; vacuum structure; cooling

**PACS:** 11.15.Ha; 12.38.Gc

**Citation:** Golubich, R.; Faber, M. A Possible Resolution to Troubles of SU(2) Center Vortex Detection in Smooth Lattice Configurations. *Universe* **2021**, *7*, 122. https://doi.org/10.3390/universe7050122

Academic Editor: Dmitri Antonov

Received: 24 March 2021
Accepted: 25 April 2021
Published: 29 April 2021

**Publisher's Note:** MDPI stays neutral with regard to jurisdictional claims in published maps and institutional affiliations.



## 1. Introduction

The center vortex model assumes that the relevant excitations of the QCD vacuum are *center vortices*, closed color magnetic flux lines evolving in time. It can explain Confinement [1] and chiral symmetry breaking [2–4]. In four-dimensional space–time, the flux lines form closed surfaces in dual space, see Figure 1. In the low-temperature phase, they percolate space–time in all dimensions.

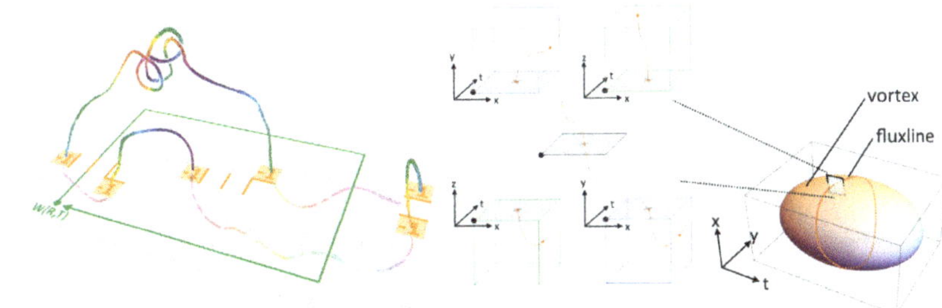

**Figure 1.** The geometric relation between piercings, the flux line and the vortex surface is schematically shown. Left: A flux line can be traced by following non-trivial plaquettes (depicted in orange with a "−1") after transformation to maximal center gauge and projection to the center degrees of freedom. Middle: Each non-trivial plaquette belongs to four elementary cubes, where the flux enters and has to leave through another plaquette. The depicted grey rectangles correspond to the same plaquette. For each cube, the three involved coordinates are indicated. Right: Due to the evolution in time, the flux line (depicted as orange line) forms a closed two-dimensional surface in four-dimensional spacetime.

Within lattice simulations, the center vortices are detected in *maximal center gauge* after projection to the center degrees of freedom. The procedure is described in more detail in Section 2. As long as the detected vortices reproduce the relevant physics, we speak of a valid *vortex finding property*. During the analysis of the color structure of vortices in smooth configurations [5] one is confronted with a loss of the vortex-finding property. Problems in detecting center vortices due to ambiguities in the gauge-fixing procedure were already found by Kovacs and Tomboulis [6]. They also point out that the thickness of vortices is of importance for the extraction of properties related to confinement. We found that this thickness can cause troubles in the vortex detection, resulting in a loss of the string tension. In search for improvements in the vortex detection, the cause of this loss is analyzed and a possible resolution discussed. We model the influence of cooling on the vortex thickness and the corresponding loss of the vortex density. An upper limit for the lattice spacing and a lower limit for the lattice size is presented. These limits are derived from measurements of the vortex density and estimates of the cross-section of flux tubes.

## 2. Materials and Methods

Our lattice simulations of the SU(2) Wilson action cover an interval of inverse coupling $\beta \in [2.1, 3.6]$ in steps of 0.05. We start with low $\beta$ values to identify discretization effects and to detect the onset of finite size effects. To check how far the compatibility of our model reaches, we expand the calculations to relatively large values of $\beta$. The lattice spacing $a$ corresponding to the respective values of $\beta$ is determined by assuming a physical string tension of $(440 \text{ MeV})^2$ via a cubic interpolation of the literature values given in Table 1. This is complemented by an extrapolation according to the asymptotic renormalization group equation for $\beta > 2.576$

$$a(\beta) = \Lambda^{-1} e^{-\frac{\beta}{8\beta_0}} \quad \text{with} \quad \beta_0 = \frac{11}{24\pi^2} \quad \text{and} \quad \Lambda = 0.015(2) \text{ fm}^{-1}, \tag{1}$$

with $\Lambda$ obtained by fitting this equation to the values of $a$ for $\beta \geq 2.6$ in Table 1.

**Table 1.** The indicted dependence of the lattice spacing $a$ in fm and the string tension $\sigma$ in lattice units on the inverse coupling $\beta$ is taken from references [7–11].

| $\beta$ | 2.3 | 2.4 | 2.5 | 2.635 | 2.74 | 2.85 |
|---|---|---|---|---|---|---|
| a [fm] | 0.165(1) | 0.1191(9) | 0.0837(4) | 0.05409(4) | 0.04078(9) | 0.0296(3) |
| $\sigma$ [lattice] | 0.136(2) | 0.071(1) | 0.0350(4) | 0.01459(2) | 0.00830(4) | 0.00438(8) |

The analysis is performed on lattices of size $8^4$ and $10^4$ with 0, 1, 2, 3, 5 and 10 Pisa-Cooling [12] steps with a cooling parameter of 0.05. We have chosen these small lattice sizes because, in bigger lattices, the finite-size effects are expected at higher values of $\beta$ and, as we will show, the detection of center vortices becomes increasingly difficult with rising values of $\beta$.

A central part of our analysis consists of identifying non-trivial center regions, regions whose is perimeter evaluated as close to non-trivial center elements, using the algorithms presented in references [13–15]. In the gauge-fixing procedure, we look for gauge matrices $\Omega$ that maximize the functional

$$R^2 = \sum_x \sum_\mu |\text{Tr}[\acute{U}_\mu(x)]|^2 \quad \text{with} \quad \acute{U}_\mu(x) = \Omega(x + e_\mu) U_\mu(x) \Omega^\dagger(x). \tag{2}$$

The non-trivial center regions are used to guide this procedure to prevent the problems found by Bornyakov et al. [16]. The detection of such non-trivial center regions is based on enlarging regions until they get as close as possible to non-trivial center elements. This quite calculation-intensive procedure is depicted in Figure 2.

As not all resulting regions are evaluated sufficiently near to a non-trivial center element, we take only those into account which are sufficiently near to non-trivial center elements.

During the gauge-fixing, only such gauge matrices $\Omega$ are allowed that preserve the sign of the non-trivial center regions. As this causes the rejection of some gauge matrices, the number of required simulated annealing steps until convergence of the gauge functional might increase.

 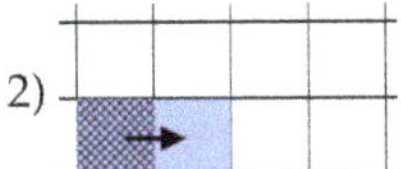 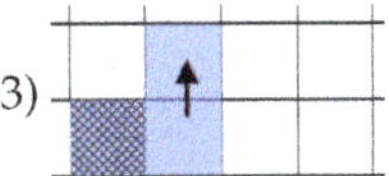

Steps 1-3: Starting with a plaquette that neither belongs to a before identified center region, nor was taken as origin for growing a region, it is tested, which enlargement by a neighboring plaquette brings the new region nearest to a center element.

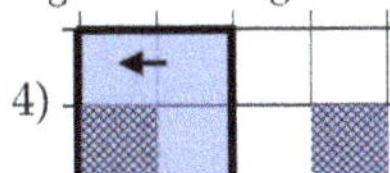 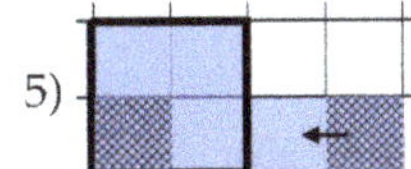 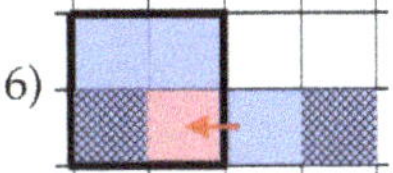

Steps 4-6: A new enlargement procedure is started with another plaquette if no enlargement leads to further improvement. During enlargement the new region could grow into an existing one. The following collision handling is used:

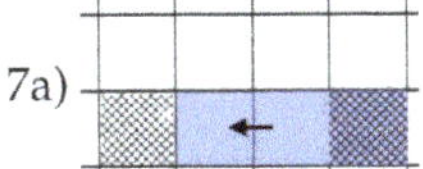 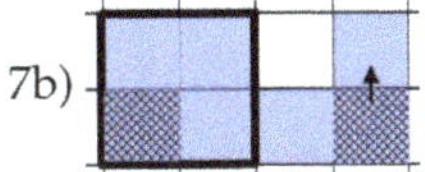

Step 7a: The evaluation of the growing region is nearer to a non-trivial center element than the evaluation of the old region: Delete the old region, only keeping the mark on its starting plaquette and allow growing.

Step 7b: The growing region deviates more from a non-trivial center element than the existing one: try other enlargements.

**Figure 2.** The non-trivial center regions, used for the gauge-fixing procedure, are detected by repeating the depicted procedure until every plaquette either belongs to an identified region or has already been used as the seed to grow a region. The direction of enlargement of the respective regions is marked by an arrow. Plaquettes that belong to a region are colored; plaquettes already used as seed are shaded. In the final determination of the non-trivial center regions enclosing a thick vortex, no collision handling is performed.

After gauge fixing and projection, plaquettes are identified that evaluate non-trivial center elements. These are dubbed *P-plaquettes* and considered to be pierced by a P-vortex.

If the number of P-plaquettes is smaller than the number of non-trivial center regions used to guide the gauge-fixing procedure, this is a clear indication of a failing vortex detection. For each value of $\beta$, the proportion of configurations where this is the case is determined. This allows to quantify the loss of the vortex-finding property besides quantifying it directly via the string tension of the center-projected configurations.

The further analysis is performed in the full SU(2) configurations. For each P-plaquette, a non-trivial center region that encloses the P-plaquette is identified. This center region is considered to be pierced by the thick vortex detected by the P-vortex. Figure 3 depicts the relation between P-vortices, thick vortices and the non-trivial center regions.

**Figure 3.** Two-dimensional slices through a four-dimensional lattice are depicted. The vortex detection starts as a best-fit procedure of P-Vortices to thick vortices, indicated by the first arrow. Then, starting from the detected P-plaquettes, non-trivial center regions are identified to reconstruct the thick vortex. These non-trivial center regions are, in general, not rectangular.

The cross-section of the flux building thethick vortex, $A_{\text{vort}}$, is measured by counting the plaquettes that build up the non-trivial center regions enclosing the corresponding thick vortex. In each configuration, we determine minimal, average and maximal cross-sections.

The string tension $\sigma$ is determined via Creutz ratios $\chi$ calculated in the center projected configurations

$$\sigma \approx \chi(R,T) = -\ln \frac{\langle W(R+1,T+1)\rangle \, \langle W(R,T)\rangle}{\langle W(R,T+1)\rangle \, \langle W(R+1,T)\rangle}, \tag{3}$$

with $R \times T$ Wilson loops $W(R,T)$. As the Coloumb-part of the potential is strongly suppressed after projecting to the center degrees of freedom, the linear part corresponding to a non-vanishing string tension is already reproduced with small loop sizes, as we saw in references [14,15,17]. Symmetric Creutz ratios are used and the average of $\chi(1,1)$ and $\chi(2,2)$ is taken to determine the string tension. Our study is based on the data generated in reference [5], where we did not save a sufficiently wide range of Wilson loop data.

Assuming independence of vortex piercings, the string tension can also be related to the vortex density $\varrho_{vort}$, the number of P-plaquettes per unit volume, via

$$\sigma \approx -\ln(1 - 2 \times \varrho_{\text{vort}}). \tag{4}$$

The requirement of uncorrelated piercing is only fulfilled if the vortex surface is strongly smoothed, otherwise this simple equation overestimates the string tension.

The working hypothesis is that the loss of the vortex-finding property, observed via a loss of the string tension, when cooling is applied, can be related to a thickening of the vortices. We will try to model the loss of the vortex density based on an analysis of the geometric structure of center vortices.

## 3. Results

The different measurements are performed for a lot of different values of $\beta$ and several cooling steps. So as not to overload the visualizations only a part of the intermediate results is depicted, showing only specific numbers of cooling steps and restricting to a smaller interval of $\beta$-values. Those parts of the data that are dominated by finite size effects are identified and excluded from the further analysis.

Starting with the quantification of the vortex-finding property presented in Figure 4, some troubles are brought to light. The proportion of configurations where fewer P-plaquettes have been identified than non-trivial center regions exist, rises rapidly when passing a specific value of $\beta$. This specific value depends on the lattice size and the number of cooling steps.

When reducing the lattice size or increasing the number of cooling steps, the loss of the vortex-finding property occurs at lowered values of $\beta$. The proportion depicted seems to saturate at about 30%, except for 10 cooling steps at a lattice of size $8^4$, where it reaches higher values. The fact that some non-trivial center regions have no corresponding P-plaquettes after gauge fixing and projection to the center degrees of freedom hints at a possible explanation for part of the lost string tension. The gauge functional given in Equation (2) is local in the sense that each gauge matrix $\Omega$ is solely based on the eight gluonic links connected to the specific lattice point. Farther distances than a single lattice spacing are not directly taken into account. In contrast, the detection of the non-trivial center regions is, in a sense, more physical as it is based solely on gauge-independent quantities, that is, the evaluation of arbitrary big Wilson loops. When detecting P-vortices in smooth configurations and high lattice resolutions, the center flux can be distributed over many link variables. Each of these links can evaluate arbitrarily close to the trivial center element, although a Wilson loop build by the links can evaluate arbitrarily near to the non-trivial center element. In such a scenario, a gauge-fixing procedure, only taking the vicinity of lattice points into account, will likely fail and result in an underestimated string tension.

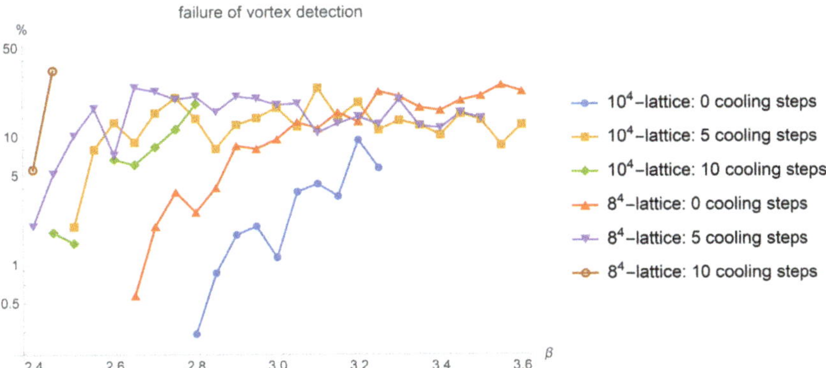



**Figure 4.** The proportion of configurations is depicted where less non-trivial plaquettes have been identified than non-trivial regions exist. The datapoints are joined to guide the eye. Due to the logarithmic scaling of axes, only non-vanishing values are depicted: all lines start with 0% at lower values of $\beta$. The interruption of the green line corresponding to the lattice of size $10^4$ at 10 cooling steps at $\beta = 2.55$ results from a vanishing percentage at the respective $\beta$-value. Observe that the curves rise at different values of $\beta$ for different number of cooling steps and different lattice sizes.

Looking at the Creutz ratios depicted in Figure 5 two possibly intertwined effects can be observed.

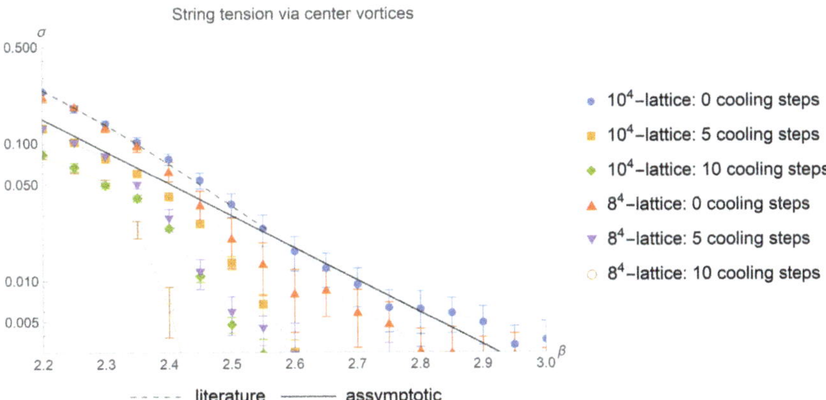



**Figure 5.** The string tension $\sigma$ is estimated via an average of the Creutz ratios $\chi(1,1)$ and $\chi(2,2)$ calculated in center-projected configurations for different numbers of cooling steps and lattice sizes. The datapoints are joined by lines to guide the eye. The literature values correspond to those listed in Table 1; the asymptotic line is given by Equation (1). Observe that, in the low $\beta$-regime, an underestimation of the string tension correlates to the number of cooling steps. This underestimation is independent of the lattice size. At higher values of $\beta$, finite size effects set in.

At sufficiently low values of $\beta$, the string tension is independent of the lattice size, but decreases with an increasing number of cooling steps. Of interest is that, for sufficiently small values of $\beta$, the deviation from the asymptotic prediction decreases with a rising value of $\beta$—for example, the $10^4$-lattice starts at 10 cooling steps with an underestimation of the asymptotic string tension of $50\% \pm 1\%$ at $\beta = 2.1$, improving to $40.8\% \pm 0.4\%$ at $\beta = 2.25$. At higher values of $\beta$, the independence from the lattice size no longer holds. For different lattice sizes, a sudden decrease in the string tension occurs at different values of $\beta$. The respective $\beta$-values are compatible for different numbers of cooling steps. The dependency on the lattice size and the independence on the number of cooling steps hint at finite size effects, but finite size effects do not give a direct explanation of the reduction in the string tension at lower values of $\beta$: We do not observe a dependency on the lattice size in the low $\beta$-regime. Based on the deviations of the string tensions for different lattice sizes, we expect finite size effects to occur at length scales around 1.3 fm, independent of cooling: observing that the lattice of size $8^4$ deviates from the $10^4$-lattice at $\beta \approx 2.3$, corresponding to a lattice spacing of $a \approx 0.165$, we acquire a physical lattice extend around 1.32 fm for the smaller lattice. The finite size effects on the bigger lattice set in at $\beta$ between approximately 2.35 and 2.4, resulting in a length scale between approximately 1.2 fm and 1.4 fm. This length scales are compatible with the findings of Kovacs and Tomboulis [18]. In Ref. [5] we also found color-homogeneous regions embedded in the vortex surface with roughly the same diameters. Similar distances can also be found between neighbouring piercings of a Wilson loop, extracted from the vortex density, as will be seen in Table 4.

A relation to the thickness $A_{\text{vort}}$ of center vortices is suspected and points towards possible further analysis. The possibility of a thick vortex expanding due to a spreading of the center flux was already suggested by Kovacs and Tomboulis in [19]. Assuming a circular cross-section of the flux tube, its diameter can be calculated as

$$d_{\text{flux}} = 2 \times \underbrace{\sqrt{\frac{A_{\text{vort}}}{\pi}}}_{r_{\text{flux}}}, \tag{5}$$

with $A_{\text{vort}}$ being the area of the flux cross-section. That flux lines are closed requires that within each two-dimensional slice through the lattice at least two vortex piercings can find place. This give a criteria on the lattice extent $L$

$$L > 2 * d_{\text{flux}}. \tag{6}$$

If $A_{\text{vort}}$ measured by a plaquette count exceeds 19 for a lattice of size $10^4$, or 12 for a lattice of size $8^4$, we can expect finite size effects to step in. These thresholds are of relevance for the average, minimal and maximal flux tube cross-section depicted in Figures 6–8. The mean flux tube cross-section presented in Figure 6 shows that we have to restrict to relative low values of $\beta$ to stay away from finite size effects.

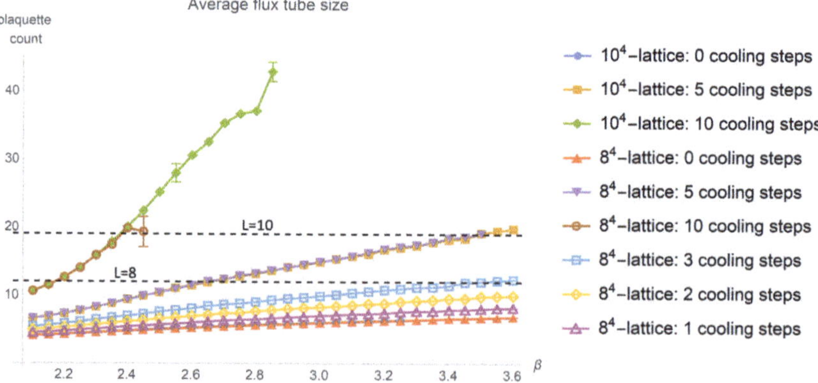


**Figure 6.** The average cross-sections of the flux tubes, measured by counting plaquettes, increases when cooling is applied. It reaches a threshold at which finite size effects are expected to become problematic, shown as a dashed line for the two lattice sizes. Measurements performed on lattices of different size have good compatibility.

Taking a look at the maximal flux tube cross-section depicted in Figure 7, we can expect finite size effects at even lower values of $\beta$: None of the data with 10 cooling steps can be expected to be free of finite size effects.

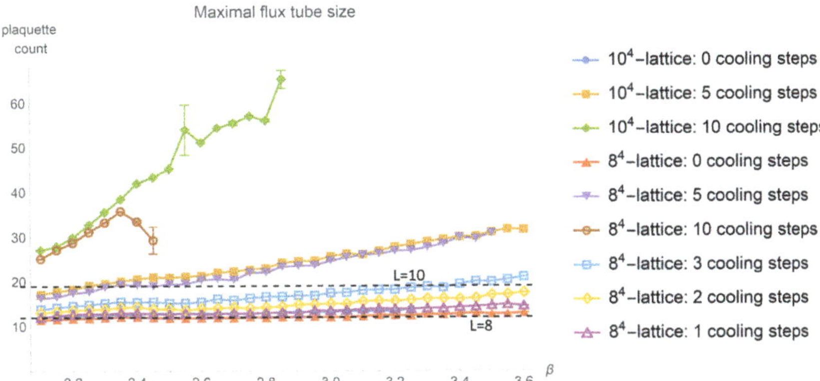

**Figure 7.** The maximal cross-sections of the flux tubes hint at finite size effects. Within our $\beta$-interval, only the lattice of size $10^4$ stays below the threshold when cooling is applied. With cooling, the different lattice sizes become more and more incompatible.

The lattice of size $8^4$ could be too small even without any cooling applied. With cooling and increasing $\beta$ the different lattice sizes become more and more incompatible. This may be caused by finite size effects and insufficient statistics. Still, the overall behaviour with cooling is qualitatively reproduced and allows the gain of another estimate on the growth rate.

Looking at the minimal tube size depicted in Figure 8 an even more sudden rise in the cross-section can be observed.

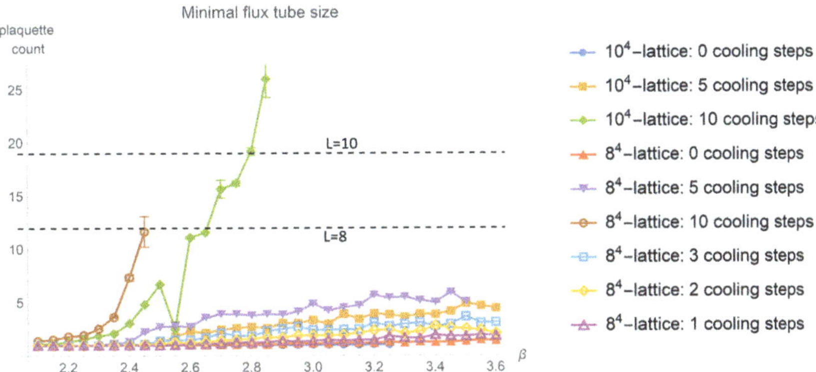

**Figure 8.** The minimal size of the flux tubes cross-sections shows a strong dependency on the lattice size. This dependency becomes even stronger when cooling is applied. Only with, at most, three cooling steps applied, the data seem thrust-worthy for $\beta < 2.3$.

We expect that the minimal flux tube cross-sections starts to grow with a certain $\beta$, where the high action density of non-trivial plaquettes leads to a suppression within the path integral. This causes a dependency on the lattice size due to the reduced statistics. For sufficiently low values of $\beta$ and sufficiently low numbers of cooling steps, the minimal flux tube cross-section is given by exactly one plaquette, independent of $\beta$ and the number of cooling steps. We restrict further analysis to $10^4$-lattices with $\beta \leq 2.3$ and, at most, five cooling steps. Nevertheless, we depict the full data in all relevant figures to allow for a check of the plausibility of our model by looking at the specific deviations of the data from our prediction.

Assuming an exponential growth in the flux tubes' cross-section with an increase in the number of cooling steps, a model of the form

$$A_{\text{vort}}(N_{\text{cool}}) = A_{\text{vort}}(0)\, e^{N_{\text{cool}}\,(g_{\text{cool}} + g_{\text{discret}}\, a)} \tag{7}$$

is fit to the data, with $N_{\text{cool}}$ being the number of cooling steps and $a$ the lattice spacing. The fit-parameter $g_{\text{cool}}$ corresponds to the exponential growth in the flux tube with cooling. As the tube size is measured by counting plaquettes, we have to account for discretization effects. This is done by adding another fit-parameter $g_{\text{discret}}$ in the exponent, related to the lattice spacing and the number of cooling steps. The two parameters are not necessarily constant as they can depend on the specific structure of interest. We restrain from carrying along another index: In the following, the values of these two parameters are to be considered only with respect to the specific context. They differ for the average cross-sections and the maximal cross-sections of flux tubes. The fit of this model to the average flux tube sizes is shown in Figure 9 in physical units. The fit is done for small $\beta$ and cooling steps indicated by black points.

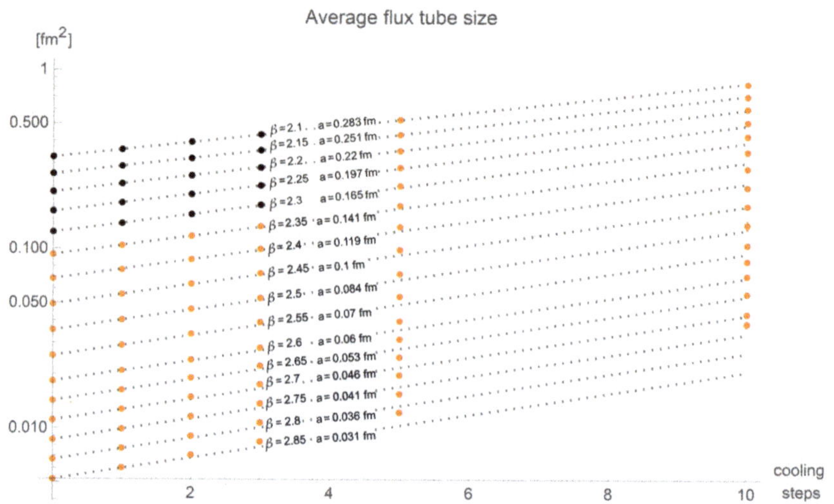


**Figure 9.** The measured data of the average flux tube cross-section for various numbers of cooling steps and several $\beta$ are shown by black and orange points. The dashed lines depict the fits according to Equation (7), where only the black datapoints were used. The corresponding fit parameters are given in Table 2. Deviations of the data from the fits can be related to finite size effects.

The fit, dashed lines reproduce the data well until the expected onset of finite size effects for cross-sections, increasing with the number of cooling steps and $\beta$. This onset is compatible to the estimates in Equation (6) and will be discussed later. At present, we concentrate on the growth in the flux tube cross-section described by the fit parameters given in Table 2. The suspected exponential growth of $A_{\text{vort}}$ is confirmed by the good quality of the fit for positive $g_{\text{cool}}$, even for larger values of $\beta$ and cooling steps.

**Table 2.** The parameters of the model described by Equation (7) and depicted in Figure 9 for average cross-sections are shown.

| Average Cross-Sections | Estimate | t-Statistic | p-Value |
| --- | --- | --- | --- |
| $g_{\text{cool}}$ | 0.14(1) | 13.6393 | $6.3 \times 10^{-11}$ |
| $g_{\text{discret}}$ | $-0.17(5)$ fm$^{-1}$ | $-3.62376$ | $1.9 \times 10^{-3}$ |

The negative value of $g_\text{discret}$ reflects the decreasing slope of the dashed lines with increasing $\beta$, indicating an influence of the lattice resolution: a coarser lattice reduces the growth of $A_\text{vort}$. The overall behaviour of $A_\text{vort}$ is qualitatively reproduced by the maximal cross-sections, as depicted in Figure 10.

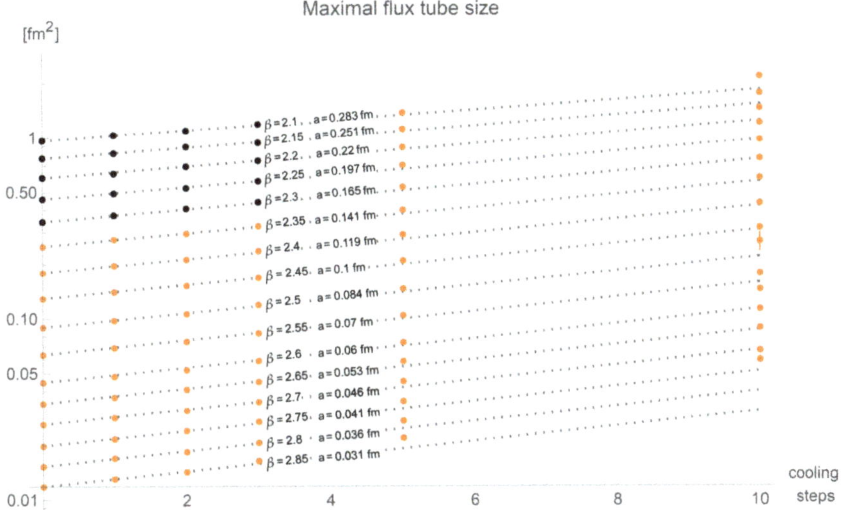


**Figure 10.** The measured data of the maximal flux tube cross-section for various numbers of cooling steps and several $\beta$ are shown by black and orange points. The dashed lines depict the fits according to Equation (7), where only the black datapoints were used. The corresponding fit parameters are given in Table 3. Deviations of the data from the fits can be related to finite size effects.

Only the growth has slowed down, as can be seen in the values given in Table 3.

**Table 3.** The parameters of the model described by Equation (7) and depicted in Figure 10 for maximal cross-sections are shown.

| Maximal Cross-Sections | Estimate | t-Statistic | p-Value |
| --- | --- | --- | --- |
| $g_\text{cool}$ | 0.0999(10) | 9.1369 | $3.5 \times 10^{-8}$ |
| $g_\text{discret}$ | $-0.13(5)$ fm$^{-1}$ | $-2.61939$ | $1.7 \times 10^{-2}$ |

This implies that the growth in $A_\text{vort}$ with increased cooling is limited.

A further influence of cooling is a smoothing of the vortex surface. We will now model this smoothing and show that the vortex flux tubes can be thickened without pushing each other apart. The vortex density $\varrho_\text{vort}$ allows to gain information about the distance of the vortex centers. Here, we have to take into account that some of the P-plaquettes belong to correlated piercings and can be attributed to short-range fluctuations. We define the quantity $A_\text{max}$ as the non-overlapping area around vortex centers.

The vortex density $\varrho_\text{vort}$ is usually calculated by dividing the number P-plaquettes by the total plaquette number. Given enough statistics, it can be determined by counting the number of piercings $N_\text{vort}$ within a sufficiently large Wilson loop of Area $A_\text{loop}$ build by $N_\text{loop}$ plaquettes

$$\varrho_\text{vort} = \frac{N_\text{vort}}{N_\text{loop}} = \frac{N_\text{vort}}{A_\text{loop} * a^{-2}} = \frac{N_\text{vort}}{(A_\text{free} + N_\text{vort} * A_\text{max}) * a^{-2}}. \tag{8}$$

In the last identity, we have split the area of the loop into two non-overlapping parts: each piercing is enclosed by circular area given by $A_{\mathrm{max}}$ and $A_{\mathrm{free}}$ covers the remaining part of the loop. When cooling is applied, we have to take into account that $A_{\mathrm{max}}$ grows.

$$\varrho_{\mathrm{vort}}(N_{\mathrm{cool}}) = \frac{N_{\mathrm{vort}}}{(A_{\mathrm{free}} + N_{\mathrm{vort}} * (A_{\mathrm{max}}(0) + \delta A_{\mathrm{max}}(N_{\mathrm{cool}}))) * a^{-2}}. \tag{9}$$

Using $A_{\mathrm{loop}} = A_{\mathrm{free}} + N_{\mathrm{vort}} * A_{\mathrm{max}}(0)$ and a model of the form given in Equation (7) for $A_{\mathrm{max}}(N_{\mathrm{cool}})$ we attain $\delta A_{\mathrm{max}} = A_{\mathrm{max}}(0)(e^{N_{\mathrm{cool}}\,(g_{\mathrm{cool}}+g_{\mathrm{discrete}}\,a)} - 1)$. It follows

$$\varrho_{\mathrm{vort}}(N_{\mathrm{cool}}) = \frac{\varrho_{\mathrm{vort}}(0)}{1 + \varrho_{\mathrm{vort}}(0)\,A_{\mathrm{max}}(0)a^{-2}\,(e^{N_{\mathrm{cool}}\,(g_{\mathrm{cool}}+g_{\mathrm{discrete}}\,a)} - 1)}. \tag{10}$$

We fit $g_{\mathrm{cool}}$, $g_{\mathrm{discrete}}$ and $A_{\mathrm{max}}(0)$ to the measurements of $\varrho_{\mathrm{vort}}$. The measured data and the fit are shown in Figure 11.

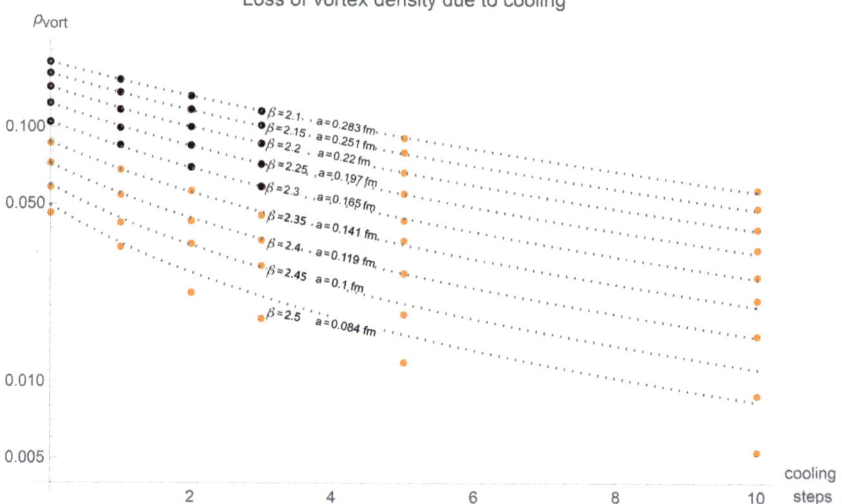


**Figure 11.** The vortex density is depicted for different values of $\beta$ and different numbers of cooling steps. For the model prediction, shown as dashed lines, only the black datapoints were used. That the datapoints fall below the model prediction at specific numbers of cooling steps for different values of $\beta$ can be explained by finite size effects. The corresponding parameters of the model are given in Table 4.

The respective fit parameters are listed in Table 4.

**Table 4.** The parameters of the model described by Equation (10) for the loss of the vortex density during cooling.

| Vortex Density | Estimate | t-Statistic | p-Value |
| --- | --- | --- | --- |
| $g_{\mathrm{cool}}$ | 0.035(1) | 26.5368 | $2.8 \times 10^{-15}$ |
| $g_{\mathrm{discret}}$ | 0.066(2) fm$^{-1}$ | 27.6254 fm$^{-1}$ | $1.5 \times 10^{-15}$ |
| $A_{\mathrm{max}}(0)$ | 1.41(5) fm$^2$ | 25.8937 fm$^2$ | $4.2 \times 10^{-15}$ |

The value of $A_{\mathrm{max}}(0)$ is larger than the flux tube cross-sections depicted in Figure 9. This, and the fact that the value of $g_{\mathrm{cool}}$, for the vortex density is smaller than those of the vortex flux tube cross-sections indicate that the majority of piercings remain separated from one another even when cooling is applied. Assuming circular geometry, we can calculate the minimal possible distance between vortex centers

$$d_{\text{center}}(N_{\text{cool}}) = 2\sqrt{\frac{A_{\max}(N_{\text{cool}})}{\pi}}. \tag{11}$$

To determine how many cooling steps are possible, we need to know how much the vortices can grow by cooling without getting into conflict. We estimate the minimal available separation by

$$s_{\text{flux}}(N_{\text{cool}}) = \underbrace{2\sqrt{\frac{A_{\max}(0)}{\pi}}}_{d_{\text{center}}(0)} - \underbrace{2\sqrt{\frac{A_{\text{vort}}(N_{\text{cool}})}{\pi}}}_{d_{\text{flux}}(N_{\text{cool}})}. \tag{12}$$

We use $d_{\text{center}}(0)$, the average distance between piercings when no loss of the vortex density occurred, and subtract the average diameter of the flux tubes $d_{\text{flux}}(N_{\text{cool}})$ with cooling applied. If $s_{\text{flux}}(N_{\text{cool}})$ becomes smaller than one lattice spacing, our methods of center vortex detection are likely to fail: we can no longer find two non-overlapping non-trivial center regions enclosing the thick vortex flux tubes. This allows for a limit for the lattice spacing to be derived, $a$, given in Equation (13) together with a limit on $L$ based on Equation (6)

$$a < s_{\text{flux}} \quad \text{and} \quad L > \text{Max}(2\overline{d_{\text{flux}}}, \text{Max}(d_{\text{flux}})). \tag{13}$$

The requirement for the lattice extent $L$ is based on the fact that two vortex piercings have to fit in every two-dimensional slicing through the lattice. Assuming a vanishing minimal flux tube size, the limit is given either by two times the average diameter $\overline{d_{\text{flux}}}$ or one times the maximal diameter $\text{Max}(d_{\text{flux}})$—whatever is bigger. The assumption of a vanishing minimal flux tube size is an approximation: on the lattice, the minimal size is given by exactly one plaquette, which is normally negligible in comparison to the lattice extent.

Using what we learned so far, we can evaluate these inequalities and find numerical values for the upper limit of $a$ and the lower limit of $L$. These are depicted in Figure 12 and will now be discussed. Discretization effects are neglected by setting $g_{\text{discret}} = 0$. Fitting the average flux tube cross-sections for configurations without cooling for $2.1 \leq \beta \leq 2.3$, see Figure 9, by a polynom up to quadratic order with respect to the lattice spacing $a$ gives

$$\overline{A_{\text{vort}}}(0) \approx 3.367(38)\ a^2 + 0.200(9)\ \text{fm}\ a, \tag{14}$$

compatible with the values we found in [20]. A fit to the maximal cross-sections without cooling for $2.1 \leq \beta \leq 2.3$, see Figure 10, results in higher fit parameters

$$Max(A_{\text{vort}}(0)) \approx 11.3(2)\ a^2 + 0.224(37)\ \text{fm}\ a. \tag{15}$$

Using this fit and Equation (12) with $A_{\max}(0)$ from Table 4 we obtain an upper limit for the lattice spacing that depends on the number of cooling steps and $g_{\text{cool}}$. With this limit, we can determine a lower limit for the required lattice extent. Both limits are shown in Figure 12 for the two different values of $g_{\text{cool}}$ resulting from average and maximal flux tube sizes from Tables 2 and 3. Let us remember how these limits were derived. Closed flux lines require sufficient room for two piercings within each two dimensional slice through the lattice—a lower limit for the lattice extent arises.

Taking the stronger limits with $g_{\text{cool}} = 0.14$, we determine the corresponding limits of $\beta$ for given lattice size and number of cooling steps. In Table 5 some numerical values are shown.

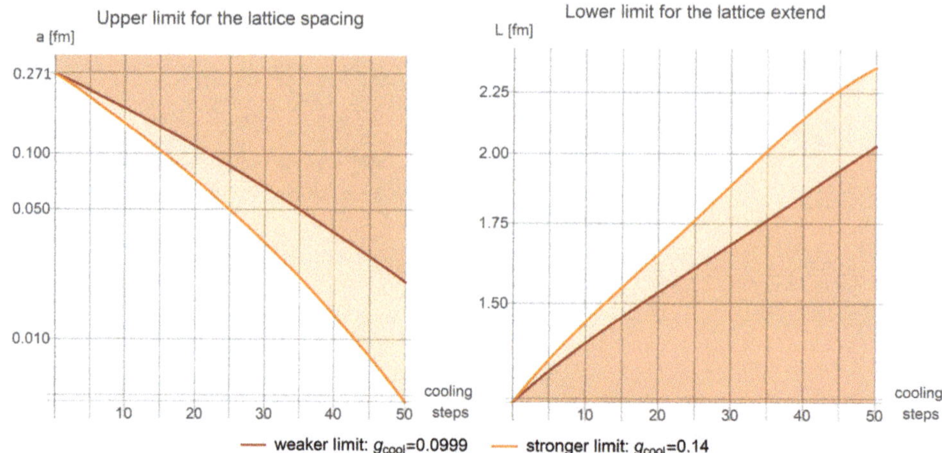



**Figure 12.** Based on the growth in the flux tubes and the reduction in the vortex density in dependency of the number of cooling steps, an upper limit for the lattice spacing (left) and a lower limit for the lattice extent (right) can be derived, as given in Equation (13). The weaker limit depicted in red is based on the slower growth in the maximal sized flux tubes with $g_{cool} = 0.0999$ (see Table 3), the stronger limit, depicted in orange, is based on the faster growth in average-sized flux tubes with $g_{cool} = 0.14$ (see Table 2).

**Table 5.** For different numbers of cooling steps and different lattice extents, the table gives a lower and an upper limit for $\beta$. "None" indicates that the limits exclude one another.

| $N_{cool} \setminus L$ | 8 | 10 | 14 | 20 | 30 | 40 | 50 |
|---|---|---|---|---|---|---|---|
| 0 | 2.12<br>2.32 | 2.12<br>2.39 | 2.12<br>2.48 | 2.12<br>2.58 | 2.12<br>2.73 | 2.12<br>2.84 | 2.12<br>2.92 |
| 1 | 2.14<br>2.31 | 2.14<br>2.38 | 2.14<br>2.48 | 2.14<br>2.58 | 2.14<br>2.73 | 2.14<br>2.83 | 2.14<br>2.91 |
| 2 | 2.16<br>2.31 | 2.16<br>2.38 | 2.16<br>2.47 | 2.16<br>2.58 | 2.16<br>2.72 | 2.16<br>2.83 | 2.16<br>2.91 |
| 3 | 2.19<br>2.3 | 2.19<br>2.37 | 2.19<br>2.47 | 2.19<br>2.57 | 2.19<br>2.71 | 2.19<br>2.82 | 2.19<br>2.9 |
| 5 | 2.23<br>2.29 | 2.23<br>2.36 | 2.23<br>2.46 | 2.23<br>2.56 | 2.23<br>2.7 | 2.23<br>2.81 | 2.23<br>2.89 |
| 10 | None | 2.34<br>2.34 | 2.34<br>2.44 | 2.34<br>2.54 | 2.34<br>2.67 | 2.34<br>2.78 | 2.34<br>2.86 |
| 15 | None | None | None | 2.44<br>2.52 | 2.44<br>2.65 | 2.44<br>2.76 | 2.44<br>2.84 |
| 20 | None | None | None | None | 2.54<br>2.63 | 2.54<br>2.73 | 2.54<br>2.82 |
| 25 | None | None | None | None | None | 2.66<br>2.71 | 2.66<br>2.79 |

We now look at the meaning of these limits for the string tension. In Figure 5, we observe that the deviation from the asymptotic prediction decreases with increasing $\beta$ in the low $\beta$-regime. We believe that this behaviour holds within the $\beta$-intervals of Table 5. The upper limit of $\beta$ can be extended by increasing the lattice size. It would be interesting to see if this alone suffices to restore full compatibility with the asymptotic string tension with modest cooling, but the required computational power might exceed our present capabilities.

## 4. Discussion

Using non-trivial center regions we analyzed how Pisa-cooling influences the cross-sections of thick center vortices. We found an exponential growth that slows down with increasing cross-sections. By geometric arguments, we derived an upper limit for the lattice spacing above which discretization effects trouble the vortex detection and a lower limit for the lattice extent where finite size effects set in. This window gets smaller with cooling and decreasing lattice extent. Cooling results in deviations from the asymptotic behaviour: an underestimation of the string tension occurs. Within the window, increasing $\beta$ leads to better agreement with the asymptotic behaviour. It would be interesting to see whether the string tension calculated on the projected lattice is, in fact, fully restored with sufficiently large $\beta$ or if only a partial restoration occurs.

By improving the method of center vortex detection, it might be possible to soften the aforementioned limits. The method of vortex detection used in this work was based on the direct maximal center gauge guided by non-trivial center regions [13–15]: we identify regions whose perimeter evaluates to the non-trivial center element and preserve their evaluation during gauge-fixing and center projection. This approach comes with three possibilities of improvement.

The growth in the flux tube due to cooling results in the non-trivial center factors within the evaluation of Wilson loops being spread over more and more links. In the original direct maximal center gauge the contribution to the gauge functional at a given site x is determined by its attached links only. By taking farther links into account, the troubles arising from the spread of the center flux may be counteracted.

When two thick vortices are not separated by at least one lattice spacing, the identification of non-trivial center regions enclosing the single piercings might fail. The original method used for the detection of non-trivial center regions is based on enlarging the perimeter of Wilson loops while preventing overlaps of the resulting regions: if overlaps occurred, the region that evaluates to a higher trace is deleted. By allowing overlaps, an improvement might be possible: more non-trivial center regions are kept to guide the further gauge-fixing procedure.

With rising number of cooling steps, more non-trivial center regions than P-plaquettes were found: The direct maximal center gauge failed to preserve some of the non-trivial center regions. This could be counteracted by inserting non-trivial factors before starting the simulated annealing procedure used to maximize the gauge functional. These non-trivial factors should guarantee that each non-trivial center region evaluates to the non-trivial center element when evaluated in the center projected configuration.

**Author Contributions:** Conceptualization, R.G.; methodology, R.G.; software, R.G. and M.F.; writing—original draft preparation, R.G.; writing—review and editing, R.G. and M.F.; supervision, M.F.; project administration, M.F.; All authors have read and agreed to the published version of the manuscript.

**Funding:** This research received no external funding.

**Institutional Review Board Statement:** Not applicable.

**Informed Consent Statement:** Not applicable.

**Data Availability Statement:** The data presented in this study are available on request from the corresponding author.

**Acknowledgments:** We thank the company *Huemer-Group* (www.huemer-group.com, accessed on 3 March 2021) and Dominik Theuerkauf for providing the computational resources speeding up our calculations.

**Conflicts of Interest:** The authors declare no conflict of interest.

# References


1. Del Debbio, L.; Faber, M.; Giedt, J.; Greensite, J.; Olejnik, S. Detection of center vortices in the lattice Yang-Mills vacuum. *Phys. Rev.* **1998**, *D58*, 094501. [CrossRef]
2. Faber, M.; Höllwieser, R. Chiral symmetry breaking on the lattice. *Prog. Part. Nucl. Phys.* **2017**, *97*, 312–355. [CrossRef]
3. Höllwieser, R.; Schweigler, T.; Faber, M.; Heller, U.M. Center Vortices and Chiral Symmetry Breaking in SU(2) Lattice Gauge Theory. *Phys. Rev.* **2013**, *D88*, 114505. [CrossRef]
4. Höllwieser, R.; Schweigler, T.; Faber, M.; Heller, U.M. Center vortices and topological charge. In Proceedings of the Xth Quark Confinement and the Hadron Spectrum—PoS(Confinement X), Munich, Germany, 21 May 2013. [CrossRef]
5. Golubich, R.; Golubich, R.; Faber, M. Properties of SU(2) Center Vortex Structure in Smooth Configurations. *Particles* **2021**, *4*, 93–105. [CrossRef]
6. Kovacs, T.G.; Tomboulis, E.T. On P vortices and the Gribov problem. *Phys. Lett. B* **1999**, *463*, 104–108. [CrossRef]
7. Bali, G.S.; Schlichter, C.; Schilling, K. Observing long color flux tubes in SU(2) lattice gauge theory. *Phys. Rev. D* **1995**, *51*, 5165–5198. [CrossRef] [PubMed]
8. Booth, S.P.; Hulsebos, A.; Irving, A.C.; McKerrell, A.; Michael, C.; Spencer, P.S.; Stephenson, P.W. SU(2) potentials from large lattices. *Nucl. Phys.* **1993**, *B394*, 509–526. [CrossRef]
9. Michael, C.; Teper, M. Towards the Continuum Limit of SU(2) Lattice Gauge Theory. *Phys. Lett.* **1987**, *B199*, 95–100. [CrossRef]
10. Perantonis, S.; Huntley, A.; Michael, C. Static Potentials From Pure SU(2) Lattice Gauge Theory. *Nucl. Phys.* **1989**, *B326*, 544–556. [CrossRef]
11. Bali, G.S.; Fingberg, J.; Heller, U.M.; Karsch, F.; Schilling, K. The Spatial string tension in the deconfined phase of the (3+1)-dimensional SU(2) gauge theory. *Phys. Rev. Lett.* **1993**, *71*, 3059–3062. [CrossRef] [PubMed]
12. Campostrini, M.; Di Giacomo, A.; Maggiore, M.; Panagopoulos, H.; Vicari, E. Cooling and the String Tension in Lattice Gauge Theories. *Phys. Lett. B* **1989**, *225*, 403–406. [CrossRef]
13. Golubich, R.; Faber, M. The Road to Solving the Gribov Problem of the Center Vortex Model in Quantum Chromodynamics. *Acta Phys. Pol. B Proc. Suppl.* **2020**, *13*, 59–65. [CrossRef]
14. Golubich, R.; Faber, M. Center Regions as a Solution to the Gribov Problem of the Center Vortex Model. *Acta Phys. Pol. B Proc. Suppl.* **2021**, *14*, 87. [CrossRef]
15. Golubich, R.; Faber, M. Improving Center Vortex Detection by Usage of Center Regions as Guidance for the Direct Maximal Center Gauge. *Particles* **2019**, *2*, 491–498. [CrossRef]
16. Bornyakov, V.G.; Komarov, D.A.; Polikarpov, M.I.; Veselov, A.I. P vortices, nexuses and effects of Gribov copies in the center gauges. Quantum chromodynamics and color confinement. In Proceedings of the International Symposium, Confinement 2000, Osaka, Japan, 7–10 March 2000; pp. 133–140.
17. Dehghan, Z.; Deldar, S.; Faber, M.; Golubich, R.; Höllwieser, R. Influence of Fermions on Vortices in SU(2)-QCD. *Preprints* **2021**, 2021040233. [CrossRef]
18. Kovacs, T.G.; Tomboulis, E.T. Vortex structure of the vacuum and confinement. *Nucl. Phys. B Proc. Suppl.* **2001**, *94*, 518–521. [CrossRef]
19. Kovács, T.G.; Tomboulis, E. Bound on the string tension by the excitation probability for a vortex. *Nuclear Phys. B Proc. Suppl.* **2000**, *83–84*, 553–555. [CrossRef]
20. Golubich, R.; Faber, M. Thickness and Color Structure of Center Vortices in Gluonic SU(2) QCD. *Particles* **2020**, *3*, 444–455. [CrossRef]

*Review*

# From Center-Vortex Ensembles to the Confining Flux Tube

**David R. Junior** [1,2,*,†], **Luis E. Oxman** [1,†] **and Gustavo M. Simões** [1,†]

1 Instituto de Física, Universidade Federal Fluminense, Avenida Litorânea, s/n, Niterói 24210-340, Brazil; leoxman@id.uff.br (L.E.O.) ; gustavomoreirasimoes@id.uff.br (G.M.S.)
2 Institut für Theoretische Physik, Universität Tübingen, Auf der Morgenstelle 14, 72076 Tübingen, Germany
* Correspondence: davidjunior@id.uff.br
† These authors contributed equally to this work.

**Abstract:** In this review, we discuss the present status of the description of confining flux tubes in SU(N) pure Yang–Mills theory in terms of ensembles of percolating center vortices. This is based on three main pillars: modeling in the continuum the ensemble components detected in the lattice, the derivation of effective field representations, and contrasting the associated properties with Monte Carlo lattice results. The integration of the present knowledge about these points is essential to get closer to a unified physical picture for confinement. Here, we shall emphasize the last advances, which point to the importance of including the non-oriented center-vortex component and non-Abelian degrees of freedom when modeling the center-vortex ensemble measure. These inputs are responsible for the emergence of topological solitons and the possibility of accommodating the asymptotic scaling properties of the confining string tension.

**Keywords:** confinement; ensembles and effective fields; topological solitons

**Citation:** Junior, D.R.; Oxman, L.E.; Simões, G.M. From Center-Vortex Ensembles to the Confining Flux Tube. *Universe* **2021**, *7*, 253. https://doi.org/10.3390/universe7080253

Academic Editor: Dmitri Antonov

Received: 7 June 2021
Accepted: 15 July 2021
Published: 21 July 2021

**Publisher's Note:** MDPI stays neutral with regard to jurisdictional claims in published maps and institutional affiliations.

## 1. Introduction

Our knowledge about the elementary particles, as well as three of the four known fundamental interactions, is successfully described by the standard model of particle physics. In particular, the quantitative behavior of the electromagnetic, weak, and strong interactions is encoded in the common language of gauge theories. In the strong sector, an important and intriguing phenomenon regarding the possible asymptotic particle states takes place. When quarks and gluons are created in a collision, they cannot move apart. Instead, they give rise to jets of colorless particles (hadrons) formed by confined quark and gluon degrees of freedom. Although confinement is key for the existence of protons and neutrons, a first-principles understanding of the mechanism underlying this phenomenon is still lacking. At high energies, the detailed scattering properties between quarks and gluons are successfully reproduced by QCD perturbative calculations in the continuum, which are possible thanks to asymptotic freedom. This is in contrast with the status at low-energies, where the validity of quantum chromodynamics (QCD) is well-established from computer simulations of the hadron spectrum which successfully make contact with the observed masses. This review focuses on this type of non-perturbative problem in pure $SU(N)$ Yang–Mills (YM) theory, which is a challenging open problem in contemporary physics. Here again, Monte Carlo simulations provide a direct way to deal with the large quantum fluctuations and compute averages of observables such as the Wilson loop, which is an order parameter for confinement in pure YM theories. As usual, the lattice calculations, as well as the center-vortex ensembles we shall discuss, consider an Euclidean (3d or 4d) spacetime. Unless explicitly stated, this is the metric that will be used throughout this work. For heavy quark probes in an irreducible representation D, the Wilson loop is given by:

$$\mathcal{W}_{\mathrm{D}}(\mathcal{C}_{\mathrm{e}}) = \frac{1}{\mathscr{D}} \operatorname{tr} \mathrm{D}\left( P\left\{ e^{i \int_{\mathcal{C}_{\mathrm{e}}} dx_\mu A_\mu(x)} \right\} \right), \tag{1}$$

211

where $\mathscr{D}$ is the dimensionality of D. The closed path $\mathcal{C}_e$ can be thought of as associated to the creation, propagation, and annihilation of a pair of quark/antiquark probes. From a rectangular path with sides $T$ and $R$, information about the static interquark potential was obtained from the large $T$ behavior $\langle \mathcal{W}_D(\mathcal{C}_e) \rangle \sim e^{-T V_D(R)}$. An area law, given by the propagation time $T$ multiplied by the interquark distance $R$, corresponds to a linear confining potential [1] (for a review, see [2]).

There are many model-independent facts that point to the importance of the center of the group $SU(N)$ to describe the confining properties of YM theory. In this regard, the first ideas relating the possible phases to the $Z(N)$ properties of the vacuum were developed in [3]. There, disorder vortex field and string field operators were introduced in (2 + 1)d and (3 + 1)d Minkowski spacetime, respectively. At equal time, they satisfy

$$\hat{\mathcal{W}}_F(\mathcal{C}_e)\,\hat{V}(\mathbf{x}) = e^{i2\pi\,L(\mathbf{x},\mathcal{C}_e)/N}\,\hat{V}(\mathbf{x})\,\hat{\mathcal{W}}_F(\mathcal{C}_e)\,, \quad \text{in } (2+1)\text{d}, \tag{2}$$

$$\hat{\mathcal{W}}_F(\mathcal{C}_e)\,\hat{V}(\mathcal{C}) = e^{i2\pi\,L(\mathcal{C},\mathcal{C}_e)/N}\,\hat{V}(\mathcal{C})\,\hat{\mathcal{W}}_F(\mathcal{C}_e)\,, \quad \text{in } (3+1)\text{d}, \tag{3}$$

where the subindex F denotes the fundamental representation, $\mathbf{x} \in \mathbb{R}^2$ ($\mathcal{C} \in \mathbb{R}^3$) is a point (curve) in real space where a thin pointlike (looplike) thin center vortex is created in three (four) dimensional spacetime. $L(\mathbf{x}, \mathcal{C}_e)$ and $L(\mathcal{C}, \mathcal{C}_e)$ are the corresponding linking numbers. An explicit realization of $\hat{V}$ was given by the action $\hat{V}|A\rangle = |A^S\rangle$, where $|A\rangle$ are quantum states with well-defined shape $A_0 = 0, A_i$ ($i = 1, 2, 3$) at a given time. The field $A_\mu^S$ has the form of a gauge transformation, but performed with a singular phase $S \in SU(N)$. To define the operator $\hat{V}(\mathbf{x})$ (respectively $\hat{V}(\mathcal{C})$), $S$ must change by a center element when going around any spatial closed loop that links $\mathbf{x}$ (respectively $\mathcal{C}$). Spurious singularities may be eliminated by using the adjoint representation $\mathrm{Ad}(S)$, which leaves a physical effect only at the point $\mathbf{x}$, or closed path $\mathcal{C}$, where $\mathrm{Ad}(S)$ is multivalued. Arguments in favor of characterizing confinement as a *magnetic* $Z(N)$ spontaneous symmetry breaking phase (center-vortex condensate),

$$\langle \hat{V}(\mathbf{x}) \rangle \neq 0, \qquad \langle \hat{V}(\mathcal{C}) \rangle \sim e^{-\mu \mathrm{Perimeter}(\mathcal{C})}\,, \tag{4}$$

were also given in that work.

The lattice also provides direct information about the role played by the center of $SU(N)$ in the confinement/deconfinement phase transition. This is observed in the properties of the Polyakov loops $P_\mathbf{x}(\mathscr{A})$, which are given by Equation (1) computed on a straight path located at a spatial coordinate $\mathbf{x}$ and extending along the Euclidean time-direction. Due to the finite-temperature periodicity conditions, these segments can be thought of as circles. By considering the fundamental representation, $P_\mathbf{x}(\mathscr{A})$ was analyzed in the lattice [4]. When changing from higher to lower temperatures, the distribution of the phase factors of $P_\mathbf{x}(\mathscr{A})$, for typical Monte Carlo configurations, shows a phase transition. At higher temperatures, for most $\mathbf{x}$, the phase factors are close to one of the center elements $e^{i2\pi k/N}$, $k = 0, \ldots, N - 1$. On the other hand, below the transition, they are equally distributed on $Z(N)$, as a function of the spatial site $\mathbf{x}$. As a result, the Monte Carlo calculation gives a transition from a non-vanishing to a vanishing gauge-field average $\langle P_\mathbf{x} \rangle$, which is in fact $\mathbf{x}$-independent, where the *electric* $Z(N)$ symmetry is not broken. This corresponds to a transition from a deconfined phase at higher $T$, where the quark free energy is finite, to a confined phase below $T_c$, where the free energy diverges.

In the full Monte Carlo simulations, the relevance of $Z(N)$ is also manifested in general Wilson loops at asymptotic distances. In this regime, the string tension only depends on the $N$-ality $k$ of D, which determines how the center $Z(N)$ of $SU(N)$ is realized in the given quark representation [5],

$$\mathrm{D}\left(e^{i\frac{2\pi}{N}}I\right) = \left(e^{i\frac{2\pi}{N}}\right)^k I_\mathscr{D}\,. \tag{5}$$

Regarding the confinement mechanism, lattice calculations aimed at determining the relevant degrees of freedom have been performed for many years. In particular, procedures have been constructed to analyze Monte Carlo $U_\mu(x) \in SU(N)$ link-configurations and extract center projected configurations $Z_\mu(x) \in Z(N)$ [6–9] (for recent techniques to improve the detection of center vortices, see [10]). A given plaquette is then said to be pierced by a thin center vortex if the product of these center elements along the corresponding links is non-trivial. Observables may then be evaluated by considering vortex-removed and vortex-only configurations. The confining properties are only well described in the latter case [6,7,11–18]. In the lattice, the analysis and visualization of center-vortex configurations [19] led to important insights regarding the origin of the topological charge density in the YM vacuum. In 3d (4d), thin center vortices are localized on worldlines (worldsheets) $\omega$. In this case, the Wilson loop in Equation (1) yields a center element

$$\mathcal{W}_{\mathrm{D}}(\mathcal{C}_{\mathrm{e}}) = \mathcal{Z}_{\mathrm{D}}(\mathcal{C}_{\mathrm{e}}) = \frac{1}{\mathscr{D}} \, \mathrm{tr} \left[ \mathrm{D}\!\left( e^{i\frac{2\pi}{N}} I \right) \right]^{\mathrm{L}(\omega,\mathcal{C}_{\mathrm{e}})} , \tag{6}$$

where $\mathrm{L}(\omega, \mathcal{C}_{\mathrm{e}})$ is the total linking number between $\omega$ and $\mathcal{C}_{\mathrm{e}}$. This result also applies to thick center vortices, when their cores are completely linked by $\mathcal{C}_{\mathrm{e}}$. In this case, $\omega$ refers to the thick center vortex guiding centers. In the scaling limit, where the lattice calculations make contact with the continuum, the density of thin center vortices detected at low temperatures is finite [7,20]. Furthermore, center vortices percolate and have positive stiffness [21,22], while the fundamental Wilson loop average over $Z_\mu(x)$ displays an area law. This is in accordance with center-vortex condensation and the Wilson loop confinement criteria. For $SU(2)$, a model based on the projected thin center-vortex ensemble captures 97.7% of the fundamental string tension. On the other hand, the percentage drops to $\sim$62% for $SU(3)$ [23]. One of the most important features of the center-vortex scenario is that it naturally explains asymptotic $N$-ality: the center element contribution in Equation (6) only depends on the $N$-ality of D. For these reasons, it is believed that the confinement mechanism should involve these degrees of freedom. For a recent discussion about this area of research, see [24].

When it comes to accommodating the model-independent full Monte Carlo calculations, some questions arise. In 3d, the full asymptotic string tension dependence on D is very well fitted by the Casimir law [25]

$$\sigma_k^{(3)} = \frac{k(N-k)}{N-1} , \tag{7}$$

which is proportional to the lowest quadratic Casimir among those representations with the same $N$-ality $k$ of D, which corresponds to the antisymmetric representation. In addition, it is precisely at asymptotic interquark distances where a model based on an ensemble of thin objects should be more reliable. This is different at intermediate distances, where finite-size effects allowed for an explanation of the observed scaling with the Casimir of D [26,27]. Then, one question is: how to capture the asymptotic law in Equation (7) from an average over percolating thin center-vortices? In 4d, where the available data cannot tell between a Casimir or a Sine law [28]

$$\sigma_k^{(4)} = \frac{k(N-k)}{N-1} \quad \text{vs.} \quad \sigma_k^{(4)} = \frac{\sin k\pi/N}{\sin \pi/N} , \tag{8}$$

is there any ensemble based on center-vortices that could reproduce one of these behaviors? More importantly, how can one explain this together with the formation of the confining flux tube observed in the lattice? This means reproducing the Lüscher term [29–31] and the observed transverse field distributions (see [32–34], and references therein). Here, we shall review some developments aimed at providing a possible answer to these questions.

In Section 2, we shall discuss the simplest Abelian center-vortex ensembles. In Section 3, we summarize, from different points of view, additional non-Abelian information and correlations that could be natural ingredients to be taken into account. In Section 4, we review ensembles of percolating oriented and non-oriented center vortices in 3d and 4d, their effective field description, as well as the possibility to accommodate the asymptotic properties of the confining string. Finally, in Section 5, we discuss recent lattice results in the light of our effective description, and present some perspectives.

## 2. Center-Vortex Ensembles

The idea that center vortices are the dominant degrees of freedom in the infrared regime means, in practice, that the Wilson loop average at asymptotic distances may well be captured by modeling the average of the center-elements in Equation (6). This line of research was mainly explored in the lattice [35] by considering an ensemble of fluctuating worldlines (in 3d) or worldsurfaces (in 4d) with tension and stiffness (see also the discussion at the beginning of Section 3.2). For example, in 4d, a theory of fluctuating center-vortex worldsurfaces in four dimensions was introduced by considering the lattice action [35]

$$S_{\text{latt}}(\omega) = \mu \mathcal{A}(\omega) + c N_p \, , \tag{9}$$

where $\mathcal{A}(\omega)$ is the area of the vortex closed worldsurface $\omega$, formed by a set of plaquettes, and $N_p$ is the number of pairs of neighboring plaquettes of the surface lying on different planes. The latter term, as well as the lattice regularization, contribute to the stiffness of the vortices. This model, initially introduced for $SU(2)$, and then generalized for $SU(3)$ [36], is able to describe important features, such as the confining string tension for fundamental quarks and the order of the deconfinement transition. This type of model can be also formulated in the continuum. The objective is the same, that is, looking for natural ensemble measures to compute center-element averages and compare them with the asymptotic information extracted from the full Monte Carlo average $\langle \mathcal{W}_{\text{D}}(\mathcal{C}_e) \rangle$. A successful comparison is expected to give important clues about the underlying mechanism of confinement. When computing center-element averages in the continuum, the simplest model has the form:

$$\langle \mathcal{Z}_{\text{D}}(\mathcal{C}_e) \rangle = \mathcal{N} \sum_{\omega} e^{-S(\omega)} \frac{1}{\mathscr{D}} \, \text{tr} \left[ \text{D}\left( e^{i \frac{2\pi}{N} I} \right) \right]^{\text{L}(\omega, \mathcal{C}_e)} , \tag{10}$$

where $\sum_{\omega}$ represents the sum over different configurations in a diluted gas of closed worldlines (in 3d) or worldsurfaces (in 4d). The weight factor $e^{-S(\omega)}$ implements the effect of center-vortex tension ($\mu$) and stiffness ($1/\kappa$) observed in the lattice [21,22]. More precisely, $S(\omega)$ contains a term proportional to the length or area of $\omega$, and another one proportional to a power of the absolute value of the curvature of $\omega$. See Equation (A3) for an explicit formula in 3 dimensions. $S(\omega)$ could also contain interactions with a scalar field $\psi$ that, when integrated with a corresponding weight $W(\psi)$, generates interactions among the variables $\omega$.

Extended models can also be introduced where the defining elements are not only given by $\omega$ but also by additional labels. At the level of the gauge field variables $A_\mu$, the center-vortex sectors can be characterized by different mappings $S_0 \in SU(N)$ containing defects (see Section 4.2). A center vortex with guiding center $\omega$ and magnetic weight $\beta$ is characterized by $S_0 = e^{-i\chi \beta \cdot T}$, $\beta \cdot T \equiv \beta|_q T_q$, where $\chi$ is a multivalued angle that changes by $2\pi$ when going around $\omega$, and $T_q, q = 1, \ldots, N-1$ are the Cartan generators. As they carry a single weight, these vortices are known as oriented (in the Cartan subalgebra). For elementary center vortices, the tuple $\beta$ is one of the magnetic weights $\beta_i$ ($i = 1, \ldots, N$) of the fundamental representation. In the region outside the vortex cores, $A_\mu$ is locally a pure gauge configuration constructed with $S_0$. Then, for fundamental quarks, the contribution to a large loop contained in that region is $i$-independent and given by the elementary center-element $(1/N) \, \text{tr}\left( e^{-i2\pi \beta_i \cdot T} \right) = e^{i2\pi/N}$ to the power $\text{L}(\omega, \mathcal{C}_e)$. Different elementary fluxes may join to form more complex configurations, provided this is done in a way that conserves the flux. For example, $N$ center-vortex guiding centers associated with different magnetic weights $\beta_i$ can be matched. For simplicity, let us consider the $SU(3)$ case in three dimensions and a configuration characterized by $S_0 = e^{i\chi_1\beta_1 \cdot T} e^{i\chi_2\beta_2 \cdot T}$, where $\chi_1$ and $\chi_2$ are multivalued when going around the closed worldlines $\omega_1$ and $\omega_2$, respectively. These worldlines could meet at a point, then follow a common open line $\gamma$, and again bifurcate to close the corresponding loops. In this case, we would have a pair of fluxes entering the initial point, carrying the fundamental weights $\beta_1, \beta_2$, and a flux leaving along $\gamma$, carrying the weight $\beta_1 + \beta_2$. In $SU(3)$, this sum is an antifundamental weight $-\beta_3$. In other words, there are three fluxes entering the initial point, which carry the three different fundamental weights $\beta_1, \beta_2, \beta_3$. This can be readily generalized to $SU(N)$, where $N$ fluxes carrying the different fundamental weights can meet at a point, as these weights satisfy $\sum_i \beta_i = 0$. Vortices may also be non-oriented [37], in the sense that they may not be described by a single weight. In this case, the center-vortex components with different fundamental weights are interpolated by instantons in 3d and monopole worldlines in 4d. These lower dimensional junctions, which carry a flux of the form $\beta_i - \beta_j$, should be weighted with additional phenomenological terms in $S(\omega)$. Furthermore, in the 4d case, three monopole worldlines carrying fluxes $\beta_i - \beta_j, \beta_j - \beta_k, \beta_k - \beta_i$ can be matched at a spacetime point. Similar higher-order matching rules are also possible. In what follows, we shall discuss the different ensembles, starting with the simplest possibilities in 3d and 4d.

## 3. Abelian Effective Description of Center Vortices

In this section, we shall briefly discuss center-vortex ensembles formed by diluted closed worldlines in 3d (Section 3.1) or worldsurfaces in 4d (Section 3.2), characterized by no other properties than tension, stiffness, and vortex–vortex interactions. No additional degrees of freedom, matching rules or correlations with lower dimensional objects will be considered here.

### 3.1. Three Dimensions

In a planar system, thin center vortices are localized on points, so they are created or annihilated by a field operator $\hat{V}(x)$. The emergence of this order parameter can be clearly seen by applying polymer techniques to center-vortex worldlines [38]. In [39], the center-element average for fundamental quarks, over all possible diluted loops, was initially represented in the form

$$\langle \mathcal{Z}_{\mathrm{F}}(\mathcal{C}_e) \rangle = \mathcal{N} \int [D\psi]\, e^{-W[\psi]}\, e^{\int_0^\infty \frac{dL}{L} \int dx \int du\, Q(x,u,x,u,L)} , \tag{11}$$

where $Q(x, u, x_0, u_0, L)$ is the integral over all paths with length $L$, starting (ending) at $x_0$ ($x$) with unit tangent vector $u_0$ ($u$), in the presence of scalar and vector sources $\psi$ and $\frac{2\pi}{N}s_\mu$, and weighted by tension and stiffness. The factor $W[\psi] = \frac{\zeta}{2} \int d^3x\, \psi^2(x)$ generates, upon integration of the auxiliary scalar field $\psi$, repulsive contact interactions between the loops with strength given by the parameter $\frac{1}{\zeta}$. Indeed, as in the exponential we have $x = x_0$, $u = u_0$, its expansion generates the diluted loop ensemble. As usual, the factor $1/L$ is to avoid loop overcounting when choosing $x_0$ on a given loop. The external source $s_\mu$ is localized on a surface $S(\mathcal{C}_e)$ whose border is the Wilson loop. As a consequence, it generates the intersection numbers between the loop-variables in $Q$ and $S(\mathcal{C}_e)$, which coincide with the different linking-numbers. Using the large-distance behavior of $Q(x, u, x_0, u_0, L)$, which satisfies a Fokker–Planck diffusion equation (given by Equations (A1) and (A7), with $b_\mu$ Abelian, and $D(\Gamma_\gamma[b_\mu])$ being the complex number $\Gamma_\gamma[b_\mu]$) we then showed that the ensemble average of center elements becomes represented by a complex scalar field $V(x)$,

$$\langle \mathcal{Z}_{\mathrm{F}}(\mathcal{C}_e) \rangle \approx \mathcal{N} \int [DV][D\bar{V}]\, e^{-\int d^3x \left[ \frac{1}{3\kappa} \overline{D_\mu V} D_\mu V + \frac{1}{2\zeta} (\bar{V}V - v^2)^2 \right]} ,$$

$$v^2 \propto -\mu\kappa > 0, \qquad D_\mu = \partial_\mu - i\frac{2\pi}{N}s_\mu . \tag{12}$$

This was obtained for small (positive) stiffness $1/\kappa$ and repulsive contact interactions. The scalar field $V$ is originated due to the approximate behavior of $Q(x, u, x_0, u_0, L)$ in Equation (A12), which turns the exponential in Equation (11) into a functional determinant. The squared mass parameter of this field is proportional to $\kappa\mu$, where $\mu$ is the center-vortex tension. For percolating objects ($\mu < 0$), the $U(1)$ symmetry of the effective field theory is spontaneously broken ($\kappa\mu < 0$). Among the consequences, we have:

1. In the center-vortex condensate, the effective description is dominated by the soft Goldstone modes, $V(x) \sim v\, e^{i\phi(x)}$. Then, the calculation of the center-element average is neither Gaussian nor dominated by a saddle-point, as it involves a compact scalar field $\phi$ and large fluctuations;
2. This is better formulated in the lattice, where the Goldstone mode sector is governed by a 3d XY model with frustration

$$S_{\text{latt}}^{(3)} = \tilde{\beta} \sum_{x,\mu} \text{Re}\left[1 - e^{i\gamma(x+\hat{\mu})}e^{-i\gamma(x)}e^{-i\alpha_\mu(x)}\right] .\tag{13}$$

The external source in Equation (12) translates into the frustration $e^{i\alpha_\mu(x)} = e^{i\frac{2\pi}{N}}$ if $S(\mathcal{C}_e)$ is crossed by the link and is trivial otherwise;
3. In the expansion of the partition function, due to the measure $\prod_x \int_{-\pi}^{\pi} d\gamma(x)$, the terms that contribute contain products of the composite $e^{i\gamma(x+\hat{\mu})}e^{-i\gamma(x)}$ (or its conjugate) over links organized forming loops. Otherwise, the integrals over the site variables at the line edges vanish (see Figure 1);
4. Due to frustration, every time $\mathcal{C}_e$ is linked, a center element is generated. Then, in the lattice, the closed center-vortex worldlines in the initial ensemble, which led to Equation (11) and gave origin to the effective description (12), are represented by the loops of item 3.

This point of view will be useful to propose other ensemble measures relying on lattice models, as in the case where the derivation of the effective description is not known, see for example Sections 3.2 and 5.3. It is also interesting to see that the initial ensemble properties encoded in Equation (11) are recovered close to the 3d XY model critical point, as expected. Indeed, using the same techniques reviewed in [40] for the case without frustration, the partition function may be formulated in terms of integer-valued divergenceless currents, originated after using the Fourier decomposition

$$e^{\beta\cos\gamma} = \sum_{b=-\infty}^{\infty} I_b(\beta)e^{ib\gamma} ,\tag{14}$$

at every lattice link. The resulting expression turns out to be equivalent to a grand canonical ensemble of non-backtracking closed loops formed by currents of strength $|b_\mu| = 1$. In the model without frustration, close to the critical point $\beta_c \approx 0.454$ (continuum limit), the relevant configurations are known to be formed by large loops rather than by multiple small loops, and multiple occupation of links is disfavored, thus making contact with the initial properties parametrized in the ensemble (see Table 1 below).

**Table 1.** The correspondence between the effective field and 3d XY model representations of the Abelian center−vortex ensemble.

| 3d XY | Effective Fields |
| --- | --- |
| large loops are favored | negative tension $\mu$ |
| multiple small loops are disfavored | positive stiffness $1/\kappa$ |
| multiple occupation of links is disfavored | repulsive interactions |

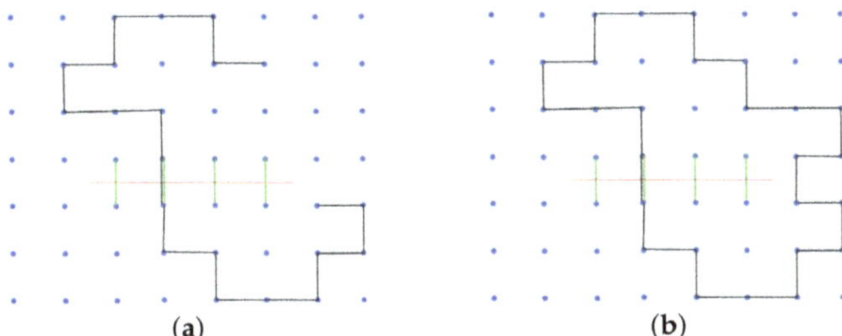

**Figure 1.** The Wilson loop and the frustration are represented in red and green, respectively. Configurations of type (**a**), which involve sites joined by open lines, do not contribute to the partition function. Only site configurations joined by loops, like the one in (**b**), contribute (with a center-element).

*3.2. Four Dimensions*

Regarding the effective description of 4d ensembles based on random surfaces, as in the $3 + 1$ dimensional world center vortices are one-dimensional objects spanning closed worldsurfaces, the emergent order parameter would be a *string field*. However, unlike the 3d case, a derivation starting from the ensemble of closed worldsurfaces with stiffness is still lacking. Such generalization should initially describe a growth process where a surface is generated, and then derive a Fokker–Planck equation for the lattice loop-to-loop probability. Similarly to what happens with end-to-end probabilities for polymers, where stiffness is essential to get a continuum limit when the monomer size goes to zero [41,42], curvature effects are expected to be essential for the continuum limit of triangulated random surfaces. Indeed, ensembles of surfaces which consider only the Polyakov (or Nambu-Goto) action leads to a phase of branched polymers [43,44]. On the other hand, in [45], the phase fluctuations of an Abelian string field with frozen modulus were approximated by a lattice *field* theory: the $U(1)$ gauge-invariant Abelian Wilson action. In other words, the Goldstone modes for a condensate of one-dimensional objects are gauge fields. Motivated by this enormous simplification and by an analogy with the 3d case, in [46] we proposed a Wilson action with frustration as a starting point to define a measure for percolating center vortices in four dimensions. This proposal will be discussed in Section 5.3. For the time being, we summarize the main initial steps, which are analogous to items 1–4 in Section 3.1:

1. In the center-vortex condensate, the effective theory is dominated by the soft Goldstone modes, which are represented by an emergent compact Abelian gauge field $V_\mu \in U(1)$. In the center-vortex context, we proposed another natural one based on $V_\mu \in SU(N)$ (see Section 5.3);
2. The lattice version of the Goldstone mode sector is given by a Wilson action with frustration;
3. In the expansion of the partition function, the relevant configurations to compute the gauge model correspond to link-variables on the edges of plaquettes organized on closed surfaces (see Figure 2);
4. The frustration is non-trivial on plaquettes $x, \mu, \nu$ that intersect $S(\mathcal{C}_e)$. Every time a closed surface links $\mathcal{C}_e$, a center-element for quarks in the representation D is generated.

Thus, the main simplification in 4d is that, in a condensate, the effective description can be captured by a local field. Similarly to 3d, where the soft modes can be read in the phase of the vortex field $V(x) \sim v\,e^{i\gamma(x)}$, the natural soft modes in 4d are given by a compact gauge field,

$$V(C) \sim v\,e^{i\gamma_\Lambda(C)}, \qquad \gamma_\Lambda(C) = \oint_C dx_\mu\, \Lambda_\mu \,. \tag{15}$$

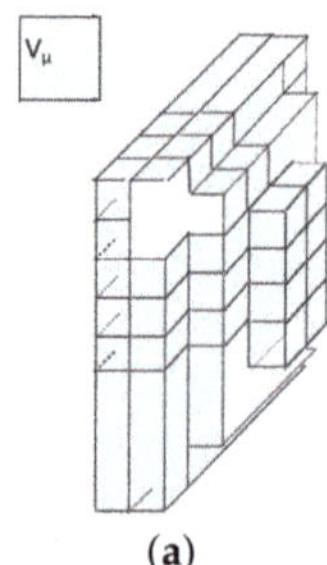


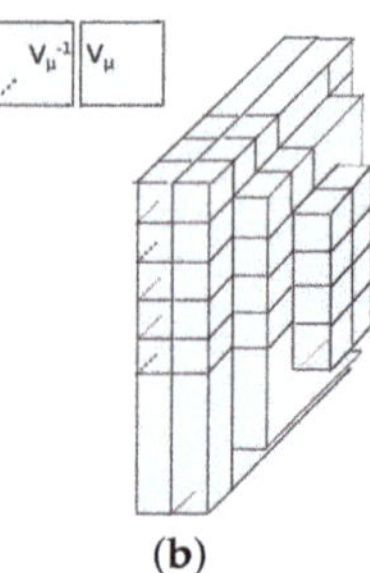


(a)      (b)

**Figure 2.** (**a**) Configurations formed by link variables distributed on plaquettes organized on an open surface do not contribute, as the $V_\mu$ link−variables at the surface edges cannot form singlets; (**b**) when they are organized on closed surfaces, singlets can be formed and the group−integral is non−trivial.

## 4. Center-Vortex Gauge Fields, Matching Rules, and Correlations

The simplest center-vortex ensembles discussed in Section 3 could provide an important basis to understand the confinement mechanism at asymptotic distances. However, they do not contain enough ingredients to reproduce more intricate properties. In this section, we shall discuss the center-vortex gauge fields and typically non-Abelian elements that could characterize the associated ensembles.

### 4.1. Thick Center Vortices and Intermediate Casimir Scaling

Before discussing generalized center-vortex ensembles with matching rules and non-oriented components, let us recall how the consideration of center-vortex thickness and the natural non-Abelian orientations in the gauge group can account for the observed Casimir scaling at intermediate distances. Some ideas along this line were initially pursued in [47]. In [26,27] (see also [48]), a simple model was put forward in the lattice, where the contribution to a planar Wilson loop along a curve $\mathcal{C}_e$ was modeled. The starting point is to postulate an ensemble of thick center vortices whose total flux, as measured by a fundamental holonomy, have different possibilities $z^j = e^{i2\pi j/N}, j = 1, \ldots, N-1$. When a thick center vortex is partially linked, the contribution to the Wilson loop is given by the insertion of a group element $G_j(x, S)$ that depends on the location ($x$) of the center-vortex midpoint (or guiding center) with respect to $\mathcal{C}_e$. It also depends on a group orientation $S$,

$$G_j(x, S) = S\mathcal{G}_j(x)S^\dagger , \tag{16}$$

where $\mathcal{G}_j = \exp\left[i\,\alpha_j \cdot T\right]$ is in the Cartan subgroup and the tuples $\alpha_j$ are formed by model-dependent scalar profiles. These profiles implement the natural condition that $G_j(x, I) = z^j I_N$, if the thick center vortex is fully enclosed by $\mathcal{C}_e$, it is $I_N$ if it is not enclosed at all, and it gives an interpolating value otherwise. After averaging over random group orientations in [26,27], they arrived at

$$\sigma_{\mathcal{C}_e}(\mathrm{D}) \equiv -\sum_x \frac{1}{A} \ln\!\left(1 - \sum_{j=0}^{N-1} f_j\!\left(1 - \frac{1}{\mathscr{D}}\mathrm{Tr}\,\mathrm{D}(\mathcal{G}_j)\right)\right) , \tag{17}$$

where $f_j$ is the probability that a given plaquette of the planar surface enclosed by $\mathcal{C}_e$ be pierced by the midpoint of a center-vortex of type $j$, $\sigma_{\mathcal{C}_e}(\mathrm{D})$ is the string tension in representation D, and $A$ is the minimal area of $\mathcal{C}_e$. At intermediate distances, after some natural approximations, an appropriate choice of profiles, and using the key formula

$$\mathrm{Tr}\left(\mathrm{D}(T_q)\mathrm{D}(T_p)\right) = \mathscr{D}\,\delta_{qp}\frac{C_2(\mathrm{D})}{N^2 - 1} , \tag{18}$$

the Casimir Scaling

$$\frac{\sigma_{\mathrm{I}}(\mathrm{D})}{\sigma_{\mathrm{I}}(\mathrm{F})} = \frac{C_2(\mathrm{D})}{C_2(\mathrm{F})} \tag{19}$$

was obtained. In [26,27], based on a specific choice of probabilities and profiles, it was also possible to reproduce different asymptotic behaviors, such as the Casimir and the Sine law. In Section 5, we shall review a different line based on oriented and non-oriented center vortices, which naturally lead to an asymptotic Casimir law. As these models are generated from weighted center-element averages, they are expected to be applicable in the asymptotic region.

### 4.2. Center-Vortex Sectors in Continuum YM Theory

Center vortex correlations were considered for the first time in [3]. In (2 + 1)d Minkowski spacetime, the order–disorder algebra in Equation (2) says that the action of $\hat{V}(\mathbf{x})$ on $|A\rangle$ gives

$$\hat{W}_{\mathrm{F}}(\mathcal{C}_{\mathrm{e}})\left(\hat{V}(\mathbf{x})|A\rangle\right) = e^{i\frac{2\pi}{N}}\, W_{\mathrm{F}}(\mathcal{C}_{\mathrm{e}})\left(\hat{V}(\mathbf{x})|A\rangle\right), \tag{20}$$

if $\mathbf{x}$ is encircled by $\mathcal{C}_{\mathrm{e}}$, and it leaves the state $|A\rangle$ unaltered otherwise. Here, $|A\rangle$ is a state with well-defined shape in the Weyl gauge $A_0 = 0$, That is, $\hat{V}(\mathbf{x})|A\rangle$ is a state where a thin center-vortex is created on top of $A_i$. In particular, the action of $\hat{V}^N(\mathbf{x})$ is trivial. Then, the possible phases were effectively described by a model with magnetic $Z(N)$ symmetry

$$\mathcal{L} = \partial^\mu \bar{V}\, \partial_\mu V + m^2\, \bar{V}V + \frac{\lambda}{2}\left(\bar{V}V\right)^2 + \xi\left(V^N + \bar{V}^N\right). \tag{21}$$

This includes quadratic and quartic correlations, as well as the $N$-th order terms that capture the possibility that $N$ vortices may annihilate. The case $m^2 > 0$ would correspond to a Higgs phase where center vortices are in the spectrum of asymptotic states. The case $m^2 < 0$ corresponds to a center-vortex condensate, with $N$ degenerate classical vacua, so that $Z(N)$ is spontaneously broken. For a detailed analysis of this effective description, see [49,50]. In [3], based on the center-vortex operator definition $\hat{V}(\mathbf{x})|A\rangle = |A^S\rangle$, discussed in Section 1, 3d Euclidean vortex Green's functions $\langle \bar{V}(y)V(x)\rangle$ were defined. This was done by considering the YM path-integral over configurations $A_\mu$ with boundary conditions around the pair of points $x, y \in \mathbb{R}^3$, such that a vortex is created at $x$, it is then propagated, and finally annihilated at $y$. When $|x - y| \to \infty$, an exponential decay would correspond to a Higgs phase and $\langle V\rangle = 0$, because of the clustering property. This agrees with the discussion above, where the Higgs phase $m^2 > 0$ is characterized by a $Z(N)$ symmetric vacuum. On the other hand, a condensate would correspond to a Green's function that tends to a constant.

Now, from the definition of the operator $\hat{V}(\mathbf{x})$, it is clear that it introduces singularities in the gauge fields. If $A$ is smooth, the configuration $A^S$ is singular, with a field strength containing a delta-singularity at the center vortex location $\mathbf{x}$. As pointed out by 't Hooft, the operator's definition could be made more precise by smearing the singularities over an infinitesimal region around $\mathbf{x}$. Otherwise, we would be working with singular infinite action gauge fields. Although this direction was not pursued in that work, the smeared Green's functions could depend on the choice of boundary conditions, for the mapping $S \in SU(N)$, around $x$ and $y$. In other words, the vortex field $\hat{V}$ could hide non-Abelian degrees of freedom which are not evidenced by the algebra in Equation (2), which only depends on properties with respect to the Wilson loop.

In [51], we proposed a partition of the full configuration space of *smooth* gauge fields $\{A_\mu\}$ into sectors $\mathcal{V}(S_0) \subset \{A_\mu\}$ characterized by topological labels $S_0$. For this objective, we introduced $N_{\mathrm{f}}$ auxiliary adjoint scalar fields $\psi_I$ by means of an identity in the YM path integral, which constrain them to be a solution to a classical equation of motion for the minimization of an auxiliary action $S_{\mathrm{aux}}(\psi_I, A)$. Imposing regularity and boundary conditions, the solution $\psi_I(A)$ is unique, and can be decomposed by means of a generalized polar decomposition $\psi_I(A) = Sq_I S^{-1}$, where $S(x) \in SU(N)$ and $(q_1, \ldots, q_{N_f})$ is a "modulus" tuple. The phase defects cannot be eliminated by regular gauge transformations $U$, which act on the left $S \to US$. A gauge field is then said to belong to a given sector $\mathcal{V}(S_0)$ if $S(A)$ is equivalent to a class representative $S_0$. The continuum of possible labels $S_0$ are characterized by the location of oriented and non-oriented center-vortex guiding centers, with all possible matching rules (see the discussion in Section 2). Although a possible label for an oriented center-vortex would be $S_0 = e^{i\chi\beta\cdot T}$, a typical non-oriented configuration is characterized by $S_0 = e^{i\chi\beta\cdot T} W$. In 3d, close to some points (instantons) on the center-vortex worldline generated by $e^{i\chi\beta\cdot T}$, the mapping $W$ behaves as a Weyl transformation that changes the fundamental weight $\beta$ to $\beta'$. Similarly, in 4d, the change occurs at some monopole worldlines on the center-vortex worldsurfaces generated by $e^{i\chi\beta\cdot T}$ (see [46]). The full YM partition function and averages of observables were then represented by a sum over partial contributions,

$$Z_{\text{YM}} = \sum_{S_0} Z_{(S_0)}, \quad \langle O \rangle_{\text{YM}} = \frac{1}{Z_{\text{YM}}} \sum_{S_0} \int_{\mathcal{V}(S_0)} [D\mathcal{A}_\mu] \, O \, e^{-S_{\text{YM}}} . \tag{22}$$

Here, $\sum_{S_0}$ is a short-hand notation for the contribution originated from the continuum of labels $S_0$. These ideas provided a glimpse of a path connecting first principles Yang–Mills theory to an ensemble containing all possible center-vortex configurations. In addition to addressing this important conceptual issue, the partition into sectors may circumvent the well-known Gribov problem when fixing the gauge in non-Abelian gauge theories, as Singer's no go theorem [52] only applies to global gauges in configuration space (see [53] for a detailed discussion). In [51], the gauge was locally fixed by a regular gauge transformation that rotates $S$ to the reference $S_0$, which is imposed by a sector dependent condition $f_{S_0}(\psi) = 0$. Furthermore, the theory was shown to be renormalizable in the vortex-free sector [54]. The extension of the renormalization proof to sectors labeled by center vortices is under way, and will be presented elsewhere. An interesting consequence of this construction is that a new label may be generated by the right multiplication, $S_0 \to S_0 \tilde{U}^{-1}$, with regular $\tilde{U}$, which is not necessarily connected to $S_0$ by a regular gauge transformation. That is, given a center-vortex sector, there is a continuum of physically inequivalent sectors characterized by non-Abelian d.o.f. where the defects are located at the same spacetime points. In the context of effective Yang–Mills–Higgs models, which describe the confining string as a smooth topological classical vortex solution, the presence of similar internal d.o.f. was previously noted in a large class of color-flavor symmetric theories [55–64].

## 5. Mixed Ensembles of Oriented and Non-Oriented Center Vortices

The general properties of center vortices discussed so far motivate the search for a natural ensemble that captures all the asymptotic properties of confinement. Among them, the formation of a confining flux tube is the most elusive one in this scenario. The formation of this object would also explain the Lüscher term, which has not been observed in projected center-vortex ensembles. Furthermore, the asymptotic Casimir law (cf. Equation (7)) should be reproduced in 3d, while in 4d we would like to understand the coexistence of $N$-ality with the Abelian-like flux tube profiles [32–34]. It is clear that a confining flux tube requires an ensemble whose effective description contains topological solitons, namely, a confining domain wall in $(2 + 1)$d and a vortex in $(3 + 1)$d. However, the simple models of oriented and uncorrelated center vortices discussed in Section 3 do not have the conditions to support these topological objects[1]. In what follows, we shall review how the inclusion of the center-vortex matching rules and correlations with lower dimensional defects (see Sections 2 and 4.2) could fill the gap between center-vortex ensembles and the formation of a flux tube. In [65,66], lattice studies showed that the 4d Abelian-projected lattice is not represented by a monopole Coulomb gas, but rather by collimated fluxes attached to the monopoles. In the continuum, these configurations correspond to the previously discussed non-oriented center vortices. While in 4d the lower dimensional defects on center-vortex worldsurfaces are monopole worldlines, in 3d they are instantons. The relevance of non-oriented center vortices to generate a non-vanishing Pontryagin index was shown in [37]. Now, although oriented and non-oriented center vortices, located at the same place, would contribute to a large Wilson loop with the same center-element, it is natural to weight them with different effective actions. In the second case, the measure should also depend on the location of the lower-dimensional defects.

### 5.1. 3d Ensemble with Asymptotic Casimir Law

In this section, we review the mixed ensembles formed by oriented and non-oriented center-vortices with $N$-line matching rules introduced in [67]. In that reference, to prepare the formalism so as to include the different correlations, we initially wrote the contribution to the Wilson loop of a thin center-vortex loop $l$ as

$$\mathcal{W}_D(\mathcal{C}_e)|_{\text{loop}} = \frac{1}{N} \operatorname{Tr} \Gamma_l[b_\mu^{\mathcal{C}_e}], \qquad \Gamma_\gamma[b_\mu] = P\{e^{i \int_\gamma dx_\mu b_\mu}\}, \tag{23}$$

where $b_\mu^{\mathcal{C}_e} = 2\pi\beta_e \cdot T s_\mu^{\mathcal{C}_e}$, $\beta_e$ is the highest magnetic weight of D, and $s_\mu^{\mathcal{C}_e}$ is a source localized on $\mathcal{C}_e$. Here, we use the notation $\beta_e \cdot T = \beta_e|_q T_q$, with $T_q$, $q = 1 \ldots, N-1$ being the Cartan generators of $SU(N)$. Then, after weighting each loop with a phenomenological factor $e^{-S(l)}$ accounting for tension and stiffness (cf. Equation (10)), and summing over all possible diluted loops, we obtained the center-element average

$$\langle \mathcal{Z}_D(\mathcal{C}_e) \rangle = e^{\int_0^\infty \frac{dL}{L} \int dx \int du \operatorname{tr} Q(x,u,x,u,L)}, \tag{24}$$

where $Q(x, u, x_0, u_0, L)$ is the integral over all the paths with length $L$ that begin at $x_0$ with unit tangent vector $u_0$, and end at $x$ with orientation $u$. This is given by Equation (A1), using as D the fundamental representation. This object satisfies a non-Abelian diffusion equation whose large $\kappa$-limit (small stiffness) solution (cf. Equation (A12)) led to approximate Equation (24) by

$$\langle \mathcal{Z}_D(\mathcal{C}_e) \rangle \approx Z_{\text{loops}} = \mathcal{N} \int [d\phi]\, e^{-\int d^3x\, \phi^\dagger \mathcal{O} \phi}, \qquad \mathcal{O} = -\frac{1}{3\kappa}(I_N \partial_\mu - ib_\mu^{\mathcal{C}_e})^2 + \mu I_N, \tag{25}$$

where $\phi$ is an emergent complex scalar field in the fundamental representation.

One basic defining property of center vortices is that $N$ such objects can be virtually created out of the vacuum at $x_0$ and then annihilated at $x$. At the level of the gauge fields, this is related to the possibility of matching $N$ guiding centers each one carrying a different fundamental magnetic weight $\beta_i$, $i = 1, \ldots, N$, which satisfy $\beta_1 + \ldots + \beta_N = 0$. Then, to incorporate all possible oriented center-vortex line matchings (see Section 5.1.1), we expanded the loop ensemble in Equation (25) considering the $N$ types of weights, each one represented by a fundamental field $\phi_i$, $i = 1, \ldots, N$. At this point, the center-element average over loops was generated from the partition function

$$Z_{\text{loops}}^N = \int [D\Phi^\dagger][D\Phi]\, e^{-\int d^3x \left[\frac{1}{3\kappa} \operatorname{Tr}\left((D_\mu\Phi)^\dagger D_\mu\Phi\right) + \mu \operatorname{Tr}(\Phi^\dagger\Phi)\right]}, \tag{26}$$

where $\Phi$ is a complex $N \times N$ matrix with components $\Phi_{ij} = \phi_j|_i$.

#### 5.1.1. Including $N$-Vortex Matching

The contribution to the Wilson loop of $N$ center-vortex worldlines starting at $x_0$ and ending at $x$, and carrying different weights, was rewritten as

$$\mathcal{W}_D(\mathcal{C}_e)|_{N-\text{lines}} = \frac{1}{N!} \epsilon_{i_1 \ldots i_N} \epsilon_{i_1' \ldots i_N'} \Gamma_{\gamma_1}[b_\mu^{\mathcal{C}_e}]|_{i_1 i_1'} \cdots \Gamma_{\gamma_N}[b_\mu^{\mathcal{C}_e}]|_{i_N i_N'}. \tag{27}$$

By weighting each line in Equation (27) with the factor $e^{-S(\gamma_i)}$, and integrating over paths with fixed endpoints and over all the lengths $L_i$ (cf. Equation (A13)), we obtained

$$C_N \propto \int d^3x\, d^3x_0\, \epsilon_{i_1 \dots i_N} \epsilon_{j_1 \dots j_N}\, G(x,x_0)_{i_1 j_1} \dots G(x,x_0)_{i_N j_N} \, , \qquad (28)$$

where $G(x,x_0)$ is the Green's function of the operator $\mathcal{O}$. In this manner, the $N$-line contribution in Equation (28) and similar processes were generated by adding a term $\propto (\det \Phi + \det \Phi^\dagger)$. The effective description thus obtained is separately invariant under local and global $SU(N)$ symmetries $S_c(x), S_f \in SU(N)$

$$\Phi \to S_c(x)\Phi, \quad b_\mu \to S_c(x)b_\mu S_c^{-1}(x) + iS_c(x)\partial_\mu S_c^{-1}(x) \, ,$$
$$\Phi \to \Phi S_f \, . \qquad (29)$$

In the effective description, other natural terms compatible with these symmetries, like the vortex–vortex interaction $\mathrm{Tr}(\Phi^\dagger \Phi)^2$, should also be included, thus leading to the center-element average $\langle \mathcal{Z}_D(\mathcal{C}_e) \rangle = Z_v[b_\mu^{\mathcal{C}_e}]/Z_v[0]$,

$$Z_v[b_\mu^{\mathcal{C}_e}] = \int [D\Phi^\dagger][D\Phi] e^{-\int d^3x \left[ \frac{1}{3\kappa}\mathrm{Tr}\left((D_\mu\Phi)^\dagger D_\mu\Phi\right) + \mu\mathrm{Tr}(\Phi^\dagger\Phi) + \frac{\lambda_0}{2}\mathrm{Tr}(\Phi^\dagger\Phi)^2 - \zeta_0(\det\Phi + \det\Phi^\dagger)\right]} \, . \qquad (30)$$

This effective description has some similarities with the 't Hooft model (cf. Equation (21)). More specifically, they coincide for configurations of the type $\Phi = V I_N$. However, there is no reason for the path-integral to favor this type of restricted configuration. Up to this point, in the percolating phase ($\mu < 0$), the quadratic and quartic terms tend to produce a manifold of classical vacua labeled by $U(N)$, while the addition of the $\det \Phi$-interaction reduces this manifold to $SU(N)$. Then, unlike the 't Hooft model, in the SSB phase this effective description has a continuum set of classical vacua which precludes the formation of the stable domain wall. It is interesting to formulate the Goldstone modes $V(x) \in SU(N)$ in the lattice, which leads to

$$S_{\text{latt}}^{(3)}(b_\mu^{\mathcal{C}_e}) = \tilde{\beta} \sum_{x,\mu} \mathrm{Re} \left[ \mathbb{I} - \bar{U}_\mu V(x+\hat{\mu})V^\dagger(x)) \right] \, , \qquad (31)$$

where $U_\mu(x) = e^{i2\pi\beta_e \cdot T} \in Z(N)$, if the link $x,\mu$ crosses $S(\mathcal{C}_e)$, and it is the identity otherwise. As expected, in the expansion of the partition function, besides the contribution of sites distributed on links that form loops, there is also one originated from $N$ lines that start or end at a common site $x$. In the former case, the singlets are included in $N \otimes \bar{N}$, while in the latter they are in the products of $N$ $V(x)$ or $V^\dagger(x)$ (compare with the Abelian case in Section 3.1). In this way, the rules originating Equation (30) can be recovered in the lattice. This type of cross-checking is useful to better understand proposals of lattice ensemble measures in situations where it is harder to derive the effective field description, like in 4d spacetime.

### 5.1.2. Including Non-Oriented Center Vortices in 3d

In terms of Gilmore–Perelemov group coherent-states (see [68,69] for a complete discussion or [46] for a summary of the main ideas) $|g,\omega\rangle = g|\omega\rangle$, $g \in SU(N)$, Equations (23) and (27) became

$$\mathcal{W}_D(\mathcal{C}_e)|_{\text{loop}} \propto \int d\mu(g)\, \langle g,\omega|\Gamma_l[b_\mu^{\mathcal{C}_e}]|g,\omega\rangle \, ,$$

$$\mathcal{W}_D(\mathcal{C}_e)|_{N-\text{lines}} \propto \int d\mu(g)d\mu(g_0)\, \langle g,\omega_1|\Gamma_{\gamma_1}[b_\mu^{\mathcal{C}_e}]|g_0,\omega_1\rangle \dots \langle g,\omega_N|\Gamma_{\gamma_N}[b_\mu^{\mathcal{C}_e}]|g_0,\omega_N\rangle \, . \qquad (32)$$

The first contribution can be thought of as associated to the creation of a center-vortex with initial fundamental weight $\omega$ and group orientation $g$, which is propagated along the closed worldline $l$, and is then annihilated. The second corresponds to $N$ vortices with different magnetic weights $\beta_i = 2N\,\omega_i$, $i = 1, \ldots, N$, created out of the vacuum at a spacetime point $x_0$, that follow separate worldlines $\gamma_i$ and then annihilate at $x_f$. Following a similar interpretation, and recalling that the center-vortex weights change at the instantons, we introduced non-oriented center vortices. When a closed object is formed by n parts $\gamma_1 \circ \gamma_2 \circ \ldots \circ \gamma_n$ with $n$ instantons at points $x_1 \ldots x_n$, we considered the contribution

$$C_n = \int d\mu(g_1) \ldots \int d\mu(g_n) \langle g_1, \omega | g_2, \omega' \rangle \langle g_2, \omega | g_3, \omega' \rangle \ldots \langle g_n, \omega | g_1, \omega' \rangle \times \tag{33}$$

$$\times \langle g_1, \omega' | \Gamma_{\gamma_n}[b_\mu^{\mathcal{C}_e}] | g_n, \omega \rangle \ldots \langle g_3, \omega' | \Gamma_{\gamma_2}[b_\mu^{\mathcal{C}_e}] | g_2, \omega \rangle \langle g_2, \omega' | \Gamma_{\gamma_1}[b_\mu^{\mathcal{C}_e}] | g_1, \omega \rangle .$$

Here, a center vortex is propagated along $\gamma_1$ from $x_1$, with orientation $g_1$ and weight $\omega$, up to $x_2$, with orientation $g_2$ and weight $\omega'$. At $x_2$, keeping the orientation $g_2$, the weight changes to $\omega'$, and then $\gamma_2$ is followed, etc. This precisely characterizes a non-oriented center vortex, where the flux orientation along the Cartan subalgebra changes. Additionally, notice that $|\omega'\rangle\langle\omega|$ is the root vector $E_\alpha$, which is in line with the presence of pointlike defects carrying adjoint charge. Moreover, when the chain configuration links the Wilson loop $\mathcal{C}_e$, one of the holonomies $\Gamma_{\gamma_1}, \ldots, \Gamma_{\gamma_n}$ gives a center element, while all the others are trivial, thus leading to the expected center-element for a chain, up to a positive and real weight factor. Performing the integrals on the group, we arrived at an additional vertex and the final formula for the ensemble average of $\mathcal{W}_D(\mathcal{C}_e)$, incorporating all the configurations discussed so far,

$$\langle \mathcal{Z}_D(\mathcal{C}_e) \rangle = \frac{Z[b_\mu^{\mathcal{C}_e}]}{Z[0]}, \qquad Z[b_\mu] = \int [D\Phi]\, e^{-S_{\text{eff}}(\Phi, b_\mu)} , \tag{34}$$

$$S_{\text{eff}}(\Phi, b_\mu) = \int d^3x \left( \text{Tr}(D_\mu \Phi)^\dagger D_\mu \Phi + V(\Phi, \Phi^\dagger) \right), \qquad D_\mu = \partial_\mu - i b_\mu , \tag{35}$$

$$V(\Phi, \Phi^\dagger) = \frac{3}{2}\lambda_0 \kappa \text{Tr}(\Phi^\dagger \Phi + \frac{\mu}{\lambda_0} I_N)^2 - \zeta_0 (3\kappa)^{\frac{N}{2}} (\det \Phi + \det \Phi^\dagger) - 3\vartheta_0 \kappa \text{Tr}\left(\Phi^\dagger T_A \Phi T_A\right), \tag{36}$$

where $\lambda_0, \zeta_0, \vartheta_0 > 0$, and we have made the redefinition $\Phi \to \sqrt{3\kappa}\Phi$ of the field. When vortices with positive stiffness percolate ($1/\kappa > 0$, $\mu < 0$), a condensate is formed. In the parameter region $\lambda_0\,, \zeta_0 \gg \vartheta_0$, the most relevant fluctuations will be parametrized by $\Phi \propto S$, $S \in SU(N)$. It is interesting to check in the lattice how the different configuration types are recovered. The additional non-oriented component in the discretized theory is generated from the product of an adjoint variable arising from the new term

$$\text{Tr}\left(\Phi^\dagger T_A \Phi T_A\right) \sim \text{const.}\,\text{Tr}(\text{Ad}(S)) , \tag{37}$$

at a lattice site $x$, with the adjoint contribution in $N \otimes \bar{N}$ associated with $V(x)$ and $V^\dagger(x)$.

*5.2. Saddle-Point Analysis in 3d*

For non-trivial $\vartheta$, the $SU(N)$ classical vacua degeneracy is lifted, and the possible global minima become discrete:

$$\Phi = v\mathcal{Z}_N, \qquad \mathcal{Z}_N = \{e^{i\frac{2\pi n}{N}} ; n = 0, 1, \ldots, N-1\} , \tag{38}$$

$$6\lambda_0 \kappa N \left( v^2 + \frac{\mu}{\lambda_0} \right) - 2\zeta_0 (3\kappa)^{\frac{N}{2}} N v^{N-2} - 3\kappa\vartheta_0(N^2 - 1) = 0 . \tag{39}$$

Thus, the presence of instantons opens the possibility of stable domain walls that interpolate the different vacua. In this case, the calculation may be approximated by a saddle-point expansion. Considering a large circular Wilson Loop $\mathcal{C}_e$ centered at the origin of the $x_2 - x_3$ plane, the effect of the source is simply to impose the boundary conditions

$$\lim_{x_1 \to -\infty} \Phi(x_1, x_2, x_3) = v I_N, \qquad \lim_{x_1 \to \infty} \Phi(x_1, x_2, x_3) = v e^{i2\pi\beta_e \cdot T}, \qquad (0, x_2, x_3) \in S(\mathcal{C}_e) . \tag{40}$$

In [67], we showed that the Ansatz

$$\Phi = (\eta I_N + \eta_0 \beta \cdot T)e^{i\theta\beta\cdot T}e^{i\alpha}$$
(41)

closes the equations of motion, yielding scalar equations for the profiles $\eta, \eta_0, \theta, \alpha$. Due to the relation $e^{i2\pi\beta_e\cdot T} = e^{-i\frac{2k\pi}{N}}$, the boundary conditions (40) may be imposed either by a solution where $\alpha$ varies with $\theta$ constant, or vice versa. The first possibility is closely related to the 't Hooft model (cf. Equation (21)). In the second case, the $\theta$ variation is governed by the Sine-Gordon equation

$$\partial_{x_1}^2 \theta = \frac{3\kappa\vartheta_0}{2}\sin(\theta) .$$
(42)

In this manner, for quarks with $N$-ality $k$, we obtained the asymptotic Casimir Law

$$\epsilon_k = \frac{k(N-k)}{N-1}\epsilon_1 ,$$
(43)

where $\epsilon_1$ is proportional to the Sine-Gordon parameter $3\kappa\vartheta_0$.

*5.3. A 4d Ensemble with Asymptotic Casimir Law*

Here, we review the ensembles of oriented and non-oriented center vortices in four dimensions as proposed in [46]. In that study, instead of deriving the effective description of center-vortex ensembles with negative tension and positive stiffness, we started the discussion from the natural Goldstone modes defined on the lattice (see also Section 3.2). The missing steps are expected to be implemented by deriving diffusion loop equations including the effect of stiffness. The lattice description of an Abelian ensemble of worldsurfaces coupled to an external Kalb–Ramond field in the form

$$\int d\sigma_1 d\sigma_2\, B_{\mu\nu}(X(\sigma_1,\sigma_2))\Sigma^{\mu\nu}(X(\sigma_1,\sigma_2)), \quad \Sigma^{\mu\nu} = \frac{\partial X^\mu}{\partial\sigma_1}\frac{\partial X^\nu}{\partial\sigma_2} - \frac{\partial X^\nu}{\partial\sigma_1}\frac{\partial X^\mu}{\partial\sigma_2} ,$$
(44)

where $X^\mu(\sigma_1,\sigma_2)$ is a parametrization of the worldsurface, was obtained in [45]. This was done in terms of a complex-valued string field $V(C)$, where $C$ is a closed loop formed by a set of lattice links. The associated action is

$$S_V = -\sum_C \sum_{p\in\eta(C)} \left[\bar{V}(C+p)U_p V(C) + \bar{V}(C-p)\bar{U}_p V(C)\right] + \sum_C m^2 \bar{V}(C)V(C) .$$
(45)

$\eta(C)$ is the set of plaquettes that share at least one common link with $C$, while $C+p$ is the path that follows $C$ until the initial site of the common link, then detours through the other three links of $p$, and continues along the remaining part of $C$. In addition, the coupling (44) originates the plaquette field $U_p = e^{ia^2 B_{\mu\nu}(p)}$. Then, the following polar decomposition was considered

$$V(C) = w(C)\prod_{l\in C} V_l, \quad V_l \in U(1) ,$$
(46)

with a phase factor that has a "local" character, as it was written in terms of the holonomy along $C$ of gauge field link-variables $V_l$. Finally, when a condensate is formed ($m^2 < 0$), it was argued that the modulus is practically frozen[2], so that $w(C) \approx w > 0$. By using this fact in Equation (45), the only links whose contribution do not cancel are those belonging to $p$:

$$\bar{V}(C+p)U_p V(C) = w^2 \prod_{l\in C+p}\prod_{l'\in C} \bar{V}_l U_p V_{l'} = w^2 U_p \prod_{l\in p}\bar{V}_l .$$
(47)

Thus,

$$S_{\text{latt}}^{(4)}(\alpha_p) = \tilde{\beta} \sum_p \text{Re} \left[ \mathbb{I} - \bar{U}_p \prod_{l \in p} V_l \right] . \tag{48}$$

where the sum is over all plaquettes $p$ and a constant was added such that the action vanishes for a trivial plaquette. Then, the description of a loop condensate, where loops are expected to percolate, is much simpler than that associated with a general phase. The string field parameter gives place to simpler gauge field Goldstone variables $V_\mu = e^{i\Lambda_\mu(l)}$, governed by a Wilson action with frustration $U_p$. This was the starting input used in [46]. An external Kalb–Ramond field that generates the center elements when the simplest center-vortex worldsurface link $\mathcal{C}_e$ is obtained by replacing $B_{\mu\nu} \to \frac{2\pi k}{N} s_{\mu\nu}$, where $k$ is the $N$-ality of the quark representation D and

$$s_{\mu\nu} = \int_{S(\mathcal{C}_e)} d^2\tilde{\sigma}_{\mu\nu} \delta^{(4)}(x - X(\sigma_1, \sigma_2)) , \tag{49}$$

$$d^2\tilde{\sigma}_{\mu\nu} = \frac{1}{2}\epsilon_{\mu\nu\alpha\beta}\left(\frac{\partial X^\alpha}{\partial\sigma_1}\frac{\partial X^\beta}{\partial\sigma_2} - \frac{\partial X^\beta}{\partial\sigma_1}\frac{\partial X^\alpha}{\partial\sigma_2}\right)d\sigma_1 d\sigma_2 \tag{50}$$

is localized on $S(\mathcal{C}_e)$. In the lattice, this localized source corresponds to a frustration $U_p = e^{i\alpha_p}$, where $\alpha_p = -2\pi k/N$ if $p$ intersects $S(\mathcal{C}_e)$ and it is trivial otherwise. Similarly to the 3d case, we can check a posteriori that the lattice expansion involves an average of center elements over closed worldsurfaces (see Section 3.2). This is a consequence of the properties of $U(1)$ group integrals. This also applies to the non-Abelian extension $V_\mu \in SU(N)$, governed by

$$S_V^{\text{latt}}(\alpha_{\mu\nu}) = \tilde{\beta} \sum_{x,\mu<\nu} \text{Re tr}\left[ I - \bar{U}_{\mu\nu} V_\mu(x) V_\nu(x + \hat{\mu}) V_\mu^\dagger(x + \hat{\nu}) V_\nu^\dagger(x) \right] ,$$

where plaquettes are denoted as usual. The closed surfaces are generated because $N \otimes \bar{N}$ contain a singlet. Interestingly, the $SU(N)$ version has additional configurations where $N$ open worldsurfaces meet at a loop formed by a set of links. This is due to the presence of a singlet in the product of $N$ link variables. Therefore, the associated normalized partition function

$$\frac{Z_V^{\text{latt}}[\alpha_{\mu\nu}]}{Z_V^{\text{latt}}[0]} , \qquad Z_V^{\text{latt}}[\alpha_{\mu\nu}] = \int [\mathcal{D}V_\mu]\, e^{-S_V^{\text{latt}}(\alpha_{\mu\nu})} \tag{51}$$

is an average of the center elements generated when a Wilson loop in representation D is linked by an ensemble of oriented center-vortex worldsurfaces with matching rules.

### 5.4. Including Non-Oriented Center Vortices in 4d

Although thin oriented or non-oriented center vortices contribute with the same center-element to the Wilson loop, they are distinct gauge field configurations, with different Yang–Mills action densities. It is then important to underline that the ensemble measure could depend on the monopole component. In order to attach center vortices to monopoles, we included dual adjoint holonomies defined on a "gas" of monopole loops and fused worldlines. In this case, because of the integration properties in the group there are additional relevant configurations like those of Figure 3a,b. The use of adjoint holonomies is in line with the fact that monopoles carry weights of the adjoint representation (the difference of fundamental weights), see [46,51].

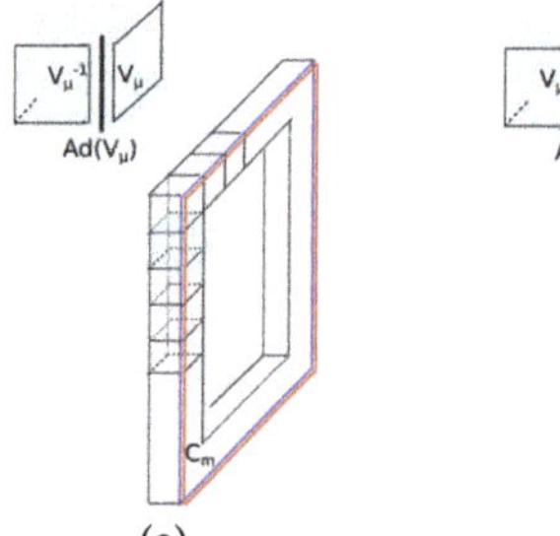
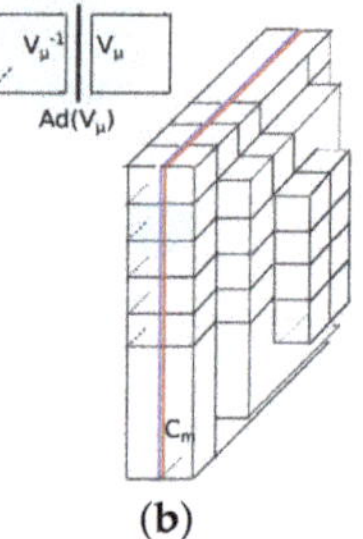
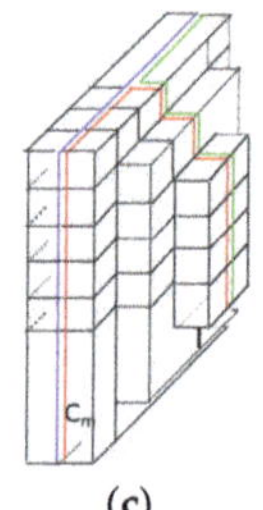



**Figure 3.** Non−oriented center vortices containing monopole worldlines. We show a configuration that contributes to the lowest order in $\tilde{\beta}$ (**a**), and one that becomes more important as $\tilde{\beta}$ is increased (**b**). A non-oriented center vortex with three matched monopole worldlines is shown in (**c**).

Then, partial contributions with $n$-loops were generated by

$$Z_{\mathrm{mix}}^{\mathrm{latt}}[\alpha_{\mu\nu}]\big|_{\mathrm{p}} \propto \int [\mathcal{D}V_\mu]\, e^{-S_V^{\mathrm{latt}}(\alpha_{\mu\nu})}\, \mathcal{W}_{\mathrm{Ad}}^{(1)}\dots \mathcal{W}_{\mathrm{Ad}}^{(n)}$$

$$\mathcal{W}_{\mathrm{Ad}}^{(k)} = \frac{1}{N^2-1}\,\mathrm{tr}\left(\prod_{(x,\mu)\in \mathcal{C}_k^{\mathrm{latt}}} \mathrm{Ad}\big(V_\mu(x)\big)\right). \tag{52}$$

In addition to the matching rules of $N$ worldsurfaces, which in the continuum occur as $N$ different fundamental magnetic weights add to zero, monopole worldlines carrying different adjoint weights (roots) can also be fused. For example, when $N \geq 3$, three worldlines carrying different roots that add up to zero can be created at a point. For this reason, we also considered partial contributions to the ensemble like

$$Z_{\mathrm{mix}}^{\mathrm{latt}}[\alpha_{\mu\nu}]\big|_{\mathrm{p}} \propto \int [\mathcal{D}V_\mu]\, e^{-S_V^{\mathrm{latt}}(\alpha_{\mu\nu})}\, D_3^{\mathrm{latt}}, \tag{53}$$

where $D_3^{\mathrm{latt}}$ is formed by combining three adjoint holonomies $\mathrm{Ad}(\Gamma_j^{\mathrm{latt}})$ (see Figure 3c). Other natural rules involve the matching of four worldlines. Then, weighting the monopole holonomies with the simplest geometrical properties (tension and stiffness), the lattice mixed ensemble of oriented and non-oriented center vortices with matching rules can be pictorially represented as

$$Z_{\mathrm{mix}}^{\mathrm{latt}}[\alpha_{\mu\nu}] = \int [\mathcal{D}V_\mu]\, e^{-S_V^{\mathrm{latt}}(\alpha_{\mu\nu})} \times \dots \tag{54}$$

where the dots represent possible combinations of holonomies as illustrated in Figure 4.

Then, noting that $e^{i2\pi k/N} = e^{-i\,2\pi\,\beta\cdot w_e}$, where $\beta$ is a fundamental magnetic weight and $w_e$ is a weight of the quark representation D, we considered the naive continuum limit, $V_\mu(x) = e^{ia\Lambda_\mu(x)}$, $\Lambda_\mu \in \mathfrak{su}(N)$,

$$Z_{\mathrm{mix}}[s_{\mu\nu}] = \int [\mathcal{D}\Lambda_\mu]\, e^{-\int d^4x\, \frac{1}{4g^2}\left(F_{\mu\nu}(\Lambda)-2\pi s_{\mu\nu}\beta_e\cdot T\right)^2} \times \dots \tag{55}$$

The dots represent all possible monopole configurations to be attached to center-vortex worldsurfaces (see Figure 5). Each contribution was obtained using the methods in the Appendix A. The first factor in Figure 5 (monopole loops) generates emergent adjoint fields coupled to the effective field $\Lambda_\mu$.

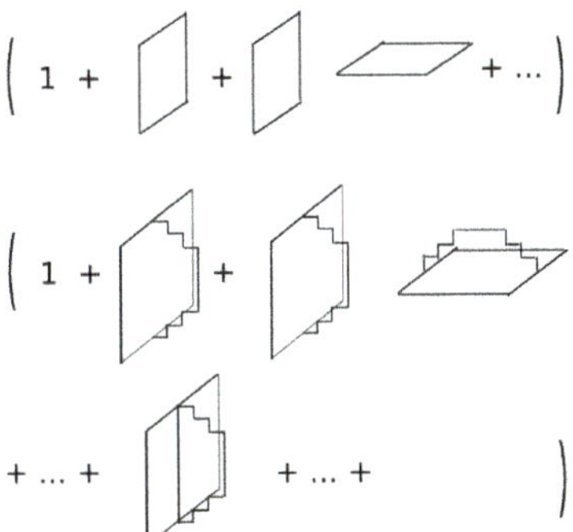

**Figure 4.** Natural combinations of holonomies that can be used to model the mixed ensemble of oriented and non-oriented center vortices. Each contribution is weighted with tension and stiffness.

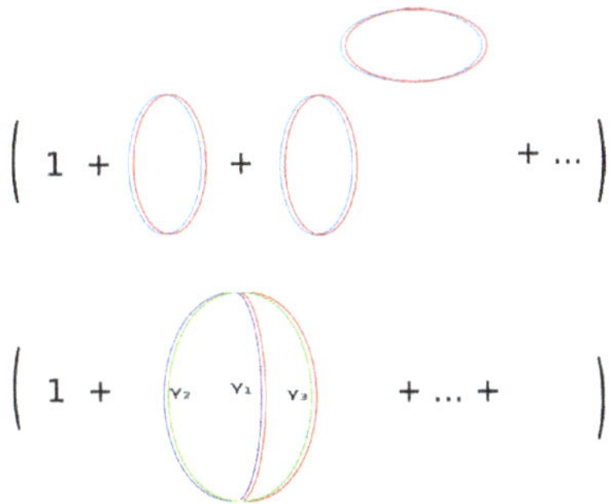

**Figure 5.** Continuum limit of the monopole sector. The worldline contributions are obtained from the solution to a Fokker–Planck diffusion equation.

For example, a diluted ensemble of a given species of monopoles, with tension $\tilde{\mu}$ and stiffness $\frac{1}{\tilde{\kappa}}$, is generated by

$$e^{\int_0^\infty \frac{dL}{L}\, \int d^4x\, du\, \mathrm{tr}\, Q(x,u,x,u,L)}\,, \tag{56}$$

where $Q$ is given by Equation (A1) and D corresponds to the adjoint representation. In the small-stiffness approximation, the non-Abelian diffusion equation for $Q$ is solved by Equation (A12), with

$$O = -\frac{\pi}{12\tilde{\kappa}}\left(\partial_\mu - i\,\mathrm{Ad}(\Lambda_\mu)\right)^2 + \tilde{\mu} I_{\mathscr{D}_{\mathrm{Ad}}}\,. \tag{57}$$

Therefore, the factor in Equation (56) was approximated by

$$e^{-\mathrm{Tr}\ln O} = \int [\mathcal{D}\zeta][\mathcal{D}\zeta^\dagger]\, e^{-\int d^4x\, \left((D_\mu\zeta^\dagger, D_\mu\zeta) + \tilde{m}^2(\zeta^\dagger,\zeta)\right)}$$
$$\tilde{m}^2 = (12/\pi)\,\tilde{\mu}\tilde{\kappa}, \quad D_\mu(\Lambda)\,\zeta = \partial_\mu\zeta - i\,[\Lambda_\mu,\zeta]\,, \tag{58}$$

where $\zeta$ is an emergent complex adjoint field, and we have introduced the Killing product between two Lie algebra elements $X, Y$ as $(X, Y) \equiv \mathrm{tr}(\mathrm{Ad}(X)\mathrm{Ad}(Y))$. In the continuum, the path-integral of $\mathrm{Ad}(\Gamma[\Lambda])$ over shapes and lengths led to the Green's function for the operator $O$, so that fusion rules like the one in Equation (53) became effective Feynman diagrams. Indeed, to differentiate the monopole lines that can be fused, the monopole loop ensemble was extended to include different species. At the end, a set of real adjoint fields

$\psi_I \in \mathfrak{su}(N)$ emerged ($I$ is a flavor index). This, together with the non-Abelian Goldstone modes (gauge fields), led to a class of effective Yang–Mills–Higgs (YMH) models,

$$Z_{\text{mix}}[s_{\mu\nu}] = \int [\mathcal{D}\Lambda_\mu][\mathcal{D}\psi]\, e^{-\int d^4x \left[ \frac{1}{4g^2} \left( F_{\mu\nu}(\Lambda) - 2\pi s_{\mu\nu}\beta_{\text{e}} \cdot T \right)^2 + \frac{1}{2}(D_\mu\psi_I, D_\mu\psi_I) + V_{\text{H}}(\psi) \right]} . \tag{59}$$

The vertex couplings weight the abundance of each fusion type. Percolating monopole worldlines (positive stiffness and negative tension) favor a spontaneous symmetry breaking phase that can easily correspond to $SU(N) \to Z(N)$ SSB. This pattern has been extensively studied in the literature (see [55–60,70–72] and references therein).

*5.5. Analysis of the Saddle Point in 4d*

In [73–75], we investigated a possible model containing $N^2 - 1$ real adjoint scalar fields $\psi_I$ and $\text{Ad}(SU(N))$ flavor symmetry,

$$V_{\text{H}}(\psi) = c + \frac{\mu^2}{2}(\psi_A, \psi_A) + \frac{\kappa}{3} f_{ABC}(\psi_A \wedge \psi_B, \psi_C) + \frac{\lambda}{4}(\psi_A \wedge \psi_B)^2 , \tag{60}$$

where $X \wedge Y \equiv -i[X, Y]$. This model includes some of the correlations previously discussed. The case $\tilde{\mu} = 0$ is specially interesting. At this point, the classical vacua are

$$\Lambda_\mu = \frac{i}{g} S \partial_\mu S^{-1}, \quad \psi_A = v S T_A S^{-1} . \tag{61a}$$

Then, the Higgs vacua manifold is $\text{Ad}(SU(N))$ and the system undergoes $SU(N) \to Z(N)$ SSB, which leads to stable confining center strings. Interestingly, at $\tilde{\mu} = 0$, we were able to find a set of BPS equations that provide vortex solutions whose energy is

$$\epsilon = 2\pi \tilde{g} v^2 \beta \cdot 2\delta , \tag{62}$$

where $\delta$ is the sum of all positive roots of the Lie algebra of $SU(N)$. Using an inductive proof based on the Young tableau properties, we showed that the smallest $\beta \cdot 2\delta$ factor is given by the $k$-A weight, the highest weight of the totally antisymmetric representation with $N$-ality $k$. Then, for a general representation $D(\cdot)$ with $N$-ality $k$, the asymptotic string tension satisfies

$$\frac{\sigma(\text{D})}{\sigma(\text{F})} = \frac{C_2(k\text{-A})}{C_2(\text{F})} = \frac{k(N-k)}{N-1} , \tag{63}$$

which is one of the possible behaviors observed in lattice simulations. Furthermore, the radial energy distribution transverse to the string is $k(N-k)$ times the distribution for a Nielsen–Olesen vortex. For $k = 1$, this agrees with the YM energy distribution of the fundamental confining string, recently obtained from lattice Monte Carlo simulations [32–34].

## 6. Discussion

We reviewed ensembles formed by oriented and non-oriented center vortices in 3d and 4d Euclidean spacetime that could capture the confinement properties of $SU(N)$ pure Yang–Mills theory. Different measures to compute center-element averages were discussed. In 3d and 4d, they include percolating oriented center-vortex worldlines and worldsurfaces that generate emergent Goldstone modes, which correspond to compact scalar and gauge fields, respectively. The models also have the natural matching rules of $N$ center vortices, as well as the non-oriented component where center-vortex worldlines (worldsurfaces) are attached to lower-dimensional defects, i.e., instantons (monopole worldlines) in 3d (4d). In addition to the weighting center vortices with tension and stiffness, it is also natural to include additional weights for the lower dimensional defects. In 4d, monopole matching rules are also included. The corresponding effective field content and the SSB pattern may lead to the formation of a confining center string, represented by a domain wall (vortex) in two-dimensional (three-dimensional) real space. The Lüscher term is originated as usual, from the string-like transverse fluctuations of the flux tube. An asymptotic Casimir law can also be accommodated. This asymptotic behavior was observed in 3d, while in 4d it is among the possibilities.

More recently, the transverse distribution of the 4d YM energy-momentum tensor $T_{\mu\nu}$ and the field profiles have been analyzed at intermediate and nearly asymptotic distances [32,34,76]. In [34], it was numerically shown that the $T_{\mu\nu}$ tensor of the Abelian Nielsen–Olesen (ANO) model cannot fit the $SU(3)$ data at the vortex guiding center for $L = 0.46$ fm (intermediate distance) and $L = 0.92$ fm (near asymptotic distance) at the same time. In fact, in [34], it was shown that the components of the energy-momentum tensor at the origin may not be accommodated for $L = 0.46$ fm. Then, on this basis, an ANO effective model to describe the fundamental string was discarded. However, while it is clear that an effective model for the confining flux tube should work at asymptotic distances, it is not that obvious that the same model could be extrapolated to intermediate distances. By intermediate distances we mean those where the string tension scales with the quadratic Casimir of the quark representation. In particular, this is the region where adjoint quarks are still confined by a linear potential, before the breaking of the adjoint string. On the other hand, in the asymptotic region, gluonic excitations around external quarks in a given irreducible representation $D(\cdot)$ may be created, so as to produce an asymptotic scaling law that only depends on the $N$-ality of $D(\cdot)$. As discussed in this review, the effective field descriptions were derived by considering the (weighted) average of center elements over oriented and non-oriented center vortices, which is expected to be applicable at asymptotic distances. In other words, we wonder if it is meaningful to discard possible effective models on the basis of the lack of adjustment to lattice data on a wide range that includes the intermediate region, where these models are not expected to fully capture the physics. Additionally, note that the known mechanism to explain intermediate Casimir scaling is based on including center-vortex thickness. In turn, these finite-size effects are not included in the ensemble definition that leads to our effective model. Interestingly, while the lattice data rule out the ANO model at intermediate distances $L = 0.46$ fm, such profiles are still among the possibilities at the nearly asymptotic distance $L = 0.92$ fm. Accordingly, the 4d $SU(N) \to Z(N)$ models we discussed in this review have a point in parameter space where the infinite flux tube profiles Abelianize, while keeping all the required $N$-ality properties. Additionally, the ideas presented in this review imply that not only an asymptotic Casimir law should be observed, but also that the transverse confining flux tube profiles for quarks in different representations should be the same, up to the asymptotic scaling law. This is true for both 3d and 4d, with the profiles being of the Sine-Gordon type in 3d. It would be interesting to test these predictions with lattice simulations.

**Funding:** This research was funded by Conselho Nacional de Desenvolvimento Científico e Tecnológico (CNPq), the Coordenação de Aperfeiçoamento de Pessoal de Nível Superior (CAPES), and the Deutscher Akademischer Austauschdienst (DAAD).

**Conflicts of Interest:** The authors declare no conflicts of interest.

## Appendix A. Non-Abelian Diffusion

Center vortices in 3 dimensions and monopoles in 4 dimensions are propagated along worldlines in Euclidean spacetime. Then, the corresponding ensembles will naturally involve the building block $Q$ associated to a worldline with length $L$ that starts at $x_0$ with orientation $u_0$ and ends at $x$ with final orientation $u$. This is given by

$$Q(x, u, x_0, u_0, L) = \int [dx(s)]_{x_0,u_0}^{x,u}\, e^{-S(\gamma)}\, D\big(\Gamma_\gamma[b_\mu]\big)\,, \tag{A1}$$

$$\Gamma_\gamma[b_\mu] = P\{e^{i \int_\gamma dx_\mu b_\mu}\}\,, \tag{A2}$$

where $S(\gamma)$ is a vortex effective action, and an interaction with a general non-Abelian gauge field $b_\mu$ was considered. We are interested in the specific form

$$S(\gamma) = \int_0^L ds \left( \frac{1}{2\kappa} \dot{u}_\mu \dot{u}_\mu + \mu \right), \quad u_\mu(s) = \frac{dx_\mu}{ds},$$
(A3)

which corresponds to tension $\mu$ and stiffness $1/\kappa$. These objects were extensively studied in [46,77]. In what follows, we review the results obtained.

For the simplest center-vortex worldlines in 3d, D is the defining $SU(N)$ representation, while for monopole worldlines in 4d, D corresponds to the adjoint. To derive a diffusion equation for this object, the paths were discretized into $M$ segments of length $\Delta L = L/M$. In this case, the path ordering was obtained from

$$P\{e^{-\int_0^L ds H(x(s),u(s))}\} = e^{-H(x_M,u_M)\Delta L} \dots e^{-H(x_1,u_1)\Delta L},$$
(A4)

where $H(x,u) = -iD(u_\mu b_\mu(x))$. The relation between the building block $Q_M$ associated to a discretized path containing $M$ segments of length $\Delta L$ and that associated with a path of length $L - \Delta L$ is given by:

$$Q_M(x,u,x_0,u_0,L) = \int d^n x' d^{n-1} u' e^{-\mu\Delta L} \psi(u - u') \times$$

$$e^{-\mu\Delta L} e^{-H(x,u)\Delta L} \delta(x - x' - u\Delta L) Q_{M-1}(x',x_0,u',u_0),$$
(A5)

with

$$\psi(u - u') = \mathcal{N} e^{-\frac{1}{2\kappa}\Delta L \left(\frac{u-u'}{\Delta L}\right)^2}$$
(A6)

arising from the discretization of the stiffness term. It acts like an angular distribution in velocity space, which tends to bring $u'$ close to $u$. Expanding Equation (A5) to first order in $\Delta L$, and taking the limit $\Delta L \to 0$, the diffusion equation

$$\left( \partial_L - \frac{\kappa\sigma}{2} \hat{L}_u^2 + \mu + u_\mu(\partial_\mu - iD(b_\mu)) \right) Q(x,u,x_0,u_0,L) = 0,$$
(A7)

was obtained, to be solved with the initial condition

$$Q(x,u,x_0,u_0,0) = \delta(x - x_0)\delta(u - u_0)I_{\mathscr{D}}.$$
(A8)

$\mathscr{D}$ is the dimension of the quark representation D and $\hat{L}_u^2$ is the Laplacian on the sphere $S^{n-1}$. The constant $\sigma$ is given, in $n$ spacetime dimensions, by

$$\sigma = \frac{\sqrt{\pi}}{2^{n-3}} \frac{\Gamma\left(\frac{n-2}{2}\right)\Gamma\left(\frac{n+1}{2}\right)}{\Gamma^2\left(\frac{n-1}{2}\right)\Gamma\left(\frac{n-3}{2}\right)} \left( \frac{4\Gamma(n-3)}{\Gamma\left(\frac{n-3}{2}\right)} - \frac{\Gamma(n-1)}{\Gamma\left(\frac{n+1}{2}\right)} \right).$$
(A9)

For the cases considered in this review ($n = 3, 4$), $\sigma = 1, 2/\pi$, respectively. In the limit of small stiffness, there is practically no correlation between $u$ and $u_0$, which allowed for a consistent solution of these equations with only the lowest angular momenta components:

$$Q(x,u,x_0,u_0,L) \approx Q_0(x,x_0,L), \quad \partial_L Q_0(x,x_0,L) = -O Q_0(x,x_0,L),$$
(A10)

$$O = -\frac{2}{(n-1)\sigma\kappa n}(\partial_\mu - iD(b_\mu))^2 + \mu, \quad Q_0(x,x_0,0) = \frac{1}{\Omega_{n-1}}\delta(x - x_0),$$
(A11)

$\Omega_{n-1}$ being the solid angle of $S^{n-1}$. This implies,

$$Q(x,u,x_0,u_0,L) \approx \langle x|e^{-LO}|x_0\rangle.$$
(A12)

Then, in this limit, we also have

$$\int_0^\infty dL\,du\,du_0 \int [Dx]_{x_0,u_0}^{x,u}\, e^{-S(\gamma)}\, \mathrm{D}(\Gamma[b]) = \int_0^\infty dL\,du\,du_0\, Q(x,u,x_0,u_0,L)$$

$$\approx \langle x|O^{-1}|x_0\rangle, \quad O\,G(x,x_0) = \delta(x-x_0)\,I_\mathscr{D}\,. \tag{A13}$$

## Notes

[1]    Namely, a SSB pattern with discrete classical vacua in $(2+1)$d and multiple connected vacua in $(3+1)$d.

[2]    Similarly to the 3d case, this phase should be stabilized by a quartic interaction.

## References


1.    Wilson, K.G. Confinement of quarks. *Phys. Rev. D* **1974**, *10*, 2445. [CrossRef]
2.    Bali, G.S. QCD forces and heavy quark bound states. *Phys. Rep.* **2001**, *343*, 1. [CrossRef]
3.    Hooft, G. On the phase transition towards permanent quark confinement. *Nucl. Phys.* **1978**, *B138*, 1. [CrossRef]
4.    Stokes, F.M.; Kamleh, W.; Leinweber, D.B. Visualizations of coherent center domains in local Polyakov loops. *Ann. Phys.* **2014**, *348*, 341. [CrossRef]
5.    Kratochvila, S.; de Forcrand, P. Observing string breaking with Wilson loops. *Nucl. Phys.* **2003**, *B671*, 103. [CrossRef]
6.    Debbio, L.D.; Faber, M.; Giedt, J.; Greensite, J.; Olejnik, S. Center dominance and $Z_2$ vortices in SU(2) lattice gauge theory. *Phys. Rev.* **1997**, *D55*, 2298.
7.    Debbio, L.D.; Faber, M.; Giedt, J.; Greensite, J.; Olejnik, S. Detection of center vortices in the lattice Yang-Mills vacuum. *Phys. Rev.* **1998**, *D58*, 094501. [CrossRef]
8.    De Forcrand, P.; Pepe, M. Center vortices and monopoles without lattice Gribov copies. *Nucl. Phys. B* **2001**, *598*, 557. [CrossRef]
9.    Faber, M.; Greensite, J.; Olejník, S. Direct Laplacian Center Gauge. *J. High Energy Phys.* **2001**, *1*, 053. [CrossRef]
10.    Golubich, R.; Faber, M. The Road to Solving the Gribov Problem of the Center Vortex Model in Quantum Chromodynamics. *Acta Phys. Polon. Suppl.* **2020**, *13*, 59. [CrossRef]
11.    Engelhardt, M.; Reinhardt, H. Center projection vortices in continuum Yang–Mills theory. *Nucl. Phys.* **2000**, *B567*, 249. [CrossRef]
12.    Alexandrou, C.; de Forcrand, P.; D'Elia, M. The role of center vortices in QCD. *Nucl. Phys. A* **2000**, *663*, 1031. [CrossRef]
13.    de Forcrand, P.; D'Elia, M. Relevance of center vortices to QCD. *Phys. Rev. Lett.* **1999**, *82*, 4582. [CrossRef]
14.    Höllwieser, R.; Schweigler, T.; Faber, M.; Heller, U.M. Center vortices and chiral symmetry breaking in SU(2) lattice gauge theory. *Phys. Rev. D* **2013**, *88*, 114505. [CrossRef]
15.    Nejad, S.M.H.; Faber, M.; Höllwieser, R. Colorful plane vortices and chiral symmetry breaking in SU(2) lattice gauge theory. *J. High Energy Phys.* **2015**, *10*, 108. [CrossRef]
16.    Deldar, S.; Dehghan, Z.; Faber, M.; Golubich, R.; Höllwieser, R. Influence of Fermions on Vortices in SU(2)-QCD. *Universe* **2021**, *7*, 130.
17.    Trewartha, D.; Kamleh, W.; Leinweber, D. Evidence that centre vortices underpin dynamical chiral symmetry breaking in SU(3) gauge theory. *Phys. Lett.* **2015**, *B747*, 373. [CrossRef]
18.    Trewartha, D.; Kamleh, W.; Leinweber, D. Connection between center vortices and instantons through gauge-field smoothing. *Phys. Rev.* **2015**, *D92*, 074507. [CrossRef]
19.    Biddle, J.C.; Kamleh, W.; Leinweber, D.B. Visualization of center vortex structure. *Phys. Rev. D* **2020**, *102*, 034504. [CrossRef]
20.    Langfeld, K.; Reinhardt, H.; Tennert, O. Confinement and scaling of the vortex vacuum of SU(2) lattice gauge theory. *Phys. Lett.* **1998**, *B419*, 317. [CrossRef]
21.    Boyko, P.Y.; Polikarpov, M.I.; Zakharov, V.I. Geometry of percolating monopole clusters. *Nucl. Phys. Proc. Suppl.* **2003**, *119*, 724. [CrossRef]
22.    Bornyakov, V.G.; Boyko, P.Y.; Polikarpov, M.I.; Zakharov, V.I. Monopole clusters at short and large distances. *Nucl. Phys.* **2003**, *B672*, 222. [CrossRef]
23.    Langfeld, K. Vortex structures in pure lattice gauge theory. *Phys. Rev.* **2004**, *D69*, 014503.
24.    Greensite, J. *An Introduction to the Confinement Problem*, 2nd ed.; Springer Nature Switzerland: Cham, Switzerland, 2020.
25.    Lucini, B.; Teper, M. Confining strings in gauge theories. *Phys. Rev.* **2001**, *D64*, 105019.
26.    Faber, M.; Greensite, J.; Olejník, S. Casimir scaling from center vortices: Towards an understanding of the adjoint string tension. *Phys. Rev.* **1998**, *D57*, 2603. [CrossRef]
27.    Greensite, J.; Langfeld, K.; Olejník, S.; Reinhardt, H.; Tok, T. Color screening, Casimir scaling, and domain structure in G(2) and SU(N) gauge theories. *Phys. Rev.* **2007**, *D75*, 034501. [CrossRef]
28.    Lucini, B.; Teper, M.; Wenger, U. Glueballs and k-strings in SU(N) gauge theories: Calculations with improved operators. *J. High Energy Phys.* **2004**, *6*, 012. [CrossRef]
29.    Lüscher, M.; Weisz, P. Quark confinement and the bosonic string. *J. High Energy Phys.* **2002**, *7*, 049. [CrossRef]
30.    Athenodorou, A.; Teper, M. SU(N) gauge theories in 2 + 1 dimensions: glueball spectra and k-string tensions. *J. High Energy Phys.* **2017**, *2*, 015. [CrossRef]
31.    Athenodorou, A.; Teper, M. On the mass of the world-sheet 'axion'in SU(N) gauge theories in 3 + 1 dimensions. *Phys. Lett.* **2017**, *B771*, 408. [CrossRef]

32. Cea, P.; Cosmai, L.; Cuteri, F.; Papa, A. Flux tubes in the QCD vacuum. *Phys. Rev.* **2017**, *D95*, 114511. [CrossRef]
33. Yanagihara, R.; Iritani, T.; Kitazawa, M.; Asakawa, M.; Hatsuda, T. Distribution of stress tensor around static quark–anti-quark from Yang–Mills gradient flow. *Phys. Lett.* **2019**, *B789*, 210. [CrossRef]
34. Yanagihara, R.; Kitazawa, M. A study of stress-tensor distribution around the flux tube in the Abelian–Higgs model. *Prog. Theor. Exp. Phys.* **2019**, *9*, 093B02; Erratum in *Prog. Theor. Exp. Phys.* **2020**, *7*, 079201. [CrossRef]
35. Engelhardt, M.; Reinhardt, H. Center vortex model for the infrared sector of Yang–Mills theory—Confinement and deconfinement. *Nucl. Phys.* **2000**, *B585*, 591. [CrossRef]
36. Engelhardt, M.; Quandt, M.; Reinhardt, H. Center vortex model for the infrared sector of SU(3) Yang–Mills theory—Confinement and deconfinement. *Nucl. Phys.* **2004**, *B685*, 227. [CrossRef]
37. Reinhardt, H. Topology of center vortices. *Nucl. Phys.* **2002**, *B628*, 133. [CrossRef]
38. de Lemos, A.L.L.; Oxman, L.E.; Teixeira, B.F.I. Derivation of an Abelian effective model for instanton chains in 3D Yang-Mills theory. *Phys. Rev.* **2012**, *D85*, 125014. [CrossRef]
39. Oxman, L.E.; Reinhardt, H. Effective theory of the D=3 center vortex ensemble. *Eur. Phys. J.* **2018**, *D78*, 177.
40. Kleinert, H. *Gauge Fields in Condensed Matter. No. Bd. 2 in Gauge Fields in Condensed Matte*; World Scientific: Singapore, 1989.
41. Kleinert, H. *Path Integrals in Quantum Mechanics, Statics, Polymer Physics, and Financial Markets*; World Scientific: Singapore, 2006.
42. Fredrickson, G.H. *The Equilibrium Theory of Inhomogeneous Polymers*, 1st ed.; Clarendon Press: Oxford, UK, 2006; p. 452.
43. Durhuus, B.; Ambjørn, J.; Jonsson, T. *Quantum Geometry: A Statistical Field Theory Approach*; Cambridge University Press: Cambridge, UK, 1997.
44. Wheater, J.F. Random surfaces: From polymer membranes to strings. *J. Phys.* **1994**, *A27*, 3323. [CrossRef]
45. Rey, S.J. Higgs mechanism for Kalb-Ramond gauge field. *Phys. Rev.* **1989**, *D40*, 3396. [CrossRef]
46. Oxman, L.E. 4D ensembles of percolating center vortices and monopole defects: The emergence of flux tubes with -ality and gluon confinement. *Phys. Rev.* **2018**, *D98*, 036018. [CrossRef]
47. Cornwall, J.M. Finding Dynamical Masses in Continuum QCD. In *Proceedings of the Workshop on Non-Perturbative Quantum Chromodynamics*; Milton, K.A., Samuel, M.A., Eds.; Birkhäuser: Stuttgart, Germany, 1983.
48. Deldar, S. Potentials between static SU(3) sources in the fat-center-vortices model. *J. High Energy Phys.* **2001**, *1*, 013. [CrossRef]
49. Fosco, C.D.; Kovner, A. Vortices and bags in dimensions. *Phys. Rev.* **2001**, *D63*, 045009.
50. Kogan, I.I.; Kovner, A. Monopoles, Vortices and Strings: Confinement and Deconfinement in 2+1 Dimensions at Weak Coupling. *arXiv* **2002**, arXiv:hep-th/0205026.
51. Oxman, L.E.; Santos-Rosa, G.C. Detecting topological sectors in continuum Yang-Mills theory and the fate of BRST symmetry. *Phys. Rev.* **2015**, *D92*, 125025. [CrossRef]
52. Singer, I.M. Commun. Some remarks on the Gribov ambiguity. *Math. Phys.* **1978**, *60*, 7. [CrossRef]
53. Fiorentini, D.; Junior, D.R.; Oxman, L.E.; Simões, G.M.; Sobreiro, R.F. Study of Gribov copies in a Yang-Mills ensemble. *Phys. Rev.* **2021**, *D103*, 114010.
54. Fiorentini, D.; Junior, D.R.; Oxman, L.E.; Sobreiro, R.F. Renormalizability of the center-vortex free sector of Yang-Mills theory. *Phys. Rev.* **2020**, *D101*, 085007.
55. Gorsky, A.; Shifman, M.; Yung, A. Non-Abelian Meissner effect in Yang-Mills theories at weak coupling. *Phys. Rev.* **2005**, *D71*, 045010. [CrossRef]
56. Hanany, A.; Tong, D. Vortices, Instantons and branes. *J. High Energy Phys.* **2003**, *307*, 037. [CrossRef]
57. Auzzi, R.; Bolognesi, S.; Evslin, J.; Konishi, K.; Yung, A. Nonabelian superconductors: Vortices and confinement in N = 2 SQCD. *Nucl. Phys. B* **2003**, *673*, 187. [CrossRef]
58. Shifman, M.; Yung, A. Non-Abelian string junctions as confined monopoles. *Phys. Rev.* **2004**, *D70*, 045004. [CrossRef]
59. Hanany, A.; Tong, D. Vortex strings and four-dimensional gauge dynamics. *J. High Energy Phys.* **2004**, *404*, 066. [CrossRef]
60. Markov, V.; Marshakov, A.; Yung, A. Non-Abelian vortices in N = 1* gauge theory. *Nucl. Phys.* **2005**, *B709*, 267. [CrossRef]
61. Balachandran, A.P.; Digal, S.; Matsuura, T. Semisuperfluid strings in high density QCD. *Phys. Rev.* **2006**, *D73*, 074009. [CrossRef]
62. Eto, M.; Isozumi, Y.; Nitta, M.; Ohashi, K.; Sakai, N. Moduli space of non-Abelian vortices. *Phys. Rev. Lett.* **2006**, *96*, 161601. [CrossRef]
63. Eto, M.; Konishi, K.; Marmorini, G.; Nitta, M.; Ohashi, K.; Vinci, W.; Yokoi, N. Non-Abelian vortices of higher winding numbers. *Phys. Rev.* **2006**, *D74*, 065021. [CrossRef]
64. Nakano, E.; Nitta, M.; Matsuura, T. Non-Abelian strings in high-density QCD: Zero modes and interactions. *Phys. Rev.* **2008**, *D78*, 045002. [CrossRef]
65. Ambjørn, J.; Giedt, J.; Greensite, J. Vortex Structure vs. Monopole Dominance in Abelian-Projected Gauge Theory. *J. High Energy Phys.* **2000**, *2*, 033. [CrossRef]
66. Greensite, J.; Höllwieser, R. Double-winding Wilson loops and monopole confinement mechanisms. *Phys. Rev.* **2015**, *D91*, 054509. [CrossRef]
67. Junior, D.R.; Oxman, L.E.; Simões, G.M. 3D Yang-Mills confining properties from a non-Abelian ensemble perspective. *J. High Energy Phys.* **2020**, *1*, 180. [CrossRef]
68. Zhang, W.; Feng, D.H.; Gilmore, R. Coherent states: theory and some applications. *Rev. Mod. Phys.* **1990**, *62*, 867. [CrossRef]
69. Perelemov, A. *Generalized Coherent States and Their Applications*; Springer: Berlin/Heidelberg, Germany, 1986.

70. De Vega, H.J.; Schaposnik, F.A. Vortices and electrically charged vortices in non-Abelian gauge theories. *Phys. Rev.* **1986**, *D34*, 3206. [CrossRef] [PubMed]
71. Auzzi, R.; Kumar, S.P. Non-Abelian k-vortex dynamics in $N = 1^*$ theory and its gravity dual. *J. High Energy Phys.* **2008**, *12*, 77. [CrossRef]
72. Oxman, L.E. Confinement of quarks and valence gluons in SU(N) Yang-Mills-Higgs models. *J. High Energy Phys.* **2013**, *3*, 038. [CrossRef]
73. Oxman, L.E.; Vercauteren, D. Exploring center strings in and relativistic Yang-Mills-Higgs models. *Phys. Rev.* **2017**, *D95*, 025001.
74. Oxman, L.E.; Simões, G.M. $k-$strings with exact Casimir law and Abelian-like profiles. *Phys. Rev.* **2019**, *D99*, 016011. [CrossRef]
75. Junior, D.R.; Oxman, L.E.; Simões, G.M. BPS strings and the stability of the asymptotic Casimir law in adjoint flavor-symmetric Yang-Mills-Higgs models. *Phys. Rev.* **2020**, *D102*, 074005. [CrossRef]
76. Nishino, S.; Kondo, K.-I.; Shibata, A.; Sasago, T.; Kato, S. Type of dual superconductivity for the SU(2) Yang–Mills theory. *Eur. Phys. J. C* **2019**, *79*, 774. [CrossRef]
77. Oxman, L.E.; Santos-Rosa, G.C.; Teixeira, B.F.I. Coloured loops in 4D and their effective field representation. *J. Phys.* **2014**, *A47*, 305401. [CrossRef]

*Review*

# Confinement in 4D: An Attempt at Classical Understanding

**Ibrahim Burak Ilhan** [1] **and Alex Kovner** [2,*]

[1] Department of Physics, METU, Ankara 06800, Turkey; burakilhan@gmail.com
[2] Physics Department, University of Connecticut, 2152 Hillside Road, Storrs, CT 06269-3046, USA
* Correspondence: alexander.kovner@uconn.edu

**Abstract:** In this review, we revisit our approach to constructing an effective theory for Abelian and Non-Abelian gauge theories in 4D. Our goal is to have an effective theory that provides a simple classical picture of the main qualitatively important features of these theories. We set out to ensure the presence of the massless photons—Goldstone bosons in Abelian theory and their disappearance in the Non-Abelian case—accompanied by the formation of confining strings between charged states. Our formulation avoids using vector fields and instead operates with the basic degrees of freedom that are the scalar fields of a nonlinear $\sigma$-model. The Mark 1 model we study turns out to have a large global symmetry group-the 2D diffeomorphism invariance in the Abelian limit, which is isomorphic to the group of all canonical transformations in the classical two dimensional phase space. This symmetry is not present in QED, and we eliminate it by "gauging" this infinite dimensional global group. Introducing additional modifications to the model (Mark 2), we are able to prove that the "Abelian" version is equivalent to the theory of a free photon. Achieving the desired property in the "Non-Abelian" regime turns out to be tricky. We are able to introduce a perturbation that leads to the formation of confining strings in our Mark 1 model. These strings have somewhat unusual properties, in that their profile does not decay exponentially away from the center of the string. In addition, the perturbation explicitly breaks the diffeomorphism invariance. Preserving this invariance in the gauged model as well as achieving confining strings in Mark 2 model remains an open question.

**Keywords:** confinement; higher order theories; gauge theory; effective field theory; magnetic flux symmetry

**Citation:** Ilhan, I.B.; Kovner, A. Confinement in 4D: An Attempt at Classical Understanding. *Universe* **2021**, *7*, 291. https://doi.org/ 10.3390/universe7080291

Academic Editor: Dmitry Antonov

Received: 8 July 2021
Accepted: 4 August 2021
Published: 7 August 2021



## 1. Introduction

Understanding confinement in Non-Abelian gauge theories is a long standing theoretical problem. There is very little doubt that QCD is confining. One has very strong indications of that from lattice gauge theory as well as from a variety of theoretical considerations. Nevertheless, a satisfactory simple understanding of the confinement phenomenon in $3 + 1$ dimensional theories is still missing. By such an understanding, we mean a simple qualitative picture that relies on universal concepts.

In $2 + 1$ dimensions, such a picture does exist. In this low dimensionality, one is able to directly relate confinement with a universal phenomenon of spontaneous symmetry breaking. The symmetry in question is a discrete symmetry generated by the magnetic flux [1–3]. The equivalence between confinement and a spontaneous breaking of magnetic symmetry provides a simple classical picture of the formation of a confining string.

There is an additional feature of gauge theories in $2 + 1$ dimensions that very much facilitates their qualitative understanding. Namely, the effective description of confining Non-Abelian gauge theories and Abelian nonconfining differs only by simple magnetic symmetry breaking deformation. The magnetic symmetry in the Abelian case is a continuous $U(1)$ group but is a discrete group $Z_N$ in $SU(N)$ gauge theories (without fundamental matter). This reduction of symmetry is affected in the effective Lagrangian by the presence of a simple deformation. The presence of this deformation, together with the spontaneous breaking of the discreet $Z_N$ group, unambiguously ensures an area law for Wilson loops and thereby a confining potential at long distances [2–4].

This review is devoted to a recent work that aims at constructing an analogous effective theory description in 3 + 1 dimensions. The goal here is to "guess" an effective description that would display features similar to the 2 + 1 dimensional case [5,6]. We design a model that embodies features of the transition between the Abelian and Non-Abelian regimes, similar to 2 + 1 dimensions. Although it is not derived from QCD per se and therefore is not a *bona fide* QCD effective theory, amusingly, it does have some properties that have appeared before in the QCD context. In particular, the model has clear similarities with the Faddeev–Niemi model, which has been proposed as an effective theory of glueballs [7–11]. We note, however, that our perspective here is completely different, and we are not concentrating on the interpretation of knots as glueballs [7–11].

Prior to introducing our effective model, we will give a short recap of the confining physics in 2 + 1 dimensional gauge theories. Consider the simplest Abelian gauge theory—QED with scalar Higgs fields. In addition to electric charge, it has a continuous magnetic global symmetry. The generator of this $U_\mu(1)$ group is the total magnetic flux through 2D, $\Phi = \int d^2x B(x)$. As any proper global symmetry, $U_\mu(1)$ has an order parameter. In the present case, this is a complex field $V$, whose physical meaning is a field associated with creation and annihilation of point-like magnetic vortices. In the Coulomb phase, its expectation value does not vanish, $\langle V \rangle = v \neq 0$, and thus, the magnetic symmetry is spontaneously broken. One can easily write down an effective low energy theory that fits this simple symmetry breaking pattern and describes the low energy dynamics. The relevant model is defined by the Lagrangian

$$L = -\partial^\mu V \partial_\mu V^* - \lambda(V^*V - \frac{e^2}{8\pi})^2 \tag{1}$$

The phase of the field $V$ appears in Equation (1) as a Goldstone boson associated with the spontaneous breaking of $U_\mu(1)$. This is nothing but the massless photon of QED. Interestingly, although the electric charge did not figure prominently in constructing Equation (1), it is indeed present in this description in the shape of the topological charge—the winding number of the field $V$

$$J_\mu = \frac{1}{e}\epsilon_{\mu\nu\lambda}\partial_\nu V^*\partial_\lambda V \tag{2}$$

A charged state of QED in the low energy description appears as a topological soliton of $V$: $V(x) = ve^{i\theta(x)}$, with $\theta = \tan^{-1}y/x$. This description is frequently called a "dual" description as the basic fields used here are dual of the fields in the original QED Lagrangian, but a more physical view is that the Lagrangian Equation (1) is merely an effective low-energy long-distance Lagrangian of QED with scalar fields.

Equation (1) is a good starting point for understanding the confinement in Non-Abelian gauge theories. Recall that in 2 + 1 dimensions, confining theories have a weakly coupled regime. For example, the $SU(N)$ Higgs model at weak coupling is confining in the weakly coupled case. The appropriate low energy description for this theory is almost identical to Equation (1), with one important difference, i.e., an additional perturbation that breaks the magnetic $U_\mu(1)$ symmetry down to $Z_N$

$$L = -\partial^\mu V \partial_\mu V^* - \lambda(V^*V - \frac{e^2}{8\pi})^2 + \mu(V^N + V^{*N}) \tag{3}$$

The presence of this additional potential has the effect of reducing the number of degenerate vacua of the Abelian theory (which is infinite) to a finite number of states connected by the $Z_N$ symmetry transformations. The effect of this reduction on the energy of a charged state is profound. A rotationally invariant "hedgehog" configuration now has an infinite energy proportional to the volume of the system. The lowest state with the nonvanishing winding number ("color charge") is not rotationally invariant anymore but instead has the winding concentrated within a quasi one-dimensional "flux tube" [2,3]. Its energy is proportional to the length of the flux tube and thus leads to linear confinement of charges.

The identification of electric (or "color") charges with topological defects in the effective theory is intuitively very appealing. Topological defects naturally have long range interactions due to their inherently nonlocal nature, which, in conjunction with spontaneous symmetry breaking, leads directly to linear confinement. Additionally, the identification of photons with Goldstone bosons in the Abelian limit furnishes a natural universal explanation for the fact that the photon is strictly massless.

The question arises if a similar description can be achieved in 3 + 1 dimensions. One would like this description to encompass the Goldstone boson nature of photons in QED as well as an interpretation of confinement in terms of topological charges in Non-Abelian theories. Of course, life in 3 + 1 dimensions is not at all that simple. First off, photons now are vector particles and thus, their interpretation as Goldstone bosons in the standard sense is questionable. Even if one successfully argues in favor of this, identification of the relevant conserved current that breaks spontaneously is far from straightforward. Clearly, this current has to be intimately tied with the dual field strength $\tilde{F}_{\mu\nu}$ since the photon is a spin one particle [12]. The identification of photons as Goldstone bosons of this higher form symmetry was achieved a while ago in [12] and was revived recently in [13]. The dual field strength, however, is an object of a very different nature than an ordinary vector current since no local order parameter that carries its charge can be defined even in principle. One might hope that a more conventional picture of symmetry breaking coexists with the "generalized symmetry" explanation, and it would be useful to clarify this. Another significant stumbling block is that we do not know of weakly coupled confining theories in 3 + 1 dimensions. QCD is certainly strongly interacting while a classical effective description of the type described before is directly applicable only for a weakly interacting theory.

These are hard problems to solve, much too hard for the present modest attempt. Instead of addressing them head on here, we will largely ignore them and instead will simply try to construct a model that encompasses the basic properties described above:

1. The degrees of freedom of the model must be scalar fields, and no fundamental gauge fields should be involved.

2. A well-defined "Abelian regime" should be clearly definable. In this regime, two massless degrees of freedom should exist. These massless particles should be Goldstone bosons and as far as possible must have the properties of photons.

3. The Abelian regime should allow for the existence of classical topological solitons associated with the nontrivial topology of the manifold of vacua. These solitons represent electrically charged particles. More precisely, we would like the topological charge of the solitons to be associated with the mapping of the spatial infinity onto the manifold of vacua and thus be identified with $\Pi_2(M)$. Charged particles in QED are excitations of finite energy, and thus, the classical energy of the solitons must be infrared finite, and more precisely, the energy density of a soliton solution away from the position of the soliton must decrease as the fourth power of the distance. This is nontrivial in 3 + 1 dimensions since our model has no gauge fields, while scalar fields that contribute to $\Pi_2$ have to be long range.

4. A "Non-Abelian regime" of the model is achieved by adding a perturbation that breaks explicitly the symmetry, which leads to the appearance of Goldstone bosons in the Abelian case. The Goldstone bosons now disappear from the spectrum or, more precisely, acquire a finite mass. In addition, in this Non-Abelian situation, the solitons do not disappear on small spatial scales, but they must become confined by a linear potential. The linear potential should arise due to the formation of a "string" or "flux tube" with finite linear energy density between the solitons.

In the first part of this review, we discuss a model (Mark 1) that exhibits all the above features. The Abelian version of the model has, in fact, been studied some years ago from a completely different perspective in [14] as a possible variation of Maxwell's theory.

The properties of this model turn out to be a little unusual. In particular, as we will see, requiring the energy of a soliton in the Abelian regime to be finite puts a very strong restriction on possible forms of the kinetic term for the scalar fields. This noncanonical kinetic term results in rather unusual properties of confining strings once the symmetry breaking perturbation is introduced. In particular, the "Non-Abelian string" is forced into having an infinite number of zero modes. This infinite degeneracy can be avoided, but the price one has to pay is adding another perturbation that does not have a natural place in the paradigm described above.

Although the model has many nice features, it does not perfectly emulate many properties of gauge theories. Most importantly, in the Abelian regime, it has more classical solutions than allowed by the structure of Abelian gauge theories; in particular, some of them carry nonvanishing magnetic charge density. Thus, the field playing the role of the dual field strength tensor is not conserved in Mark1. A related problem is that we are not able to find classical solutions that can represent arbitrary multiphoton states. Although solutions of equations of motion that behave as single photons can be constructed, we show that there are no solutions that correspond to a two-photon state with arbitrary photon polarization.

This is partly due to the fact that the global symmetry group of the model turns out to be much larger than naively anticipated. The global symmetry group turns out to be isomorphic to diffeomorphism symmetry in two dimensions. These diffeomorphism transformations act nontrivially on the Hilbert space even though the fields that we identify with the electric and magnetic fields of QED are invariant under their action. QED does not possess such a large global symmetry.

We then discuss an approach devised to eliminate this global symmetry, which amounts to "gauging" it. The framework we work in is very different from the usual gauge theories, where "gauging" amounts to eliminating a set of local degrees of freedom. In our case, gauging applies only to global group of transformations and therefore does not change the number of local degrees of freedom.

Unfortunately, although we are able to eliminate the global diffeomorphisms from the model, it turns out not to be enough to bring it into full conformity with QED. We, therefore, take a different track and discuss a modification (Mark 2), which circumvents this obstacle. We show that the the model Mark 2, which shares many features with Mark 1, is indeed equivalent to the theory of a free Maxwell field in 3 + 1 dimensions. However, even though we are able to reproduce the Abelian limit, introducing a reasonable Non-Abelian perturbation turns out to be quite tricky. We make some comments on how this can be achieved, but the implementation is left for the future.

## 2. The Abelian Model: Mark 1

### 2.1. The Field Space and the Lagrangian

As explained in the introduction, our aim is to devise a model containing scalar fields only with two massless degrees of freedom, which allows for finite energy solitons. We, thus, zero in on a theory of two scalar fields. In order to have a chance to get Goldstone bosons, we endow it with $SU(2)$ symmetry. The simplest option open for us is an $O(3)$ nonlinear $\sigma$-model.

$$\phi^a, \ a = 1, 2, 3; \quad \phi^2 = 1 \tag{4}$$

This moduli space allows for a nontrivial topology $\Pi_2(S^2)$. We will identify the corresponding topological charge with the electric charge of QED

$$Q = \frac{e}{4\pi^2} \int d^3x \, \epsilon_{abc} \epsilon_{ijk} \partial^i \phi^a \partial_j \phi^b \partial_k \phi^c \tag{5}$$

This identification when extended to current density suggests the following representation of the electric current and, by extension, of the electromagnetic field in the effective description:

$$J^\mu = \frac{e}{4\pi^2} \epsilon_{abc} \epsilon^{\mu\nu\lambda\sigma} \partial_\nu \phi^a \partial_\lambda \phi^b \partial_\sigma \phi^c \tag{6}$$

$$F^{\mu\nu} = \epsilon^{abc}\epsilon^{\mu\nu\lambda\sigma}\phi^a\partial_\lambda\phi^b\partial_\sigma\phi^c \tag{7}$$

Our initial challenge is the following potential problem. A standard two derivative kinetic term would lead to infrared divergent energy of a soliton carrying a nonvanishing topological charge of Equation (5). Consider the simplest topologically nontrivial field configuration:

$$\phi_h^a(x) = \frac{r^a}{|r|}f(|r|); \qquad f(|r|) \to_{r\to\infty} 1 \tag{8}$$

The derivatives of the field decrease as the first power of the distance away from the soliton core, and therefore, the standard two derivative kinetic term gives energy, which diverges linearly in the infrared. The only way to cure this divergence is to not allow a two derivative kinetic term but instead consider a kinetic term with more than two derivatives.

Some reflection shows that there is a unique four derivative term that would do the job, which is also the most natural choice from another point of view. Since our goal is to approach the QED as close as possible, the natural choice for the kinetic term is the square of the field strength tensor in Equation (7). When written in terms of the scalar field $\phi^a$, this is just the well-known Skyrme term.

We, thus, consider a somewhat unusual nonlinear $\sigma$-model, which is defined by the Lagrangian:

$$L = \frac{1}{16e^2}F^{\mu\nu}F_{\mu\nu} + \lambda(\phi^2 - 1)^2 \tag{9}$$

with $F^{\mu\nu}$ defined in Equation (7).

At first sight, it may seem strange that the sign of the $F^2$ term in the Lagrangian Equation (9) is positive, while in QED, the same term enters with the negative sign. The sign in Equation (9) is determined by the requirement of positive definiteness of the Hamiltonian and is thus nonnegotiable. However, the reversal of the sign of the kinetic term is a staple of dual models. The $2 + 1$ dimensional models described in the introduction exhibit the same feature. In the Lagrangian of the effective theory, the kinetic term is the standard $|\partial_\mu V|^2$, while in QED, it is of course $-\tilde{F}_\mu^2$. With the identification of $V^*\partial_\mu V \propto \tilde{F}_\mu$, the signs of the two kinetic terms are again opposite. The reason for this inversion is that, while in QED, the electric field is proportional to the time derivative of the basic field (in this case, $A_\mu$). In the effective dual description, it is the magnetic field that contains the time derivative of the vertex field $V$. Thus, in order for the Hamiltonian of the two models to be the same, the kinetic terms in the respective Lagrangians must have opposite signs.

In Equation (9), we have introduced a coupling $\lambda$. The role of this coupling is easy to understand. In the strong coupling limit $\lambda \to \infty$, the isovector $\phi$ is forced to have a unit length, and we are back to Equation (4). In this limit, the field strength is trivially conserved

$$\partial_\nu F^{\mu\nu} = 0 \tag{10}$$

which means that electric current vanishes. Therefore, the strong coupling limit should correspond to the pure Maxwell theory: the energy of the soliton Equation (8) at strong coupling diverges linearly in the ultraviolet. At finite $\lambda$, the radial component of the field $\phi^a$ can vary in space and vanishes in the soliton core. This eliminates the UV divergence of the energy, and the soliton has a finite mass. The finite value $\lambda$, therefore, corresponds to an Abelian theory with charged matter of finite mass. We again stress that the energy of the soliton is also finite in the infrared, thanks to our choice of the four derivative action. For the hedgehog configuration in Equation (8), the "electric field" decreases as $E_i(x) \propto \frac{\hat{r}_i}{r^2}$, and the energy density away from the soliton core decreases as $1/r^4$. This is the same as the behavior of the Coulomb energy of a static electric charge in the electrodynamics.

Interestingly, the same model was discussed a while ago in [14] with an entirely different motivation.

### 2.2. The Equations of Motion

Given the Lagrangian, we can now write down equations of motion for our effective theory. We work in the strong coupling limit and define two independent degrees of freedom as

$$\phi_3 = z, \qquad \psi = \phi_1 + i\phi_2 = \sqrt{1 - z^2}\,e^{i\chi} \tag{11}$$

With this definition, we have

$$F^{\mu\nu} = \epsilon^{\mu\nu\alpha\beta}\epsilon_{abc}\phi^a\partial_\alpha\phi^b\partial_\beta\phi^c = -2\epsilon^{\mu\nu\alpha\beta}\partial_\alpha z\partial_\beta\chi \tag{12}$$

The Lagrangian becomes

$$L = \frac{1}{4e^2}(\partial_\mu z\partial_\nu\chi - \partial_\mu\chi\partial_\nu z)^2 \tag{13}$$

and the equations of motion are

$$\partial^\mu\left[\frac{1}{e^2}\partial^\nu\chi(\partial_\mu z\partial_\nu\chi - \partial_\nu z\partial_\mu\chi)\right] = 0$$
$$\partial^\mu\left[\frac{1}{e^2}\partial^\nu z(\partial_\mu z\partial_\nu\chi - \partial_\nu z\partial_\mu\chi)\right] = 0 \tag{14}$$

Interestingly, these equations can be written as

$$\frac{1}{e^2}\partial_\nu G(z,\chi)\partial_\mu(\partial_\mu z\partial_\nu\chi - \partial_\nu z\partial_\mu\chi) = \frac{1}{e^2}\partial_\nu\left[G(z,\chi)\partial_\mu(\partial_\mu z\partial_\nu\chi - \partial_\nu z\partial_\mu\chi)\right] = 0 \tag{15}$$

with $G(z,\chi)$ being an arbitrary function of two variables. These can be thought of as an infinite number of conservation equations, where the conserved current corresponding to a given function $G(z,\chi)$ is defined as

$$J_\nu^G = G(z,\chi)\partial^\mu(\partial_\mu z\partial_\nu\chi - \partial_\nu z\partial_\mu\chi) \tag{16}$$

We will see later that the existence of an infinite number of conserved currents is a very important feature.

### 2.3. The Symmetries of the Model and Correspondence to Electrodynamics

Given that we have identified and infinite number of conserved currents from Equation (16), we see that the choice of the Skyrme term as the kinetic term in the Lagrangian allows a very large global symmetry group of the model. The global symmetry group of Equation (9) is not just the $SO(3)$ group that we required from the outset but is isomorphic to the group of diffeomorphisms in two dimensions.

This is easy to understand. The field strength in Equation (7) is related to an infinitesimal area element on a configuration space. A given field configuration $\phi^a(x)$ defines a map from space-time to a sphere $S^2$. Consider a given component, such as the field strength tensor, say $F_{12}$ at some point $x$. To express it in terms of $\phi$, we consider three infinitesimally close points $A \equiv x^\mu$, $B \equiv x^\mu + \delta^{\mu 1}a$, and $C \equiv x^\mu + \delta^{\mu 2}a$. These three points in space-time map into three infinitesimally close points on the sphere $\phi^a(A)$, $\phi^a(B)$, $\phi^a(C)$. The field strength $F_{12}$ is proportional (up to the factor $a^{-2}$) to the area of the infinitesimal triangle on $S^2$ defined by these three points. Since our Lagrangian depends only on $F^{\mu\nu}$, clearly an arbitrary area preserving reparametrization of the sphere leaves our action unchanged.

The original $SO(3)$ global symmetry is only a *small* subgroup of the area preserving diffeomorphisms of $S^2$—the group we will denote $Sdiff(S^2)$ [15]. As an aside, we note that this group is also isomorphic to the group of canonical transformations on a classical two-dimensional phase space. The infinitesimal $Sdiff(S^2)$ transformation when acting on $z$ and $\chi$ is written as

$$z \to z + \frac{\partial G}{\partial\chi}; \quad \chi \to \chi - \frac{\partial G}{\partial z} \tag{17}$$

The Noether currents are those given by Equation (16), and the equations of motion are equivalent to conservation equations of these currents.

An intriguing point is that the symmetry Equation (17) is reminiscent of the world sheet diffeomorphism invariance of the Nambu-Gotto string. Indeed, if the fields $z$ and $\chi$ are thought as the world sheet string coordinates, the world sheet diffeomorphism invariance is precisely Equation (17). The model discussed here is not motivated by a string theory and *a priori* has nothing to do with a string theory. Nevertheless, the similarities may run deeper than just a coincidence since the fundamental "order parameters" of the magnetic symmetry in $3 + 1$ dimensions are indeed magnetic vortex strings [12]. The $S^2$ topology of the world sheet would then ask for closed string loops. The analogy is indeed intriguing and would be worthwhile pursuing further, but since this is not the goal of our exploratory efforts here, we will return to the field theoretical approach in the rest of this review.

The fact that the global symmetry group of the model is so large means that the moduli space (space of all vacuum configurations) is much larger than $S^2$, which corresponds to a symmetry breaking pattern $SO(3) \to SO(2)$. Consider an arbitrary field configuration $\phi^a(x)$ that maps the configuration space into *any* **one-dimensional** curve on $S^2$. Such a configuration has a vanishing action and therefore is a classical vacuum. The full moduli space is, therefore, the union of maps $\phi^a(x)$ that map $R^4$ to $L$, where $L$ is an arbitrary point or a one-dimensional curve on $S^2$.

Still the important question for us is whether the topology of this moduli space is right to support classically quantized topological charge. Indeed, from its definition, it is clear that the topological charge $Q$ is quantized on any smooth classical configuration of fields $\phi(x)$. The catch is that there are many more degenerate soliton configurations than just the rotationally invariant hedgehog of Equation (8). Any $Sdiff(S^2)$ transformation with an arbitrary (regular) function $G$ of Equation (17) applied to Equation (8) generates a configuration $\phi_h^{aG}(x)$, which carries the same charge $Q$ as $\phi_h^a(x)$ and is degenerate with it in energy. However, although these are different field configurations, they all have the same electric field $E_i = \epsilon_{ijk}\epsilon^{abc}\phi^a \partial_j \phi^b \partial_k \phi^c$ since the field strength is invariant under the action of $Sdiff(S^2)$. Thus, if one is physically only allowed to measure electromagnetic fields, all these solitons look identical.

*2.4. The Photon States-a.k.a. Plane Waves*

We have constructed the Lagrangian Equation (9) so that it has the maximal similarity to QED when written in terms of the putative electromagnetic fields. This does not yet ensure that the content of the theory is the same as that of electrodynamics. We do know that the field strength $F^{\mu\nu}$ identified in Equation (7) satisfies half of Maxwell's equations—the Coulomb law and the evolution equations for electric field. The other half of Maxwell's equations (dynamics of magnetic fields) have to follow from the equations of motions of our model. Indeed, there is clear similarity between Equation (15) and the Maxwell's equations. Equation (15) can be rewritten in terms of $F^{\mu\nu}$ as

$$[\partial_\nu G(z,\chi)]\partial_\mu \tilde{F}^{\mu\nu} = 0 \tag{18}$$

This ensures that for any configuration of the fields $z$, the $\chi$ that satisfies $\partial_\mu \tilde{F}^{\mu\nu} = 0$ also satisfies the equations of motion of our model. However, the converse is not assured. In Appendix A, we give an example of a solution of Equation (15) that does not satisfy the equations of motion of electrodynamics.

Thus, there is no full equivalence between the model Equation (9) and electrodynamics. Nevertheless, we can ask to what extent the spectrum of solutions of Equation (9) contains basic excitations of QED. The excitations of particular interest in the present context are of course the photons. (Although we are dealing with a classical theory and not a quantum theory, we will, with a slight abuse of language, refer to plane wave configurations of $F^{\mu\nu}$ with light-like momentum as photons).

We aim now to show that free wave excitations are indeed solutions of Equation (15). Consider the following configuration

$$\chi(x) = A\epsilon^{\mu}x_{\mu}; \qquad z(x) = \sin k^{\mu}x_{\mu} \tag{19}$$

where the vector $\epsilon^{\mu}$ is normalized as $\epsilon^{\mu}\epsilon_{\mu} = -1$. Calculating the field strength, we find

$$\tilde{F}^{\mu\nu} = A(\epsilon^{\mu}k^{\nu} - \epsilon^{\nu}k^{\mu})\cos k \cdot x \tag{20}$$

Thus,

$$\partial_{\mu}\tilde{F}^{\mu\nu} = -A\left[(\epsilon \cdot k)k^{\nu} - k^{2}\epsilon^{\nu}\right]\sin k \cdot x \tag{21}$$

If the momentum vector is light-like and the polarization vector $\epsilon$ is perpendicular to $k$, this vanishes:

$$k^{2} = 0; \quad \epsilon \cdot k = 0 \tag{22}$$

For a fixed light-like momentum $k_{\mu}$, Equation (22) has three independent solutions for $\epsilon^{\mu}$, one of which is proportional to $k_{\mu}$. For $\epsilon_{\mu} \propto k_{\mu}$, the field strength tensor Equation (19) vanishes, and so, there are two independent polarization vectors $\epsilon_{\lambda}^{\mu}$, $\lambda = 1, 2$ that yield plane wave solutions for $F^{\mu\nu}$. We can always choose the polarization vectors so that their zeroth component vanishes $\epsilon_{\lambda}^{\mu} = (0, \epsilon_{\lambda}^{i})$, just like in electrodynamics. The square of the overall amplitude of wave $A$ in the quantum case is proportional to the number of photons with a given momentum and a given polarization vector.

Note that the freedom in the choice of the independent polarization vectors is exactly the same as in electrodynamics

$$\epsilon^{\mu} \to \epsilon^{\mu} + ak^{\mu} \tag{23}$$

This change of polarization vector is generated by the transformation

$$\chi \to \chi + a \arcsin z \tag{24}$$

which is a particular element of the $Sdiff(S^{2})$ group from Equation (17). More generally, the field configuration in Equation (19) can be transformed by any element of $Sdiff(S^{2})$ without causing a change in $F^{\mu\nu}$.

The solution Equations (19)–(22) describe a state that in all respects is equivalent to the freely propagating photon, and we will refer to it as such. The solution Equation (19) suggests an intuitive interpretation for the properties of the photon states in terms of the effective theory. The momentum of the photon is simply the momentum associated with the variation of the third component of the isovector $\phi^{a}$, while the direction of the photon polarization vector is the direction of the spatial variation of the phase $\chi$.

We again note that the present formulation is easier to interpret in terms of quantities dual to those normally used in QED. One usually introduces the vector potential $A_{\mu}$ via $F^{\mu\nu} = \partial_{\mu}A_{\nu} - \partial_{\nu}A_{\mu}$, which potentiates the homogeneous Maxwell's equation $\partial_{\mu}\tilde{F}^{\mu\nu} = 0$. However, in the absence of electric charges, one can alternatively potentiate the other half of Maxwell's equation by introducing the dual vector potential via $\tilde{F}^{\mu\nu} = \partial_{\mu}\tilde{A}_{\nu} - \partial_{\nu}\tilde{A}_{\mu}$. In the absence of charges, the dynamics of the dual vector potential $\tilde{A}_{\mu}$ is identical to that of $A_{\mu}$, and it can be expanded in exactly the same polarization basis as $A_{\mu}$. In this dual formulation, QED possesses a dual gauge symmetry $\tilde{A}_{\mu} \to \tilde{A}_{\mu} + \partial_{\mu}\lambda(x)$.

To make the correspondence to our model more obvious we can introduce a "dual vector potential" by

$$\tilde{A}_{\mu} = z\partial_{\mu}\chi \tag{25}$$

As opposed to the field strength tensor itself, this object is not invariant under the $Sdiff(S^{2})$ group transformation from Equation (17):

$$\tilde{A}_{\mu} \to \tilde{A}_{\mu} + \partial_{\mu}[G - z\frac{\partial G}{\partial z}] \tag{26}$$

This is similar to the dual gauge transformation in electrodynamics with the gauge function $\lambda(x) = G - z\frac{\partial G}{\partial z}$.

The analogy of Equation (25) is suggestive, but one has to keep in mind that this is not at all an equivalence. First, the transformation Equation (26) is not a gauge transformation but rather the action of a global symmetry transformation of the Lagrangian on $\tilde{A}_\mu$ of Equation (25). More importantly, an arbitrary vector field cannot be expressed in terms of two scalars by a relation of the type Equation (25), even allowing for a possible gauge ambiguity. Thus, Equation (25) cannot be considered merely a convenient parametrization of the dual potential of electrodynamics. For this reason, the variation of the Lagrangian Equation (9) with respect to such a constrained vector potential does not lead to directly to homogeneous Maxwell's equations but instead to Equation (18).

We have thus determined that monochromatic plane wave $\tilde{F}_{\mu\nu}$ solves the equations of motion of our effective model. In QED, which is a linear theory of $F_{\mu\nu}$, the immediate consequence is that a superposition of such waves is a solution as well. However, Equation (18) is not linear in the basic field variables, and thus, a superposition of two such solutions is not assured to be a solution as well. Let us try to construct a "two photon state" by slightly extending the ansatz Equation (19).

$$\chi = \lambda_\mu x_\mu; \quad z = a\sin k^\mu x_\mu + b\sin p^\mu x_\mu \tag{27}$$

with $k^\mu$ and $p^\mu$—both light-like vectors, $\lambda^\mu k_\mu = \lambda^\mu p_\mu = 0$ and $\lambda^\mu \lambda_\mu = -1$. The latter two conditions can be satisfied by taking

$$\lambda^\mu = \alpha\left[\epsilon^\mu - \frac{\epsilon \cdot k}{k \cdot p}p_\mu - \frac{\epsilon \cdot p}{k \cdot p}k_\mu\right] \tag{28}$$

with an arbitrary vector $\epsilon^\mu$ and an appropriate normalization constant $\alpha$.

The dual field strength tensor is now:

$$\tilde{F}_{\mu\nu} = a(k_\mu \epsilon_\nu^k - k_\nu \epsilon_\mu^k)\cos k \cdot x + b(p_\mu \epsilon_\nu^p - p_\nu \epsilon_\mu^p)\cos p \cdot x \tag{29}$$

with

$$\epsilon_\mu^k = \lambda_\mu - \frac{\lambda_0}{k_0}k_\mu; \quad \epsilon_\mu^p = \lambda_\mu - \frac{\lambda_0}{p_0}p_\mu; \tag{30}$$

This looks like a bona fide two-photon state. However, our ansatz does not yield a generic two-photon state with arbitrary polarization vectors: both the polarization vectors $\epsilon^k$ and $\epsilon^p$ above have equal components in the perpendicular direction to the plane spanned by $p^i$; $k^i$. Thus, we are one degree of freedom short and cannot construct a two-photon state with arbitrary polarizations of both photons. Although this might look merely like a limitation of our particular ansatz, we show in Appendix A that this problem is not restricted to the ansatz Equation (27) but is unfortunately a genuine limitation of our effective model.

## 3. Going Non-Abelian: The "Confining String"

Our main goal in this project is to have a model representation of the confinement phenomenon in Non-Abelian theories. We, therefore, take the same trek as in 2 + 1 dimensions. Namely, we will add to the Lagrangian Equation (9) a simple perturbation that explicitly breaks the global symmetry of the model. This modification of low energy description is meant to get rid of the multiple vacuum structure inherent to spontaneous symmetry breaking and therefore eliminate massless excitations. For convenience, we will choose a potential that (classically) sets the vacuum expectation value of the field $z$ to unity.

With the above considerations, we are led to consider the Lagrangian

$$L = \frac{1}{16e^2}F^{\mu\nu}F_{\mu\nu} - \frac{2}{e^2}\Lambda^2(z-1)^2 \tag{31}$$

The potential we have added of course breaks the $SO(3)$ symmetry, but in addition, it is also not invariant under a general $Sdiff(S^2)$ transformation. However, the $Sdiff(S^2)$ is not broken completely but only up to the subgroup

$$\chi \to \chi - \frac{dG(z)}{dz} \tag{32}$$

We keep this in mind throughout the discussion of this section.

The equations of motion of the perturbed model are

$$\partial^\mu \left[ \frac{1}{e^2} \partial^\nu \chi (\partial_\mu z \partial_\nu \chi - \partial_\nu z \partial_\mu \chi) \right] = \frac{4}{e^2} \Lambda^2 (z - 1)$$
$$\partial^\mu \left[ \frac{1}{e^2} \partial^\nu z (\partial_\mu z \partial_\nu \chi - \partial_\nu z \partial_\mu \chi) \right] = 0 \tag{33}$$

These equations do not have static topologically stable solutions of finite energy. However, one can still ask what is the energy of a configuration of a soliton and antisoliton separated far in space. As the answer to this question, we expect to find a (approximately) translationally invariant string-like configuration that connects the soliton and the antisoliton and produces a linear confining potential between the two. Consider a static field configuration translationally invariant in the third direction. The only components of $F_{\mu\nu}$ that do not vanish then are:

$$F^{03} = 2\epsilon^{ij} \partial_i z \partial_j \chi \tag{34}$$

Let us take the following ansatz, which preserves rotational symmetry in the $x_1 - x_2$ plane:

$$\chi(x) = \theta(x); \qquad z(x) = z(r) \tag{35}$$

Here, $r$ and $\theta$ are the polar coordinates in the $x_1, x_2$ plane. This ansatz parametrizes a configuration with a unit winding in the $x_1, x_2$ plane, which should be the case for a string connecting a soliton and an antisoliton. The soliton partner of the pair is located at a very large negative value of $x_3$. At even more negative $x_3$, the field must relax into the vacuum $\phi^1 = \phi^2 = 0$; $z = 1$. Therefore, the topological charge calculated on a surface enclosing the soliton (but not its antisoliton partner) should be given by the two dimensional topological charge—the winding number of the phase $\chi$ on any surface crossed by the string. The same argument applies for the antisoliton, which resides at large positive value of $x_3$. Our ansatz, therefore, describes a confining string connecting a widely separated soliton–antisoliton pair.

Interestingly, the equation of motion for the field $\chi$ is automatically satisfied for Equation (35). The only nontrivial equation is that for $z$:

$$4z'' = 4\Lambda^2 (z - 1) \tag{36}$$

with $z' \equiv \frac{dz}{d(r^2)}$

In order for the solution to have finite linear energy density, $z$ must satisfy the boundary conditions:

$$z(0) = -1, \; z(\infty) = 1 \tag{37}$$

The solution with these boundary conditions is

$$z(r^2) = 1 - 2\exp\{-\Lambda r^2\} \tag{38}$$

Some of the properties of this solution are intuitively appealing. It has a finite width governed by the only dimensional parameter $\Lambda$. Outside of this width, the fields relax to vacuum. Inside the string, the potential energy is finite, and thus, the string carries finite linear energy density. The string tension is found to be

$$\sigma = 8\pi\frac{\Lambda}{e^2} \tag{39}$$

One feature of the solution, however, is rather peculiar. Away from the string core, the fields do not approach the vacuum configuration exponentially but rather as a Gaussian in transverse distance. The string profile is, therefore, unusual as it has a very sharp boundary, outside of which the vacuum is reached very quickly. Such a behavior is unusual and, in fact, is not possible in a local field theory with a finite mass gap and a finite number of massive excitations. We can trace the origins of this behavior back to the non canonical kinetic term in Equation (9), which has four derivatives. For simple dimensional reasons, the kinetic energy for a rotationally invariant configuration is given by the second derivative with respect to $r^2$ rather than $r$, which results in a Gaussian rather than exponential decay of the solution fields.

### 4. The $Z_N$ Preserving Perturbation

The potential of Equation (31) breaks the $SO(3)$ as well as the $Sdiff(S^2)$ symmetries but leaves an $O(2)$ subgroup of $SO(3)$ and a large subgroup $Sdiff(S^2)$ (Equation (32)) unbroken. On the other hand, if we follow a direct analogy with $2 + 1$ dimensions, we expect the effective theory in the Non-Abelian regime to preserve only a $Z_N$ subgroup of $SO(3)$. We can easily implement such a perturbation in the effective theory. Let us modify the Lagrangian to

$$L = \frac{1}{16e^2}F^2 - \frac{2}{e^2}\Lambda^2(z-1)^2\left[1 - \mu(\psi^N + \psi^{\star N})\right] = \frac{1}{16e^2}F^2 - \frac{2}{e^2}\Lambda^2(z-1)^2\left[1 - 2\mu(1-z^2)^{N/2}\cos N\chi\right] \tag{40}$$

For large enough $\mu$, the additional perturbation shifts the lowest energy value away from $z = 1$. For simplicity, we will only consider values

$$\mu < \frac{1}{2} \tag{41}$$

for which the vacuum configuration remains at $z = 1$.

We will now study the effect of the additional perturbation on the structure of the "confining string".

Assuming a long string in the direction $x_3$, the energy per unit length can be written as

$$E = \int d^2x \frac{1}{2e^2}(\epsilon_{ij}\partial_i z\partial_j\chi)^2 + \frac{2}{e^2}\Lambda^2(z-1)^2\left[1 - 2\mu(1-z^2)^{N/2}\cos N\chi\right] \tag{42}$$

### 4.1. Perturbative Solution

For small values $\mu \ll 1$, we can find corrections to the string solution perturbatively. Let us take the following ansatz for the perturbative solution:

$$z(r,\theta) = z(r); \qquad \chi = \theta + \chi_1(r,\theta) = \theta + f(r)\sin N\theta \tag{43}$$

where $z(r)$ is given by Equation (38). Although this is not the most general possible form of the perturbative correction, it nevertheless yields a solution to first order in $\mu$, as we now show.

The leading order perturbative equation for $f$ is

$$\frac{1}{e^2}8N^2(z')^2 f \sin N\theta = \frac{1}{e^2}N\mu\Lambda^2(z-1)^2(1-z^2)^{N/2}\sin N\theta \tag{44}$$

which is solved by

$$f(r^2) = \frac{\mu}{N}\left[2e^{-\Lambda r^2}(1 - e^{-\Lambda r^2})\right]^{N/2} \tag{45}$$

In principle, we have to consider also the minimization equation for $z(r)$. It reads

$$\frac{1}{e^2}8N\left[2z''f + z'f'\right] = \frac{1}{e^2}4\mu\Lambda\left[2(z-1)(1-z^2)^{N/2} - Nz(z-1)^2(1-z^2)^{N/2-1}\right] \quad (46)$$

One can explicitly check that this equation is satisfied by the perturbative expression of Equation (45) and $z(r)$ of Equation (38).

The longitudinal electric field inside the string is given by

$$F^{03} = -4\Lambda e^{-\Lambda r^2}\left[1 + \mu\left(2e^{-\Lambda r^2}(1 - e^{-\Lambda r^2})\right)^{N/2}\cos N\theta\right] \quad (47)$$

As before, the electric field is concentrated within the radius $\Lambda^{1/2}$ in the transverse plane, with an angular modulation of the transverse profile due to the additional $Z_N$ invariant potential.

*4.2. General Solution*

We now demonstrate the string solution beyond the simple perturbative approximation discussed above. Minimizing the energy functional Equation (42) yields the equations:

$$\frac{1}{e^2}\epsilon_{ij}\partial_j\chi\partial_iF = \frac{\partial U}{\partial z}$$
$$\frac{1}{e^2}\epsilon_{ij}\partial_jz\partial_iF = -\frac{\partial U}{\partial \chi} \quad (48)$$

with

$$F \equiv \frac{1}{2}F^{03} = \epsilon_{ij}\partial_iz\partial_j\chi\,, \quad (49)$$

where $U$ is the potential of Equation (42).

These equations can be combined into:

$$\frac{1}{2e^2}\partial_k(F^2) = \partial_k U \quad (50)$$

Requiring that the electric field vanishes at transverse infinity, as should be the case for any finite energy density configuration, we find

$$F^2 = 2e^2U; \quad F = \sqrt{2e^2U} \quad (51)$$

Let us work in the modified polar coordinates $(\tau = r^2, \theta)$. We then have

$$\partial_\tau z\partial_\theta\chi - \partial_\theta z\partial_\tau\chi = \sqrt{\frac{1}{2}e^2U} \quad (52)$$

This equation has infinite number of solutions. This infinite degeneracy results from an unusual symmetry of the energy functional Equation (42). Consider the group of area-preserving diffeomorphisms on a plane $SDiff(R^2)$

$$(z(x),\chi(x)) \rightarrow (z(x'),\chi(x')); \quad \frac{\partial(x'^1,x'^2)}{\partial(x^1,x^2)} = 1 \quad (53)$$

These transformations leave the energy functional Equation (42) invariant. Therefore, any string solution can be transformed by a transformation Equation (53), generating an infinite number of degenerate solutions. Note that the longitudinal electric field is itself invariant under Equation (53), and thus, all the degenerate solutions have identical electric field and energy density profiles.

Interestingly, the transformations $SDiff(R^2)$ of Equation (53) are diffeomorphisms on the coordinate space rather than on the field space. Thus, this is a different diffeomorphisms than $Sdiff(S^2)$, which we discussed in the previous section and is explicitly broken by the potential $U$. The symmetry $SDiff(R^2)$ is in a sense accidental since it only exists for configurations translationally invariant in one direction.

Let us now discuss two solutions related by $SDiff(R^2)$. We can utilize the large symmetry by prescribing a simple dependence of $\chi$ on the angle : $\chi = \theta$. The equation for $z$ then follows from Equation (52).

$$\partial_\tau z = \sqrt{\frac{1}{2}e^2 U} = \sqrt{\Lambda^2(z-1)^2\left[1 - 2\mu(1-z^2)^{N/2}\cos N\theta\right]} \tag{54}$$

The coordinate $\theta$ enters here as a parameter, and for a given value of $\theta$, the solution is

$$\tau = \int_{-1}^{z(\tau)} dz \frac{1}{\sqrt{\Lambda^2(z-1)^2\left[1 - 2\mu(1-z^2)^{N/2}\cos N\theta\right]}} \tag{55}$$

This has correct large distance asymptotic behavior since as $\tau \to \infty$, the function $z$ has to approach unity for the right hand side to diverge. It is easy to find the large distance asymptotics of the solution. When $z$ is close to unity, the term proportional to $\mu$ in the denominator can be neglected, and we have

$$\tau = \int_{-1}^{z(\tau)} dz \frac{1}{\sqrt{\Lambda^2(z-1)^2}} \tag{56}$$

which is solved by

$$z(\tau \to \infty) = 1 - 2e^{-\Lambda\tau} \tag{57}$$

This is identical to Equation (38), and thus, the IR asymptotics of the solution is not sensitive to the $Z_N$ perturbation.

As an example of another solution, we assume that $z$ has no angular dependence. We then have:

$$\partial_\tau z \partial_\theta \chi = \sqrt{\Lambda^2(z-1)^2\left[1 - 2\mu(1-z^2)^{N/2}\cos N\chi\right]} \tag{58}$$

This determines $\theta$ as a function of $r$:

$$\theta = \int_0^{\chi(r,\theta)} \frac{z' d\chi}{\sqrt{\Lambda^2(z-1)^2\left[1 - 2\mu(1-z^2)^{N/2}\cos N\chi\right]}} \tag{59}$$

The explicit solution is

$$\theta = \frac{2}{N}\frac{z'}{\sqrt{\Lambda^2(z-1)^2(1-2\mu(1-z^2)^{N/2})}} F\left(\frac{N\chi}{2}, \frac{4\mu(1-z^2)^{N/2}}{2\mu(1-z^2)^{N/2}-1}\right) \tag{60}$$

where $F(\phi, m)$ is the incomplete elliptic integral of the first kind:

$$F(\phi, m) = \int_0^\phi (1 - m\sin\theta^2)^{-1/2}d\theta \tag{61}$$

Imposing on the solution the boundary condition

$$\chi(\theta + 2\pi) = \chi(\theta) + 2\pi \tag{62}$$

and using $F(\frac{k\pi}{2}, m) = kK(m)$, with $K(m)$—the complete elliptic integral of the first kind, gives

$$2\pi = \frac{4z'}{\sqrt{\Lambda^2(z-1)^2(1-2\mu(1-z^2)^{N/2})}} K\left(\frac{4\mu(1-z^2)^{N/2}}{2\mu(1-z^2)^{N/2}-1}\right) \tag{63}$$

It is easy to check that for $z \to 1$, this reduces to

$$z' = \Lambda(1 - z) \qquad (64)$$

and thus, the IR asymptotics again is the same as in Equation (38).

### 5. Discussion of the Model Mark 1

In constructing our model, we have tried to follow the guide of 2 + 1 dimensional gauge theories and, based on several requirements, "guess" a theory of scalar fields that may emulate the effective theory of 3 + 1 dimensional gauge theories. The model we were led to is not quite satisfactory, but it does have several interesting and intriguing features.

First off, already in the Abelian limit, it is quite peculiar. It possesses an infinite dimensional global symmetry group, which is spontaneously broken by classical solutions of lowest energy. On the other hand, the observables that we would like to identify with physical quantities in QED turn out to be invariant under this symmetry. This may seem problematic; however, we note that a somewhat similar situation occurs in 2 + 1 dimensions and, in general, in dual type descriptions. In 2 + 1 dimensional gauge theories, the electromagnetic field is invariant under the action of the magnetic $U(1)$ symmetry, which does act nontrivially on the magnetic vortex field—the basic degree of freedom in the effective/'dual" description. In the present 3 + 1 dimensional model, likewise, the electromagnetic field does not feel the action of the (infinite) global symmetry group $Sdiff(S^2)$, which does act nontrivially on the "fundamental" scalar fields of the effective theory.

The global $Sdiff(S^2)$ symmetry is classically broken by the lowest energy configurations. This is similar to 2 + 1 dimensions, but the situation is more involved. In 2 + 1 dimensions, we had to deal with a standard symmetry breaking pattern of symmetry with a finite number of generators. In our 3 + 1 dimensional model, on the other hand, the symmetry group is infinitely dimensional, and thus, the space of vacuum configurations is very large. It includes not only translationally invariant field configurations but also configurations with nontrivial spatial dependence. These configurations break translational invariance in addition to the global $Sdiff(S^2)$ symmetry. This is not a unique situation, and in fact, such a situation is ubiquitous in condensed matter systems, but in relativistic field theories, it is quite rare. As a result, since the vacuum set has large entropy, it could well be that classical analysis fails in this model quite badly. Many of the classical vacua differ from each other only in the finite region of space. Generically in cases like this, upon quantization, these configurations become connected by tunneling transitions of finite probability. One, therefore, may expect that the quantum portrait of moduli space is very different from the classical one. This question is worth investigating on its own, but this goes far beyond the scope of the present work.

With a symmetry breaking perturbation, our model exhibits a simple classical mechanism of confinement of topological defects, such as in 2 + 1 dimensions. However, some peculiarities are present again. We have discovered that string solutions are infinitely degenerate. The static energy of configurations translationally invariant in one direction has an additional diffeomorphism invariance. This is not the same invariance as in the Abelian limit, as the diffeomorphisms in question are transformations in coordinate space and not in the field space. Nonetheless, this symmetry leads to degeneracy between an infinite number of solutions, all of which have the same electric field. As far as the electric field profile is concerned, the solution, as far as we can ascertain, is unique. This infinite degeneracy makes one wonder about the fate of such strings in a quantum theory, given that they carry large entropy associated with the degeneracy.

All the peculiar features of the model stem from the nonconventional, higher derivative kinetic term required to have finite energy of a soliton in the absence of the potential. One could add the standard two derivative kinetic term $\partial^\mu \phi^a \partial_\mu \phi^a$ as a perturbation. Although we have not explored this possibility in detail, it is clear that this would lift the degeneracy between the different string solutions. With this additional kinetic term, our model becomes identical with the model proposed by Faddeev and Niemi in [7–11] as an effective theory of QCD. Note, however, that our proposed picture of confinement is very different from and in a way complementary to that of [7–11]. The authors of [7–11] are mostly interested in closed string solutions meant to represent the glueballs, while in our way of thinking, it is the open strings, with the endpoints representing "constituent gluons" that play the main role in analogy with 2 + 1 dimensions [2,3,16]. In the Faddeev–Niemi model, stability of closed string solutions is ensured by nontrivial twisting of the phase of the scalar field along the string. Open strings, on the other hand, are not associated with twist and in principle can break into shorter strings, which is the case in QCD. The stability of classical strings solutions in a quantum theory is not absolute but is rather an approximate feature that arises dynamically provided the string endpoints are sufficiently heavy [17]. This endpoint mass suppresses quantum tunneling, which is responsible for the decay of long strings.

Finally, it is worthwhile noting that the addition of the two derivative kinetic term makes our model similar to the $CP^1$ model, which has been recently discussed in the literature in relation to effective models of confinement [18].

The large global symmetry of our model in the Abelian, which has no obvious parallel in QED, is worrisome. One can wonder if it is responsible at least partially for the absence of an arbitrary "two-photon state", as we have found here. It is, therefore, natural to try and eliminate this symmetry from the model. In the next section, we describe an approach to doing so by "gauging" this global symmetry. This amounts to restricting the Hilbert space of the model to states that are invariant under $Sdiff(S^2)$.

## 6. Gauging $Sdiff(S^2)$

In this section, we show how the global $Sdiff(2)$ symmetry can be eliminated from the theory. The standard way of going about such a task is to "gauge" the symmetry, i.e., to impose the vanishment of the appropriate charge. It is usually employed to eliminate local symmetries; however, as a matter of principle, it can also be done for global symmetries. We will now describe this procedure.

Recall that the symmetry in question is

$$z \to z + \frac{\partial G(z,\chi)}{\partial \chi}, \ \chi \to \chi - \frac{\partial G}{\partial z} \tag{65}$$

with $G$ being an arbitrary function of the two variables $z$ and $\chi$ but does not explicitly depend on space-time coordinates.

This symmetry is associated with the conserved currents

$$J_\nu^G = G(z,\chi)J_\nu = G(z,\chi)\partial^\mu(\partial_\mu z \partial_\nu \chi - \partial_\mu \chi \partial_\nu z). \tag{66}$$

where

$$J_\nu = \partial^\mu(\partial_\mu z \partial_\nu \chi - \partial_\mu \chi \partial_\nu z) \tag{67}$$

The corresponding charges are

$$Q^G = \int d^3x G(z,\chi)J_0 = \int d^3x G(z,\chi)\partial^\mu(\partial_\mu z \partial_0 \chi - \partial_\mu \chi \partial_0 z) \tag{68}$$

We note for future use that the symmetry transformation can be written as a canonical transformation on a phase space spanned by $z$ and $\chi$.

$$\delta z = \{z,G\}; \ \ \delta \chi = \{\chi,G\}; \ \ \{A,B\} \equiv \frac{\partial A}{\partial z}\frac{\partial B}{\partial \chi} - \frac{\partial A}{\partial \chi}\frac{\partial B}{\partial z} \tag{69}$$

To gauge this symmetry, we first introduce the analog of the zeroth component of vector potential $\Lambda(z,\chi,t)$. Note that $\Lambda$ is not an arbitrary function of space-time coordinates but only a function of the field variables $z$ and $\chi$ and time $t$.

We now change our definition of the "magnetic field" to

$$B_k = 2(\partial_k\chi\partial_0 z - \partial_0\chi\partial_k z) - \partial_k\Lambda = 2(\partial_k\chi\partial_0 z - \partial_0\chi\partial_k z) - \frac{\partial\Lambda}{\partial z}\partial_k z - \frac{\partial\Lambda}{\partial\chi}\partial_k\chi \tag{70}$$

Defining "covariant derivative" as

$$\begin{aligned}
\nabla_0\chi &= \partial_0\chi + \frac{1}{2}\frac{\partial\Lambda}{\partial z} \\
\nabla_0 z &= \partial_0 z - \frac{1}{2}\frac{\partial\Lambda}{\partial\chi}
\end{aligned} \tag{71}$$

we can write this as

$$B_k = 2(\partial_k\chi\nabla_0 z - \nabla_0\chi\partial_k z) \tag{72}$$

Note that this definition of covariant derivative implies for any function of $z$ and $\chi$

$$\nabla_0\Phi(z,\chi) = \frac{d}{dt}\Phi - \frac{1}{2}\{\Phi,\Lambda\} \tag{73}$$

With this altered definition of the magnetic field, and the electric field is still defined as

$$E_i = 2\epsilon_{ijk}\partial_j z\partial_k\chi \tag{74}$$

we now write the Lagrangian

$$\mathcal{L} = -\frac{1}{2}(\vec{E}^2 - \vec{B}^2) \tag{75}$$

As we show now, this Lagrangian is gauge invariant. First, let us consider time independent transformations from Equation (69). Under this transformation, we define the transformation of $\Lambda$ as

$$\delta\Lambda = -\{\Lambda,G\} = -\left[\frac{\partial\Lambda}{\partial z}\frac{\partial G}{\partial\chi} - \frac{\partial\Lambda}{\partial\chi}\frac{\partial G}{\partial z}\right] \tag{76}$$

Note, that this equation should be understood as the change in the functional form of $\Lambda$ as a function of $z$ and $\chi$. With this definition and taking into account that the values of $z$ and $\chi$ change according to Equation (69), we find

$$\Lambda'(z',\chi') = \Lambda(z,\chi) \tag{77}$$

Thus, it is easy to see that both $E_i$ and $B_i$ are invariant under the time-independent transformations Equations (69) and (76).

Now, consider time-dependent transformations, $G(z,\chi,t)$. The electric field is invariant under the time-dependent transformations as well. For the magnetic field, a straightforward calculation gives

$$\begin{aligned}
2(\partial_k\chi\partial_0 z - \partial_0\chi\partial_k z) \rightarrow &2\left[\partial_k\left(\chi - \frac{\partial G}{\partial z}\right)\partial_0\left(z + \frac{\partial G}{\partial\chi}\right) - \partial_0\left(\chi - \frac{\partial G}{\partial z}\right)\partial_k\left(z + \frac{\partial G}{\partial\chi}\right)\right] \\
= &2(\partial_k\chi\partial_0 z - \partial_0\chi\partial_k z) + 2\partial_i\partial_0 G + O(G^2)
\end{aligned} \tag{78}$$

Thus, if we define the transformation of $\Lambda$ as

$$\delta\Lambda = 2\partial_0 G - \{\Lambda,G\} \tag{79}$$

we find that the magnetic field in Equation (72) is invariant.

To summarize, we have now constructed the Lagrangian, which is invariant under arbitrary time-dependent $Sdiff(S^2)$ transformations. Physically, this gauge invariance means that the $Sdiff(S^2)$ charges are required to vanish on physical configurations. Indeed, we can see that the equation of motion for $\Lambda$ is indeed equivalent to this constraint. We note that variation with respect to $\Lambda$ should be done with care since $\Lambda$ is not an independent field. One cannot vary space-time dependence of $\Lambda$ arbitrarily; instead, one has to vary the functional form of the dependence on the field $z$ and $\chi$.

Let us derive equations of motion for Lagrangian Equation (75). Varying with respect to $z$ and $\chi$, we obtain

$$\partial_i \chi \left[ \partial_0 B_i + \epsilon_{ijk} \partial_j E_k \right] - \partial_i B_i \nabla_0 \chi = 0 \tag{80}$$

$$\partial_i z \left[ \partial_0 B_i + \epsilon_{ijk} \partial_j E_k \right] - \partial_i B_i \nabla_0 z = 0$$

or in relativistic notation

$$\nabla^\nu \chi \partial^\mu \tilde{F}_{\mu\nu} = 0; \qquad \nabla^\nu z \partial^\mu \tilde{F}_{\mu\nu} = 0 \tag{81}$$

These can be combined into a covariant conservation equation

$$\nabla^\mu J_\mu^G = 0 \tag{82}$$

with the current

$$J_\mu^G \equiv G(z, \chi) \partial^\nu \tilde{F}_{\nu\mu} = 0 \tag{83}$$

with arbitrary function $G$.

In addition, there is an equation obtained by differentiation with respect to $\Lambda$. To understand how to derive this equation, we can expand $\Lambda(z, \chi)$ in a complete basis of functions on a two-dimensional space, for example, by writing $\Lambda(z, \chi) = \int dp\, dq\, e^{ipz + iq\chi} \tilde{\Lambda}(p, q)$ and substituting it into the action, then differentiate with respect to $\tilde{\Lambda}$. The resulting equations are

$$\int d^3x J_0^G = \int d^3x G(z, \chi) \partial_i B_i = 0 \tag{84}$$

This equations are rather interesting. They put a large number of constraints on the divergence of the magnetic field. Unfortunately, the number of constraints is not large enough to ensure that magnetic monopole charge vanishes, as $G$ is only a function of two variables (at any given time), while the coordinate space is obviously three-dimensional.

One could ask whether the modification we made can help us find arbitrary two gluon states in the spectrum. Unfortunately, the answer is negative. The simplest way to see it is to realize that one can gauge fix the "vector potential" $\Lambda$ to zero—the Hamiltonian gauge of sorts. In this gauge, the dynamical equations of the model are identical with the equations of Mark 1. Thus, we do not have new solutions to the equations of motion. The gauging does eliminate those solutions that do not satisfy the constraint Equation (84), but it does not generate any new solutions to the equations of motion.

Thus, although it feels like gauging $Sdiff(S^2)$ may be a step in the right direction, it is not sufficient. In the next section, we discuss a further modification of the model-Mark 2, which starts from the same premise but successfully reproduces the theory of free photon.

## 7. The Model Mark 2

The model of [5], despite having some interesting properties, fails to describe adequately the low energy dynamics of the Abelian limit. As we have learned from the previous section, gauging the $Sdiff(S^2)$ symmetry does not solve the main problems of [5], i.e., on one hand, the constraints it imposes are not sufficient to ensure vanishing of magnetic charge density, and on the other hand, it does not allow for additional solutions of equations of motion that can be identified with multiphoton states of arbitrary polarization. Both of these deficiencies are associated with the fact that the "vector potential" $\Lambda$ is not a bona fide local degree of freedom but only a function of two variables $z$ and $\chi$. Let us extend our approach by allowing $\Lambda$ to become an independent function of space time. We, therefore, change our definition of magnetic field to [6]

$$F^{\mu\nu} = \epsilon^{\mu\nu\alpha\beta} [\epsilon^{abc} \phi_a \partial_\alpha \phi_b \partial_\beta \phi_c + (n \cdot \partial) n_\alpha \partial_\beta \Phi] \tag{85}$$

Here, $n = (1, 0, 0, 0)$ is a time-like vector of unit length and $\Phi$ is a scalar field [19]. This is a generalization of Equation (70) with $\Lambda \to \partial_0 \Phi$.

We stress that, as opposed to the discussion in the previous section, $\Phi(x)$ is now a *bona fide* field that has a general dependence on space-time coordinates.

The Lagrangian, as before, is

$$\mathcal{L} = \frac{1}{4} F_{\mu\nu} F^{\mu\nu} = -\frac{1}{2} (\vec{E}^2 - \vec{B}^2) \tag{86}$$

One may worry that since $n$ is chosen to be a time-like vector, the model is not a Lorentz invariant. Nevertheless, we will show below that the model possesses a Lorentz invariant super selection sector, and it is this sector that will turn out to be equivalent to QED.

We now have to understand what effect the modification has on the Abelian limit of the model. We will analyze its canonical structure and will demonstrate that it is identical to that of free electrodynamics. This applies to the commutators between the "electric" and "magnetic" fields and the Hamiltonian. We, thereby, demonstrate that the model is equivalent to the theory of a free noninteracting photon, even though it is not formulated in terms of a covariant vector potential field. We also derive the Lorentz transformation properties on the degrees of freedom of the model. We demonstrate that the fields $\phi^i$ are not covariant scalar fields but instead transform nontrivially and noncovariantly under the Lorentz group. We confirm that due to these modified transformation properties, the model retains Lorentz invariance.

## 8. Equations of Motion and Canonical Structure

### 8.1. Equations of Motion

As before, we use the following parametrization of the basic fields $\chi$ and $z$ via $\phi_3 = z$ and $\phi_1 + i\phi_2 = \sqrt{1 - z^2} e^{i\chi}$. The electromagnetic field can now be written as:

$$F^{\mu\nu} = \epsilon^{\mu\nu\alpha\beta} [-2 \partial_\beta \chi \partial_\alpha z + n_\alpha \partial_\beta \partial_0 \Phi] \tag{87}$$

or explicitly

$$E_i = 2\epsilon_{ijk} \partial_j z \partial_k \chi \tag{88}$$

$$B_k = [2(\partial_k \chi \partial_0 z - \partial_0 \chi \partial_k z) - \partial_k \partial_0 \Phi] \tag{89}$$

Varying the action, Equation (86) yields the following equations of motion

$$\partial_0 \partial_k \left[ F_{ij} \epsilon^{ij0k} \right] = 0 = \partial_0 \partial_k B_k \tag{90}$$

$$\partial_\beta \chi \partial_\alpha (F_{\mu\nu} \epsilon^{\mu\nu\alpha\beta}) = 0 = \partial_k \chi \partial_\alpha (F_{\mu\nu} \epsilon^{\mu\nu\alpha k}) = \partial_k \chi (\partial_0 B_k + (\partial \times E)_k) \tag{91}$$

$$\partial_\beta z \partial_\alpha (F_{\mu\nu} \epsilon^{\mu\nu\alpha\beta}) = 0 = \partial_k z \partial_\alpha (F_{\mu\nu} \epsilon^{\mu\nu\alpha k}) = \partial_k z (\partial_0 B_k + (\partial \times E)_k) \tag{92}$$

The main difference with our previous attempt is Equation (90). This equation now means that the "magnetic charge density" $\partial_k B_k$ is *locally* conserved. In the current model, therefore, the magnetic charge density is time independent at any space point. Equation (90) imposes the existence of "super selection sectors" characterized by the value of the magnetic charge density at all spatial coordinates. Clearly, most of these sectors are not translationally invariant. In order to preserve translational invariance, we limit our consideration to the trivial sector with $\partial_k B_k = 0$. The rest of our discussion pertains exclusively to this super selection sector.

Using this constraint on the magnetic field, we can invert Equations (91) and (92). (One has to be careful since there is an ambiguity in the inversion of Equations (91) and (92). In general, we find $\partial_0 B_k + (\partial \times E)_k = \alpha E_k$, where $\alpha$ is an arbitrary constant. Nonetheless, since $B$ and $\partial \times E$ are pseudovectors, while $E$ is a vector, a nonvanishing value of $\alpha$ would violate parity. Imposing parity invariance on the solution sets $\alpha = 0$.)

$$\partial_0 B_k + (\partial \times E)_k = 0 \tag{93}$$

Recall that with the field strength components given by Equation (87), the "electric" equation

$$\partial_\mu F^{\mu\nu} = 0 \tag{94}$$

is satisfied identically. Thus, the equations of motion of the model Mark 2 are the full set of Maxwell's equations.

### 8.2. The Hamiltonian

Let us now turn to the Hamiltonian description of the model. The canonical momenta as calculated from Equation (87) are given by :

$$p_z = \frac{\delta L}{\delta \partial_0 z} = F_{ij}\epsilon^{ij0k}\partial_k \chi = 2B_k \partial_k \chi = 2\partial_k \chi [2(\partial_k \chi \partial_0 z - \partial_0 \chi \partial_k z) - \partial_k \partial_0 \Phi] \tag{95}$$

$$p_\chi = \frac{\delta L}{\delta \partial_0 \chi} = F_{ij}\epsilon^{ijk0}\partial_k z = -2B_k \partial_k z = -2\partial_k z [2(\partial_k \chi \partial_0 z - \partial_0 \chi \partial_k z) - \partial_k \partial_0 \Phi] \tag{96}$$

$$p_\Phi = \frac{\delta L}{\delta \partial_0 \Phi} = \frac{1}{2}\partial_k (F_{ij}\epsilon^{ij0k}) = \partial_k B_k = \partial_k [2(\partial_k \chi \partial_0 z - \partial_0 \chi \partial_k z) - \partial_k \partial_0 \Phi] \tag{97}$$

The time derivatives of the fields can be expressed as:

$$\dot{\chi} = \frac{1}{E^2}[p_z(z\chi) + p_\chi \chi^2 + \epsilon_{ijk}\dot{\Phi}_i E_j \chi_k] \tag{98}$$

$$\dot{z} = \frac{1}{E^2}[p_z z^2 + p_\chi(z\chi) + \epsilon_{ijk}\dot{\Phi}_i E_j z_k] \tag{99}$$

Using these equations, we express $p_\Phi$ as:

$$p_\Phi = \partial_k \left[ \frac{1}{E^2}\epsilon_{klm}E_l(p_z z_m + p_\chi \chi_m) - \frac{1}{E^2}E_k E_i \dot{\Phi}_i \right] \tag{100}$$

or in terms of a "vector potential"

$$D_k = \frac{1}{E^2}\epsilon_{klm}E_l(p_z z_m + p_\chi \chi_m) \tag{101}$$

as

$$p_\Phi = \partial_k (D_k - \hat{E}_k \hat{E}_i \dot{\Phi}_i) \tag{102}$$

The Hamiltonian is then calculated as:

$$H = \int d^3x [p_z \dot{z} + p_\chi \dot{\chi} + p_\Phi \dot{\Phi} - L] = \int d^3x \frac{1}{2}\left(E^2 + B^2\right) \tag{103}$$

In arriving at this expression, we have neglected a boundary term $\int d^3x \partial_k (B_k \dot{\Phi})$.

### 8.3. Canonical Structure

In order to prove the equivalence between Mark 2 and QED, we have to show that the canonical commutation relations of $E_i$ and $B_i$ are identical in the two theories. We do this in the present section.

Since all components of the electric field in Mark 2l are functions only of coordinates and not canonical momenta, they clearly commute with each other

$$\left[E_i(x), E_j(y)\right] = 0 \tag{104}$$

In order to calculate the commutation relations between electric and magnetic fields, we restrict ourselves explicitly to the zero magnetic charge density super selection sector and set $p_\Phi = 0$. Equation (102) then becomes

$$\frac{\partial_k D_k}{E} = \hat{E}_k \partial_k \left( \frac{\hat{E}_i \Phi_i}{E} \right) \tag{105}$$

where we have used $\partial_k E_k = 0$.

Formally, the solution is written as

$$\hat{E}_i \Phi_i = E(x) \int_{-\infty}^{x} dl_C \frac{\partial_k D_k}{E} \tag{106}$$

In this expression, the integration contour $C$ starts at $x$ and ends at some point at spatial infinity. The direction along the contour at every point is parallel to the direction of the electric field at this point.

Using the definition of $D$, we have:

$$B_k = D_k - E_k \int_{-\infty}^{x} dl_C \frac{\partial_m D_m}{E} \tag{107}$$

Consider first the following auxiliary quantity

$$
\begin{aligned}
[E_i(x), D_k(y)] &= 2i \frac{E_l(y)}{E^2(y)} \epsilon_{iab} \epsilon_{klm} [\partial_a^x \delta(x-y) \chi_b(x) z_m(y) + \partial_b^x \delta(x-y) z_a(x) \chi_m(y)] \\
&= 2i \frac{E_l(y)}{E^2(y)} \epsilon_{iab} \epsilon_{klm} \partial_a^x \delta(x-y) [\chi_b(y) z_m(y) - z_b(y) \chi_m(y)] \\
&= i \hat{E}_l(y) \hat{E}_c(y) \epsilon_{iab} \epsilon_{klm} \epsilon_{cmb} \partial_a^x \delta(x-y) = i [\epsilon_{iak} - \hat{E}_b(y) \hat{E}_k(y) \epsilon_{iab}] \partial_a^x \delta(x-y)
\end{aligned}
\tag{108}
$$

Using this, we calculate

$$
\begin{aligned}
[E_i(x), B_k(y)] &= [E_i(x), D_k(y)] - E_k(y) \int_{-\infty}^{y} dl_C \frac{\partial_m^t [E_i(x), D_m(t)]}{E(t)} \\
&= [E_i(x), D_k(y)] - E_k(y) \int_{-\infty}^{y} dl_C \frac{1}{E(t)} \partial_m^t [(\epsilon_{iam} - \hat{E}_b(t) \hat{E}_m(t) \epsilon_{iab}) \partial_a^x \delta(x-t)] \\
&= [E_i(x), D_k(y)] + E_k(y) \int_{\infty}^{y} dl_C \hat{E}_m(t) \partial_m^t \left( \frac{\hat{E}_b(t)}{E(t)} \epsilon_{iab} \partial_a^x \delta(x-t) \right) \\
&= i \epsilon_{iak} \partial_a^x \delta(x-y)
\end{aligned}
\tag{109}
$$

Here, it was important that the integration contour $C$ is parallel to the electric field everywhere. In addition, we have assumed that all the fields vanish at the spatial boundary.

The commutator Equation (109) is identical to the commutator of corresponding quantities in QED.

We now consider the commutator of the magnetic fields.

It is easy to see that $[B_i(x), B_a(y)] = 0$ as long as the curve $C_x$ that enters the definition of $B_i(x)$ in Equation (107) does not pass through the point $y$, and $C_y$ does not pass through $x$. If this condition is not satisfied, direct evaluation of the commutator is not easy. Instead, we argue indirectly. One can straightforwardly obtain a number of relations involving the commutator of interest. Consider, for instance,

$$[B_i(x) \partial_i \chi(x), B_j(y) \partial_j z(y)] = [p_z(x), p_\chi(y)] = 0 \tag{110}$$

Trivially:

$$B_i(x)\partial_j z(y)[\partial_i\chi(x), B_j(y)] + B_j(y)\partial_i\chi(x)[B_i(x), \partial_j z(y)] + \partial_i\chi(x)\partial_j z(y)[B_i(x), B_j(y)] =$$

$$\left(B_i(x)\partial_j z(y)\partial_i^{(x)}\frac{\partial D_j(y)}{\partial p_\chi(x)} - B_j(y)\partial_i\chi(x)\partial_j^{(y)}\frac{\partial D_i(x)}{\partial p_z(y)}\right) + \partial_i\chi(x)\partial_j z(y)[B_i(x), B_j(y)] = \tag{111}$$

$$(B_i(y)\partial_i^{(x)}\delta(x-y) + B_i(x)\partial_i^{(y)}\delta(x-y)) + \partial_i\chi(x)\partial_j z(y)[B_i(x), B_j(y)] =$$

$$\partial_i\chi(x)\partial_j z(y)[B_i(x), B_j(y)] = 0$$

Similarly,

$$\partial_i z(x)\partial_j z(y)[B_i(x), B_j(y)] = \partial_i\chi(x)\partial_j\chi(y)[B_i(x), B_j(y)] = 0 \tag{112}$$

Using $\partial_k B_k = 0$, we obtain:

$$\partial_i z(x)\partial_j^y[B_i(x), B_j(y)] = \partial_i\chi(x)\partial_j^y[B_i(x), B_j(y)] = \partial_i^x\partial_j^y[B_i(x), B_j(y)] = 0 \tag{113}$$

Defining the matrix $M_{ij}(x,y) \equiv [B_i(x), B_j(y)]$, we, therefore, find that it is antisymmetric under the exchange $(i, x) \leftrightarrow (j, y)$ and satisfies the set of equations (Equations (111)–(113)). The general solution for these constraints can be written as

$$M_{ij}(x,y) = E_i(x)F_j(y) - E_j(y)F_i(x) \tag{114}$$

with $F_i(x)$ being an arbitrary vector function. However, we already saw that when $x$ does not belong to $C_y$ and $y$ does not belong to $C_x$, then $M_{ij}(x,y) = 0$. This determines $F_i(x) = 0$, so that finally:

$$[B_i(x), B_j(y)] = 0 \tag{115}$$

for all $x, y$.

### 9. Lorentz Transformations of the Fields

We now wish to discuss the properties of the fields $z$ and $\chi$ under Lorentz transformations. The rotational transformation properties of $z$ and $\chi$ are clearly those of a scalar field. This is obvious since the rotational invariance is represented in our model in the standard linear manner. This is not the situation with Lorentz boosts. The electric and magnetic fields are components of a covariant Lorentz tensor, and therefore, it is clear that $z$ and $\chi$ cannot be covariant scalar fields.

We take the following parametrization of infinitesimal Lorentz transformations:

$$z(x) \to z(\Lambda^{-1}x) = (1 + \beta\Delta)z(x) + a$$
$$\chi(x) \to \chi(\Lambda^{-1}x) = (1 + \beta\Delta)\chi(x) + b \tag{116}$$
$$\Theta(x) \equiv \partial_0\Phi \to \Theta(\Lambda^{-1}x) = \Theta(x) + c$$

Here, $\beta$ is the boost parameter and $\Delta \equiv \omega^\mu{}_\nu x^\nu\partial_\mu$ with $\omega^\mu_\nu$—an antisymmetric generator of Lorentz transformation. The boost in the direction of a unit vector $\hat{n}$ is generated by $\omega^i_0 = \hat{n}_i$. The terms involving $a, b$, and $c$ are not canonical, and we will determine them requiring that $F^{\mu\nu}$ transforms as a covariant tensor.

Let us consider first a boost in the direction of the first axis, $\hat{n} = (1,0,0)$. The components of the field strength tensor transform as

$$E_2(x) \to E_2(\Lambda^{-1}x) - \beta B_3(\Lambda^{-1}x) \tag{117}$$

Writing this in terms of $z, \chi$, and $\Theta$ and using Equation (116), we have:

$$E_2(x) = 2[\partial_3 z(x)\partial_1\chi(x) - \partial_1 z(x)\partial_3(x)] \to 2[\partial_3 z(\Lambda^{-1}x)\partial_1\chi(\Lambda^{-1}x) - \partial_1 z(\Lambda^{-1}x)\partial_3(\Lambda^{-1}x)] \tag{118}$$

Comparing the two we obtain:

$$-\beta\partial_3\Theta + 2[\partial_3 z\partial_1 b + \partial_3 a\partial_1\chi - \partial_1 z\partial_3 b - \partial_1 a\partial_3\chi] = 0 \tag{119}$$

Similarly, the transformation of $E_1$ yields

$$2(\partial_2 z\partial_3 b + \partial_2 a\partial_3\chi - \partial_3 z\partial_2 b - \partial_3 a\partial_2\chi) = 0 \tag{120}$$

and that of $E_3$:

$$\beta\partial_2\Theta + 2[\partial_1 z\partial_2 b + \partial_1 a\partial_2\chi - \partial_2 z\partial_1 b - \partial_2 a\partial_1\chi] = 0 \tag{121}$$

Introducing $f_i = 2(a\partial_i\chi - b\partial_i z)$, and $u_i = (0, \beta\partial_3\Theta, -\beta\partial_2\Theta)$, the above equations combine into

$$\epsilon_{ijk}\partial_j f_k = u_i \tag{122}$$

The general solution for $f$ is:

$$\begin{aligned}
f_i &= -\frac{\epsilon_{ijk}\partial_j u_k}{\partial^2} + \partial_i\tilde{\lambda} \\
&= \beta\hat{n}_i\Theta + \partial_i\lambda
\end{aligned} \tag{123}$$

where

$$\tilde{\lambda} - \beta\frac{\hat{n}_i\partial_i}{\partial^2}\Theta = \lambda \tag{124}$$

with $\lambda$ still to be determined.

Noting that Equations (123) and (89) become the same under the substitution, Equation (123) for $a$ and $b$ can be solved as

$$\begin{aligned}
\partial_0 z &\to a \\
\partial_0\chi &\to b \\
\partial_0\Phi &\to \lambda \\
B_k &\to \beta\Theta\hat{n}_k
\end{aligned} \tag{125}$$

Using Equations (98) and (99), we find:

$$\begin{aligned}
a &= \frac{1}{E^2}(\beta\Theta\hat{n}_i + \lambda_i)\epsilon_{ijk}E_j z_k \\
b &= \frac{1}{E^2}(\beta\Theta\hat{n}_i + \lambda_i)\epsilon_{ijk}E_j\chi_k
\end{aligned} \tag{126}$$

Now, Equation (123) becomes an equation for $\lambda$:

$$E_i(\beta\Theta\hat{n}_i + \partial_i\lambda) = 0 \tag{127}$$

which yields:

$$\lambda(x) = -\beta\int_\infty^x dl_C\hat{E}_i\hat{n}_i\Theta \tag{128}$$

where as before, the contour $C$ is locally parallel to the vector $E_i$.

The function $c$ can similarly be determined by considering the transformation of the magnetic field.

$$B_1(x) \to B_1(\Lambda^{-1}x) = 2[\partial_1\chi(\Lambda^{-1}x)\partial_0(\Lambda^{-1}x) - \partial_0\chi(\Lambda^{-1}x)\partial_1 z(\Lambda^{-1}x)] - \partial_1\Theta(\Lambda^{-1}x) \tag{129}$$

yields

$$2[\partial_1\chi\partial_0 a + \partial_1 b\partial_0 z - \partial_0\chi\partial_1 a - \partial_0 b\partial_1 z] - \partial_1 c + \beta\Delta\partial_1\Theta = 0 \tag{130}$$

Similarly, the transformation of $B_2$ and $B_3$ gives

$$2[\partial_2\chi\partial_0 a + \partial_0 z\partial_2 b - \partial_0\chi\partial_2 a - \partial_2 z\partial_0 b] - \partial_2 c + \beta\Delta\partial_2\Theta = 0$$
$$2[\partial_3\chi\partial_0 a + \partial_0 z\partial_3 b - \partial_0\chi\partial_3 b - \partial_3 z\partial_0 b] - \partial_3 c + \beta\Delta\partial_3\Theta = 0 \tag{131}$$

These can be combined into a single vector equation

$$\partial_0 f_i - \partial_i f_0 - \partial_i c + \beta\Delta\partial_i\Theta = 0 \tag{132}$$

Using Equation (123), we write this:

$$\partial_i[\partial_0\lambda - f_0 - c + \beta\Delta\Theta] = 0 \tag{133}$$

yielding

$$c = 2(a\partial_0\chi - b\partial_0 z) - \partial_0\lambda - \beta\Delta\Theta$$
$$= \beta\left[\frac{2}{E^2}\epsilon_{ijk}E_j(\partial_0\chi z_k - \partial_0 z\chi_k)\left[\Theta\hat{n}_i - \partial_i\int_\infty^x dl_C\hat{E}_i\hat{n}_i\Theta\right] + \partial_0\int_\infty^x dl_C\hat{E}_i\hat{n}_i\Theta - \Delta\Theta\right] \tag{134}$$

To summarize, the fields $z$, $\chi$, and $\Phi$ under Lorentz boost transform according to Equation (116) with $a,b$, and $c$ given in Equations (126), (128), and (134). These somewhat complicated transformation properties ensure that electromagnetic fields are components of the covariant Lorentz tensor.

## 10. Discussion of Model MARK 2

Our amended model (Mark 2) is equivalent to the theory of a free photon. We were led to this model by our wish to eliminate the global $Sdiff(S^2)$ symmetry but had to go further from the original model in order to achieve equivalence with QED. What is the fate of $Sdiff(S^2)$ in Mark 2? It is indeed easy to see that this symmetry is gauged. In order to see that, let us write

$$\partial_0\Phi = \Lambda(z,\chi,t) + \partial_0\bar{\Phi} \tag{135}$$

Assigning to $\Lambda$ the same transformation properties under $Sdiff(S^2)$ as before and requiring $\bar{\Phi}$ to be invariant, we see that the Lagrangian Equation (86) is indeed invariant under the $Sdiff(S^2)$ global gauge transformation. Note that the decomposition Equation (135) is always possible, given that $\Phi$ is an arbitrary function of space-time coordinates. It is important that we have been able to obtain the theory of a free photon. Our main goal, however, was (and remains) to understand confinement in the Non-Abelian case. Here, the road is still very long and winding, and at this point, there are mainly questions. We need to generalize our model in several directions. First, charged states have not been included in the model. This should be relatively straightforward to mitigate. As suggested in [5], we should relax the constraint of constant length of the sigma model field $\phi^a$ and instead endow the modulus field $\phi^2$ with nontrivial dynamics. This will soften the classical behavior of the model in UV and will lead to UV finite energy of charged states. The configuration space of our model is $SO(3) \times R$, with the $SO(3)$ symmetry broken spontaneously to $O(2)$. The moduli space should, therefore, have a nontrivial homotopy group $\Pi_2(M) = Z$ and allow for a nontrivial topological charge, which is identified with the electric charge. (There may be some subtlety in this argument related to the fact that the global gauge group $Sdiff(S^2)$ has to be modded out. However, since the gauge transformation is global, we do not anticipate any problems.)

The more complicated question is how to extend this model into the Non-Abelian regime. Following the logic of [5], we should add a perturbation thats explicitly breaks the global symmetry of the model and via this breaking generates a linear potential between the charges. Here, first of all, we need to understand whether this perturbation should preserve the $Sdiff(S^2)$ gauge symmetry or should break it explicitly. Such a global gauge symmetry was not present in 2 + 1 dimensional models [2,3], and we lack guidance on this question from 2 + 1 d. It seems likely that the $Sdiff(S^2)$ should be preserved by the perturbation. If that is the case, the type of perturbations considered in [5] do not fit the bill. Perhaps one should deal directly with the breaking of the generalized magnetic symmetry—the symmetry generated by the magnetic flux [12,13] in terms of its order parameter—the t Hooft loop [1].

A rough idea of how this can work is the following. Let us try to define an operator that breaks the generalized magnetic symmetry. This should be an analog of a t Hooft loop operator, except it should have end points, so rather a t Hooft line operator. We write the following bilocal expression

$$V(x,y) = \exp\left(i \int_{t=-\infty}^{t=0} dt A_0(x,t)\right) \exp\left(i \int_x^y A_i(z, t=0)dz^i\right) \exp\left(i \int_{t=0}^{t=-\infty} dt A_0(y,t)\right) \tag{136}$$

Here, the components of dual vector potential are chosen as $A_i = z\partial_i\chi$ and $A_0 = z\partial_0\chi + \frac{1}{2}\partial_0\Phi(x)$. Given the $Sdiff(S^2)$ transformation properties of the various operators, we have

$$A_i \to A_i + \partial_i\left[G - z\frac{\partial G}{\partial z}\right] \tag{137}$$

$$A_0 \to A_0 + \frac{d}{dt}\left[G(z,\chi;t) - z\frac{\partial G}{\partial z}\right] \tag{138}$$

Under the assumption that the fields vanish at infinity, it is easy to see that the operator Equation (136) is invariant under $Sdiff(S^2)$. In terms of its quantum numbers, this operator essentially creates a monopole–antimonopole pair at points $x$ and $y$. For infinitesimally close points $y = x + \epsilon$, this becomes

$$V(x,\epsilon_i) = 1 + i\epsilon_i\left[z\partial_i\chi - \partial_i\Phi - \partial_i\int_{-\infty}^0 dtz\partial_0\chi\right] + \epsilon_i\epsilon_j\left[z\partial_i\chi - \frac{1}{2}\partial_i\Phi - \partial_i\int_{-\infty}^0 dtz\partial_0\chi\right]\left[z\partial_j\chi - \frac{1}{2}\partial_i\Phi - \partial_j\int_{-\infty}^0 dtz\partial_0\chi\right] \tag{139}$$

We could contemplate averaging this operator over the direction of the point splitting vector $\epsilon$, which would kill the linear in $\epsilon$ term and would result in a term reminiscent of the gauge invariant Stueckelberg mass for the dual vector potential [20]. Adding an $n$-th power of such an operator as a perturbation to the Lagrangian would seem a reasonable way to proceed in order to break the generalized magnetic symmetry to the $Z_N$ subgroup.

Unfortunately, Equation (139) contains a term that is nonlocal in time. Thus, adding it to the Lagrangian would lead to nonlocal in time theory, which amounts to adding extra degrees of freedom in disguise. Although this may turn out to be necessary, it is clearly outside the rather tight framework that we have set out to ourselves from the beginning. Thus, before taking this route, a better understanding is necessary. We hope to be able to make progress in this approach in the future.

**Author Contributions:** Both authors contributed equally. Both All authors have read and agreed to the published version of the manuscript.

**Funding:** I.B.I. is partially supported by TUBITAK (The Scientific and Technological Research Council of Turkey) under the grant no 117F203. A.K. is supported by the NSF Nuclear Theory grant 1913890.

**Institutional Review Board Statement:** Not applicable.

**Informed Consent Statement:** Not applicable.

**Data Availability Statement:** Not applicable.

**Conflicts of Interest:** The authors declare no conflict of interest.

## Appendix A

In this appendix, we show that the model considered in this paper does not admit two-photon solutions with arbitrary polarizations. We are looking for two-photon solutions for which the electromagnetic tensor is of the form:

$$\tilde{F}_{\mu\nu} = \partial_{[\mu}z\partial_{\nu]}\chi = A(k_\mu\epsilon_\nu^1 - k_\nu\epsilon_\mu^1)\cos kx + B(p_\mu\epsilon_\nu^2 - p_\nu\epsilon_\mu^2)\cos px \tag{A1}$$

For simplicity, we choose the case when the first photon has momentum $k$ in $x$-direction and polarization $a$ in $y$-direction, while the second photon has momentum $p$ in $y$-direction and polarization $b$ in $z$-direction. Note that this case is not covered by our construction of two-photon states in the body of the paper.

Now, for components of $\tilde{F}_{\mu\nu}$, we have:

$$\partial_{[0}z\partial_{1]}\chi = 0 = \partial_{[1}z\partial_{3]}\chi = 0 \tag{A2}$$

$$\partial_{[0}z\partial_{2]}\chi = ka\cos kx = -\partial_{[1}z\partial_{2]}\chi \tag{A3}$$

$$\partial_{[0}z\partial_{3]}\chi = pb\cos px = -\partial_{[2}z\partial_{3]}\chi \tag{A4}$$

Introducing new coordinates $(x, y, z, t) \to (\bar{x} = t - x, \bar{y} = t - y, \bar{t} = t, \bar{z} = z)$ and using unbarred symbols for notational simplicity, we have:

$$\partial_{[t}z\partial_{y]}\chi = \partial_{[t}z\partial_{z]}\chi = \partial_{[x}z\partial_{z]}\chi = 0 \tag{A5}$$

$$\partial_{[t}z\partial_{x]}\chi = \partial_{[x}z\partial_{y]}\chi = -ka\cos kx \tag{A6}$$

$$\partial_{[y}z\partial_{z]}\chi = pb\cos py \tag{A7}$$

These equations have no solutions. Assuming $\partial_t z \neq 0$, the first two equations in Equation (A5) imply $\partial_y z\partial_z\chi - \partial_z z\partial_y\chi = 0$, which contradicts Equation (A7). Alternatively, assuming $\partial_t z = 0$, implies vanishing of either $\partial_t\chi$ or two other partial derivatives of $z$. It is then easy to see that both these options are in conflict with the rest of the equations. The result is that a two-photon state with this polarization pattern cannot be constructed in this model.

The model also contains solutions that do not satisfy the homogeneous Maxwell equation. As an example of such a solution consider the configuration

$$\chi = \sin p \cdot x; \quad z = \sin k \cdot x \tag{A8}$$

It is easy to see that this configuration satisfies equations of motion provided

$$(p \cdot k)^2 - p^2 k^2 = 0 \tag{A9}$$

A simple example is a light-like momentum $k^\mu$ and a space-like momentum $p^\mu$ satisfying $p \cdot k = 0$. This yields the dual field strength

$$\tilde{F}_{\mu\nu} \propto (k_\mu p_\nu - k_\nu p_\mu)[\cos(p + k) \cdot x + \cos(p - k) \cdot x] \tag{A10}$$

which is not conserved

$$\partial^\mu \tilde{F}_{\mu\nu} \propto p^2 k_\nu [\sin(p + k) \cdot x + \sin(p - k) \cdot x] \tag{A11}$$

In fact, both momenta $k + p$ and $k - p$ are space-like, and thus, $\tilde{F}_{\mu\nu}$ looks tachyonic. However, as mentioned in the Discussion, since the model classically has many degenerate vacua with broken translational invariance, the interpretation of classical solutions as excitations is not so clear.

## References


1. Hooft, G. On the Phase Transition Towards Permanent Quark Confinement. *Nucl. Phys. B* **1978**, *138*, 1–25. [CrossRef]
2. Kovner, A. *At the Frontier of Particle Physics*; Shifman, M., Ed.; World Scientific: Singapore, 2001; Volume 3, pp. 1777–1825;
3. IKogan, I.; Kovner, A. *At the Frontier of Particle Physics*; Shifman, N., Ed.; World Scientific: Singapore, 2002; Volume 4, pp. 2335–2407.
4. Altes, C.K.; Kovner, A. Magnetic Z(N) symmetry in hot QCD and the spatial Wilson loop. *Phys. Rev. D* **2000**, *62*, 096008. [CrossRef]
5. Kovner, A.; Ilhan, I.B. Curious case of an effective theory. *Phys. Rev. D* **2013**, *88*, 125004.
6. Kovner, A.; Ilhan, I.B. Photons without vector fields. *Phys. Rev. D* **2016**, *93*, 025015.
7. Faddeev, L.D.; Niemi, A.J. Stable knot-like structures in classical field theory. *Nature* **1997**, *387*, 58–61. [CrossRef]
8. Faddeev, L.D.; Niemi, A.J. Partially Dual Variables in SU(2) Yang-Mills Theory. *Phys. Rev. Lett.* **1999**, *82*, 1624. [CrossRef]
9. Faddeev, L.D.; Niemi, A.J. Partial duality in SU(N) Yang-Mills theory. *Phys. Lett. B* **1999**, *449*, 214–218. [CrossRef]
10. Faddeev, L.D.; Niemi, A.J. Decomposing the Yang-Mills Field. *Phys. Lett. B* **1999**, *464*, 90–93. [CrossRef]
11. Faddeev, L.D.; Niemi, A.J. Spin-Charge Separation, Conformal Covariance and the SU(2) Yang-Mills Theory. *Nucl. Phys. B* **2007**, *776*, 38–65. [CrossRef]
12. Kovner, A.; Rosenstein, B. New Look at $QED_4$: The Photon as a Goldstone Boson and the Topological Interpretation of Electric Charge. *Phys. Rev. D* **1994**, *49*, 5571–5581. [CrossRef]
13. Gaiotto, D.; Kapustin, A.; Seiberg, N.; Willett, B. Generalized Global Symmetries. *JHEP* **2015**, *2*, 172. [CrossRef]
14. Guendelman, E.I.; Nissimov, E.; Pacheva, S. Volume-Preserving Diffeomorphisms' versus Local Gauge Symmetry. *Phys. Lett. B* **1995**, *360*, 57–64. [CrossRef]
15. Takasaki, K. *Turku 1991, Proceedings, Topological and Geometrical Methods in Field Theory*; Mickelsson, J., Pekonen, O., Eds.; World Scientific: River Edge, NJ, USA, 1992; pp. 383–397.
16. Kovner, A.; Rosenstein, B. Strings and string breaking in 2+1 dimensional nonabelian theories. *JHEP* **1998**, *9809*, 003. [CrossRef]
17. Greensite, J. An introduction to the confinement problem. *Lect. Notes Phys.* **2011**, *821*, 1–211.
18. Gorsky, A.; Shifman, M.; Yung, A. Revisiting the Faddeev-Skyrme Model and Hopf Solitons. *Phys. Rev. D* **2013**, *88*, 045026. [CrossRef]
19. Gliozzi, F. String-like topological excitations of the electromagnetic field. *Nucl. Phys. B* **1978**, *141*, 379–390. [CrossRef]
20. Stueckelberg, E.C.G. Interaction energy in electrodynamics and in the field theory of nuclear forces. *Helv. Phys. Acta* **1938**, *11*, 225.

*Article*

# Axial Anomaly in Galaxies and the Dark Universe

Janning Meinert * and Ralf Hofmann *

Institut für Theoretische Physik, Universität Heidelberg, Philosophenweg 16, D-69120 Heidelberg, Germany
* Correspondence: j.meinert@thphys.uni-heidelberg.de (J.M.); r.hofmann@thphys.uni-heidelberg.de (R.H.)

**Abstract:** Motivated by the $SU(2)_{CMB}$ modification of the cosmological model $\Lambda$CDM, we consider isolated fuzzy-dark-matter lumps, made of ultralight axion particles whose masses arise due to distinct SU(2) Yang–Mills scales and the Planck mass $M_P$. In contrast to $SU(2)_{CMB}$, these Yang–Mills theories are in confining phases (zero temperature) throughout most of the Universe's history and associate with the three lepton flavours of the Standard Model of particle physics. As the Universe expands, axionic fuzzy dark matter comprises a three-component fluid which undergoes certain depercolation transitions when dark energy (a global axion condensate) is converted into dark matter. We extract the lightest axion mass $m_{a,e} = 0.675 \times 10^{-23}$ eV from well motivated model fits to observed rotation curves in low-surface-brightness galaxies (SPARC catalogue). Since the virial mass of an isolated lump solely depends on $M_P$ and the associated Yang–Mills scale the properties of an $e$-lump predict those of $\mu$- and $\tau$-lumps. As a result, a typical $e$-lump virial mass $\sim 6.3 \times 10^{10} \, M_\odot$ suggests that massive compact objects in galactic centers such as Sagittarius A* in the Milky Way are (merged) $\mu$- and $\tau$-lumps. In addition, $\tau$-lumps may constitute globular clusters. $SU(2)_{CMB}$ is always thermalised, and its axion condensate never has depercolated. If the axial anomaly indeed would link leptons with dark matter and the CMB with dark energy then this would demystify the dark Universe through a firmly established feature of particle physics.

**Keywords:** galaxy rotation curves; low surface brightness; dark matter; dark energy; ultralight axion particles; cores; halos; mass-density; profiles; pure Yang–Mills theory

**Citation:** Meinert, J.; Hofmann, R. Axial Anomaly in Galaxies and the Dark Universe. *Universe* **2021**, *7*, 198. https://doi.org/10.3390/universe7060198

Academic Editor: Dmitry Antonov

Received: 10 May 2021
Accepted: 3 June 2021
Published: 13 June 2021

**Publisher's Note:** MDPI stays neutral with regard to jurisdictional claims in published maps and institutional affiliations.



## 1. Introduction

Dark matter was introduced as an explanation for the anomalous, kinematic behavior of luminous test matter in comparison with the gravity exerted by its luminous surroundings, e.g., virialised stars within a galaxy [1] or a virialised galaxy within a cluster of galaxies [2]. That luminous matter can be segregated from dark matter is evidenced by the bullet cluster in observing hot intergalactic plasma (X-ray) in between localised dark-mass distributions (gravitational lensing) [3,4].

The present Standard Model of Cosmology (SMC) $\Lambda$CDM posits a spatially flat Universe [5] with about 70% dark energy, inducing late-time acceleration [6,7]. This model requires a substantial contribution of about 26% cold dark matter to the critical density and allows for a contribution of baryons of roughly 4%.

To determine all parameters of $\Lambda$CDM at a high accuracy, cosmological distance scales can be calibrated by high-redshift data (inverse distance ladder, global cosmology), coming from precision observations of the Cosmic Microwave Background (CMB) or from large-scale structure surveys probing Baryon Acoustic Oscillations (BAO). Alternatively, low-redshift data (direct distance ladder, local cosmology) can be used by appeal to standard or standardisable candles such as cepheids, TRGB stars, supernovae Ia, and supernovae II. Recently, a comparison between global and local cosmology has revealed tensions [8] in some of the cosmological parameter values (e.g., $H_0$ [9–12] and $\sigma_8 - \Omega_m$ [13–15], see also [16] for the context of a high-redshift modification of $\Lambda$CDM).

These interesting discrepancies motivate modifications of $\Lambda$CDM [17]. A cosmological model aiming to resolve these tensions should target high-redshift radiation and the dark sector. In particular, models which are in principle falsifiable by terrestrial experiments and which pass such tests could lead to a demystification of the dark Universe. However, searches for weakly interacting, massive and stable particles (WIMPS) [18], whose potential existence is suggested by certain extensions of the Standard Model of Particle Physics (SMPP), so far have not produced any detection [19,20].

An attractive possibility to explain the feebleness of a potential interaction between the dark sector of the SMC and SMPP matter in terms of the large hierarchy between particle-physics scales and the Planck mass is the theoretically [21–23] and experimentally [24] solidly anchored occurrence of an axial anomaly, which is induced by topological charge densities [25] in the ground states of pure Yang–Mills theories [26]. The axial anomaly acts on top of a dynamical chiral symmetry breaking mediated by a force of hierarchically large mass scale compared to the scales of the Yang–Mills theories. To enable the axial anomaly throughout the Universe's entire history chiral fermions, which acquire mass through gravitational torsion and which can be integrated out in a Planck-scale de Sitter background [27], need to be fundamentally charged under certain gauge groups. In such a scenario gravity itself—a strong force at the Planck scale—would induce the dynamical chiral symmetry breaking [28–30]. The anomaly then generates an axion mass $m_a$ [25] for particles that a priori are chiral Nambu–Goldstone bosons. Working in natural units $c = \hbar = k_B = 1$, one has

$$m_a = \frac{\Lambda^2}{M_P},\tag{1}$$

where $\Lambda$ denotes a Yang–Mills scale and $M_P = 1.221 \times 10^{28}$ eV the Planck mass [28,30]. The cold-dark-matter (CDM) paradigm is successful in explaining large-scale structure in the $\Lambda$CDM context but exhibits problems at small scales, e.g., galactic and lower [31]: While N-body simulations within $\Lambda$CDM reveal matter-density profiles of the galactic DM halos that are characterised by a central cusp of the Navarro-Frenk-White (NFW) type [32], $\rho_{\text{NFW}} \propto r^{-1}$ [33] ($r$ the radial distance to the center of the galaxy), observations suggest a core or soliton profile $\rho_{\text{sol}}(r)$ subject to a constant central matter density $\rho_c = \rho_{\text{sol}}(r = 0)$, see, e.g., [34–40]. A model of fuzzy dark matter (FDM) [34,36,41–48], according to the ground-state solution of the Schrödinger–Poisson system embedded into cosmological simulations [47], posits a condensate of free axion particles within the galactic core. For the radial range

$$r_{200} > r > r_e > 3\,r_c\tag{2}$$

the associated central matter densities $\rho_{\text{sol}}(r)$ gives way to a selfgravitating cloud of effective, nonrelativistic particles of mass $\sim \lambda_{\text{deB}}^3 \times \rho_{\text{NFW}}(r)$. Here $r_{200}$ denotes the virial radius defined such that

$$\rho_{\text{NFW}}(r_{200}) = 200\,\frac{3\,M_P^2}{8\pi}H_0^2,\tag{3}$$

where $H_0$ is the Hubble constant, and $\lambda_{\text{deB}} = \lambda_{\text{deB}}(r)$ indicates the de-Broglie wavelength of an axionic particle for $r_e < r < r_{200}$ where the NFW model applies. Note that within the core region $r < r_e$ the correlation length in the condensate is given by the reduced Compton wave length $\lambda_C = 1/m_a$. In what follows, we will refer to such a system— condensate core plus NFW-tail—as a *lump*. In [49], FDM fits to the rotations curves of low-surface-brightness galaxies, which are plausibly assumed to be dominated by dark matter, have produced an axion mass of $m_a = 0.554 \times 10^{-23}$ eV. Note also that the cosmological simulation of [47] associates the axionic scalar field with dark-matter *perturbations* only but not with the *background* dark-matter density which is assumed to be conventional CDM.

Another potential difficulty with $\Lambda$CDM, which FDM is capable of addressing, is the prediction of too many satellite galaxies around large hosts like the Milky Way or Andromeda [50], see, however [51] for a cosmological simulation within CDM. A recent match of observed satellite abundances with cosmological simulations within the FDM context yields a stringent bound on the axionic particle mass $m_a$ [51]: $m_a > 2.9 \times 10^{-21}$ eV.

This bound is consistent with $m_a = 2.5^{+3.6}_{-2.0} \times 10^{-21}$ eV derived from an analysis of the Milky Way rotation curve in [39].

There is yet another indication that $\Lambda$CDM may face a problem in delaying the formation of large galaxies of mass $M \sim 10^{12} M_\odot$ due to their hierarchical formation out of less massive ones. This seems to contradict the high-redshift observation of such galaxies [52] and suggests that a component of active structure formation is at work.

Assuming axions to be a classical ideal gas of non-relativistic particles the mass $m_a$ can be extracted from CMB simulations of the full Planck data subject to scalar adiabatic, isocurvature, and tensor-mode initial conditions [53] ($10^{-25}$ eV$\leq m_a \leq 10^{-24}$ eV with a 10% contribution to DM and a 1% contribution of isocurvature and tensor modes) and from a modelling of Lyman-$\alpha$ data [54] with conservative assumptions on the thermal history of the intergalactic medium. For the XQ-100 and HIRES/MIKE quasar spectra samples one obtains, respectively, $m_a \geq 7.12 \times 10^{-22}$ eV and $m_a \geq 1.43 \times 10^{-21}$ eV.

In our discussion of Section 5 we conclude that three axion species of hierarchically different masses could determine the dark-matter physics of our Universe. When comparing the results of axion-mass extractions with FDM based axion-mass constraints obtained in the literature it is important to observe that a *single* axion species always is assumed. For example, this is true of the combined axion-mass bound $m_a > 3.8 \times 10^{-21}$ eV, derived from modelling the Lyman-$\alpha$ flux power spectrum by hydrodynamical simulations [54], and it applies to the cosmological evolution of scalar-field based dark-matter perturbations yielding an axion mass of $m_a \sim 8 \times 10^{-23}$ eV in [47].

In the present article we are interested in pursuing the consequences of FDM for the physics of dark matter on super-galactic and sub-galactic scales within a cosmological model which deviates from $\Lambda$CDM in three essential points: (i) FDM is subject to three instead of one nonthermal axionic particle species, whose present cosmological mass densities are nearly equal, (ii) axion lumps (condensate core plus halo of fluctuating density granules) cosmologically originate from depercolation transitions at distinct redshifts $z_{p,i}$ out of homogeneous condensates [16], and (iii) the usual, nearly scale invariant spectrum of adiabatic curvature fluctuations imprinted as an initial condition for cosmological cold-dark-matter evolution, presumably created by inflation, does not apply.

Point (i) derives from the match of axion species with the three lepton families of the Standard Model of particle physics. These leptons emerge in the confining phases of SU(2) Yang–Mills theories [55]. According to Equation (1) axion masses are then determined by the universal Peccei-Quinn scale $M_P$ and the distinct Yang–Mills scales $\Lambda_e$, $\Lambda_\mu$, and $\Lambda_\tau$.

Point (ii) is suggested by a cosmological model [16] which is induced by the postulate that the CMB itself is described by an SU(2) gauge theory [26] and which fits the CMB power spectra TT, TE, and EE remarkably well except for low $l$. The according overshoot in TT at large angular scales may be due to the neglect of the nontrivial, SU(2)-induced photon dispersion at low frequencies.

Point (iii) relates to the fact that a condensate does not maintain density perturbations on cosmological scales and that $z_{p,e} \sim 53$. As a consequence, constraints on axion masses from cosmological simulations by confrontation with the observed small-scale structure should be repeated based on the model of [16]. This, however, is beyond the scope of the present work.

To discuss point (ii) further, we refer to [16], where a dark sector was introduced as a deformation of $\Lambda$CDM. This modification models a sudden transition from dark energy to dark matter at a redshift $z_p = 53$. Such a transition is required phenomenologically to reconcile high-$z$ cosmology (well below the Planckian regime but prior to and including recombination), where the dark-matter density is reduced compared to $\Lambda$CDM, with well-tested low-$z$ cosmology. That a reduced dark-matter density is required at high $z$ is as a result of an SU(2)$_{\text{CMB}}$-induced temperature-$z$ relation [56]. Depercolation of a formely spatially homogeneous axion condensate, which introduces a change of the equation of state from $\rho = -P$ to $P = 0$, is a result of the Hubble radius $r_H$—the spatial scale of causal connectedness in a Friedmann–Lemaitre–Robertson–Walker (FLRW) Universe—exceeding by far the gravitational Bohr radius $r_B$ of an isolated, spherically symmetric system of selfgravitating axion particles. The value of the ratio $r_H/r_B$ at depercolation so far is subject to phenomenological extraction, but should intrinsically be computable in the future by analysis of the Schrödinger–Poisson system in a thus linearly perturbed background cosmology whose dark sector is governed by axion fields subject to their potentials.

Roughly speaking, at depercolation from an equation of state $\rho = -P$ the quantum correlations in the axionic system become insufficient to maintain the homogeneity of the formerly homogeneously Bose-condensed state. The latter therefore decays or depercolates into selfgravitating islands of axionic matter whose central regions continue to be spatially confined Bose condensates but whose peripheries are virialised, quantum correlated particle clouds of an energy density that decays rapidly in the distance $r$ to the gravitational center to approach the cosmological dark-sector density. On cosmological scales, each of these islands (lumps) can be considered a massive (nonrelativistic) particle by itself such that the equation of state of the associated ensemble becomes $P = 0$: The density of lumps then dilutes as $a^{-3}$ where $a$ denotes the cosmological scale factor.

For the entire dark sector we have

$$\Omega_{ds}(z) = \Omega_\Lambda + \Omega_{pdm,0}(z+1)^3 + \Omega_{edm,0}\begin{cases} (z+1)^3, & z < z_{p,e} \\ (z_{p,e}+1)^3, & z \geq z_{p,e} \end{cases}. \tag{4}$$

Fits of this model to the TT, TE, and EE CMB power spectra reveal that $\Omega_{edm,0} \sim \frac{1}{2}\Omega_{pdm,0}$. Here $\Omega_{pdm,0}$ denotes a primordial contribution to the present dark-matter density parameter $\Omega_{dm,0} = \Omega_{edm,0} + \Omega_{pdm,0}$ while $\Omega_{edm,0}$ refers to the emergence of dark matter due to the depercolation of a formerly homogeneous Bose–Einstein condensate into isolated lumps once their typical Bohr radius is well covered by the horizon radius $r_H$. One may question that depercolation occurs suddenly at $z_{p,e}$, the only justification so far being the economy of the model. If a first-principle simulation of the Schrödinger–Poisson system plus background cosmology reveals that the transition from dark energy to dark matter during depercolation involves a finite $z$-range then this has to be included in the model of Equation (4).

After depercolation has occurred, a small dark-energy residual $\Omega_\Lambda$ persists to become the dominant cosmological constant today. As we will argue in Section 5, the primordial dark-matter density $\Omega_{pdm,0}$ could originate from the stepwise depercolation of former dark energy in the form of super-horizon sized $\mu$- and $\tau$-lumps. Therefore, dark energy dominates the dark sector at sufficiently high $z$. However, due to radiation dominance dark energy then was a marginal contribution to the expansion rate. The model of [16] was shown to fit the CMB anisotropies with a low baryon density, the local value for the redshift of re-ionisation [57], and the local value of $H_0$ from supernovae Ia distance-redshift extractions [10,11].

The purpose of the present work is to propose a scenario which accommodates $\Omega_{edm,0}$, $\Omega_{pdm,0}$, and $\Omega_\Lambda$. At the same time, we aim at explaining the parameters $\Omega_{edm,0}$ and $\Omega_{pdm,0}$ in terms of axial anomalies subject to a Planck-mass Peccei-Quinn scale and three SU(2) Yang–Mills theories associated with the three lepton families. In addition, an explanation of parameter $\Omega_\Lambda$ is proposed which invokes the SU(2) Yang–Mills theory underlying the CMB. Hence, the explicit gauge-theory content of our model is: $SU(2)_e \times SU(2)_\mu \times SU(2)_\tau \times SU(2)_{CMB}$.

We start with the observation in [36] that ultralight bosons necessarily need to occur in the form of selfgravitating condensates in the cores of galaxies. Because these cores were separated in the course of nonthermal depercolation halos of axion particles, correlated due to gravitational virialisation on the scale of their de Broglie wavelength, were formed around the condensates. Such a halo reaches out to a radius, say, of $r_{200}$ where its mass density starts to fall below 200 times the critical cosmological energy density of the spatially flat FLRW Universe. A key concept in describing such a system—a lump—is the gravitational Bohr radius $r_B$ defined as

$$r_B \equiv \frac{M_{\mathrm{P}}^2}{M m_a^2} , \tag{5}$$

where $M$ is the mass of the lump which should coincide with the viral mass, say $M_{200}$. We use two FDM models of the galactic mass density $\rho(r)$ to describe low-surface-brightness galaxies and to extract the axion mass $m_a$: The Soliton-NFW model, see [44] and references therein, and the Burkert model [58,59].

Rather model independently, we extract a typical value of $m_{a,e} \sim 0.7 \times 10^{-23}$ eV which confirms the value obtained in [49]. With Equation (1) this value of $m_{a,e}$ implies a Yang–Mills scale of $\Lambda_e \sim 287$ eV. This is smaller than $\Lambda_e = 511 \, \mathrm{keV}/118.6 = 4.31 \, \mathrm{keV}$ found in [55] where a link to an SU(2) Yang–Mills theory governing the first lepton family is made: SU(2)$_e$. Note that the larger value of $\Lambda_e$ was extracted in the *deconfining* phase [55] while the smaller value, obtained from the axion mass $m_{a,e}$, relates to the confining phase. The suppression of Yang–Mills scale is plausible because topological charges, which invoke the axial anomaly, are less resolved in the confining as compared to the deconfining phase. The gravitational Bohr radius associated with a typical $e$-lump mass of $M_e \sim 6.3 \times 10^{10} \, M_\odot$ turns out to be $r_{B,e} \sim 0.26$ kpc.

Having fixed the scales of SU(2)$_{\mathrm{CMB}}$, SU(2)$_e$ and linked their lumps to dark energy and the dark-matter halos of low-surface-brightness galaxies, respectively, we associate the lumps of SU(2)- and SU(2)$_\emptyset$ with $\Omega_{\mathrm{pdm},0}$ of the dark-sector cosmological model in Equation (4). Within a galaxy, each individual $\mu$- and $\tau$-lump provides a mass fraction of $(m_e/m_\mu)^2 \sim 2.3 \times 10^{-5}$ and $(m_e/m_\tau)^2 \sim 8.3 \times 10^{-8}$, respectively, of the mass $M_e$ of an $e$-lump, see Equation (11).

This paper is organised as follows. In Section 2 we discuss features of lumps in terms of a universal ratio between reduced Compton wavelength and gravitational Bohr radius. As a result, a typical lump mass can be expressed solely in terms of Yang–Mills scale and Planck mass. The rotation curves of galaxies with low surface brightness (SPARC library) are analysed in Section 3 using two models with spherically symmetric mass densities: the Soliton–Navarro–Frenk–White (SNFW) and the Burkert model. Assuming that only one Planck-scale axion species dominates the dark halo of a low-surface-brightness galaxy in terms of an isolated, unmerged $e$-lump, we extract the typical axion mass $m_{a,e}$ in Section 3.2. In Section 3.3 we demonstrate the consistency of axion-mass extraction between the two models: The gravitational Bohr radius, determined in SNFW, together with the lump mass, obtained from the Burkert-model-fit, predicts an axion mass which is compatible with the axion mass extracted from the soliton-core density of the SNFW model. The typical value of the axion mass suggests an association with SU(2) Yang–Mills dynamics responsible for the emergence of the first lepton family. In Section 4, this information is used to discuss the cosmological origin and role of lumps played in the dark Universe in association with the two other lepton families and the SU(2) gauge theory propounded to describe the CMB [16,56]. As a result, on subgalactic scales the $\mu$-lumps could explain the presence of massive compact objects in galactic centers such as Sagittarius A$^*$ in the Milky Way [60,61] while $\tau$-lumps may relate to globular clusters [62]. On super-galactic scales and for $z < z_{p,e}$, however, lumps from all axion species act like CDM. On the other hand, the CMB-lump's extent always exceeds the Hubble radius by many orders of magnitude and therefore should associate with dark energy. Finally, in Section 5 we discuss in more detail how certain dark structures of the Milky Way may have originated in terms of $\mu$- and $\tau$-lumps. We also provide a summary and an outlook on future work. We work in natural units $\hbar = c = k_B = 1$.

## 2. Gravitational Bohr Radius and Reduced Compton Wave Length of a Planck–Scale Axion

We start by conveying some features of basic axion lumps, cosmologically originated by depercolation transitions, that we wish to study. Let

$$\lambda_{C,i} \equiv \frac{1}{m_{a,i}}$$ (6)

denote the reduced Compton wavelength and

$$d_{a,i} \equiv \left(\frac{m_{a,i}}{\bar{\rho}_i}\right)^{1/3}$$ (7)

the mean distance between axion particles within the spherically symmetric core of the lump of mean dark-matter mass density $\bar{\rho}_i$. One has

$$\bar{\rho}_i \sim \frac{M_{200,i}}{\frac{4\pi}{3}r_{B,i}^3}.$$ (8)

The energy densities $\rho_i$ of each of the three dark-energy like homogeneous condensates of axionic particles prior to lump depercolation are assumed to arise due to Planckian physics [30]. Therefore, each $\bar{\rho}_i$ may only depend on $M_P$ and $m_{a,i}$ ($i = e, \mu, \tau$). Finite-extent, isolated, unmerged lumps self-consistently are characterised by a fixed ratio between the reduced Compton wavelength $\lambda_{C,i}$—the correlation length in the condensate of free axion particles at zero temperature—and the Bohr radius $r_{B,i}$.

Let us explain this. Causal lump segregation due to cosmological expansion (depercolation), which sets in when the Hubble radius $r_H$ becomes sufficiently larger than $r_B$, is adiabatically slow and generates a sharply peaked distribution of lump masses (and Bohr radii) in producing typically sized condensate cores. These cores are surrounded by halos of axion particles that represent regions of the dissolved condensate and nonthermally are released by the mutual pull of cores during depercolation. In principle, we can state that for an isolated, unmerged lump

$$\frac{r_{B,i}}{\lambda_C} = \kappa(\delta_i),$$ (9)

where $\kappa$ is a smooth dimensionless function of its dimensionless argument $\delta_i \equiv m_{a,i}/M_P$ with the property that $\lim_{\delta_i \to 0} \kappa_i(\delta_i) < \infty$. This is because the typical mass $M_i \sim M_{200,i}$ of an isolated, unmerged lump, which enters $r_{B,i}$ via Equation (5), is, due to adiabatically slow depercolation, by itself only a function of the two mass scales $m_{a,i}$ and $M_P$ mediating the interplay between quantum and gravitational correlations that give rise to the formation of the lump. Since $\delta_i$ is much smaller than unity, we can treat the right-hand side of Equation (9) as a *universal* constant. In practice, we will in Section 3 derive the values of $r_{B,e}$ and $m_{a,e}$ by matching dark-matter halos of low surface-brightness galaxies with well motivated models of a lump's mass density. As a result, we state a value of $\kappa \sim 314$ in Equation (25) of Section 4.

Equation (9) together with Equations (1), (5) and (6) imply for the mass $M_i$ of the isolated, unmerged lump

$$M_i = \frac{1}{\kappa}\frac{M_P^3}{\Lambda_i^2}.$$ (10)

Equation (10) is important because it predicts that the ratios of lump masses solely are determined by the squares of the ratios of the respective Yang–Mills scales or, what is the same [55], by the ratios of charged lepton masses $m_e$, $m_\mu$, and $m_\tau$. One has

$$\frac{M_\tau}{M_\mu} = \left(\frac{m_\tau}{m_\mu}\right)^2 \sim 283, \quad \frac{M_\mu}{M_e} = \left(\frac{m_e}{m_\mu}\right)^2 \sim 2.3 \times 10^{-5},$$

$$\frac{M_\tau}{M_e} = \left(\frac{m_e}{m_\tau}\right)^2 \sim 8.3 \times 10^{-8}. \tag{11}$$

Moreover, Equations (1), (6)–(8) and (10) fix the ratio $\xi_i \equiv \frac{d_{a,i}}{\lambda_{C,i}}$ as

$$\xi_i = \left(\frac{4\pi}{3}\right)^{1/3}\left(\kappa\frac{\Lambda_i}{M_P}\right)^{4/3}. \tag{12}$$

Since $\Lambda_i \ll M_P$ we have $\xi_i \ll 1$, and therefore a large number of axion particles are covered by one reduced Compton wave length. This assures that the assumption of a condensate core is selfconsistent. A thermodynamical argument for the necessity of axion condensates throughout the Universe's expansion history is given in Section 4. In [36], the non-local and non-linear (integro-differential) Schrödinger equation, obtained from a linear Schrödinger equation and a Poisson equation for the gravitational potential, see, e.g., [63], governing the lump, was analysed. An excitation of such a lump in terms of its wavefunction $\psi_i$ containing radial zeros was envisaged in [36,49]. Here instead, we assume the isolated, unmerged lump to be in its ground state, parameterised by a phenomenological mass density $\rho_i(r) \propto |\psi_i|^2(r) > 0$ which represents the lump well [47].

Finally, Equation (5) together with Equations (1) and (10) yield for the gravitational Bohr radius

$$r_{B,i} = \kappa\frac{M_P}{\Lambda_i^2}. \tag{13}$$

## 3. Analysis of Rotation Curves

In this section, we extract the axion mass $m_{a,e}$ from observed RCs of low-surface-brightness galaxies which fix the lump mass $M_e$ and a characterising length scale—the gravitational Bohr radius $r_{B,e}$. This, in turn, determines the (primary, see Section 4) Yang–Mills scale $\Lambda_e$ associated with the lump. We analyse RCs from the SPARC library [64].

### 3.1. Fuzzy Dark Matter: Soliton–Navarro–Frenk–White vs. Burkert Model

To investigate, for a given galaxy and RC, the underlying spherically symmetric mass density $\rho(r)$ it is useful to introduce the orbit-enclosed mass

$$M(r) = 4\pi \int_0^r dr'\, r'^2\rho(r'). \tag{14}$$

Assuming virialisation, spherical symmetry, and Newtonian gravity the orbital velocity $V(r)$ of a test mass (a star) is given as

$$V(r) = \sqrt{\frac{GM(r)}{r}}, \tag{15}$$

where $M(r)$ is defined in Equation (14), and $G \equiv M_P^{-2}$ denotes Newton's constant. The lump mass $M$ is defined to be $M_{200} \equiv M(r_{200})$ where $r_{200}$ is given by Equation (3).

For an extraction of $m_{a,e}$ and therefore the associated Yang–Mills scale governing the mass of a lump according to Equation (10), we use the Soliton–Navarro—Frenk–White

(SNFW) and the Burkert model. The mass-density profile of the NFW-part of the SNFW-model is given as [33]

$$\rho_{\text{NFW}}(r) = \frac{\rho_s^{\text{NFW}}}{\frac{r}{r_s}\left(1 + \frac{r}{r_s}\right)^2}, \tag{16}$$

where $\rho_s^{\text{NFW}}$ associates with the central mass density, and $r_s$ is a scale radius which represents the onset of the asymptotic cubic decay in distance $r$ to the galactic center. Note that profile $\rho_{\text{NFW}}$ exhibits an infinite cusp as $r \to 0$ and that the orbit-enclosed mass $M(r)$ diverges logarithmically with the cutoff radius $r$ for the integral in Equation (14). In order to avoid the cuspy behavior for $r \to 0$, an axionic Bose–Einstein condensate (soliton density profile) is assumed to describe the soliton region $r \leq r_\epsilon$. From the ground-state solution of the Schrödinger–Poisson system for a single axion species one obtains a good analytic description of the soliton density profile as [49]

$$\rho_{\text{sol}}(r) = \frac{\rho_c}{(1 + 0.091(r/r_c)^2)^8}, \tag{17}$$

where $\rho_c$ is the core density [47]. On the whole, the fuzzy dark matter profile can than be approximated as

$$\rho_{\text{FDM}}(r) = \Theta(r_\epsilon - r)\rho_{\text{sol}} + \Theta(r - r_\epsilon)\rho_{\text{NFW}}. \tag{18}$$

For the Burkert model one assumes a mass-density profile of the form [58,59]

$$\rho_{\text{Bu}}(r) = \frac{\rho_0\, r_0^3}{(r + r_0)(r^2 + r_0^2)} \tag{19}$$

where $\rho_0$ refers to the central mass density and $r_0$ is a scale radius.

*3.2. Analysis of RCs in the SNFW Model*

Using Equations (14), (15) and (18), we obtain the orbital velocity $V_{\text{SNFW}}$ of the SNFW model [65] [Equation (17)] which is fitted to observed RCs. This determines the parameters $r_\epsilon$, $r_s$, and $\rho_c$. The density $\rho_s$ relates to these fit parameters by demanding continuity of the SNFW mass density at $r_\epsilon$ [49]. As a result, one has

$$\rho_s(\rho_c, r_c, r_\epsilon, r_s) = \rho_c\frac{(r_\epsilon/r_s)(1 + r_\epsilon/r_s)^2}{(1 + 0.091(r_\epsilon/r_c)^2)^8}. \tag{20}$$

Examples of good fits with $\chi^2/\text{d.o.f.} < 1$ are shown in Figure 1, see Tables 1 and 2 for the corresponding fit parameters. The derived quantity $m_{a,\epsilon}$ is extracted from the following equation [47]

$$\rho_c \equiv 1.9 \times 10^9 (m_{a,\epsilon}/10^{-23}\text{eV})^{-2}(r_c/\text{kpc})^{-4}M_\odot\text{kpc}^{-3} \tag{21}$$

The other derived quantities $r_{200}$ and $M_{200}$ are obtained by employing Equations (3) and (14) with $M(r = r_{200}) \equiv M_{200}$, respectively. In Figure 2, a frequency distribution of $m_{a,\epsilon}$ is shown, based on a sample of 17 best fitting galaxies, see Figure 1 for the fits to the RCs. The maximum of the smooth-kernel-distribution (solid line) is at

$$m_{a,\epsilon} = (0.72 \pm 0.5) \times 10^{-23}\,\text{eV}. \tag{22}$$

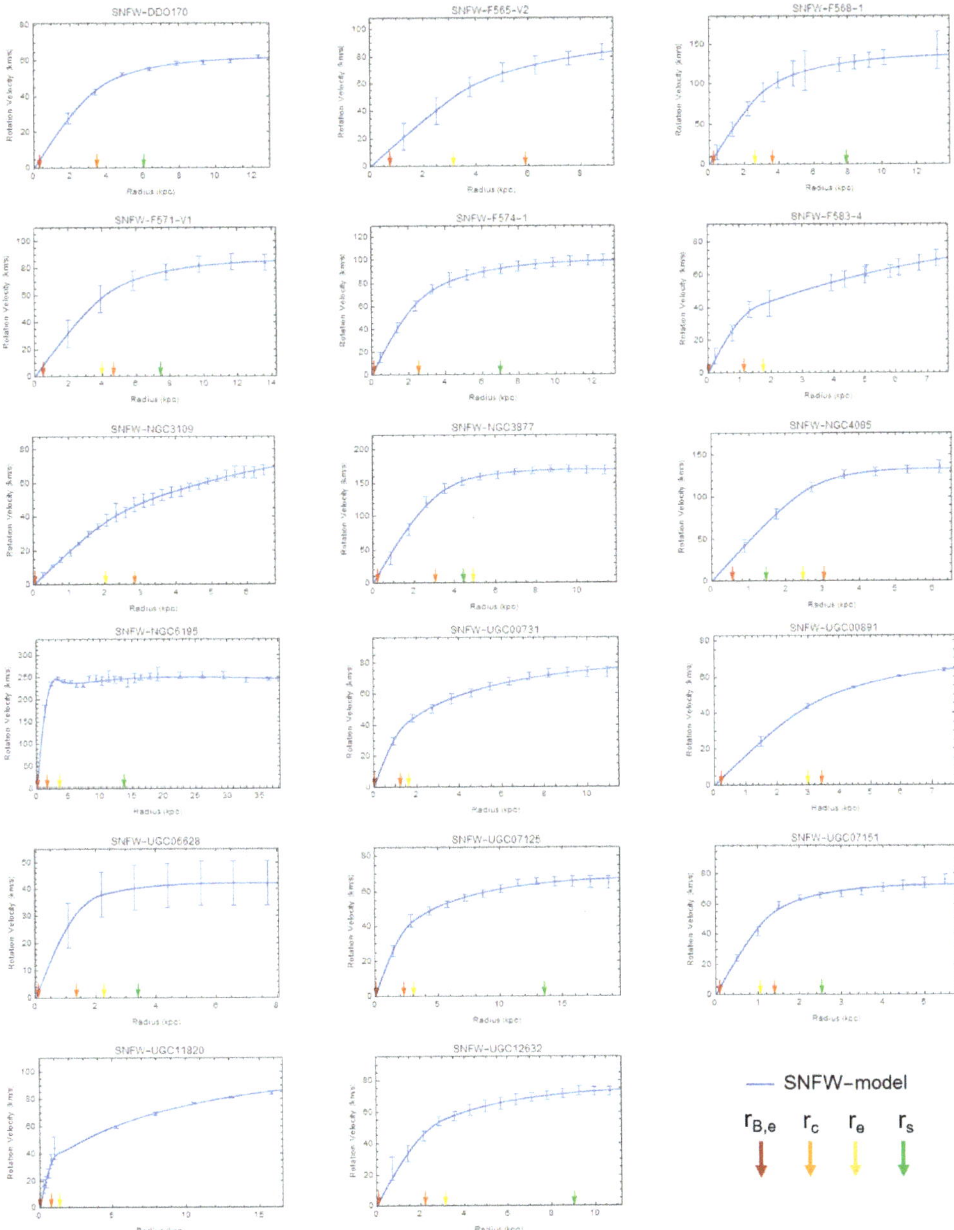

**Figure 1.** Best fits of SNFW to RCs of 17 SPARC galaxies. The arrows indicate the Bohr radius of the $e$-lump, $r_{B,e}$ (red), the core radius of the soliton $r_c$ (orange), the transition radius from the soliton model to the NFW model $r_e$ (yellow), and the scale radius of the NFW model $r_s$ (green).

**Table 1.** Fits of RCs to SNFW model: Galaxy name, Hubble Type, $\chi^2$/d.o.f., luminosity, axion mass $m_a$, $r_{200}$, virial mass $M_{200}$, central density $\rho_c$, scale radius $r_s$, transition radius $r_e$, core radius $r_c$, and $r_e/r_c$. The fitting constraints are heuristic and motivated by the results of [49], $r_{200} < 200$ kpc, $r_e, r_c < 6$ kpc, and $r_e/r_c > 0.1$.

| Galaxy | Hub. Type | Lum. $[L_\odot/\text{pc}^2]$ | $\chi^2$ /d.o.f. | $m_{a,e}$ $[\text{eV} \times 10^{-23}]$ | $r_{200}$ [kpc] | $M_{200}$ $[M_\odot \times 10^{10}]$ |
|---|---|---|---|---|---|---|
| DDO170 | Im | 73.93 | 0.73 | $1.00 \pm 0.82$ | $36.90 \pm 8.16$ | $2.81 \pm 1.95$ |
| F565-V2 | Im | 40.26 | 0.02 | $0.31 \pm 0.47$ | $60.71 \pm 11.52$ | $11.65 \pm 5.05$ |
| F568-1 | Sc | 57.13 | 0.02 | $0.41 \pm 0.24$ | $70.06 \pm 7.82$ | $22.59 \pm 6.67$ |
| F571-V1 | Sd | 64.39 | 0.03 | $0.50 \pm 0.30$ | $49.71 \pm 6.70$ | $7.12 \pm 2.83$ |
| F574-1 | Sd | 128.48 | 0.02 | $0.93 \pm 0.15$ | $53.96 \pm 3.09$ | $9.79 \pm 1.44$ |
| F583-4 | Sc | 83.34 | 0.13 | $3.87 \pm 1.37$ | $75.08 \pm 32.44$ | $17.88 \pm 13.83$ |
| NGC3109 | Sm | 140.87 | 0.18 | $1.14 \pm 0.39$ | $77.09 \pm 18.07$ | $19.85 \pm 5.45$ |
| NGC3877 | Sc | 3410.59 | 0.17 | $0.39 \pm 0.04$ | $69.37 \pm 12.49$ | $26.91 \pm 6.58$ |
| NGC4085 | Sc | 5021.46 | 0.07 | $0.41 \pm 0.18$ | $46.07 \pm 7.09$ | $9.26 \pm 2.78$ |
| NGC6195 | Sb | 174.11 | 0.47 | $0.48 \pm 0.03$ | $121.72 \pm 9.61$ | $122.40 \pm 24.56$ |
| UGC00731 | Im | 82.57 | 0.19 | $3.39 \pm 0.87$ | $53.27 \pm 8.51$ | $7.62 \pm 2.90$ |
| UGC00891 | Sm | 113.98 | 0.01 | $0.93 \pm 0.10$ | $44.92 \pm 1.42$ | $4.78 \pm 0.36$ |
| UGC06628 | Sm | 103.00 | 0.00 | $3.61 \pm 0.10$ | $23.35 \pm 0.87$ | $0.77 \pm 0.05$ |
| UGC07125 | Sm | 103.00 | 0.25 | $1.81 \pm 0.65$ | $47.55 \pm 10.38$ | $4.97 \pm 2.68$ |
| UGC07151 | Scd | 965.67 | 0.72 | $1.94 \pm 2.32$ | $32.32 \pm 5.65$ | $2.53 \pm 1.58$ |
| UGC11820 | Sm | 34.11 | 0.82 | $5.99 \pm 4.75$ | $75.83 \pm 49.15$ | $18.10 \pm 34.72$ |
| UGC12632 | Sm | 66.81 | 0.09 | $1.47 \pm 0.23$ | $46.91 \pm 6.10$ | $5.56 \pm 1.48$ |

**Table 2.** Fits of RCs to SNFW model: Galaxy name, central density $\rho_c$, soliton density $\rho_s$, scale radius $r_s$, transition radius $r_e$, core radius $r_c$, and $r_e/r_c$. The fitting constraints are heuristic and motivated by the results of [49], $r_{200} < 200$ kpc, $r_e, r_c < 6$ kpc, and $r_e/r_c > 0.1$.

| Galaxy | $\rho_c \times 10^7$ $[M_\odot/\text{kpc}^3]$ | $\rho_s \times 10^7$ $[M_\odot/\text{kpc}^3]$ | $r_s$ [kpc] | $r_e$ [kpc] | $r_c$ [kpc] | $r_e/r_c$ |
|---|---|---|---|---|---|---|
| DDO170 | $1.34 \pm 0.38$ | $0.95 \pm 0.78$ | $6.02 \pm 1.51$ | $3.48 \pm 2.99$ | $3.45 \pm 1.39$ | 1.01 |
| F565-V2 | $1.6 \pm 0.21$ | $0.51 \pm 0.33$ | $12.46 \pm 4.33$ | $3.14 \pm 0.8$ | $5.89 \pm 4.38$ | 0.53 |
| F568-1 | $6.61 \pm 0.56$ | $2.67 \pm 1.04$ | $7.87 \pm 1.36$ | $2.59 \pm 0.6$ | $3.62 \pm 1.06$ | 0.72 |
| F571-V1 | $1.61 \pm 0.24$ | $1.22 \pm 0.68$ | $7.42 \pm 1.71$ | $3.96 \pm 1.21$ | $4.65 \pm 1.4$ | 0.85 |
| F574-1 | $5.41 \pm 0.26$ | $1.81 \pm 0.32$ | $6.98 \pm 0.46$ | $2.52 \pm 1.04$ | $2.52 \pm 0.21$ | 1.00 |
| F583-4 | $8.05 \pm 1.34$ | $0.12 \pm 0.13$ | $27.1 \pm 19.49$ | $1.73 \pm 0.41$ | $1.12 \pm 0.19$ | 1.55 |
| NGC3109 | $2.28 \pm 0.09$ | $0.14 \pm 0.08$ | $25.88 \pm 11.92$ | $2.03 \pm 0.56$ | $2.83 \pm 0.48$ | 0.71 |
| NGC3877 | $15. \pm 0.91$ | $13.56 \pm 15.7$ | $4.4 \pm 2.43$ | $4.87 \pm 0.91$ | $3.03 \pm 0.15$ | 1.61 |
| NGC4085 | $13.49 \pm 1.04$ | $103.33 \pm 134.83$ | $1.46 \pm 0.82$ | $2.46 \pm 0.41$ | $3.02 \pm 0.65$ | 0.81 |
| NGC6195 | $128.1 \pm 7.47$ | $2.67 \pm 0.6$ | $13.68 \pm 0.96$ | $3.55 \pm 0.12$ | $1.59 \pm 0.05$ | 2.23 |
| UGC00731 | $7.41 \pm 0.87$ | $0.39 \pm 0.16$ | $12.17 \pm 1.85$ | $1.62 \pm 0.34$ | $1.22 \pm 0.15$ | 1.32 |
| UGC00891 | $1.63 \pm 0.05$ | $0.57 \pm 0.06$ | $8.85 \pm 0.39$ | $2.97 \pm 0.19$ | $3.42 \pm 0.19$ | 0.87 |
| UGC06628 | $4.34 \pm 0.08$ | $1.3 \pm 0.21$ | $3.4 \pm 0.28$ | $2.27 \pm 0.07$ | $1.35 \pm 0.02$ | 1.68 |
| UGC07125 | $2.2 \pm 0.39$ | $0.22 \pm 0.12$ | $13.55 \pm 2.2$ | $3.06 \pm 0.94$ | $2.27 \pm 0.39$ | 1.35 |
| UGC07151 | $13.48 \pm 2.86$ | $7.49 \pm 5.62$ | $2.52 \pm 0.66$ | $1.05 \pm 0.46$ | $1.39 \pm 0.83$ | 0.75 |
| UGC11820 | $15.04 \pm 4.43$ | $0.11 \pm 0.2$ | $28.64 \pm 6.01$ | $1.35 \pm 0.66$ | $0.77 \pm 0.3$ | 1.76 |
| UGC12632 | $3.77 \pm 0.3$ | $0.61 \pm 0.24$ | $9.01 \pm 1.82$ | $3.12 \pm 0.49$ | $2.2 \pm 0.16$ | 1.42 |

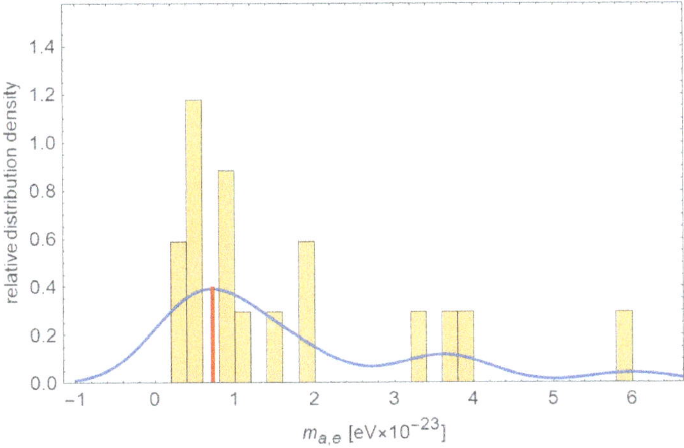

**Figure 2.** Frequency distribution of axion mass $m_{a,e}$ as extracted from the SNFW model for 17 best fitting galaxies. The maximum of the smooth-kernel-distribution (solid, blue line) is at $m_{a,e} = (0.72 \pm 0.5) \times 10^{-23}$ eV (red, vertical line).

In Figure 3 a frequency distribution of $M_{200}$ is shown for the 17 best-fitting galaxies. Figure 4 depicts the distribution of these galaxies in the $M_{200}$—surface-brightness plane. The maximum of the smooth-kernel-distribution is at $M_{200} = (6.3 \pm 3) \times 10^{10}\, M_\odot$. With Equation (5) this implies a mean Bohr radius of

$$
r_{B,e} = \frac{M_p^2}{(6.3 \pm 4) \times 10^{10}\, M_\odot \, ((0.72 \pm 0.5) \times 10^{-23}\, \text{eV})^2}
$$
$$
= (0.26 \pm 0.1)\, \text{kpc} \,. \tag{23}
$$

This value of $r_{B,e}$ is used in the Burkert-model analysis of Section 3.3 to extract the frequency distribution of $m_{a,e}$ via the frequency distribution of $M_{200}$.

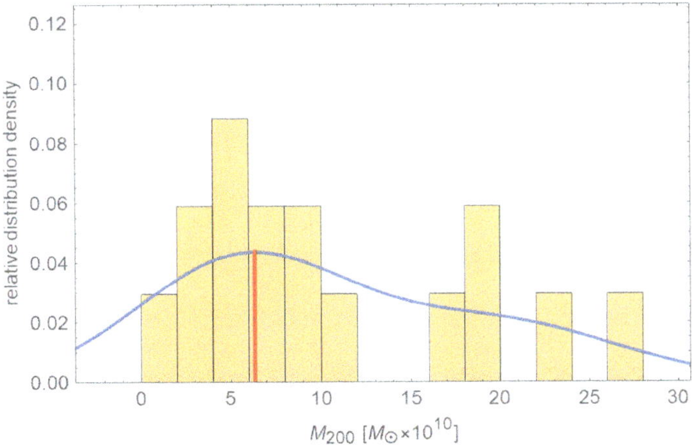

**Figure 3.** Frequency distribution of the virial mass $M_{200}$ in units of solar masses $M_\odot$ from the 17 best-fitting galaxies in the SNFW model. The maximum of the smooth-kernel-distribution (solid line) is at $M_{200} = (6.3 \pm 4) \times 10^{10}\, M_\odot$.

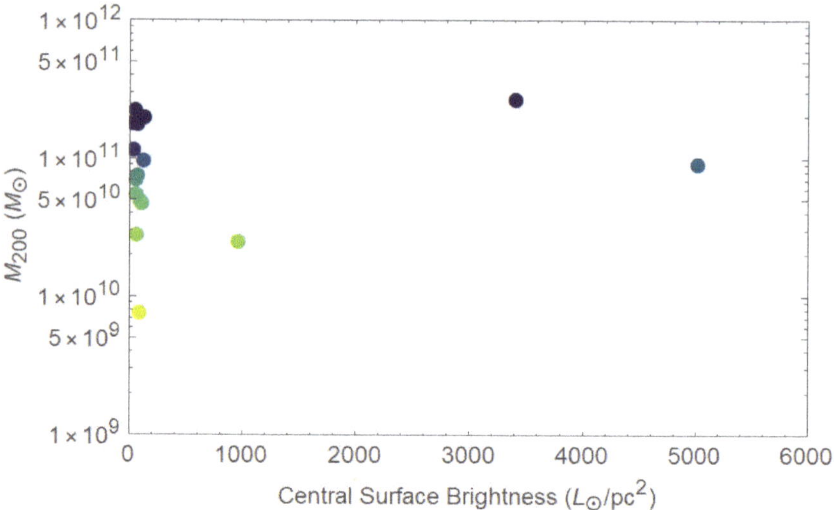

**Figure 4.** Extracted virial masses $M_{200}$ in units of solar masses $M_\odot$ from sample of the 17 best-fitting galaxies in the SNFW model.

### 3.3. Analysis of RCs in the Burkert Model

Figure 5 depict the fits of the Burkert model to the 17 RCs used in the SNFW fits. Tables 3 and 4 indicate that three out of these 17 RCs are fitted with a $\chi^2/\text{d.o.f.} > 1$. Therefore, we resort to a sample of 80 galaxies which fit with $\chi^2/\text{d.o.f.} < 1$.

Our strategy to demonstrate independence of the mean value of $m_{a,e}$ on the details of the two realistic models SNFW and Burkert is to also determine it from Equation (5). To do this, we use the value of the gravitational Bohr radius $r_{B,e}$ in Equation (23) and the values of $M_{200}$ extracted from RC fits within an ensemble of 80 SPARC galaxies to the Burkert model. The results are shown in Tables 3 and 4, Figures 6 and 7. This yields a frequency distribution of $m_{a,e}$ shown in Figure 8. Obviously, the maximum of the smooth-kernel distribution, $m_{a,e} = (0.65 \pm 0.4) \times 10^{-23}$ eV, is compatible with that in the SNFW model $m_{a,e} = (0.72 \pm 0.5) \times 10^{-23}$ eV. Notice how $M_{200}$ clusters around the value $M_{200} \sim 5 \times 10^{10} \, M_\odot$.

In our treatment of cosmological and astrophysical implications, we appeal to the mean value of $m_{a,e}$-extractions in the SNFW and the Burkert model as

$$m_{a,e} = 0.675 \times 10^{-23} \, \text{eV} \,. \tag{24}$$

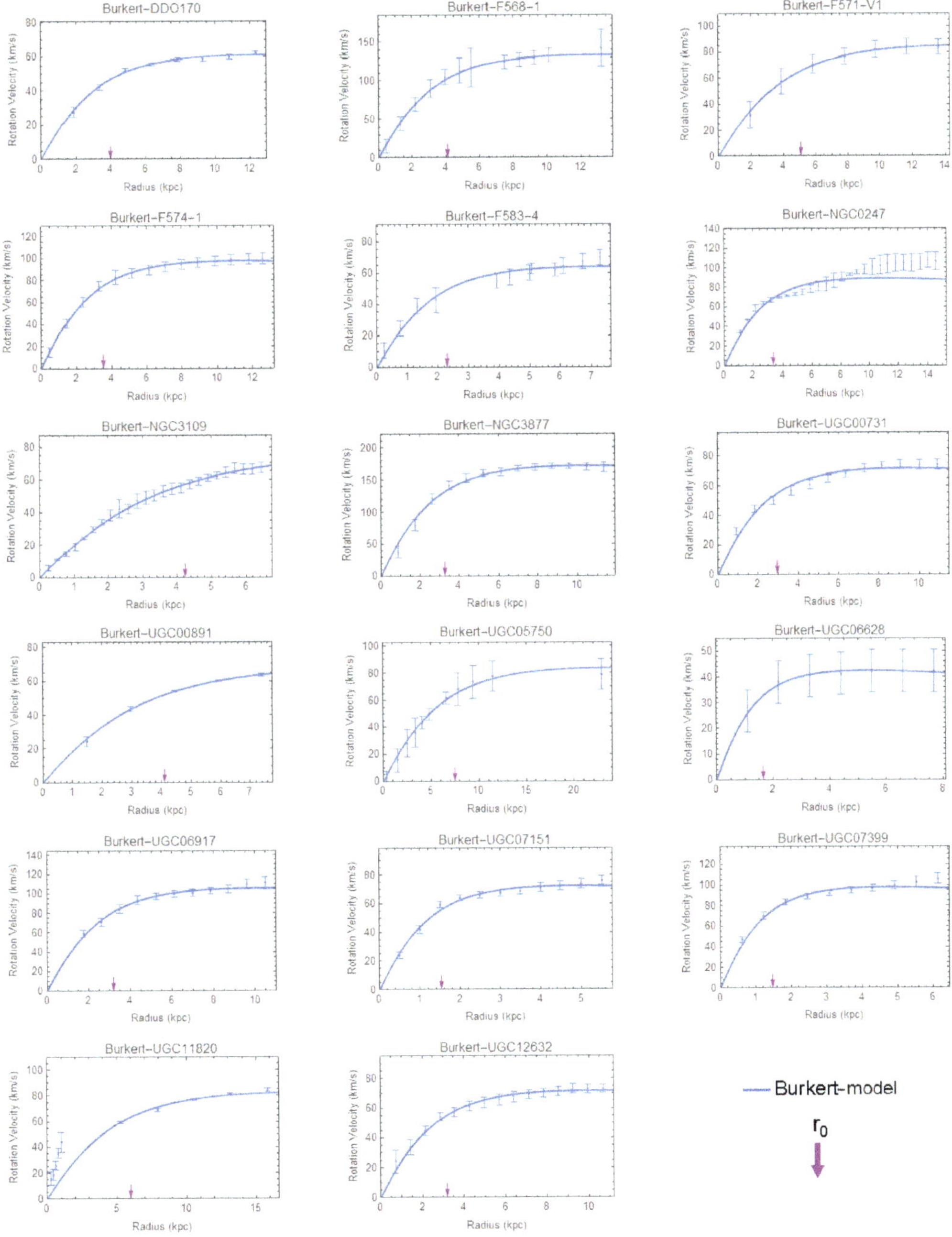

**Figure 5.** Burkert-model fits to the 17 best fitting SNFW-model galaxies. The purple arrow indicates the value of $r_0$.

**Table 3.** Burkert model: Galaxy name, Hubble Type, $\chi^2$/d.o.f., luminosity, axion mass $m_a$, $r_{200}$, virial mass $M_{200}$, core density $\rho_0$, and core radius $r_0$. In the red frame the 17 galaxies used for the SNFW fit are highlighted.

| Galaxy | Hub. Type | Lum. $[L_\odot/\mathrm{pc}^2]$ | $\chi^2$/d.o.f. | $m_{a,e}$ $[\mathrm{eV}\times10^{-23}]$ | $r_{200}$ [kpc] | $M_{200}$ $[M_\odot \times 10^{10}]$ | $\rho_0 \times 10^7$ $[M_\odot/\mathrm{kpc}^3]$ | $r_0$ [kpc] |
|---|---|---|---|---|---|---|---|---|
| DDO170 | Im | 73.93 | 0.74 | $1.18 \pm 0.24$ | $33.22 \pm 1.64$ | $2.34 \pm 0.28$ | $2.03 \pm 0.14$ | $3.98 \pm 0.16$ |
| F565-V2 | Im | 40.26 | 0.04 | $0.68 \pm 0.14$ | $47.44 \pm 3.11$ | $7.06 \pm 1.03$ | $2.39 \pm 0.16$ | $5.39 \pm 0.29$ |
| F568-1 | Sc | 57.13 | 0.06 | $0.47 \pm 0.09$ | $57.09 \pm 2.82$ | $15.19 \pm 1.47$ | $9.13 \pm 0.46$ | $4.10 \pm 0.13$ |
| F571-V1 | Sd | 64.39 | 0.03 | $0.74 \pm 0.15$ | $44.84 \pm 2.51$ | $6.03 \pm 0.69$ | $2.45 \pm 0.15$ | $5.06 \pm 0.20$ |
| F574-1 | Sd | 128.48 | 0.10 | $0.70 \pm 0.14$ | $43.92 \pm 1.74$ | $6.62 \pm 0.45$ | $6.69 \pm 0.26$ | $3.51 \pm 0.08$ |
| F583-4 | Sc | 83.34 | 0.51 | $1.33 \pm 0.34$ | $28.84 \pm 3.82$ | $1.85 \pm 0.61$ | $6.66 \pm 1.26$ | $2.30 \pm 0.25$ |
| NGC3109 | Sm | 140.87 | 0.17 | $0.91 \pm 0.18$ | $38.94 \pm 0.91$ | $3.93 \pm 0.23$ | $2.68 \pm 0.06$ | $4.23 \pm 0.09$ |
| NGC3877 | Sc | 3410.59 | 0.46 | $0.38 \pm 0.08$ | $63.43 \pm 3.48$ | $23.22 \pm 3.28$ | $23.82 \pm 1.88$ | $3.26 \pm 0.15$ |
| NGC4085 | Sc | 5021.46 | 0.63 | $0.50 \pm 0.12$ | $52.85 \pm 5.37$ | $13.34 \pm 3.53$ | $22.57 \pm 2.90$ | $2.77 \pm 0.25$ |
| NGC6195 | Sb | 174.11 | 50.97 | $0.21 \pm 0.06$ | $91.02 \pm 16.07$ | $74.49 \pm 32.29$ | $37.13 \pm 9.18$ | $4.07 \pm 0.56$ |
| UGC00731 | Im | 82.57 | 1.55 | $1.08 \pm 0.24$ | $33.71 \pm 3.07$ | $2.83 \pm 0.66$ | $5.15 \pm 0.68$ | $2.92 \pm 0.23$ |
| UGC00891 | Sm | 113.98 | 0.14 | $1.04 \pm 0.20$ | $35.94 \pm 0.96$ | $3.03 \pm 0.19$ | $2.34 \pm 0.07$ | $4.09 \pm 0.09$ |
| UGC06628 | Sm | 103.00 | 0.01 | $2.40 \pm 0.48$ | $19.65 \pm 0.92$ | $0.57 \pm 0.07$ | $5.82 \pm 0.41$ | $1.63 \pm 0.06$ |
| UGC07125 | Sm | 103.00 | 0.83 | $1.06 \pm 0.22$ | $36.11 \pm 2.61$ | $2.90 \pm 0.51$ | $1.63 \pm 0.17$ | $4.67 \pm 0.28$ |
| UGC07151 | Scd | 965.67 | 0.94 | $1.32 \pm 0.28$ | $27.68 \pm 1.93$ | $1.87 \pm 0.34$ | $19.42 \pm 1.91$ | $1.52 \pm 0.09$ |
| UGC11820 | Sm | 34.11 | 11.39 | $0.73 \pm 0.20$ | $46.42 \pm 7.09$ | $6.17 \pm 2.34$ | $1.63 \pm 0.34$ | $6.00 \pm 0.80$ |
| UGC12632 | Sm | 66.81 | 0.22 | $1.05 \pm 0.21$ | $34.69 \pm 1.37$ | $3.01 \pm 0.30$ | $4.40 \pm 0.25$ | $3.18 \pm 0.10$ |
| CamB | Im | 66.20 | 0.02 | $1.10 \pm 0.31$ | $36.66 \pm 6.71$ | $2.74 \pm 1.15$ | $0.91 \pm 0.04$ | $5.81 \pm 1.06$ |
| D512-2 | Im | 93.94 | 0.33 | $2.52 \pm 0.76$ | $19.25 \pm 3.55$ | $0.52 \pm 0.24$ | $4.68 \pm 0.99$ | $1.73 \pm 0.29$ |
| D564-8 | Im | 21.13 | 0.02 | $4.15 \pm 0.82$ | $14.37 \pm 0.50$ | $0.19 \pm 0.02$ | $2.10 \pm 0.08$ | $1.70 \pm 0.05$ |
| DDO064 | Im | 151.65 | 0.40 | $1.47 \pm 0.45$ | $27.02 \pm 4.99$ | $1.51 \pm 0.71$ | $6.87 \pm 1.10$ | $2.12 \pm 0.37$ |
| F563-1 | Sm | 41.77 | 0.54 | $0.59 \pm 0.13$ | $50.11 \pm 4.84$ | $9.55 \pm 2.26$ | $5.90 \pm 0.83$ | $4.16 \pm 0.32$ |
| F563-V2 | Im | 146.16 | 0.15 | $0.60 \pm 0.13$ | $47.48 \pm 3.69$ | $9.24 \pm 1.70$ | $14.65 \pm 1.42$ | $2.89 \pm 0.18$ |
| F567-2 | Sm | 46.65 | 0.25 | $1.64 \pm 0.56$ | $26.26 \pm 6.02$ | $1.23 \pm 0.69$ | $2.66 \pm 0.82$ | $2.88 \pm 0.56$ |
| F568-3 | Sd | 132.08 | 0.80 | $0.49 \pm 0.10$ | $58.93 \pm 5.29$ | $13.95 \pm 2.73$ | $2.76 \pm 0.23$ | $6.39 \pm 0.47$ |
| F568-V1 | Sd | 90.54 | 0.05 | $0.60 \pm 0.12$ | $47.54 \pm 2.15$ | $9.09 \pm 0.82$ | $12.07 \pm 0.57$ | $3.10 \pm 0.09$ |
| F579-V1 | Sc | 201.76 | 0.47 | $0.71 \pm 0.17$ | $41.3 \pm 4.54$ | $6.59 \pm 1.75$ | $25.8 \pm 4.07$ | $2.08 \pm 0.17$ |
| F583-1 | Sm | 60.93 | 0.11 | $0.73 \pm 0.14$ | $45.23 \pm 1.46$ | $6.23 \pm 0.43$ | $2.73 \pm 0.08$ | $4.90 \pm 0.13$ |
| KK98-251 | Im | 52.10 | 0.57 | $1.86 \pm 0.46$ | $24.45 \pm 3.06$ | $0.95 \pm 0.29$ | $2.25 \pm 0.22$ | $2.83 \pm 0.34$ |
| NGC0024 | Sc | 1182.58 | 0.61 | $0.81 \pm 0.16$ | $36.96 \pm 1.41$ | $5.03 \pm 0.51$ | $51.79 \pm 3.18$ | $1.46 \pm 0.05$ |
| NGC0055 | Sm | 391.59 | 0.32 | $0.70 \pm 0.14$ | $46.22 \pm 1.19$ | $6.62 \pm 0.42$ | $2.80 \pm 0.08$ | $4.94 \pm 0.12$ |
| NGC0100 | Scd | 1193.52 | 0.10 | $0.81 \pm 0.16$ | $40.65 \pm 1.10$ | $5.04 \pm 0.33$ | $5.73 \pm 0.18$ | $3.40 \pm 0.08$ |
| NGC0300 | Sd | 437.35 | 0.74 | $0.76 \pm 0.15$ | $41.89 \pm 1.63$ | $5.71 \pm 0.58$ | $7.45 \pm 0.43$ | $3.20 \pm 0.11$ |
| NGC2366 | Im | 113.98 | 0.69 | $1.61 \pm 0.32$ | $25.86 \pm 1.22$ | $1.27 \pm 0.15$ | $4.98 \pm 0.29$ | $2.27 \pm 0.10$ |
| NGC2915 | BCD | 313.93 | 0.30 | $1.04 \pm 0.22$ | $32.66 \pm 2.31$ | $3.03 \pm 0.58$ | $17.25 \pm 2.02$ | $1.87 \pm 0.11$ |
| NGC2976 | Sc | 1502.55 | 0.62 | $0.51 \pm 0.12$ | $52.34 \pm 5.22$ | $12.77 \pm 3.27$ | $20.73 \pm 1.02$ | $2.82 \pm 0.28$ |
| NGC3917 | Scd | 1226.96 | 0.87 | $0.43 \pm 0.09$ | $61.13 \pm 2.97$ | $18.06 \pm 2.19$ | $8.28 \pm 0.54$ | $4.51 \pm 0.18$ |
| NGC3949 | Sbc | 185.71 | 0.85 | $0.45 \pm 0.11$ | $54.76 \pm 6.26$ | $16.38 \pm 4.96$ | $50.95 \pm 7.60$ | $2.18 \pm 0.22$ |
| NGC3953 | Sbc | 1999.08 | 0.20 | $0.27 \pm 0.05$ | $77.09 \pm 3.05$ | $44.03 \pm 4.44$ | $37.19 \pm 2.18$ | $3.41 \pm 0.11$ |
| NGC3972 | Sbc | 1587.93 | 0.51 | $0.5 \pm 0.10$ | $53.96 \pm 3.61$ | $13.22 \pm 2.25$ | $13.02 \pm 1.14$ | $3.41 \pm 0.20$ |
| NGC3992 | Sbc | 3257.09 | 0.95 | $0.19 \pm 0.04$ | $99.24 \pm 8.40$ | $88.76 \pm 19.86$ | $23.2 \pm 3.34$ | $5.15 \pm 0.34$ |
| NGC4068 | Im | 261.11 | 0.07 | $1.03 \pm 0.28$ | $35.59 \pm 5.60$ | $3.09 \pm 1.19$ | $3.22 \pm 0.24$ | $3.63 \pm 0.56$ |
| NGC4088 | Sbc | 3988.69 | 0.65 | $0.36 \pm 0.08$ | $64.61 \pm 4.96$ | $24.74 \pm 5.03$ | $25.43 \pm 3.19$ | $3.25 \pm 0.20$ |
| NGC4157 | Sb | 23813.87 | 0.85 | $0.33 \pm 0.07$ | $68.59 \pm 5.23$ | $30.54 \pm 6.20$ | $32.73 \pm 4.20$ | $3.17 \pm 0.19$ |
| NGC4217 | Sb | 4373.51 | 0.55 | $0.36 \pm 0.07$ | $64.62 \pm 2.90$ | $25.96 \pm 3.00$ | $37.45 \pm 2.57$ | $2.85 \pm 0.10$ |
| NGC4389 | Sbc | 322.72 | 0.15 | $0.41 \pm 0.10$ | $63.93 \pm 7.79$ | $19.69 \pm 5.94$ | $5.91 \pm 0.55$ | $5.29 \pm 0.61$ |
| NGC4559 | Scd | 1602.62 | 0.76 | $0.54 \pm 0.11$ | $52.00 \pm 2.59$ | $11.35 \pm 1.47$ | $9.78 \pm 0.78$ | $3.62 \pm 0.15$ |
| UGC00634 | Sm | 126.13 | 0.87 | $0.51 \pm 0.11$ | $57.24 \pm 4.75$ | $12.59 \pm 2.54$ | $2.70 \pm 0.31$ | $6.22 \pm 0.43$ |
| UGC01230 | Sm | 69.32 | 0.19 | $0.56 \pm 0.12$ | $51.30 \pm 3.78$ | $10.35 \pm 1.81$ | $6.27 \pm 0.67$ | $4.18 \pm 0.23$ |

**Table 4.** Burkert model with $\chi^2/\text{d.o.f.} < 1$: Galaxy name, Hubble Type, $\chi^2/\text{d.o.f.}$, luminosity, axion mass $m_a$, $r_{200}$, virial mass $M_{200}$, core density $\rho_0$, and core radius $r_0$ (ordered as in the SPARC library).

| Galaxy | Hub. Type | Lum. $[L_\odot/\text{pc}^2]$ | $\chi^2$ /d.o.f. | $m_{a,e}$ $[\text{eV} \times 10^{-23}]$ | $r_{200}$ [kpc] | $M_{200}$ $[M_\odot \times 10^{10}]$ | $\rho_0 \times 10^7$ $[M_\odot/\text{kpc}^3]$ | $r_0$ [kpc] |
|---|---|---|---|---|---|---|---|---|
| UGC01281 | Sdm | 135.78 | 0.17 | $1.16 \pm 0.24$ | $32.78 \pm 1.89$ | $2.45 \pm 0.35$ | $3.56 \pm 0.2$ | $3.23 \pm 0.17$ |
| UGC02023 | Im | 121.57 | 0.03 | $0.36 \pm 0.16$ | $73.23 \pm 24.56$ | $24.82 \pm 19.87$ | $1.92 \pm 0.21$ | $8.95 \pm 2.97$ |
| UGC04305 | Im | 88.07 | 0.76 | $3.42 \pm 0.9$ | $15.39 \pm 2.06$ | $0.28 \pm 0.1$ | $7.2 \pm 1.55$ | $1.19 \pm 0.14$ |
| UGC04325 | Sm | 213.22 | 0.26 | $1. \pm 0.2$ | $32.71 \pm 1.31$ | $3.29 \pm 0.34$ | $31.75 \pm 1.83$ | $1.53 \pm 0.05$ |
| UGC04483 | Im | 82.57 | 0.26 | $6.58 \pm 1.48$ | $9.51 \pm 0.84$ | $0.08 \pm 0.02$ | $19.32 \pm 2.23$ | $0.52 \pm 0.04$ |
| UGC04499 | Sdm | 127.30 | 0.09 | $1.05 \pm 0.21$ | $34. \pm 1.01$ | $2.96 \pm 0.21$ | $5.97 \pm 0.24$ | $2.8 \pm 0.07$ |
| UGC05005 | Im | 65.59 | 0.06 | $0.52 \pm 0.1$ | $59.27 \pm 2.86$ | $12.09 \pm 1.21$ | $1.1 \pm 0.06$ | $8.85 \pm 0.32$ |
| UGC05414 | Im | 127.30 | 0.11 | $1.25 \pm 0.25$ | $30.31 \pm 1.42$ | $2.09 \pm 0.24$ | $5.9 \pm 0.3$ | $2.51 \pm 0.11$ |
| UGC05750 | Sdm | 124.98 | 0.08 | $0.68 \pm 0.14$ | $49.77 \pm 3.23$ | $7.13 \pm 0.97$ | $1.07 \pm 0.06$ | $7.5 \pm 0.4$ |
| UGC05829 | Im | 63.22 | 0.33 | $0.93 \pm 0.23$ | $38.9 \pm 4.8$ | $3.81 \pm 1.16$ | $2.21 \pm 0.28$ | $4.52 \pm 0.52$ |
| UGC05918 | Im | 24.94 | 0.06 | $2.35 \pm 0.48$ | $19.9 \pm 1.15$ | $0.59 \pm 0.09$ | $6.01 \pm 0.47$ | $1.64 \pm 0.08$ |
| UGC05999 | Im | 51.62 | 0.35 | $0.53 \pm 0.12$ | $57.19 \pm 5.87$ | $11.71 \pm 2.81$ | $1.72 \pm 0.22$ | $7.29 \pm 0.63$ |
| UGC06399 | Sm | 311.05 | 0.03 | $0.8 \pm 0.16$ | $40.82 \pm 1.2$ | $5.16 \pm 0.36$ | $6.2 \pm 0.22$ | $3.32 \pm 0.08$ |
| UGC06446 | Sd | 86.46 | 0.78 | $1.05 \pm 0.23$ | $32.95 \pm 2.75$ | $3.01 \pm 0.66$ | $13.01 \pm 1.64$ | $2.08 \pm 0.15$ |
| UGC06667 | Scd | 614.94 | 0.11 | $0.81 \pm 0.16$ | $40.55 \pm 1.3$ | $5.03 \pm 0.39$ | $5.97 \pm 0.23$ | $3.34 \pm 0.09$ |
| UGC06917 | Sm | 261.11 | 0.35 | $0.66 \pm 0.13$ | $45.28 \pm 2.24$ | $7.48 \pm 0.92$ | $9.48 \pm 0.64$ | $3.19 \pm 0.13$ |
| UGC06923 | Im | 347.40 | 0.4 | $0.97 \pm 0.26$ | $34.69 \pm 4.92$ | $3.46 \pm 1.28$ | $11.64 \pm 2.17$ | $2.28 \pm 0.29$ |
| UGC06930 | Sd | 189.16 | 0.21 | $0.62 \pm 0.13$ | $47.7 \pm 3.06$ | $8.45 \pm 1.38$ | $7.41 \pm 0.72$ | $3.65 \pm 0.19$ |
| UGC06983 | Scd | 121.57 | 0.68 | $0.66 \pm 0.14$ | $45.05 \pm 3.3$ | $7.54 \pm 1.42$ | $11.27 \pm 1.24$ | $2.99 \pm 0.18$ |
| UGC07089 | Sdm | 520.99 | 0.22 | $0.86 \pm 0.18$ | $40.29 \pm 2.82$ | $4.41 \pm 0.75$ | $2.82 \pm 0.23$ | $4.31 \pm 0.27$ |
| UGC07261 | Sdm | 566.02 | 0.33 | $1.28 \pm 0.31$ | $28.32 \pm 3.03$ | $2. \pm 0.57$ | $18.97 \pm 3.11$ | $1.57 \pm 0.14$ |
| UGC07323 | Sdm | 283.68 | 0.8 | $0.71 \pm 0.17$ | $44.67 \pm 5.34$ | $6.55 \pm 1.98$ | $5.06 \pm 0.62$ | $3.9 \pm 0.43$ |
| UGC07524 | Sm | 106.86 | 0.57 | $0.86 \pm 0.17$ | $39.92 \pm 1.58$ | $4.45 \pm 0.44$ | $3.66 \pm 0.19$ | $3.89 \pm 0.13$ |
| UGC07559 | Im | 55.06 | 0.03 | $2.93 \pm 0.6$ | $17.5 \pm 0.93$ | $0.38 \pm 0.05$ | $4.15 \pm 0.23$ | $1.63 \pm 0.08$ |
| UGC07577 | Im | 54.55 | 0.04 | $3.14 \pm 1.07$ | $18.08 \pm 4.38$ | $0.33 \pm 0.19$ | $1.02 \pm 0.12$ | $2.76 \pm 0.66$ |
| UGC07603 | Sd | 520.99 | 0.22 | $1.66 \pm 0.33$ | $23.7 \pm 1.$ | $1.19 \pm 0.13$ | $21.7 \pm 1.3$ | $1.26 \pm 0.05$ |
| UGC07608 | Im | 46.65 | 0.07 | $1. \pm 0.22$ | $35.39 \pm 2.78$ | $3.27 \pm 0.65$ | $5.23 \pm 0.44$ | $3.05 \pm 0.22$ |
| UGC07866 | Im | 97.46 | 0.1 | $3.78 \pm 0.92$ | $14.34 \pm 1.63$ | $0.23 \pm 0.07$ | $7.72 \pm 1.11$ | $1.08 \pm 0.11$ |
| UGC08490 | Sm | 576.54 | 0.95 | $1.24 \pm 0.25$ | $28.16 \pm 1.51$ | $2.14 \pm 0.31$ | $38.07 \pm 3.44$ | $1.23 \pm 0.05$ |
| UGC08837 | Im | 77.42 | 0.29 | $0.54 \pm 0.18$ | $57.39 \pm 13.51$ | $11.12 \pm 6.14$ | $1.26 \pm 0.11$ | $8.13 \pm 1.89$ |
| UGC09037 | Scd | 841.07 | 0.61 | $0.34 \pm 0.07$ | $72.32 \pm 3.54$ | $28.35 \pm 2.75$ | $5.11 \pm 0.28$ | $6.33 \pm 0.21$ |
| UGC09992 | Im | 73.25 | 0.24 | $3.68 \pm 1.34$ | $14.37 \pm 3.31$ | $0.24 \pm 0.15$ | $10.76 \pm 3.86$ | $0.97 \pm 0.19$ |
| UGC11557 | Sdm | 337.93 | 0.28 | $0.73 \pm 0.17$ | $44.75 \pm 4.74$ | $6.16 \pm 1.6$ | $3.16 \pm 0.4$ | $4.6 \pm 0.43$ |
| UGC12506 | Scd | 5608.28 | 0.69 | $0.2 \pm 0.04$ | $96.54 \pm 6.95$ | $80.52 \pm 12.65$ | $17.8 \pm 1.77$ | $5.53 \pm 0.26$ |
| UGCA281 | BCD | 12.05 | 0.34 | $4.71 \pm 1.08$ | $11.71 \pm 1.1$ | $0.15 \pm 0.04$ | $27.12 \pm 2.79$ | $0.58 \pm 0.05$ |

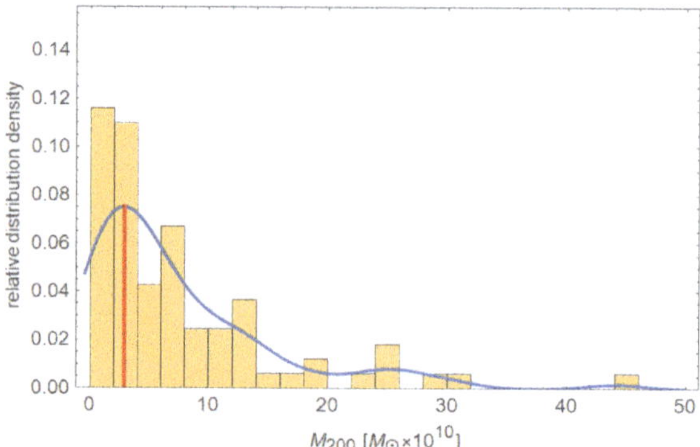

**Figure 6.** Frequency distribution of the virial mass $M_{200}$ in units of solar masses $M_\odot$ from the 80 best-fitting galaxies in the Burkert model. The maximum of the smooth-kernel-distribution (solid line) is at $M_{200} = (2.9 \pm 4) \times 10^{10}\, M_\odot$.

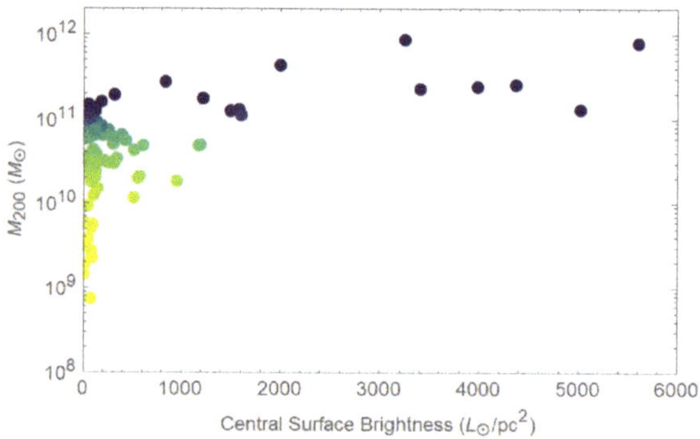

**Figure 7.** Extracted virial masses $M_{200}$ in units of solar masses from Burkert-model fits to 80 RCs with $\chi^2/\text{d.o.f.} < 1$ vs. the respective galaxy's central surface brightness in units of $L_\odot/\text{pc}^2$.

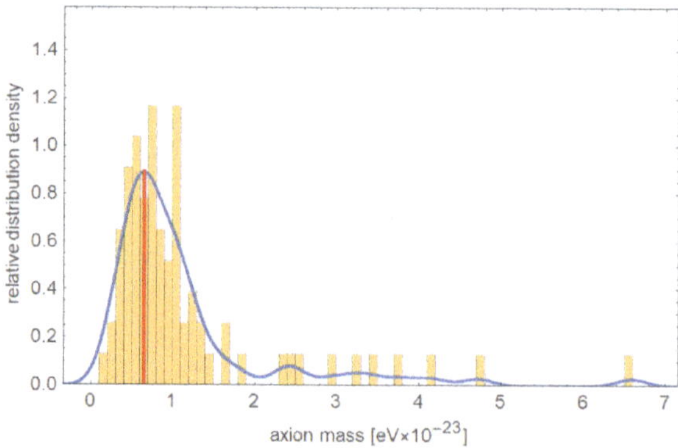

**Figure 8.** Frequency distribution of 80 axion masses $m_{a,e}$, extracted from the Burkert-model fits of $M_{200}$ to the RCs of galaxies with a $\chi^2/\text{d.o.f.} < 1$. The maximum of the smooth-kernel distribution (solid, blue line) is $m_{a,e} = (0.65 \pm 0.4) \times 10^{-23}\,\text{eV}$ (red, vertical line).

## 4. Galactic Central Regions and the Dark Sector of the Universe

Interpreting the dark-matter structure of a typical low-surface-brightness galaxy as an *e*-lump, we have $r_{B,e} = 0.26\,\mathrm{kpc}$ from the SNFW model, see Section 3.2. Therefore, the value of $\kappa$ in Equation (9) is

$$\kappa = 314 \,. \tag{25}$$

With $m_{a,e} = 0.675 \times 10^{-23}\,\mathrm{eV}$ Equation (1) yields

$$\Lambda_e = 287\,\mathrm{eV} \,. \tag{26}$$

This is by only a factor 15 smaller than the scale $\Lambda_e = m_e/118.6$ ($m_e = 511\,\mathrm{keV}$ the mass of the electron) of an SU(2) Yang–Mills theory proposed in [55] to originate the electron's mass in terms of a fuzzy ball of deconfining phase. There the deconfining region is immersed into the confining phase and formed by the selfintersection of a center-vortex loop. Considering an undistorted Yang–Mills theory for simplicity[1], the factor of 15 could be explained by a stronger screening of topological charge density—the origin of the axial anomaly—in the confining ground state, composed of round, pointlike center-vortex loops, versus the deconfining thermal ground state, made of densely packed, spatially extended (anti)caloron centers subject to overlapping peripheries [26]. The factor of 15 so far is a purely phenomenological result (it could be expected to be O(100) or higher) which is plausible qualitatively because of the reduced topological charge density in the confining phase where overlapping magnetic monopoles and antimonopoles, aligned within hardly resolved center vortices, are the topological charge carriers. The complex interplay between the would-be Goldstone nature of the axion, as prescribed by fermion interaction at the Planck scale, and the topological charge density of an SU(2) Yang–Mills theory deeply in its confining phase is anything but understood quantitatively so far. One may hope that simulations of the axion potential in a center-vortex model of the confining phase, such a proposed in [66], will yield more quantitative insights in the future.

The link between the masses of the three species of ultralight axions, whose fuzzy condensates form lumps of typical masses $M_e$, $M_\mu$, and $M_\tau$, with the three lepton families via the Planck-scale originated axial anomaly within confining phases of SU(2) Yang–Mills theories is compelling. In particular, $M_e = M_{200}$ can be determined by mild modelling of direct observation, as done in Section 3, while $M_\mu$ and $M_\tau$ are predicted by an appeal to Equations (11). Such a scenario allows to address two questions: (i) the implication of a given lump's selfgravity for its stability and (ii) the cosmological origin of a given species of isolated lumps.

Before we discuss question (i) we would like to provide a thermodynamical argument, based on our knowledge gained about axion and lump masses in terms of Yang–Mills scales and the Planck mass, why Planck-scale axions associated with the lepton families always occur in the form of fuzzy or homogeneous condensates. Namely, the Yang–Mills scales $\Lambda_e$, $\Lambda_\mu = \Lambda_e\, m_\mu/m_e$, and $\Lambda_\tau = \Lambda_e\, m_\tau/m_e$ together with Equations (1) and (26), yield axion masses as

$$\begin{aligned}
m_{a,e} &\sim 0.675 \times 10^{-23}\,\mathrm{eV}\,, \\
m_{a,\mu} &\sim 2.89 \times 10^{-19}\,\mathrm{eV}\,, \\
m_{a,\tau} &\sim 8.17 \times 10^{-17}\,\mathrm{eV}\,.
\end{aligned} \tag{27}$$

The critical temperature $T_c$ for the Bose–Einstein condensation of a quantum gas of free bosons of mass $m_a$ and (mean) number density $n_a \sim M/(m_a \frac{4}{3}\pi r_B^3)$ is given as

$$T_c = \frac{2\pi}{m_a}\left(\frac{n_a}{\zeta(3/2)}\right)^{2/3} \,. \tag{28}$$

We conclude from Equations (11), (13), (27) and (28) that

$$
\begin{aligned}
T_{c,e} &\sim 9.7 \times 10^{30}\,\text{GeV}\,, \\
T_{c,\mu} &\sim 7.7 \times 10^{39}\,\text{GeV}\,, \\
T_{c,\tau} &\sim 6.1 \times 10^{42}\,\text{GeV}\,.
\end{aligned}
\tag{29}
$$

All three critical temperatures are comfortably larger than the Planck mass $M_P = 1.22 \times 10^{19}\,\text{GeV}$ such that throughout the Universe's expansion history and modulo depercolation, which generates a nonthermal halo of particles correlated on the de Broglie wave length around a condensate core, the Bose-condensed state of $e$-, $\mu$-, and $\tau$-axions is guaranteed and consistent with $\xi \ll 1$, compare with Equation (12).

We now turn back to question (i). Explicit lump masses can be obtained from Equation (11) based on the typical mass $M_e = 6.3 \times 10^{10}\,M_\odot$ of an $e$-lump. One has

$$
\begin{aligned}
M_\mu &= 1.5 \times 10^6\,M_\odot\,, \\
M_\tau &= 5.2 \times 10^3\,M_\odot\,.
\end{aligned}
\tag{30}
$$

For the computation of the respective gravitational Bohr radii according to Equation (5) both quantities, axion mass $m_{a,i}$ and lump mass $M_i$, are required. To judge the gravitational stability of a given isolated and unmerged lump throughout its evolution a comparison between the typical Bohr radius $r_{B,i}$ and the typical Schwarzschild radius $r_{\text{SD},i}$, defined as

$$
r_{\text{SD,i}} \equiv \frac{2M_i}{M_P^2}\,,
\tag{31}
$$

is in order. Using $M_e = 6.3 \times 10^{10}\,M_\odot$, Figure 9 indicates the implied values of the Bohr radii $r_{B,e}$, $r_{B,\mu}$, and $r_{B,\tau}$ by dots on the curves of all possible Bohr radii as functions of their lump masses when keeping the axion mass $m_{a,i}$ fixed. Notice that for all three cases, $e$-lumps, $\mu$-lumps, and $\tau$-lumps, typical Bohr radii are considerably larger than their Schwarzschild radii. Indeed, from Equations (1), (10), and (31) it follows that

$$
\frac{r_B}{r_{\text{SD}}} = \frac{1}{2}\kappa^2\,.
\tag{32}
$$

With $\kappa = 314$ we have $r_B/r_{\text{SD}} = 4.92 \times 10^4$. An adiabatic pursuit of the solid lines in Figure 9 down to their intersections with the dashed line reveals that an increase of lump mass by a factor $\sim 222$ is required to reach the critical mass for black-hole formation. While this is unlikely to occur through mergers of $e$-lumps within their peers it is conceivable for merging $\mu$- and $\tau$-lumps, see below.

The mean mass density of a lump scales with the fourth power of the Yang–Mills scale, see Eqsuation (8), (10) and (13). With the hierarchies in Yang–Mills scales $\Lambda_\tau/\Lambda_\mu \sim 17$ or $\Lambda_\mu/\Lambda_e \sim 200$ it is conceivable that sufficiently large number of lumps of a higher Yang–Mills scale, embedded into a lump of a lower scale, catalyse the latter's gravitational compaction to the point of collapse, see, however, discussion in Section 5.1.

With Equation (11) we have $M_\mu/M_e \sim 2.3 \times 10^{-5}$ such that a dark mass of the self-gravitating dark-matter disk of the Milky Way, exhibiting a radial scale of $(7.5 \cdots 8.85)$ kpc and a mass of $M_{\text{MW}} = (2 \cdots 3) \times 10^{11}\,M_\odot$ [62], would contain a *few* previously isolated but now merged $e$-lumps. This implies with Equation (11) a $\mu$-lump mass of

$$
M_\mu = (4.7 \cdots 7) \times 10^6\,M_\odot\,.
\tag{33}
$$

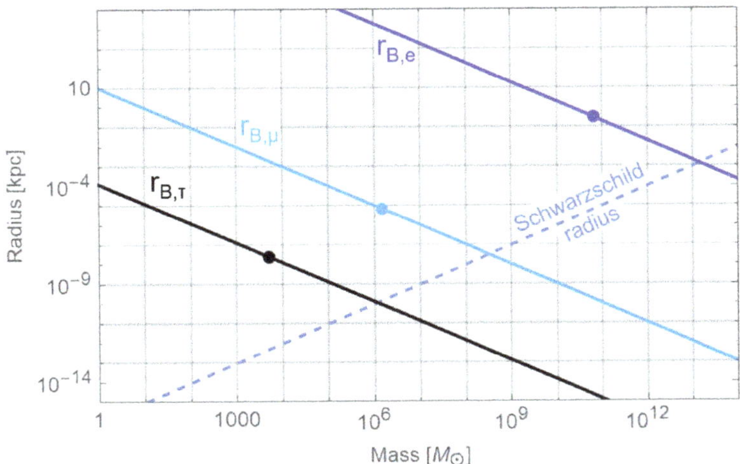


**Figure 9.** Schwarzschild radius (dashed line) and gravitational Bohr radii (solid lines; dark blue: $e$-lump, turquoise: $\mu$-lump, and black: $\tau$-lump) as functions of lump mass in units of solar mass $M_\odot$. The dots indicate lump masses which derive from the typical mass of an isolated $e$-lump $M_e = 6.3 \times 10^{10}\, M_\odot$ suggested by the analysis of the RCs of low-surface-brightness galaxies performed in Section 3.

In [62], the mass of the dark halo of the Milky Way, which is virialised up to $r \sim 350$ kpc, is determined as $1.8 \times 10^{12}\, M_\odot$. In addition to the halo and the disk, there is a ringlike dark-matter structure within $(13 \cdots 18.5)$ kpc of mass $(2.2 \cdots 2.8) \times 10^{10}\, M_\odot$. Since these structures probably are, judged within the here-discussed framework, due to contaminations of a seeding $e$-lump by the accretion of $\tau$- and $\mu$-lumps we ignore them in what follows. In any case, a virialised dark-matter halo of 350 kpc radial extent easily accomodates the dark mass ratio $\sim 0.1$ between the selfgravitating dark-matter disk and the dark halo in terms of accreted $\tau$- and $\mu$-lumps.

Interestingly, the lower mass bound of Equation (33) is contained in the mass range $(4.5 \pm 0.4) \times 10^6\, M_\odot$ [61] or $(4.31 \pm 0.36) \times 10^6\, M_\odot$ [60] of the central compact object extracted from orbit analysis of S-stars.

Next, we discuss question (ii). Consider a situation where the gravitational Bohr radius $r_B$ exceeds the Hubble radius $r_H(z) = H^{-1}(z)$ at some redshift $z$. Here $H(z)$ defines the Hubble parameter subject to a given cosmological model. In such a situation, the lump acts like a homogeneous energy density (dark energy) within the causally connected region of the Universe roughly spanned by $r_H$. If $r_B$ falls *sizably* below $r_H$ then formerly homogeneous energy density may decay into isolated lumps. In order to predict at which redshift $z_p$ such a depercolation epoch has taken place we rely on the extraction of the epoch $z_{p,e} = 53$ in [16] for the depercolation of $e$-lumps. To extract the depercolation redshifts $z_{p,\mu}$ and $z_{p,\tau}$ we use the cosmological model SU(2)$_{\text{CMB}}$ proposed in [16] with parameters values given in column 2 of Table 1 of that paper. In Figure 10 the relative density parameters of the cosmological model SU(2)$_{\text{CMB}}$ are depicted as functions of $z$, and the point of $e$-lump depercolation $z_{p,e} = 53$ is marked by the cusps in dark energy and matter.

The strategy to extract $z_{p,\mu}$ and $z_{p,\tau}$ out of information collected at $z_{p,e} = 53$ is to determine the ratio $\alpha_e$ of $r_H = 16.4$ Mpc at $z_{p,e} = 53$ and $r_{B,e} = 0.26$ kpc for a typical, isolated, and unmerged $e$-lump as

$$\alpha_e \equiv \left. \frac{r_H}{r_{B,e}} \right|_{z=z_{p,e}} = 55{,}476\,. \tag{34}$$

It is plausible that $\alpha_e$ can be promoted to a universal (that is, independent of the Yang–Mills scale and temperature) constant $\alpha$, again, because of the large hierarchy between all Yang–Mills scales to the Planck mass $M_P$. Moreover, the ratio of radiation temperature to the Planck mass $M_P$ remains very small within the regime of redshifts considered in typical CMB simulations. Using the cosmological model SU(2)$_{\text{CMB}}$, Equation (13), and demanding $\alpha$ to set the condition for $\mu$- and $\tau$-lump depercolation ($r_H \equiv \alpha\, r_{B,i}$), one obtains

$$z_{p,\mu} = 40,000, \quad z_{p,\tau} = 685,000. \tag{35}$$

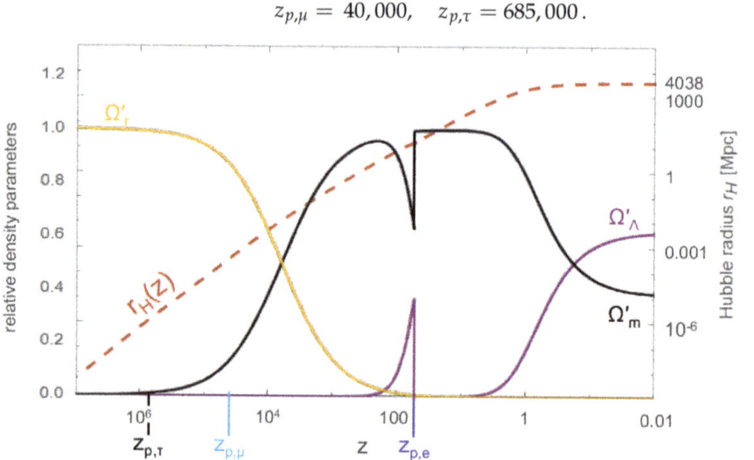


**Figure 10.** Cosmological model SU(2)$_{\text{CMB}}$ of [16] (with parameter values fitted to the TT, TE, and EE CMB Planck power spectra and taken from column 2 of Table 1 of that paper) in terms of relative density parameters as functions of redshift $z$. Normalised density parameters refer to dark energy ($\Omega'_\Lambda$), to total matter (baryonic and dark, $\Omega'_m$), and to radiation (three flavours of massless neutrinos and eight relativistic polarisations in a CMB subject to SU(2)$_{\text{CMB}}$, $\Omega'_r$). The dotted red line represents the Hubble radius of this model. The redshifts of $e$-lump, $\mu$-lump, and $\tau$-lump depercolations are indicated by vertical lines intersecting the $z$-axis. Only $e$-lump depercolation is taken into account explicitly within the cosmological model SU(2)$_{\text{CMB}}$ since at $z_{p,\mu} = 40,000$ and $z_{p,\tau} = 685,000$ the Universe is radiation dominated.

In Figure 10 the relative density parameters $\Omega'_\Lambda$ (dark energy), $\Omega'_m$ for total matter (baryonic and dark), $\Omega'_r$ (total radiation), and the Hubble radius $r_H$ are depicted as functions of $z$. Moreover, the redshifts of $e$-lump, $\mu$-lump, and $\tau$-lump depercolations—$z_{p,e}$, $z_{p,\mu}$, and $z_{p,\tau}$—are indicated by vertical lines intersecting the $z$-axis. The depercolation epochs for $\mu$- and $\tau$-lumps at redshifts $z_{p,\mu} = 40,000$, and $z_{p,\tau} = 685,000$ are not modelled within SU(2)$_{\text{CMB}}$ because the Universe then is radiation dominated.

In Figure 11 a schematic evolution of the Universe's dark sector, subject to the SU(2) Yang–Mills theories SU(2)$_\tau$, SU(2)$_\mu$, SU(2)$_e$, and SU(2)$_{\text{CMB}}$ invoking Planck-scale induced axial anomalies, is depicted.

After a possible epoch of Planck-scale inflation and reheating the temperature of the radiation dominated Universe is close to the Planck mass $M_P$, and $r_H \sim M_P^{-1}$. In this situation, the Bohr radii of the various hypothetical lump species (Peccei-Quinn scale $M_P$, SU(2)$_\tau$, SU(2)$_\mu$, SU(2)$_e$, and SU(2)$_{\text{CMB}}$ Yang–Mills dynamics) are much larger than $r_H$, and the (marginal) dark sector of the model then solely contains dark energy.

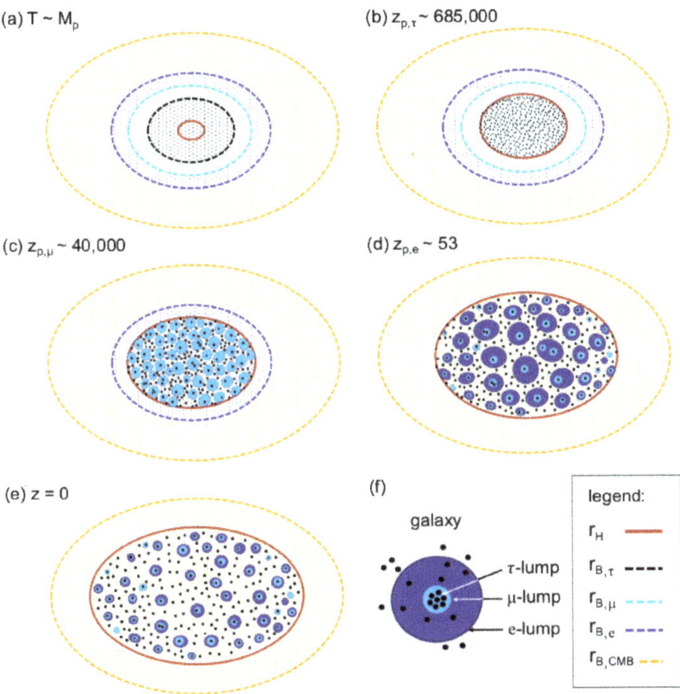



**Figure 11.** The evolution of the Universe's dark sector according to SU(2) Yang–Mills theories of scales $\Lambda_e = m_e/(15 \times 118.6)$, $\Lambda_\mu = m_\mu/(15 \times 118.6)$, $\Lambda_\tau = m_\tau/(15 \times 118.6)$ (confining phases, screened), and $\Lambda_{CMB} \sim 10^{-4}$ eV (deconfining phase, unscreened) invoking Planck-scale induced axial anomalies. The horizon size, set by the Hubble radius $r_H$ at various epochs (**a–e**), is shown by a red circumference. At epoch (**a**) gravity induced chiral symmetry breaking at the Planck scale creates a would-be-Goldstone boson which, due to the axial anomaly, gives rise to four ultralight axionic particle species. Their gravitational Bohr radii $r_{B,\tau}$, $r_{B,\mu}$, $r_{B,e}$, and $r_{B,CMB}$ are much larger than $r_H$. Therefore, the associated energy densities should be interpreted as dark energy. (**b**) As the radiation dominated Universe expands the smallest Bohr radius $r_{B,\tau}$ falls below $r_H$. Once the ratio $\alpha \equiv r_H/r_{B,\tau}$ is sufficiently large ($\alpha = 55{,}500$) $\tau$-lumps depercolate ($z_{p,\tau} = 685{,}000$). (**c**) As the Universe expands further the Bohr radius $r_{B,\mu}$ falls below $r_H$. When the ratio of $r_H$ and $r_{B,\mu}$ again equals about $\alpha = 55{,}500$ $\mu$-lumps derpercolate ($z_{p,\mu} = 40{,}000$). The cosmological matter densities of $\tau$ and $\mu$-lumps are comparable [16]. Since the mass of an isolated, unmerged $\tau$-lump is by a factor of about $(m_\tau/m_\mu)^2 \sim 283$ smaller than the mass of an isolated, unmerged $\mu$-lump it then follows that the number density of $\tau$-lumps is by this factor larger compared to the number density of $\mu$-lumps. (**d**) Upon continued expansion down to redshift $z_{p,e} = 53$ e-lumps depercolate. Their number density is by a factor of $(m_\mu/m_e)^2 \sim 42{,}750$ smaller than the number density of $\mu$-lumps. (**e**) The value of $r_{B,CMB}$ is vastly larger than $r_H(z = 0)$: $r_{B,CMB} = 2.4 \times 10^{10}$ Mpc vs. $r_H(z = 0) = 4038$ Mpc. Therefore, a depercolation of CMB-lumps up to the present is excluded. As a consequence, the condensate of CMB-axions *is* dark energy. (**f**) Possible dark-matter configuration of a galaxy including $\tau$-lumps and a single $\mu$-lump inside an e-lump.

  Around $z_{p,\tau} = 685{,}000$ (radiation domination) the depercolation of $\tau$-lumps occurs for $\alpha \equiv r_H/r_{B,\tau} \sim 55{,}500$. Once released, they evolve like pressureless, non-relativistic particles and, cosmologically seen, represent dark matter.

  As the Universe expands further, the ratio $\alpha \equiv r_H/r_{B,\mu} \sim 55{,}500$ is reached such that $\mu$-lumps start to depercolate at $z_{p,\mu} = 40{,}000$. Since they contribute to the cosmological dark-matter density roughly the same amount like $\tau$-lumps, see [16] for a fit of so-called primordial and emergent dark-matter densities to TT, TE, and EE power spectra of the 2015 Planck data, one concludes from Equation (11) that their number density is by a factor $(m_\tau/m_\mu)^2 \sim 283$ smaller than that of $\tau$-lumps. For a first estimate this assumes a neglect of local gravitational interactions. That is, at $\mu$-lump depercolation there are roughly 300

$\tau$-lumps inside one $\mu$-lump. Each of these $\tau$-lumps possesses a mass of $M_\tau = 5.2 \times 10^3 \, M_\odot$. The implied accretion process involving additional $\tau$-lumps may catalyse the gravitational compaction of the thus contaminated $\mu$-lump, see discussion in Section 5.1.

At $z_{p,e} = 53$ $e$-lumps depercolate [16]. Again, disregarding local gravitational binding, we conclude from Equation (11) and a nearly equal contribution of each lump species to the cosmological dark-matter density [16] that the number densities of $\mu$- and $\tau$-lumps are by factors of $(m_\mu/m_e)^2 \sim 42{,}750$ and $(m_\tau/m_e)^2 \sim 283\times 42{,}750$, respectively, larger than the number density of $e$-lumps. At $e$-lump depercolation we thus have $42{,}750$ $\mu$-lumps and $42{,}750 \times 283 \sim 1.2 \times 10^7$ $\tau$-lumps within one $e$-lump.

Again, ignoring local gravitational binding effects, the dilution of $\tau$- and $\mu$-lump densities by cosmological expansion predicts that today we have $42{,}750/(z_{p,e}+1)^3 = 0.27$ $\mu$-lumps and $42{,}750 \times 283/(z_{p,e}+1)^3 = 77$ $\tau$-lumps within one $e$-lump. Local gravitational binding should correct these numbers to higher values but the orders of magnitude—$O(1)$ for $\mu$-lumps and $O(100)$ for $\tau$-lumps—should remain unaffected. It is conspicuous that the number of globular clusters within the Milky Way is in the hundreds [67], with typical masses between ten to several hundred thousand solar masses [61]. With $M_\tau = 5.2 \times 10^3 \, M_\odot$ it is plausible that the dark-mass portion of these clusters is constituted by a single or a small number of merged $\tau$-lumps. In addition, in the Milky Way there is one central massive and dark object with about $(4.5 \pm 0.4) \times 10^6$ [61] or $(4.31 \pm 0.36) \times 10^6$ solar masses [60]. If, indeed, there is roughly one isolated $\mu$-lump per isolated $e$-lump today then the mass range of the Milky Way's dark-matter disk, interpreted as a merger of few isolated $e$-lumps, implies the mass range of Equation (33) for the associated $\mu$-lump merger. This range contains the mass of the central massive and dark object determined in [60,61].

## 5. Discussion, Summary, and Outlook

### 5.1. Speculations on Origins of Milky Way's Structure

The results of Section 5 on mass ranges of $\tau$-lumps, $\mu$-lumps, and $e$-lumps being compatible with typical masses of globular clusters, the mass of the central compact Galactic object [60,61], and the mass of the selfgravitating dark-matter disk of the Milky Way, respectively, is compelling. We expect that similar assignments can be made to according structures in other spiral galaxies.

Could the origin of the central compact object in Milky Way be the result of $\tau$- and $\mu$-lump mergers? As Figure 9 suggests, a merger of $n \geq 222$ isolated $\tau$- or $\mu$-lumps is required for black hole formation. Since we know that the mass of the central compact object is $\sim 4 \times 10^6 M_\odot$ a merger of $n \geq 222$ $\mu$-lumps is excluded for Milky Way. Thus, only a merger of $n \geq 222$ $\tau$-lumps, possibly catalysed by the consumption of a few $\mu$-lumps, is a viable candidate for black-hole formation in our Galaxy. Such a process—merging of several hundred $\tau$-lumps within the gravitational field of a few merging $\mu$-lumps down to the point of gravitational collapse—would be consistent with the results of [60,61] who fit stellar orbits around the central massive object of Milky Way extremely well to a single-point-mass potential. Indeed, the gravitational Bohr radius of a $\mu$-lump is $7 \times 10^{-6}$ kpc while the closest approach of an S2 star to the gravitational center of the central massive object of Milky Way is $17\,\mathrm{lh} = 5.8 \times 10^{-7}$ kpc [60]. Therefore, $\mu$-lumps need to collapse in order to be consistent with a point-mass potential.

The Milky Way's contamination with baryons, its comparably large dark-disk mass vs. the mass of the low-surface-brightness galaxies analysed in Section 3, and possibly tidal shear from the dark ring and the dark halo during its evolution introduce deviations from the simple structure of a typical low-surface-brightness galaxy. Simulations, which take all the here-discussed components into account, could indicate how typical such structures are, rather independently of primordial density perturbations.

Isolated $\tau$-, $\mu$-, and $e$-lumps, which did not accrete sufficiently many baryons to be directly visible, comprise dark-matter galaxies that are interspersed in between visible galaxies. The discovery of such dark galaxies, pinning down their merger-physics, and determinations of their substructure by gravitational microlensing and gravitational-wave astronomy could support the here-proposed scenario of active structure formation on sub-galactic scales.

*5.2. Summary and Outlook*

In this paper, we propose that the dark Universe can be understood in terms of axial anomalies [21–23] which are invoked by screened Yang–Mills scales in association with the leptonic mass spectrum. This produces three ultra-light axion species. Such pseudo Nambu-Goldstone bosons are assumed to owe their very existence to a gravitationally induced chiral symmetry breaking with a universal Peccei–Quinn scale [25] of order the Planck mass $M_P = 1.22 \times 10^{19}$ GeV [30]. We therefore refer to each of these particle species as *Planck-scale* axions. Because of the relation $m_{a,i} = \Lambda_i^2/M_P$ the screened Yang–Mills scale $\Lambda_i$ derives from knowledge of the axion mass $m_{a,i}$. Empirically, the here-extracted screened scale $\Lambda_e = 287$ eV points to the first lepton family, compare with [55]. This enables predictions of typical lump and axion masses in association with two additional SU(2) Yang–Mills theories associating with $\mu$ and $\tau$ leptons.

Even though the emergence of axion mass [25] and the existence of lepton families [55] are governed by the same SU(2) gauge principle, the interactions between these ultra-light pseudo scalars and visible leptonic matter is extremely feeble. Thus, the here-proposed relation between visible and dark matter could demystify the dark Universe. An important aspect of Planck-scale axions is their Bose–Einstein, yet non-thermal, condensed state. A selfgravitating, isolated fuzzy condensate (lump) of a given axion species $i = e, \mu, \tau$ is chiefly characterised by the gravitational Bohr radius $r_{B,i}$ [36] given in terms of the axion mass $m_{a,i}$ and the lump mass $M_i = M_{200,i}$ (virial mass), see Equation (5). As it turns out, for $i = e$ the information about the latter two parameters is contained in observable rotation curves of low-surface-brightness galaxies with similar extents. Realistic models for the dark-matter density profiles derive from ground-state solutions of the spherically symmetric Poisson–Schrödinger system at zero temperature and for a single axion species. These solutions describe selfgravitating fuzzy axion condensates, compare with [47]. Two such models, the Soliton-NFW and the Burkert model, were employed in our present extractions of $m_{a,e}$ and $M_e$ under the assumption that the dark-matter density in a typical low-surface brightness galaxy is dominated by a single axion species. Our result $m_{a,e} = 0.675 \times 10^{-23}$ eV is consistent with the result of [49]: $m_{a,e} = 0.554 \times 10^{-23}$ eV. Interestingly, such an axion mass is close to the result $10^{-25}$ eV $\leq m_a \leq 10^{-24}$ eV [53] obtained by treating axions as a classical ideal gas of non-relativistic particles—in stark contrast to the Bose condensed state suggested by Equation (28) or the gas surrounding it with intrinsic correlations governed by large de-Broglie wavelengths. This value of the axion mass is considerably lower then typical lower bounds obtained in the literature: $m_a > 2.9 \times 10^{-21}$ eV [51], $m_a = 2.5^{+3.6}_{-2.0} \times 10^{-21}$ eV [39], $m_a > 3.8 \times 10^{-21}$ eV [54], and $m_a \sim 8 \times 10^{-23}$ eV in [47]. We propose that this discrepancy could be due to the omission of the other two axion species with a mass spectrum given by Equation (27). For example, the dark-matter and thus baryonic density variations along the line of sight probed by a Lyman-$\alpha$ forest do not refer to gravitationally bound systems and therefore should be influenced by all *three* axion species.

Once axions and their lumps are categorised, questions about (i) the cosmological origin of lumps and (ii) their role in the evolution of galactic structure can be asked. Point (i) is addressed by consulting a cosmological model (SU(2)$_{\text{CMB}}$ [16]) which requires the emergence of dark matter by lump depercolation at defined redshifts, see also [68]. Depercolation of $e$-lumps at redshift $z_{p,e} = 53$ anchors the depercolations of the two other lump species. One obtains $z_{p,\mu} = 40{,}000$ and $z_{p,\tau} = 685{,}000$.

The critical temperature $T_{c,e}$ of SU(2)$_e$ for the deconfining-preconfining phase transition (roughly equal to the temperature of the Hagedorn transition to the confining phase [26]) is $T_{c,e} = 9.49$ keV [55]. A question arises whether this transition could affect observable small-scale angular features of the CMB. In the SU(2)$_{\text{CMB}}$ based cosmological model of [16] $T_{c,e} = 9.49$ keV corresponds to a redshift of $z_{c,e} = 6.4 \times 10^7$. (Typically, CMB simulation are initialised at $z = 10^9$ [69]). Traversing the preconfining-deconfining phase transition at $z_{c,e}$ an already strongly radiation dominated Universe receives additional radiation density and entropy. However, we expect that the horizon crossing of curvature perturbation at $z > z_{c,e}$, which may influence small-scale matter perturbations, will affect CMB anisotropies on angular scales $l > 3000$ only. Therefore, Silk damping would reduce the magnitudes of these multipoles to below the observational errors.

Up to the present, lump depercolation does not occur for the Planck-scale axion species associated with $SU(2)_{CMB}$: here, the gravitational Bohr radius of the axion condensate always exceeds the Hubble radius by many orders of magnitude. As for point (ii), the masses and Bohr radii of $\mu$- and $\tau$-lumps seem to be related with the central massive compact object of the Milky Way [60,61] and globular clusters [62], respectively. Within a given galaxy such active components of structure formation possibly originate compact stellar streams through tidal forces acting on $\tau$-lumps. Whether this is supported by observation could be decided by a confrontation of N-body simulations (stars) in the selfgravitating background of the externally deformed lump.

Apart from cosmological and astrophysical observation, which should increasingly be able to judge the viability of the here-proposed scenario, there are alternative terrestrial experiments which can check the predictions of the underlying SU(2) gauge-theory pattern. Let us quote two examples: First, there is a predicted low-frequency spectral black-body anomaly at low temperatures ($T \sim 5\,\mathrm{K}$) [70] which could be searched for with a relatively low instrumental effort. Second, an experimental link to $SU(2)_e$ would be the detection of the Hagedorn transition in a plasma at electron temperature 9.49 keV and the stabilisation of a macroscopically large plasma ball at a temperature of $1.3 \times 9.49$ keV [55]. Such electron temperatures should be attainable by state-of-the-art nuclear-fusion experiments such as ITER or by fusion experiments with inertial plasma confinement.

**Author Contributions:** Conceptualization, J.M. and R.H.; methodology, J.M. and R.H.; software, J.M. and R.H.; validation, J.M. and R.H.; formal analysis, J.M. and R.H.; investigation, J.M. and R.H.; resources, J.M. and R.H.; data curation, J.M. and R.H.; writing—original draft preparation, J.M. and R.H.; writing—review and editing, J.M. and R.H.; visualization, J.M. and R.H.; supervision, J.M. and R.H.; project administration, J.M. and R.H.; funding acquisition, J.M. and R.H. All authors have read and agreed to the published version of the manuscript.

**Funding:** This research received no external funding.

**Institutional Review Board Statement:** Not applicable.

**Informed Consent Statement:** Not applicable.

**Data Availability Statement:** The SPARC library was analysed in support of this research [64]. The processed data and program underlying this article will be shared on request to the corresponding author.

**Conflicts of Interest:** The authors declare no conflict of interest.

## Note

1.   The chiral dynamics at the Planck scale, which produces the axion field, to some extent resolves the ground states of Yang–Mills theories: axions become massive by virtue of the anomaly because of this very resolution of topological charge density.

## References

1. Rubin, V.C.; Ford, W.K., Jr. Rotation of the Andromeda Nebula from a Spectroscopic Survey of Emission Regions. *Astrophys. J.* **1970**, *159*, 379–403. [CrossRef]
2. Zwicky, F. On the Masses of Nebulae and of Clusters of Nebulae. *Astrophys. J.* **1937**, *86*, 217. [CrossRef]
3. Tucker, W.; Blanco, P.; Rappoport, S.; David, L.; Fabricant, D.; Falco, E.E.; Forman, W.; Dressler, A.; Ramella, M. 1e0657-56: A contender for the hottest known cluster of galaxies. *Astrophys. J.* **1998**, *496*, L5. [CrossRef]
4. Clowe, D.; Bradač, M.; Gonzalez, A.H.; Markevitch, M.; Randall, S.W.; Jones, C.; Zaritsky, D. A Direct Empirical Proof of the Existence of Dark Matter. *Astrophys. J.* **2006**, *648*, L109. [CrossRef]
5. de Bernardis, P.; Ade, P.A.R.; Bock, J.J.; Bond, J.R.; Borrill, J.; Boscaleri, A.; Coble, K.; Crill, B.P.; De Gasperis, G.; Farese, P.C.; et al. A flat Universe from high-resolution maps of the cosmic microwave background radiation. *Nature* **2000**, *404*, 955. [CrossRef]

6. Riess, A.G.; Filippenko, A.V.; Challis, P.; Clocchiatti, A.; Diercks, A.; Garnavich, P.M.; Gilliland, R.L.; Hogan, C.J.; Jha, S.; Kirshner, R.P.; et al. Observational Evidence from Supernovae for an Accelerating Universe and a Cosmological Constant. *Astron. J.* **1998**, *116*, 1009. [CrossRef]

7. Perlmutter, S.; Aldering, G.; Goldhaber, G.; Knop, R.A.; Nugent, P.; Castro, P.G.; Deustua, S.; Fabbro, S.; Goobar, A.; Groom, D.E.; et al. Measurements of $\Omega$ and $\Lambda$ from 42 high redshift supernovae. *Astrophys. J.* **1999**, *517*, 565–586. [CrossRef]

8. Verde, L.; Treu, T.; Riess, A.G. Tensions between the early and late Universe. *Nat. Astron.* **2019**, *3*, 891. [CrossRef]

9. Aghanim, N.; Akrami, Y.; Ashdown, M.; Aumont, J.; Baccigalupi, C.; Ballardini, M.; Banday, A.J.; Barreiro, R.B.; Bartolo, N.; Basak, S.; et al. Planck 2018 results. VI. Cosmological parameters. *Astron. Astrophys.* **2020**, *641*, A6. [CrossRef]

10. Reid, M.J.; Pesce, D.W.; Riess, A.G. An Improved Distance to NGC 4258 and Its Implications for the Hubble Constant. *Astrophys. J.* **2019**, *886*, L27. [CrossRef]

11. Riess, A.G.; Casertano, S.; Yuan, W.; Macri, L.; Anderson, J.; MacKenty, J.W.; Bowers, J.B.; Clubb, K.I.; Filippenko, A.V.; Jones, D.O.; et al. New Parallaxes of Galactic Cepheids from Spatially Scanning the Hubble Space Telescope: Implications for the Hubble Constant. *Astrophys. J.* **2018**, *855*, 136. [CrossRef]

12. Wong, K.C.; Suyu, S.H.; Auger, M.W.; Bonvin, V.; Courbin, F.; Fassnacht, C.D.; Halkola, A.; Rusu, C.E.; Sluse, D.; Sonnenfeld, A.; et al. H0LiCOW IV. Lens mass model of HE 0435-1223 and blind measurement of its time-delay distance for cosmology. *Mon. Not. R. Astron. Soc.* **2017**, *465*, 4895–4913. [CrossRef]

13. Abbott, T.; Abdalla, F.; Alarcon, A.; Aleksić, J.; Allam, S.; Allen, S.; Amara, A.; Annis, J.; Asorey, J.; Avila, S.; et al. Dark Energy Survey year 1 results: Cosmological constraints from galaxy clustering and weak lensing. *Phys. Rev. D* **2018**, *98*. [CrossRef]

14. Troxel, M.; MacCrann, N.; Zuntz, J.; Eifler, T.; Krause, E.; Dodelson, S.; Gruen, D.; Blazek, J.; Friedrich, O.; Samuroff, S.; et al. Dark Energy Survey Year 1 results: Cosmological constraints from cosmic shear. *Phys. Rev. D* **2018**, *98*. [CrossRef]

15. Tröster, T.; Sánchez, A.G.; Asgari, M.; Blake, C.; Crocce, M.; Heymans, C.; Hildebrandt, H.; Joachimi, B.; Joudaki, S.; Kannawadi, A.; et al. Cosmology from large-scale structure. *Astron. Astrophys.* **2020**, *633*, L10. [CrossRef]

16. Hahn, S.; Hofmann, R.; Kramer, D. SU(2)$_{\mathrm{CMB}}$ and the cosmological model: angular power spectra. *Mon. Not. R. Astron. Soc.* **2019**, *482*, 4290. [CrossRef]

17. Krishnan, C.; Mohayaee, R.; Colgáin, E.Ó.; Sheikh-Jabbari, M.M.; Yin, L. Does Hubble Tension Signal a Breakdown in FLRW Cosmology? *arXiv* **2021**, arXiv:2105.09790.

18. Kolb, E.W.; Turner, M.S. *The Early Universe, Taylor and Francis*; Westview Press: Boulder, CO, USA, 1990.

19. Akerib, D.; Araújo, H.M.; Bai, X.; Bailey, A.J.; Balajthy, J.; Bedikian, S.; Bernard, E.; Bernstein, A.; Bolozdynya, A.; Bradley, A.; et al. First results from the LUX dark matter experiment at the Sanford Underground Research Facility. *Phys. Rev. Lett.* **2014**, *112*, 091303. [CrossRef] [PubMed]

20. Akerib, D.S.; Akerlof, C.W.; Alsum, S.K.; Araújo, H.M.; Arthurs, M.; Bai, X.; Bailey, A.J.; Balajthy, J.; Balashov, S.; Bauer, D.; et al. Projected WIMP sensitivity of the LUX-ZEPLIN dark matter experiment. *Phys. Rev. D* **2020**, *101*, 052002. [CrossRef]

21. Adler, S.L.; Bardeen, W.A. Absence of higher order corrections in the anomalous axial vector divergence equation. *Phys. Rev.* **1969**, *182*, 1517–1536. [CrossRef]

22. Bell, J.; Jackiw, R. A PCAC puzzle: $\pi^0 \rightarrow \gamma\gamma$ in the $\sigma$ model. *Nuovo Cim. A* **1969**, *60*, 47–61. [CrossRef]

23. Fujikawa, K. Path Integral Measure for Gauge Invariant Fermion Theories. *Phys. Rev. Lett.* **1979**, *42*, 1195–1198. [CrossRef]

24. Atherton, H.W.; Bovet, C.; Coet, P.; Desalvo, R.; Doble, N.; Maleyran, R.; Anderson, E.W.; Von Dardel, G.; Kulka, K.; Boratav, M.; et al. Direct measurement of the lifetime of the neutral pion. *Phys. Lett. B* **1985**, *158*, 81–84. [CrossRef]

25. Peccei, R.; Quinn, H.R. Constraints Imposed by CP Conservation in the Presence of Instantons. *Phys. Rev. D* **1977**, *16*, 1791–1797. [CrossRef]

26. Hofmann, R. *The Thermodynamics of Quantum Yang–Mills Theory: Theory And Applications*, 2nd ed.; World Scientific: Singapore, 2016; p. 1.

27. Candelas, P.; Raine, D.J. General-relativistic quantum field theory: An exactly soluble model. *Phys. Rev. D* **1975**, *12*, 965. [CrossRef]

28. Frieman, J.A.; Hill, C.T.; Stebbins, A.; Waga, I. Cosmology with ultralight pseudo Nambu-Goldstone bosons. *Phys. Rev. Lett.* **1995**, *75*, 2077–2080. [CrossRef] [PubMed]

29. Gross, D.J.; Wilczek, F. Ultraviolet Behavior of Nonabelian Gauge Theories. *Phys. Rev. Lett.* **1973**, *30*, 1343–1346. [CrossRef]

30. Giacosa, F.; Hofmann, R.; Neubert, M. A model for the very early Universe. *J. High Energy Phys.* **2008**, *2*, 077. [CrossRef]

31. Weinberg, D.H.; Bullock, J.S.; Governato, F.; Kuzio de Naray, R.; Peter, A.H.G. Cold dark matter: Controversies on small scales. *Proc. Natl. Acad. Sci. USA* **2015**, *112*, 12249–12255. [CrossRef]

32. Bullock, J.S.; Kolatt, T.S.; Sigad, Y.; Somerville, R.S.; Kravtsov, A.V.; Klypin, A.A.; Primack, J.R.; Dekel, A. Profiles of dark haloes. Evolution, scatter, and environment. *Mon. Not. R. Astron. Soc.* **2001**, *321*, 559–575. [CrossRef]

33. Navarro, J.F.; Frenk, C.S.; White, S.D. A Universal density profile from hierarchical clustering. *Astrophys. J.* **1997**, *490*, 493–508. [CrossRef]

34. Baldeschi, M.R.; Ruffini, R.; Gelmini, G.B. On massive fermions and bosons in galactic halos. *Phys. Lett. B* **1983**, *122*, 221–224. [CrossRef]

35. Membrado, M.; Pacheco, A.; Sañudo, J. Hartree solutions for the self-Yukawian boson sphere. *Phys. Rev. A* **1989**, *39*, 4207. [CrossRef] [PubMed]

36. Ji, S.; Sin, S.J. Late-time Phase transition and the Galactic halo as a Bose Liquid: (II) the Effect of Visible Matter. *Phys. Rev. D* **1994**, *50*. [CrossRef] [PubMed]

37. de Blok, W.J.G.; McGaugh, S.S.; Rubin, V.C. High-Resolution Rotation Curves of Low Surface Brightness Galaxies. II. Mass Models. *Astron. J.* **2001**, *122*, 2396–2427. [CrossRef]

38. Kuzio de Naray, R.; McGaugh, S.S.; de Blok, W.J.G. Mass Models for Low Surface Brightness Galaxies with High Resolution Optical Velocity Fields. *Astrophys. J.* **2008**, *676*, 920–943. [CrossRef]

39. Maleki, A.; Baghram, S.; Rahvar, S. Constraint on the mass of fuzzy dark matter from the rotation curve of the Milky Way. *Phys. Rev. D* **2020**, *101*, 103504. [CrossRef]

40. Pawlowski, M.S.; Kroupa, P. The Vast Polar Structure of the Milky Way Attains New Members. *Astrophys. J.* **2014**, *790*, 74. [CrossRef]

41. Martinez-Medina, L.A.; Robles, V.H.; Matos, T. Dwarf galaxies in multistate scalar field dark matter halos. *Phys. Rev. D* **2015**, *91*, 023519. [CrossRef]

42. Magana, J.; Matos, T. A brief Review of the Scalar Field Dark Matter model. *J. Phys. Conf. Ser.* **2012**, *378*, 012012. [CrossRef]

43. Suárez, A.; Robles, V.H.; Matos, T. A Review on the Scalar Field/Bose–Einstein Condensate Dark Matter Model. *Astrophys. Space Sci. Proc.* **2014**, *38*, 107–142. [CrossRef]

44. Matos, T.; Robles, V.H. Scalar Field (Wave) Dark Matter. 2016. Available online: arXiv:astro-ph.GA/1601.01350 (accessed on 10 September 2020).

45. Marsh, D.J.E. Axion Cosmology. *Phys. Rep.* **2016**, *643*, 1–79. [CrossRef]

46. Hui, L.; Ostriker, J.P.; Tremaine, S.; Witten, E. Ultralight scalars as cosmological dark matter. *Phys. Rev. D* **2017**, *95*, 043541. [CrossRef]

47. Schive, H.Y.; Chiueh, T.; Broadhurst, T. Cosmic Structure as the Quantum Interference of a Coherent Dark Wave. *Nat. Phys.* **2014**, *10*, 496–499. [CrossRef]

48. Amorisco, N.C.; Loeb, A. First Constraints on Fuzzy Dark Matter from the Dynamics of Stellar Streams in the Milky Way. **2018**. Available online: arXiv:astro-ph.GA/1808.00464 (accessed on 30 October 2020).

49. Bernal, T.; Fernández-Hernández, L.M.; Matos, T.; Rodríguez-Meza, M.A. Rotation curves of high-resolution LSB and SPARC galaxies with fuzzy and multistate (ultralight boson) scalar field dark matter. *Mon. Not. R. Astron. Soc.* **2018**, *475*, 1447–1468. [CrossRef]

50. Pawlowski, M.S. The Vast Polar Structure of the Milky Way and Filamentary Accretion of Sub-Halos. In Proceedings of the 13th Marcel Grossmann Meeting on Recent Developments in Theoretical and Experimental General Relativity, Astrophysics, and Relativistic Field Theories, Stockholm, Sweden, 1–7 July 2015; pp. 1724–1726. [CrossRef]

51. Nadler, E.O.; Wechsler, R.H.; Bechtol, K.; Mao, Y.Y.; Green, G.; Drlica-Wagner, A.; McNanna, M.; Mau, S.; Pace, A.B.; Simon, J.D.; et al. Milky Way Satellite Census—II. Galaxy-Halo Connection Constraints Including the Impact of the Large Magellanic Cloud. *Astrophys. J.* **2020**, *893*, 48. [CrossRef]

52. Caputi, K.I.; Ilbert, O.; Laigle, C.; McCracken, H.J.; Le Fèvre, O.; Fynbo, J.; Milvang-Jensen, B.; Capak, P.; Salvato, M.; Taniguchi, Y. Spitzer bright, UltraVISTA faint sources in COSMOS: the contribution to the overall population of massive galaxies at z = 3–7. *Astrophys. J.* **2015**, *810*, 73. [CrossRef]

53. Hložek, R.; Marsh, D.J.E.; Grin, D. Using the full power of the cosmic microwave background to probe axion dark matter. *Mon. Not. R. Astron. Soc.* **2018**, *476*, 3063–3085. [CrossRef]

54. Iršič, V.; Viel, M.; Haehnelt, M.G.; Bolton, J.S.; Becker, G.D. First constraints on fuzzy dark matter from Lyman-$\alpha$ forest data and hydrodynamical simulations. *Phys. Rev. Lett.* **2017**, *119*, 031302. [CrossRef]

55. Hofmann, R. The isolated electron: De Broglie's "hidden" thermodynamics, SU(2) Quantum Yang–Mills theory, and a strongly perturbed BPS monopole. *Entropy* **2017**, *19*, 575. [CrossRef]

56. Hofmann, R. Relic photon temperature versus redshift and the cosmic neutrino background. *Ann. Phys.* **2015**, *527*, 254. [CrossRef]

57. Becker, R.H.; Fan, X.; White, R.L.; Strauss, M.A.; Narayanan, V.K.; Lupton, R.H.; Gunn, J.E.; Annis, J.; Bahcall, N.A.; Brinkmann, J.; et al. Evidence for Reionization at z $\sim$ 6: Detection of a Gunn-Peterson Trough in a z = 6.28 Quasar. *Astron. J.* **2001**, *122*, 2850–2857. [CrossRef]

58. Salucci, P.; Burkert, A. Dark matter scaling relations. *Astrophys. J. Lett.* **2000**, *537*, L9–L12. [CrossRef]

59. Burkert, A. The Structure of Dark Matter Halos in Dwarf Galaxies. *Astrophys. J.* **1995**, *447*. [CrossRef]

60. Gillessen, S.; Eisenhauer, F.; Trippe, S.; Alexander, T.; Genzel, R.; Martins, F.; Ott, T. Monitoring stellar orbits around the Massive Black Hole in the Galactic Center. *Astrophys. J.* **2009**, *692*, 1075. [CrossRef]

61. Ghez, A.M.; Salim, S.; Weinberg, N.N.; Lu, J.R.; Do, T.; Dunn, J.K.; Matthews, K.; Morris, M.R.; Yelda, S.; Becklin, E.E.; et al. Measuring Distance and Properties of the Milky Way's Central Supermassive Black Hole with Stellar Orbits. *Astrophys. J.* **2008**, *689*, 1044–1062. [CrossRef]

62. Kalberla, P.; Dedes, L.; Kerp, J.; Haud, U. Dark matter in the Milky Way, II. the HI gas distribution as a tracer of the gravitational potential. *Astron. Astrophys.* **2007**, *469*, 511. [CrossRef]

63. Bar, N.; Blas, D.; Blum, K.; Sibiryakov, S. Galactic rotation curves versus ultralight dark matter: Implications of the soliton-host halo relation. *Phys. Rev. D* **2018**, *98*. [CrossRef]

64. Lelli, F.; McGaugh, S.S.; Schombert, J.M. SPARC: Mass Models for 175 Disk Galaxies with Spitzer Photometry and Accurate Rotation Curves. *Astron. J.* **2016**, *152*, 157. [CrossRef]

65. Robles, V.H.; Matos, T. Flat Central Density Profile and Constant DM Surface Density in Galaxies from Scalar Field Dark Matter. *Mon. Not. R. Astron. Soc.* **2012**, *422*, 282–289. [CrossRef]
66. Engelhardt, M. Center vortex model for the infrared sector of Yang–Mills theory: Topological susceptibility. *Nucl. Phys. B* **2000**, *585*, 614. [CrossRef]
67. Harris, W.E. A New Catalog of Globular Clusters in the Milky Way. 2010. Available online: arXiv:astro-ph.GA/1012.3224 (accessed on 15 December 2020).
68. Hofmann, R. An SU(2) Gauge Principle for the Cosmic Microwave Background: Perspectives on the Dark Sector of the Cosmological Model. *Universe* **2020**, *6*, 135. [CrossRef]
69. Ma, C.P.; Bertschinger, E. Cosmological Perturbation Theory in the Synchronous and Conformal Newtonian Gauges. *Astrophys. J.* **1995**, *455*, 7. [CrossRef]
70. Hofmann, R. Low-frequency line temperatures of the CMB. *Ann. Phys.* **2009**, *18*, 634. [CrossRef]

MDPI

St. Alban-Anlage 66

4052 Basel

Switzerland

Tel. +41 61 683 77 34

Fax +41 61 302 89 18

www.mdpi.com

*Universe* Editorial Office

E-mail: universe@mdpi.com

www.mdpi.com/journal/universe